Dynamical Paleoclimatology

This is Volume 80 in the
INTERNATIONAL GEOPHYSICS SERIES
A series of monographs and textbooks
Edited by RENATA DMOWSKA, JAMES R. HOLTON,
and H. THOMAS ROSSBY

A complete list of books in this series appears at the end of this volume.

Dynamical Paleoclimatology

Generalized Theory of Global Climate Change

Barry Saltzman
Department of Geology and Geophysics
Yale University
New Haven, Connecticut

San Diego San Francisco New York Boston London Sydney Tokyo

Front cover image of West Antarctic icebergs courtesy of Terence J. Hughes.

This book is printed on acid-free paper. ♾

Academic Press
A division of Harcourt, Inc.
525 B Street, Suite 1900, San Diego, California 92101-4495, USA
http://www.academicpress.com

Academic Press
Harcourt Place, 32 Jamestown Road, London NW1 7BY, UK
http://www.academicpress.com

Library of Congress Catalog Card Number: 2001091216

International Standard Book Number: 0-12-617331-1

PRINTED IN THE UNITED STATES OF AMERICA
01 02 03 04 05 06 MM 9 8 7 6 5 4 3 2 1

To Sheila,

With gratitude for all the years of loving support.

Contents

Prologue xv

Acknowledgments xix

List of Symbols xxi

PART I Foundations

1 INTRODUCTION: The Basic Challenge 3

1.1 The Climate System 3

1.2 Some Basic Observations 4

1.3 External Forcing 9

1.3.1 Astronomical Forcing 9

1.3.2 Tectonic Forcing 12

1.4 The Ice-Age Problem 14

2 TECHNIQUES FOR CLIMATE RECONSTRUCTION 17

2.1 Historical Methods 17

2.1.1 Direct Quantitative Measurements 17

2.1.2 Descriptive Accounts of General Environmental Conditions 18

2.2 Surficial Biogeologic Proxy Evidence 18

2.2.1 Annually Layered Life Forms 18

2.2.2 Surface Geomorphic Evidence 19

2.3 Conventional Nonisotopic Stratigraphic Analyses of Sedimentary Rock and Ice 20

2.3.1 Physical Indicators 21

2.3.2 Paleobiological Indicators (Fossil Faunal Types and Abundances) 22

2.4 Isotopic Methods 23

2.4.1 Oxygen Isotopes 23

2.4.2 Deuterium and Beryllium in Ice Cores 24

2.4.3 Stable Carbon Isotopes 25

2.4.4 Strontium and Osmium Isotopes 26

2.5 Nonisotopic Geochemical Methods 26

2.5.1 Cadmium Analysis 26

2.5.2 Greenhouse Gas Analysis of Trapped Air in Ice Cores 27

2.5.3 Chemical and Biological Constituents and Dust Layers in Ice Cores 27

2.6 Dating the Proxy Evidence (Geochronometry) 27

3 A SURVEY OF GLOBAL PALEOCLIMATIC VARIATIONS 30

3.1 The Phanerozoic Eon (Past 600 My) 31

3.2 The Cenozoic Era (Past 65 My) 34

3.3 The Plio-Pleistocene (Past 5 My) 35

3.4 Variations during the Last Ice Age: IRD Events 37

3.5 The Last Glacial Maximum (20 ka) 38

3.6 Postglacial Changes: The Past 20 ky 39

3.7 The Past 100 Years 40

3.8 The Generalized Spectrum of Climatic Variance 41

3.9 A Qualitative Discussion of Causes 44

4 GENERAL THEORETICAL CONSIDERATIONS 47

4.1 The Fundamental Equations 47

4.2 Time Averaging and Stochastic Forcing 51

4.3 Response Times and Equilibrium 55

4.4 Spatial Averaging 60

4.5 Climatic-Mean Mass and Energy Balance Equations 63

4.5.1 The Water Mass Balance 63

4.5.2 Energy Balance 65

5 SPECIAL THEORETICAL CONSIDERATIONS FOR PALEOCLIMATE: Structuring a Dynamical Approach 68

5.1 A Basic Problem: Noncalculable Levels of Energy and Mass Flow 69

5.2 An Overall Strategy 72

5.3 Notational Simplifications for Resolving Total Climate Variability 74

5.4 A Structured Dynamical Approach 76

5.5 The External Forcing Function, F 82

5.5.1 Astronomical/Cosmic Forcing 82

5.5.2 Tectonic Forcing 82

6 BASIC CONCEPTS OF DYNAMICAL SYSTEMS ANALYSIS: Prototypical Climatic Applications 84

6.1 Local (or Internal) Stability 84
6.2 The Generic Cubic Nonlinearity 86
6.3 Structural (or External) Stability: Elements of Bifurcation Theory 87
6.4 Multivariable Systems 92
6.4.1 The Two-Variable Phase Plane 92
6.5 A Prototype Two-Variable Model 95
6.5.1 Sensitivity of Equilibria to Changes in Parameters: Prediction of the Second Kind 97
6.5.2 Structural Stability 99
6.6 The Prototype Two-Variable System as a Stochastic-Dynamical System: Effects of Random Forcing 103
6.6.1 The Stochastic Amplitude 104
6.6.2 Structural Stochastic Stability 104
6.7 More Than Two-Variable Systems: Deterministic Chaos 108

PART II Physics of the Separate Domains

7 MODELING THE ATMOSPHERE AND SURFACE STATE AS FAST-RESPONSE COMPONENTS 113

7.1 The General Circulation Model 114
7.2 Lower Resolution Models: Statistical-Dynamical Models and the Energy Balance Model 115
7.2.1 A Zonal-Average SDM 116
7.2.2 Axially Asymmetric SDMs 117
7.2.3 The Complete Time-Average State 119
7.3 Thermodynamic Models 119
7.3.1 Radiative–Convective Models 119
7.3.2 Vertically Averaged Models (the EBM) 120
7.4 The Basic Energy Balance Model 121
7.5 Equilibria and Dynamical Properties of the Zero-Dimensional (Global Average) EBM 123
7.6 Stochastic Resonance 127
7.7 The One-Dimensional (Latitude-Dependent) EBM 129
7.8 Transitivity Properties of the Atmospheric and Surface Climatic State: Inferences from a GCM 132
7.9 Closure Relationships Based on GCM Sensitivity Experiments 134

7.9.1 Surface Temperature Sensitivity 135

7.10 Formal Feedback Analysis of the Fast-Response Equilibrium State 139

7.11 Paleoclimatic Simulations 143

8 THE SLOW-RESPONSE "CONTROL" VARIABLES: An Overview 146

8.1 The Ice Sheets 147

8.1.1 Key Variables 147

8.1.2 Observations 148

8.2 Greenhouse Gases: Carbon Dioxide 149

8.3 The Thermohaline Ocean State 151

8.4 A Three-Dimensional Phase-Space Trajectory 154

9 GLOBAL DYNAMICS OF THE ICE SHEETS 158

9.1 Basic Equations and Boundary Conditions 158

9.2 A Scale Analysis 163

9.3 The Vertically Integrated Ice-Sheet Model 166

9.4 The Surface Mass Balance 168

9.5 Basal Temperature and Melting 169

9.6 Deformable Basal Regolith 171

9.7 Ice Streams and Ice Shelves 172

9.8 Bedrock Depression 172

9.9 Sea Level Change and the Ice Sheets: The Depression-Calving Hypothesis 173

9.10 Paleoclimatic Applications of the Vertically Integrated Model 176

9.11 A Global Dynamical Equation for Ice Mass 177

10 DYNAMICS OF ATMOSPHERIC CO_2 181

10.1 The Air–Sea Flux, $Q^{\uparrow}$ 183

10.1.1 Qualitative Analysis of the Factors Affecting $Q^{\uparrow}$ 185

10.1.2 Mathematical Formulation of the Ocean Carbon Balance 189

10.1.3 A Parameterization for $Q^{\uparrow}$ 191

10.2 Terrestrial Organic Carbon Exchange, $W_G^{\uparrow}$ 192

10.2.1 Sea Level Change Effects 194

10.2.2 Thermal Effects 194

10.2.3 Ice Cover Effects 194

10.2.4 Long-Term Terrestrial Organic Burial, $\widehat{W}_G^{\downarrow}$ 195

10.2.5 The Global Mass Balance of Organic Carbon 196

10.3 Outgassing Processes, $V^{\uparrow}$ 196
10.4 Rock Weathering Downdraw, $W^{\downarrow}$ 197
10.5 A Global Dynamical Equation for Atmospheric CO_2 200
10.6 Modeling the Tectonically Forced CO_2 Variations, $\widehat{\mu}$: Long-Term Rock Processes 200
10.6.1 The Long-Term Oceanic Carbon Balance 201
10.6.2 The GEOCARB Model 201
10.7 Overview of the Full Global Carbon Cycle 205

11 SIMPLIFIED DYNAMICS OF THE THERMOHALINE OCEAN STATE 206

11.1 General Equations 208
11.1.1 Boundary Conditions 209
11.2 A Prototype Four-Box Ocean Model 210
11.3 The Wind-Driven, Local-Convective, and Baroclinic Eddy Circulations 211
11.3.1 The Wind-Driven Circulation: Gyres and Upwelling 211
11.3.2 Local Convective Overturnings and Baroclinic Eddy Circulations 215
11.4 The Two-Box Thermohaline Circulation Model: Possible Bimodality of the Ocean State 216
11.4.1 The Two-Box System 216
11.4.2 A Simple Model of the TH Circulation 219
11.4.3 Meridional Fluxes 221
11.4.4 Dynamical Analysis of the Two-Box Model 222
11.5 Integral Equations for the Deep Ocean State 226
11.5.1 The Deep Ocean Temperature 226
11.5.2 The Deep Ocean Salinity 228
11.6 Global Dynamical Equations for the Thermohaline State: θ and S_{φ} 229

PART III Unified Dynamical Theory

12 THE COUPLED FAST- AND SLOW-RESPONSE VARIABLES AS A GLOBAL DYNAMICAL SYSTEM: Outline of a Theory of Paleoclimatic Variation 235

12.1 The Unified Model: A Paleoclimate Dynamics Model 236
12.2 Feedback-Loop Representation 238
12.3 Elimination of the Fast-Response Variables: The Center Manifold 241

12.4 Sources of Instability: The Dissipative Rate Constants 242
12.5 Formal Separation into Tectonic Equilibrium and Departure Equations 244

13 FORCED EVOLUTION OF THE TECTONIC-MEAN CLIMATIC STATE 247

13.1 Effects of Changing Solar Luminosity and Rotation Rate 248
13.1.1 Solar Luminosity ($\mathcal{S}$) 248
13.1.2 Rotation Rate (Ω) 249
13.2 General Effects of Changing Land–Ocean Distribution and Topography (h) 249
13.3 Effects of Long-Term Variations of Volcanic and Cosmic Dust and Bolides 253
13.4 Multimillion-Year Evolution of CO_2 255
13.4.1 The GEOCARB Solution 255
13.4.2 First-Order Response of Global Ice Mass and Deep Ocean Temperature to Tectonic CO_2 Variations 259
13.5 Possible Role of Salinity-Driven Instability of the Tectonic-Mean State 260
13.6 Snapshot Atmospheric and Surficial Equilibrium Responses to Prescribed $\hat{y}$-Fields Using GCMs 261

14 THE LATE CENOZOIC ICE-AGE DEPARTURES: An Overview of Previous Ideas and Models 262

14.1 General Review: Forced vs. Free Models 262
14.1.1 Models in Which Earth-Orbital Forcing Is Necessary 263
14.1.2 Instability-Driven (Auto-oscillatory) Models 265
14.1.3 Hierarchical Classification in Terms of Increasing Physical Complexity 266
14.2 Forced Ice-Line Models (Box 1, Fig. 14-1) 266
14.3 Ice-Sheet Inertia Models 267
14.3.1 The Simplest Forms (Box 2) 267
14.3.2 More Physically Based Ice-Sheet Models: First Applications 268
14.3.3 Direct Bedrock Effects (Box 3) 269
14.3.4 Bedrock-Calving Effects (Box 4) 270
14.3.5 Basal Meltwater and Sliding (Box 5) 270
14.3.6 Ice Streams and Ice Shelf Effects 270
14.3.7 Continental Ice-Sheet Movement (Box 6) 270
14.3.8 Three-Dimensional (λ, φ, h_I) Ice-Sheet Models 271
14.4 The Need for Enhancement of the Coupled Ice-Sheet/Atmospheric Climate Models 271

14.5 Ice-Sheet Variables Coupled with Additional Slow-Response Variables 272
14.5.1 Regolith Mass, m_r (Box 7) 272
14.5.2 The Deep Ocean Temperature θ (Box 8) 273
14.5.3 The Salinity Gradient S_φ (Box 9) 274
14.6 Carbon Dioxide, μ (Box 10) 274
14.6.1 Earlier History 274
14.6.2 Quantitative Revival of the Carbon Dioxide Hypothesis 275
14.7 Summary 276

15 A GLOBAL THEORY OF THE LATE CENOZOIC ICE AGES: Glacial Onset and Oscillation 278
15.1 Specialization of the Model 279
15.2 The 100-ky Oscillation as a Free Response: Determination of the Adjustable Parameters 282
15.2.1 Nondimensional Form 283
15.2.2 Internal Stability Analysis to Locate a Free 100-ky-Period Oscillation in Parameter Space 284
15.3 Milankovitch Forcing of the Free Oscillation 286
15.4 Structural Stability as a Function of the Tectonic CO_2 Level 288
15.5 A More Complete Solution 290
15.6 Predictions 295
15.7 Robustness and Sensitivity 297
15.8 Summary: A Revival of the CO_2 Theory of the Ice Ages 298

16 MILLENNIAL-SCALE VARIATIONS 301
16.1 Theory of Heinrich Oscillations 303
16.1.1 The "Binge–Purge" Model 304
16.1.2 Scale Analysis of the Factors Influencing T_B 305
16.1.3 Diagnostic Analysis 306
16.1.4 Dynamical Analysis: A Simple Heinrich-Scale Oscillator 308
16.2 Dynamics of the D-O Scale Oscillations 311

17 CLOSING THOUGHTS: EPILOGUE 314
17.1 Toward a More Complete Theory 314
17.2 Epilogue: The "Ice Ages" and "Physics" 318

Bibliography 321
Index 343
List of Volumes in the Series 351

Prologue

"We dance round in a ring and suppose,
but the secret sits in the middle and knows"

ROBERT FROST (1945)

Like all great poetry, this insightful couplet by Robert Frost contains many levels of meaning for all areas of human experience, some of which, in my view, are especially relevant for the subject of this book. For example, from a physical viewpoint it is very likely that the "secret" of paleoclimatic variability lies in the center of the internal slow-response climate system, e.g., in the deep ocean, the behavior of which we are just beginning to unravel. From a more mathematical viewpoint, it has yet to be established whether the "rings" of oscillatory variability observed, highlighted by the major near-100-ky-period ice ages of the Late Pleistocene, are centered about stable or unstable equilibria that determine whether they are forced or free.

Although these and many other questions are far from resolution, the time seems ripe to at least expose these issues in a more formal way and attempt to provide a framework for building toward a rational explanation (i.e., a theory) of the rich and mounting array of evidence of the variability that Earth's climate has exhibited over the eons. One can therefore only view this book as a skeletal form of a more complete theory of paleoclimatic variability that will undoubtedly be fleshed out considerably in the future as part of a long-term process.

This process will involve a step-by-step merging of increasing observational evidence, which owes so much to the fields of geology, geochemistry, and paleontology, with theoretical understanding in terms of all the conservation principles that must ultimately rest on dynamical considerations. Since, as emphasized in this book, we must essentially proceed in a phenomenological, inductive manner in which we discover the nature of the "slow manifold" of the climate system, including ultra-slow rate constants needed to account for the data, it is probably fair to say that the observational side of the problem will turn out to be the more important of the two. Nonetheless, as indicated by the title, the emphasis here will be on the theoretical side, and, in particular, the aim will be to provide a foundation for a more truly time-dependent (i.e., dynamical) approach to the theory of paleoclimatic evolution. Hence the role of dynamical systems analysis will play a larger role than in most previous treatments.

On a more general level, in attempting to account for long-term paleoclimatic variations we are led to broaden our view of the climate system and restructure our approach to a fuller theory of climate. The aim of this book is to provide a basis for such a restructured approach.

We begin by describing the external forcing of the climate system and the observed response to this forcing, as represented by the proxy evidence for paleoclimatic variations (with emphasis on the ice ages of the Late Cenozoic). An approach toward a theory of the variations is proposed based on concepts from dynamical systems analysis. In this dynamical context the role of general circulation models (GCMs) is placed in a larger perspective as the "equilibrium" component governing only the "fast-response" parts of the full system (e.g., the atmosphere and surficial state, including the biosphere). In contrast, the "slow-response" parts of the system to which the fast-response variables tend to be quasi-statistically equilibrated (for example, the ice sheets and their underlying bedrock and basal states, the deep ocean, and the global carbon inventory) are governed by nonequilibrium (i.e., dynamical) equations. Separate chapters are devoted to the detailed physics operative in each of these slow-response domains. In the last part of the book, comprising five chapters, we demonstrate how all of these components can be combined in a unified theory consisting of a closed set of equations, satisfying all conservative requirements, from which the main paleoclimatic variations can be deduced with some additional predictions. Suggestions for further progress toward a more complete theory of climate evolution are made in the last chapter.

The book is intended for graduate students and research workers in all the areas of climate theory typically included under such umbrella designations as "paleoclimatology," "earth system studies," and "global change research." If there were any single groups to whom this book might be most directly aimed, it would probably be to those, like myself, whose roots lie in dynamical meteorology but whose interests have shifted markedly toward the emerging field of "climate dynamics." My sincere hope, however, is that a much broader group will find the work of value, particularly those in the disparate fields of physics and geology.

In a larger sense, we shall be crossing the spectral gap separating the major peak that represents weather variability from the next major peak that represents ice-age variability. Whereas weather variability is commonly treated by the more deductive, quantitative methods of fluid dynamics wherein the notion of internal system instability is well established, in the case of the latter paleoclimatic variability these methods and ideas have only begun to be applied in the context of the relevant coupled ice/ocean/biochemosphere that is involved. In another respect, this spectral gap also represents a sort of "cultural gap" between practitioners in all the subfields of geology who have brilliantly and painstakingly been reconstructing the paleoclimatic record and practitioners basically steeped in the methods of mathematical physics. It is again my hope that this book will help bridge this gap.

To this latter end, an extensive bibliography containing contributions to both the geological and physical/dynamical sides of the problem is included; but, in spite of our efforts to be comprehensive, even this long bibliography is undoubtedly incomplete and will not do sufficient justice to many highly worthy contributions to this vast subject area. At the least, there should be enough material to give the reader a good start toward a fuller exploration of the topic being referenced. In this regard, we must recognize that a single book cannot possibly cover all the important aspects of a subject so broad as paleoclimatology. As noted, we focus here on the overall structure and approach to a comprehensive theory based on dynamical system ideas. This work should therefore be complemented by books devoted to other aspects and more specialized treatments of topics that we could only give a rather diluted coverage. The following works are good examples of such books, listed by field: paleoclimate reconstruction techniques and observations (Hecht, 1985; Crowley and North, 1991; Frakes *et al.,* 1992; Lowe and Walker, 1997; Williams *et al.,* 1998; Bradley, 1999; Cronin, 1999); atmospheric climate physics and modeling (Washington and Parkinson, 1986; Peixoto and Oort, 1992; Hartmann, 1994; McGuffie and Henderson-Sellers, 1997); atmospheric and oceanic dynamics (Gill, 1982); cryospheric dynamics (Paterson, 1994; Oerlemans and Van der Veen, 1984; Hughes, 1998); climate system modeling (Trenberth, 1992); geophysical dynamical systems (Ghil and Childress, 1987); and biogeochemical cycles (Jacobson *et al.,* 2000).

To conclude this prologue we recall a perceptive observation by my former colleague at Yale, Richard Foster Flint (1974), that, whereas we now have good basic paradigms for the history of Earth's surface (global tectonics) and the history of life on Earth (Darwinian evolution), a similar basic paradigm for the history of climatic evolution still remains to be established. In this book we shall suggest a candidate for such a paradigm centered on the role of carbon dioxide and the potential for instability in the full slow-response climate system—but this is getting way ahead of the story. Let us first enter gently into the marvelous fantasy-land of the past Earth that many devoted scientists have created, in terms of both description and theory, ever hopeful that we are indeed truly discove ring this major piece in the history of Earth.

REFERENCES

Bradley, R. S. (1999). "Paleoclimatology: Reconstructing Climates of the Quaternary," 2nd Ed. Academic Press, San Diego.

Cronin, T. M. (1999). "Principles of Paleoclimatology." Columbia University Press, New York.

Crowley, T., and G. North (1991). *In* "Paleoclimatology," p. 339. Oxford University Press, New York.

Frakes, L. A., J. E. Francis, and J. I. Syktus (1992). "Climate Modes of the Phanerozoic." Cambridge University Press, New York.

Ghil, M., and S. Childress (1987). "Topics in Geophysical Fluid Dynamics: Atmospheric Dynamics, Dynamo Theory, and Climate Dynamics." Springer-Verlag, New York.

Gill, A. E. (1982). "Atmosphere-Ocean Dynamics." Academic Press, Orlando.

Hartmann, D. L. (1994). "Global Physical Climatology." Academic Press, San Diego.

Hecht, A. (1985). *In* "Paleoclimate Analysis and Modeling," p. 445. Wiley, New York.

Hughes, T. (1998). *In* "Ice Sheets," p. 399. Oxford University Press, New York.

Jacobson, M. C., R. J. Charlson, H. Rodhe, and G. H. Orians, eds. (2000). "Earth System Science." Academic Press, San Diego.

Lowe, J., and M. Walker (1997). *In* "Reconstructing Quaternary Environments," 2nd Ed., p. 446. Longman, New York.

McGuffie, K., and A. Henderson-Sellers (1997). *In* "A Climate Modelling Primer," p. 217. Wiley, New York.

Oerlemans, J., and C. Van der Veen (1984). *In* "Ice-Sheets and Climate," p. 217. Reidel, Dordrecht.

Paterson, W. (1994). *In* "The Physics of Glaciers," 3rd Ed., p. 480. Pergamon Press, Oxford.

Peixoto, J., and A. Oort (1992). *In* "Physics of Climate," p. 520. American Institute of Physics, New York.
Trenberth, K. E. (1992). "Climate System Modeling." Cambridge University Press, Cambridge.
Washington, W., and C. Parkinson (1986). *In* "An Introduction to Three Dimensional Climate Modelling," p. 422. University Science, Mill Valley, CA.
Williams, M., D. Dunkerley, P. de Decker, P. Kershaw, and J. Chappell (1998). "Quaternary Environments." Arnold, London.

Acknowledgments

At the outset, I express my heartfelt thanks to Kirk Maasch, whose efforts to help me see this book through to publication kept my spirits high and hopes of completion alive during my darker days of illness. It is to Kirk that I owe the final publication-ready form of the text including all of the figures.

In earlier stages, Aida Rodriguez expertly and cheerfully transformed my handwritten manuscript to serviceable typescript through many, perhaps exasperating, revisions over several years. This was followed by further transformation of Aida's work into a more mathematically friendly computer format, with further revisions, by Zavareh Kothavala. I am extremely grateful to Zav, who volunteered for this task at the expense of a good deal of his valuable postdoctoral research time and energy. I am deeply thankful to Aida, Zav, and Kirk for their indispensable technical assistance.

On the scientific side, this book can only be viewed as the end product of many years of scholarly interaction and collaborative research with many individuals. I take great pleasure in expressing my gratitude to the many associates and students with whom I have been fortunate to work and share in the development and growth of the field now termed "climate system dynamics." Of foremost significance for the special paleoclimatic emphasis of this book are my major collaborators, Alfonso Sutera, Kirk Maasch, Bob Oglesby, and Mikhail Verbitsky and also Dick Moritz, with whom I shared some early adventures in this field when he was a graduate advisee. Tony Hansen and Gianni Matteucci also made very special contributions for which I am grateful. My ongoing learning process in this highly multidisciplinary field owes a great deal to all of the above-mentioned individuals, as well as to Bob Berner (carbon cycle) and Terry Hughes (ice-sheet dynamics).

The unfailing encouragement of my wife, Sheila, who has patiently dealt with the obsessiveness needed to complete a book of this kind, has been constant throughout this writing process, as well as throughout my career. It is to her that I dedicate this work.

Although my children, Matthew and Jennifer, have been on the sidelines during this endeavor, they have always been on my mind as beautiful representatives of the next generation to whom this book is largely aimed. Thus, my whole family has participated in my work over the years.

Finally, I take pleasure in acknowledging my gratitude to the National Science Foundation for its continued generous support of our research program at Yale University over many years.

List of Symbols

Principal Symbols and Definitions

Symbol	Definition	Section or Chapter of First Use
a	(1) Radius of Earth	4.1
	(2) Characteristic snowfall rate ($\mathrm{m\,s^{-1}}$)	9.2
A	(1) Parameterization constant for longwave radiative flux	7.4
	(2) Alkalinity	10.1
$A^{\uparrow}$	Global anthropogenic flux of carbon	10
A_{I}	Ice-sheet area	9.2
A_{w}	Ocean area	9.9
$\mathcal{A}$	Net ice accumulation ($\mathrm{m\,y^{-1}}$)	9.4
b_{j}	Coefficients in dynamical equation for CO_2 variations	15.1
B, B	(1) Parameterization constant for longwave radiative flux	7.4
	(2) Subscript denoting ice-sheet basal state	9.1
	(3) Rate of carbon exchange	10.1
$\mathcal{B}$	Coefficient for log response of temperature to CO_2	7.9.1
c	Specific heat at constant volume ($\mathrm{J\,kg^{-1}\,K^{-1}}$)	4.1
C	Rate of condensation in column	4.5
$C^{\uparrow}$	Upward thermal conduction in an ice sheet	16.1.1
C_{a}	Total mass of atmospheric carbon	10
C_{G}	Total mass of organic carbon	10.2
C_{w}	Total mass of carbon in ocean	10.6
$\mathcal{C}$	Rate of condensation	4.1
$\mathcal{C}_{\mathrm{I}}$	Rate of ice-sheet loss due to calving	9.9, 9.11
$\mathcal{C}^{\uparrow}_{\theta,\mathrm{S}}$	Rate of decrease of deep ocean temperature (θ), or salinity (S), due to upward advection	11.5
d	(1) Depth of upper ocean box	10.1.2, 11.2
	(2) Index for upper ocean box property	10.1.2
d_{E}	Depth of ocean Ekman layer	11.3

Principal Symbols and Definitions

Symbol	Definition	Section or Chapter of First Use
D	(1) Vertical depth scale	4.1, 4.3
	(2) Eddy conduction coefficient ($\mathrm{W\,K^{-1}}$)	7.4
	(3) Depth of ocean floor	11.1
	(4) Index for deep ocean properties	11.4
D_B	Mean bedrock depression	5.4, 9.9
$\mathcal{D}$	Rate of frictional heating per unit area due to basal ice sliding	9.1
$\mathcal{D}^{\downarrow}_{\theta,\mathrm{S}}$	Rate of increase of deep ocean temperature (θ), or salinity (S), due to downward diffusion	11.5
e	(1) Eccentricity of Earth's orbit	1.3
	(2) Snow line or equilibrium line altitude (ELA)	9.4
E	Rate of surface evaporation ($\mathrm{kg\,m^{-2}\,s^{-1}}$)	4.5
$\mathcal{E}$	Rate of evaporation ($\mathrm{kg\,m^{-3}\,s^{-1}}$)	4.1
f	Coriolis parameter	11.1
F	External forcing function	5.4
$F^{\uparrow}$	Upward freshwater flux ($\mathrm{kg\,m^{-2}\,s^{-1}}$)	11.1
$\mathbf{F}$	Frictional force per unit volume	4.1
$\mathcal{F}^{\uparrow}$	Net upward freshwater flux ($\mathrm{m^3\,s^{-1}}$)	11.4
g	Acceleration of gravity	4.1
g_i	Equilibrium solution for fast-response variable x_i as a function of $y = (Y, F)$	7.9
$G^{\uparrow}$	Upward geothermal flux ($\mathrm{W\,m^{-2}}$)	1.3
$\mathcal{G}^{\uparrow}$	Net geothermal flux ($\mathrm{m^3\,K\,s^{-1}}$)	11.4
$\mathcal{G}^{\uparrow}_{\theta}$	Rate of deep ocean temperature increase due to geothermal flux	11.5
h	Height of the surface relative to sea level	4.4, 5.5
h_I	Thickness of ice column	4.1
h_r	Thickness of deformable regolith	5.4
H	(1) Scale thickness of ice sheet	8.1
	(2) Heinrich episode	16
$H^{(j)\uparrow}_\mathrm{S}$	Upward surface heat flux due to solar radiation ($j = 1$), terrestrial radiation ($j = 2$), and small-scale convection ($j = 3$)	4.5
$H^{(j)\downarrow}_\mathrm{T}$	Downward heat flux at the top of the atmosphere due to solar radiation ($j = 1$) and longwave radiation ($j = 2$)	4.5
$\mathcal{H}^{\uparrow}$	Net upward heat flux at ocean surface ($\mathrm{m^3\,K\,s^{-1}}$)	11.4
i, i	(1) Index for set of variables	4.2
	(2) Subscript denoting ice property	9.1

Principal Symbols and Definitions

Symbol	Definition	Section or Chapter of First Use
$\mathbf{i}$	Eastward unit vector	4.1
I, I	(1) Global ice mass	3, 8.1
	(2) Subscript denoting ice-sheet property	9.1
j	Index for set of variables	4.2
$\mathbf{j}$	Northward unit vector	4.1
J_0	Jacobian matrix	6.1
$J_{\mathrm{I},\theta}$	Sea level change per ice mass change (I) or ocean temperature change (θ)	9.9
$\mathbf{J}$	Horizontal flux vector ($J_\lambda \mathbf{i} + J_\varphi \mathbf{j}$)	4.5
$\mathcal{J}_{\mathrm{p}}$	Energy flux across latitude φ_{p}	4.5
k	Thermal diffusivity ($\mathrm{m}^2\,\mathrm{s}^{-1}$)	4.3, 9.1
$k_j^{(i)}$	Response function of variable X_i to y_j	7.9
k_{B}	Basal ice sliding coefficient	9.3
$\mathbf{k}$	Upward unit vector	4.1
K	Rheological coefficient in Glen's law	9.1
$K_{\mathrm{v,h}}^{(T,S)}$	Oceanic eddy diffusivity for temperature (T) or salinity (S), in vertical (v) or horizontal (h) direction ($\mathrm{m}^2\,\mathrm{s}^{-1}$)	11.1
L	(1) Horizontal spatial scale	4.3
	(2) Lateral ice-sheet extent	9.2
$L_{\mathrm{v,f}}$	Latent heat of vaporization (v) or fusion (f)	4.5
m	Mass per unit area	4.5
M	Melting rate per unit area ($\mathrm{kg\,m^{-2}\,s^{-1}}$)	4.5
$M(X_i)$	Nonlinear function of variables X_i	6.1
$\mathcal{M}$	Melting rate per unit volume ($\mathrm{kg\,m^{-3}\,s^{-1}}$)	4.1
n	Power in Glen's rheological law	9.1
N	Net ice-sheet surface accumulation minus ice creep perimeter loss	9.11
$N(x_i)$	Nonlinear function of variables x_i	6.1
$N^{\uparrow}$	Net upward radiative flux at top of atmosphere [$\equiv (H_T^{(1)\downarrow} - H_T^{(2)\uparrow})$]	4.5
p	Pressure	4.1
p_{s}	Surface pressure	4.4
P	(1) Period of oscillation	4.2
	(2) Precipitation rate ($\mathrm{kg\,m^{-2}\,s^{-1}}$)	4.5
	(3) Phosphate mass	10.1
P_{i}	Snowfall rate ($\mathrm{kg\,m^{-2}\,s^{-1}}$)	4.5
$\mathcal{P}$	Departure of pressure from its hydrostatic value	11.1

Principal Symbols and Definitions

Symbol	Definition	Section or Chapter of First Use
q	Rate of heat addition per unit mass (W kg^{-1})	4.1
$q^{\uparrow}$	Sea to air carbon flux	10
$q_{\psi,\phi}$	Oceanic thermohaline (ψ) or gyre (ϕ) volume exchange flux (Sv)	11.4
$Q^{\uparrow}$	Net global sea to air flux of carbon	10
Q_x	Net horizontal flux of property x	11.4
r	Subscript denoting regolith property	9.6
$r_{1,2}$	Semimajor (1) and minor (2) axes of Earth's elliptic orbit	1.3
R	(1) Gas constant for air	
	(2) Incoming solar radiation at the top of the atmosphere ($\equiv (H_{\mathrm{T}}^{(1)\downarrow})$ (W m^{-2})	5.5
$R^{(x)}$	Rate of riverine flux of property (x) to the ocean	10.1
$\mathcal{R}$	River water flux into ocean	11.1
s	Nondimensional equator to pole salinity differences, δS	11.4
S	Salinity (%)	4.1
$S_{\mathrm{w}}^{\uparrow}$	Rate of basal water thickness loss due to downward flux	9.5
$\mathcal{S}$	Solar constant (W m^{-2})	1.3
$\mathcal{S}_0$	Present value of solar constant (1368 W m^{-2})	7.5
$\mathcal{S}_\xi$	Source function for constituent (ξ)	4.1
$\mathcal{S}_{\mathrm{I}}$	Rate of ice mass loss due to surging	9.9, 9.11
t	Time	4.1
t^*	Time scale for ice-sheet surface to base advective flow	9.2
T	Temperature	4.1
T_{B}	Temperature at the base of ice sheet	9.1
T_{M}	Pressure melting-point temperature	8.4, 9.1
T_ψ	Summer sea level temperature near ice sheet	9.11, 15.1
u	Eastward speed	4.1
U	Scale of horizontal motion in ice sheet	9.2
$U^{\downarrow}$	Rate of cosmic mass influx	5.5
v	Northward speed	4.1
$\mathbf{v}$	Horizontal velocity vector ($u\mathbf{i} + v\mathbf{j}$)	4.1
V	(1) Lyapunov pseudo-potential	6.1, 7.5
	(2) Volume	8.3
$V^{\uparrow}$	Volcanic flux to the atmosphere	5.5
V_α	Set of velocity components ($\alpha = u, v, w$)	9.1
$\mathbf{V}$	Three-dimensional velocity vector	4.1

Principal Symbols and Definitions

Symbol	Definition	Section or Chapter of First Use
w	Upward speed	4.1
W	(1) Characteristic vertical ice motion	9.2
	(2) Width of an ocean basin	11.4.1
W_B	Water thickness at base of ice sheet	5.4, 9.5
$W{\downarrow}$	Net rock weathering downdraw of CO_2	10
$W_G^{\uparrow}$	Net organic weathering release of CO_2	10
$\mathcal{W}$	Carbonate–silicate weathering	5.5
x	(1) Arbitrary variable	4.2
	(2) Cartesian eastward distance	4.1
X	Set of fast-response variables	5.4
y	(1) Cartesian northward distance	4.1
	(2) Combined set of slow-response variables and forcing functions $(Y + F)$	7.9
Y	Set of slow-response variables	5.3
z	(1) Upward vertical distance	4.1, 9.1
	(2) Arbitrary variable	6.1
Z	(1) Scale height above prevailing sea level of a continent, undisturbed by the presence of an ice sheet	8, 9.9
	(2) Ocean depth below sea level	11.1
	(3) Strength of ocean circulation	12.2
$Z^{(x)}$	Oceanic flux convergence of property x	10.1.2
α	Albedo	7.4
α_P	Planetary albedo	7.5
β	(1) Sensitivity function for surface temperature response to change solar constant	6.5.1
	(2) Variation of Coriolis parameter with latitude	11.3
γ	(1) Newtonian cooling coefficient for horizontal eddy heat flux	7.4
	(2) Atmospheric lapse rate	9.1, 16.1.2
Γ	Newtonian restoring coefficient for sea to air heat flux	11.4.1
δ	Averaging time interval	4.2, 5.2
$\delta(X)$	Measure of the ratio of the isotopic form of X to its more abundant form	2.4
$\delta_{i,j}$	Kronecker delta	6.1
δ_B	Local bedrock depression	8.1, 9.1
$\delta_{C,G}$	Mean $\delta\,^{13}C$ value of all carbonate rocks (C) or all organic material (G)	10.6

Principal Symbols and Definitions

Symbol	Definition	Section or Chapter of First Use
δ_X	Difference in property X between equator and pole	11.4
δ_S	Rate of ice thickness loss due to surging	16.1.4
Δ	Depth of the surface active layer	4.4
Δx	Departure of climatic-mean value from the tectonic-mean value of x	5.3
ε	Response time	4.3
$\dot{\varepsilon}_{\alpha,\beta}$	Strain rate tensor (s^{-1})	9.1
ε_D	Response time for bedrock depression	9.11
ζ	(1) Vertical component of vorticity	11.3
	(2) Departure of ice-sheet mass from its tectonic-mean value ($\equiv \Delta I$)	12.5
η	(1) Sine of the ice-edge latitude	6.5
	(2) Thickness scale of ice-sheet basal boundary layer	9.5
θ	Bulk ocean temperature	5.4
ϑ	Departure of θ from its tectonic-mean value $\widehat{\theta}$	10.5, 12.5
κ	Rayleigh friction coefficient	11.3
λ	(1) Longitude	4.1
	(2) Conduction coefficient, i.e., thermal conductivity ($W\,m^{-1}\,K$)	9.1
$\lambda_{1,2,3}$	Basal sliding velocity parameters	9.1
λ_j	Nondimensional parameters ($j = 1, 2, \ldots$)	15.2
Λ	(1) Longitude of perihelion	1.3
	(2) Heat capacity factor ($J\,m^{-2}\,K^{-1}$)	7.4
μ	Atmospheric carbon dioxide concentration (ppmv)	8
μ_T	Coefficient of volume expansion	4.1
μ_P	Coefficient of isothermal compressibility	4.1
μ_S	Salinity coefficient	4.1
μ_i	Viscosity of ice	9.1
ν	Damping coefficient (s^{-1})	9.11
$\nu_{u,v}$	Horizontal eddy diffusivity for momentum	11.1
ν_w	Vertical eddy diffusivity for momentum	11.1
ξ	(1) Mass mixing ratio of a constituent	4.1
	(2) Sine of latitude	6.5
	(3) Sea level change due to ice sheets	9.1
	(4) Departure of atmospheric CO_2 from its tectonic-mean value ($\equiv \Delta\mu$)	10.5, 12.5

Principal Symbols and Definitions

Symbol	Definition	Section or Chapter of First Use
Ξ	Factor relating shear $(\partial\mathbf{v}/\partial z)$ to depth within an ice sheet	9.3
Π	Nondimensional freshwater flux	11.4
Π_x	Total mass of property x	4.4, 4.5
ρ	Density	4.1
ρ_φ	Poleward gradient of ocean density	11.5
σ	(1) Stefan–Boltzmann constant	7.5
	(2) Departure of salinity gradient from its tectonic-mean value $(\equiv \Delta S\varphi)$	12.5
σ_E	Surface area of globe	4.1
Σ	Composite surface climatic state	12.2
$\tau_{\alpha,\beta}$	Viscous stress in α direction on planes perpendicular to β	4.1
$\underline{\tau}_\mathrm{s}$	Surface wind stress	11.1
$\underline{\tau}_\mathrm{B}$	Basal shear stress	9.1
$\underline{\tau}_\mathrm{h}$	Three-dimensional vector stress tangent to ice surface	9.1
Υ	Sensible heat content of a column per unit area	4.5
ϕ	Stream function for wind-driven ocean circulation	11
φ	Latitude	4.1
Φ	(1) Geopotential (gz)	4.1
	(2) Nondimensional ocean gyre volume exchange flux	11.4
χ	Mass concentration (kg m^{-3})	4.1
χ_c	Mass concentration of carbon	10
χ_r	mass concentration of regolith	9.6
ψ	Stream function for the thermohaline circulation	11, 11.4
ψ_i	Set of fast-response variables	7.10
ψ_E	Vertical overturning of the ocean due to divergent part of wind-driven (Ekman) circulation	11
ψ_tot	Total overturning circulation of the ocean $(= \psi + \psi_\mathrm{E})$	11
Ψ	(1) Ice-sheet mass	8.1, 9.11
	(2) Nondimensional thermohaline volume exchange flux	11.4
ω	(1) Stochastic forcing function	4.1
	(2) Eigenvalue	6.1
Ω	Angular speed of Earth's rotation	1.3
$\mathbf{\Omega}$	Angular velocity of Earth	4.1

Averaging and Other Operator Symbols

Symbol	Definition	Section or Chapter of First Use
x^*	Reference value of x	4.1
$\bar{x}$	Time average of x	4.2
$x^\star$	Departure from time average [$\equiv (x - \bar{x})$]	4.2
$\langle x \rangle$	Zonal average of x	4.4
x_*	Departure from zonal average [$\equiv (x - \langle x \rangle)$]	4.4
$\widetilde{x}$	Global area average x	4.4
$x_\star$	Departure from $\widetilde{x}$	4.4
$\check{x}$	Vertical average of x	4.4
$\check{x}$	Global mean value of $x (\equiv \widetilde{\check{x}})$	4.4, 11.4.1
$\widehat{x}$	Tectonic (10 My) average of x	5.3
Δx	Departure from $\widehat{x}$	5.3
x_0	Equilibrium value of x	5.3
x'	Departure from x_0	5.3
∇	Three-dimensional gradient operator	4.1

Acronyms

Symbol	Definition	Section or Chapter of First Use
A-B	Alleröd–Bölling	3.6
AGCM	Atmospheric general circulation model	4.2
CCM	Community climate model (NCAR)	7.8
CLIMAP	Climate: Long-range investigation, mapping, and prediction	3.5
C-O	Climatic optimum	3.6
CSM	Climate system model	5
D-O	Dansgaard–Oeschger oscillation	3.6,14.5.3
DSDP	Deep Sea Drilling Project	15.5
EBM	Energy balance model	4.5
ELA	Equilibrium line altitude	9.4
EMIC	Earth-system model of intermediate complexity	5.4
GCM	General circulation (or global climate) model	4.1,4.3
GEOCARB	Berner (1994) model for long-term CO_2 changes	10.5
GRIP	Greenland Ice Core Project	3.4
IPCC	Intergovernmental Panel on Climate Change	3.7

Acronyms

Symbol	Definition	Section or Chapter of First Use
IRD	Ice-rafted debris	2.3.1
LGM	Last glacial maximum	3.5
MIS	Marine isotope stage	Fig. 16.1
MLM	Mixed-layer ocean model	4.3
NADW	North Atlantic deep water	8.3
NCAR	National Center for Atmospheric Research	7.8
ODP	Ocean Drilling Program	1.1
OGCM	Ocean general circulation model	5
PDM	Paleoclimate dynamics model	5.4
RCM	Radiative–convective model	7.3.1
SDM	Statistical-dynamical model	4.2,7.2
SHM	Surface hydrological model	4.3
SIM	Sea-ice model	4.3
SPECMAP	Spectral Mapping Project	3.3
SST	Sea surface temperature	3.4
TBM	Terrestrial biosphere model	4.3
TH	Thermohaline property	8.3,10.1.1,11
Y D	Younger Dryas	3.6

PART I

Foundations

1

INTRODUCTION
The Basic Challenge

The scientific enterprise proceeds in two main directions: on one hand there is a striving to amass as many raw observational facts as possible, broadening the base of phenomena to be explained, while on the other hand there is a simultaneous effort to find order in these facts by showing they are not wholly independent of one another, but are deducible and predictable, in common, from a set of statements of a relatively general and concise nature (i.e., a theory). In trying to describe and understand the major past variations of Earth's climate, we quickly become aware of the enormous difficulties and deficiencies in both of these directions, and of the necessity that progress in one direction will go hand-in-hand with progress in the other. Therefore, although our emphasis will be on the theoretical side, it is important to provide at least a brief review of paleoclimatic observational techniques and findings. Before discussing these observational aspects more fully in Chapters 2 and 3, we devote this first chapter to certain fundamental ideas that define our area of concern: the climate system; internal climatic variables and external forcing; a brief review of some of the more salient observations, including the basic signature provided by ice variations; and the theoretical challenge posed by the observations.

1.1 THE CLIMATE SYSTEM

Let us begin by defining "climate" as the roughly 100-year average state of the atmosphere and surface boundary layers where we and most other biota reside. In choosing this averaging period we are essentially defining paleoclimatic variations by excluding, *ab initio*, explicit consideration of all the variations that occur within a human lifetime (e.g., seasonal, interannual, and interdecadal variations), though the variability and covariability on these scales must be implicitly represented through parameterizations or stochastic forcing in discussing paleoclimate. This average state includes not only the mean conditions, but the variances and all higher moments of the probability distributions of all the relevant physical variables (e.g., temperature, wind velocity, water vapor concentration, clouds, precipitation). In contrast, the "climate

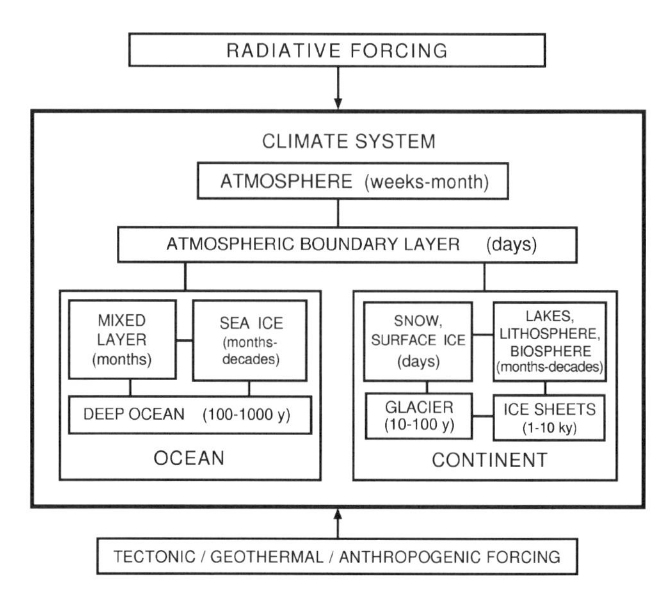

Figure 1-1 Schematic representation of the component domains of the internal climate system, showing their typical response time constants, and main sources of forcing.

system" comprises all those physical domains that interact freely with the atmosphere and boundary layers and influence their states; this includes the oceans, ice masses, and biosphere, as well as those upper portions of the solid Earth that participate in the changes of the overlying atmosphere. External "climatic forcing" is defined as all those factors that influence the climate system but are not themselves influenced by climatic behavior (e.g., solar radiation, tectonic motions of Earth's lithosphere driven by mantle convection). In Fig. 1-1 we schematically portray the internal climate system and external forcing.

1.2 SOME BASIC OBSERVATIONS

From all indications the state of all these domains of the climatic system has been changing continuously throughout Earth's history—at times, relatively rapidly. A highly idealized spectrum of atmospheric thermal variability at a midlatitude point over the age of Earth is shown in Fig. 1-2. The highest frequency fluctuations are due to phenomena in the atmosphere, where much of the total variance resides in frequencies shorter than 1 month. This reflects the presence of phenomena such as zonal wind (index) cycles, weather-producing cyclones and anticyclones, mesoscale systems, cumulus cells, and microturbulent eddy motions, as well as the diurnal cycle. The ocean tends to contribute variations of lower amplitude and frequency, compared to the atmosphere, and the glacial ice sheet variations contribute the lowest frequencies in the system. In fact, much of the variability exhibited at periods greater than 10 thousand years (ky) is known from geologic evidence to be strongly associated with planetary

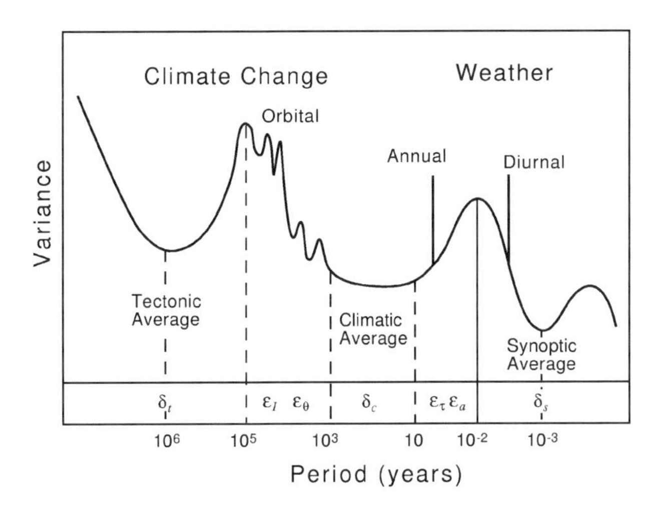

Figure 1-2 Hypothetical, highly idealized, spectrum of atmospheric thermal variance at a midlatitude point over the age of the Earth. δ_s, δ_c, and δ_t denote the synoptic, climatic, and tectonic averaging intervals, respectively, and ε_a, ε_τ, ε_θ, and ε_I denote the approximate response times for the atmosphere, oceanic mixed layer, deep ocean, and ice sheets, respectively, as will be discussed in Section 4.3.

ice mass variations, which provide a primary signature of the changing state of the climate system on longer time scales. In this regard, it is of fundamental significance that, due to Earth's special distance from the sun and its basic gaseous composition, the temperature of Earth's atmosphere and surface embraces values that admit the presence of water in all three of its phases. Thus, Earth's climate system is delicately poised near the freezing point of water, allowing relatively slow but major fluctuations in the proportion of surficial ice to liquid water that involve only small perturbations of the global energy cycle. The low levels of net energy flux involved, occurring in a complex, heterogeneous, nonlinear, nonequilibrium system, pose a class of problems that is as difficult and important as those more commonly treated in modern physics.

An estimate of the temporal variations of temperature over the age of Earth is pictured in Fig. 1-3 (Frakes, 1979), showing that there have been at least two main episodes of major cooling and glaciation over the past 600 million years (My), i.e., over the Phanerozoic. These episodes occurred about 300 My ago (Ma) during the Permo-Carboniferous, and during the past few million years (late Cenozoic). The Cretaceous (about 100 Ma) was a period of particularly warm climate. We shall be concerned with this full time range, with special emphasis on the late Cenozoic, during which time a complex set of fairly well-documented climatic changes occurred, including a dramatic sequence of glaciations. In Fig. 1-4 we show an oxygen isotope (δ^{18}O) record in a tropical western Pacific sedimentary core (ODP 806) for the past 5 My, high values measuring the combined effects of increased planetary ice mass (net global evaporation over precipitation) and more localized cooling and/or salinization of the ocean surface. (A discussion of the oxygen isotope method and other techniques of paleoclimatic reconstruction is given in Chapter 2.) Most of the δ^{18}O change shown in this record is probably due to the growth and disappearance of major ice sheets on

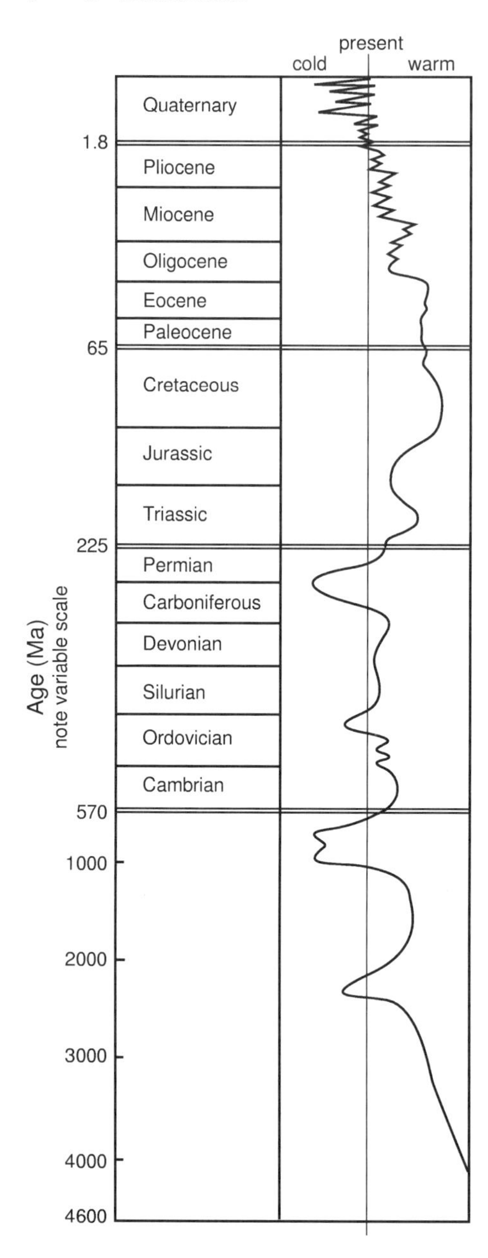

Figure 1-3 Idealized representation of the variations of mean global temperature over the age of the Earth based on geologic proxy evidence. Coldest periods are associated with large-scale glaciation. After Frakes (1979).

North America and Eurasia. Note that in Fig. 1-4, and all subsequent figures, time advances to the right as in most physics contexts but is opposite to convention in most geological contexts.

A more detailed, expanded representation of the changes over the past 1 My is shown in Fig. 1-4b. In this figure, and subsequently in this book, we denote "ky before the present" by ka. The differences between the last glacial maximum (LGM) at about

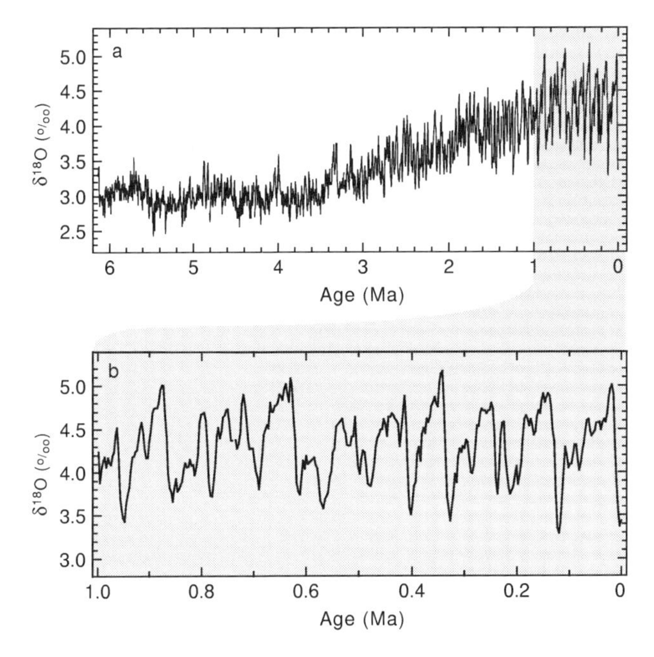

Figure 1-4 (a) δ^{18}O-Derived estimates of global ice mass variability over 6 My; (b) a more detailed view of the past 1 My. After Shackleton (1999).

20 ka and the present day minimum value represent a major change of the face of our planet and its climate, as depicted in the reconstruction of the ice sheets at 18 ka described in Chapter 3. From spectral analyses of the curves shown in Fig. 1-4 it has been determined that large-amplitude near-100-ky period oscillations occurred over roughly the past 900 thousand years (ky) (the late Pleistocene), preceded by a roughly 2-My time span during which ice levels were generally higher than at present, characterized by shorter period, mainly near-40-ky period, oscillations. At roughly 2.5 Ma, a major transition occurred from an earlier regime of minimal ice amounts and lower amplitude variations. Thus, these past 2.5 My represent a "glacial epoch," embedded in which are "ice ages" representing periods of extensive glaciation (including the major 100-ky period oscillations of the late Pleistocene). We now find ourselves at a minimum (interglacial) point of the last of these oscillations, recovering from the last ice age that reached its maximum at 20 ka. During this last glacial period detailed sedimentary core records reveal a quasi-oscillatory (6- to 10-ky period) sequence of episodes of massive iceberg discharges from the ice sheets surrounding the North Atlantic, known as Heinrich events. Moreover, during the transition to the present state, as well as during the glacial state, higher frequency (millenial-scale) variations have been recorded (e.g., the so-called Dansgaard–Oeschger oscillations). A fuller discussion of the paleoclimatic observations is given in Chapter 3.

The long-term goal is to construct a theory by which we can establish a predictive connection between the external forcing and the observed behavior of the climate system such as shown in Figs. 1-2, 1-3, and 1-4. By a "theory" we simply mean a set of

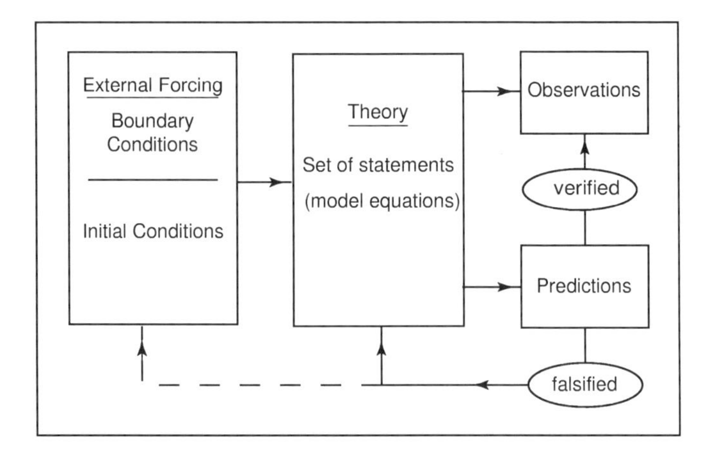

Figure 1-5 Schematic representation of the role of theory in providing, as an ongoing process, the connection of both external forcing (i.e., boundary conditions) and initial conditions with the observations of paleoclimatic variability.

statements consistent with the physical principles of conservation of mass, momentum, and energy (most conveniently expressed in a mathematical form) from which we can deduce this observed behavior and also predict (and thereby guide the search for) some as yet unobserved past or future behavior. (It would, of course, be desirable to test the long-term predictions of a paleoclimate theory for the future climatic state, but alas, verification would be possible only by generations well beyond our own.) In Fig. 1-5 we depict schematically the role of theory in providing, as an ongoing process, the connection of both boundary conditions (i.e., external forcing) and initial conditions with the observed paleoclimatic variability. Thus, a theoretical model in our sense, as represented by the central box in Fig. 1-5, is no more than our latest hypothesis for how the climate system works (in our case on paleotime scales), to be constantly tested against observations and improved by modifying the set of statements comprising the model.

Because, as noted, we hold as an article of faith that these statements must be expressions of the known conservation laws, our particular task is to isolate from the myriad of potential feedbacks those that are essential to account for the variability observed. We are in fact dealing with what may truly be termed a "complex system" involving all the physical domains schematically portrayed in Fig. 1-1 (atmosphere, hydrosphere, cryosphere, lithosphere, and biosphere), each possessing different, heterogeneous properties describable by many variables, and varying with vastly different time constants over vastly different spatial scales. Though the equations expressing the governing thermomechanical laws for each domain or subdomain may be known (see Chapter 4), these same equations may be inadequate when applied to the whole "complex" system on paleoclimatic time scales. It may therefore be desirable to develop completely new phenomenological statements that would likely take the form of a nonlinear dynamical system, which still satisfy the conservation principles.

1.3 EXTERNAL FORCING

In trying to explain variability of the kind shown in Fig. 1-4 it is natural to look first for possible changes in external forcing that can directly excite a response of the kind observed in the internal system. Before we start a brief review of the possible sources of such variable forcing, it cannot be overemphasized that a good deal (perhaps most) of observed paleoclimatic variability may owe its origin to *steady* forcing (e.g., the basic unperturbed meridional variation of insolation between equator and pole). Such steady forcing may power major "free" climatic changes through internal instability of the states that tend to be generated by the steady forcing.

The variable external factors affecting the climatic system can be divided into two main groups: (1) those of an astronomical or cosmic nature controlled largely by the processes involved in the maintenance and evolution of the solar system and beyond, and (2) those of a tectonic nature involving energy sources within the solid portion of the planet.

1.3.1 Astronomical Forcing

1.3.1.1 Solar luminosity changes. The sun's shortwave radiative output, as measured by the energy flux normal to the outer limit of the planetary atmosphere (the solar "constant" $\mathcal{S}$), varies to some extent even on shorter (secular) time scales (connected to sunspot variations, for example), but from solar–astrophysical considerations (Newman and Rood, 1977) is believed to have increased monotonically over the age of Earth at a rate of about 68×10^{-9} W m^{-2} y^{-1} (see Fig. 1-6a). Other, much less energetic variations of the solar corpuscular flux (the "solar wind"), which interacts strongly with Earth's magnetic field out to distances of more than 15 Earth radii, have undoubtedly also occurred.

1.3.1.2 Earth-orbital changes. The most regular and best known climatic forcing variations result from Earth's changing orbital relation to the sun. These are primarily the familiar diurnal and annual (i.e., seasonal) variations associated with Earth's rotation, tilt, and elliptical revolution about the sun, and the perturbations of these seasonal variations that provide the most important source of external forcing of the climate system on the same scale as the ice ages. These latter perturbations are due to the changes in Earth's orientation and position relative to the sun due to changes in the following orbital elements: (1) the eccentricity

$$e = \frac{\sqrt{r_1^2 - r_2^2}}{r_2}$$

where r_1 and r_2 are the semi major and minor axes of the elliptic orbit; (2) the obliquity or angular tilt of Earth's axis relative to the plane of the ecliptic, ε; and (3) the precessional index ($e \sin \Lambda$), where Λ is the longitude of perihelion measured from a fixed point along the orbit, the vernal equinox. These variations, caused by interactions of Earth (particularly its oblate form) with the sun, moon, and the other planets and

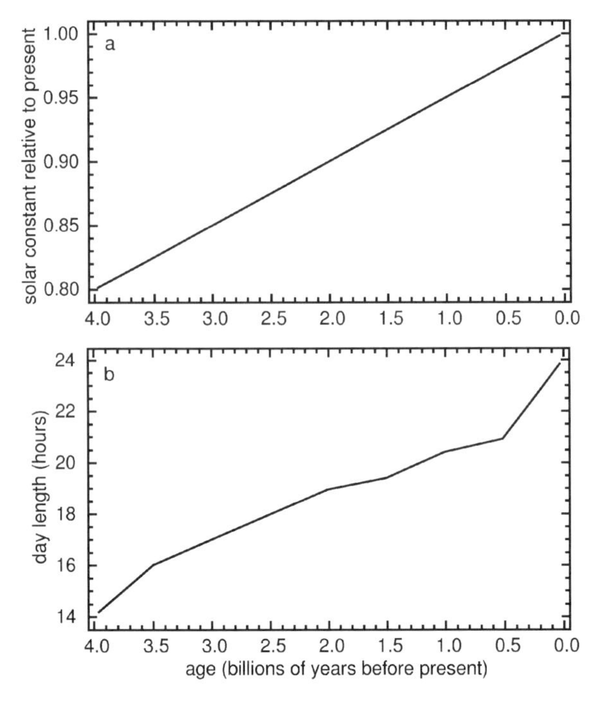

Figure 1-6 (a) Evolution of the solar constant and (b) of the length of the day. After Kuhn *et al.* (1989).

calculable to fairly high accuracy by the methods of celestial mechanics (Stockwell, 1875; Laskar, 1988), are shown in Fig. 1-7 for the past 2 My (Berger, 1978) [see Ghil and Childress (1987, pp. 412–419) and Bradley (1999, pp. 35–46) for good general discussions]. To a first approximation eccentricity varies with dominant periods near 100 and 400 ky; obliquity varies with split dominant periods near 41 ky; and the precession index varies with a split period near 19 and 23 ky modulated by the near-100- and 400-ky eccentricity periods. Henceforth, we shall often refer to these latter split 19/23-ky periods simply as the "near-20-ky" precession index period. All of these orbital features appear to prevail with some small modifications over at least the past 1.5 My (Berger *et al.*, 1998a); beyond about 35–50 Ma the instability of the orbital motions limits the possibility for accurate calculation of the Earth-orbital parameters (Laskar, 1999).

It was suggested qualitatively by Adhémar (1842) and Croll (1875) and later determined quantitatively by Milankovitch (1930, 1941) that these systematic variations of Earth-orbital elements (e, ε, Λ) would lead to varying meridional distributions of insolation that could force climatic changes on Earth and possibly account for the late Cenozoic ice ages (see also Vernekar, 1972; Berger and Loutre, 1991). Henceforth we shall use the terms "Earth-orbital forcing" and "Milankovitch forcing" interchangeably. Of the orbital periods, only the near-100- and 400-ky eccentricity variations (which have an amplitude of less than 0.03) can lead to a change in the net annual irradiance of the Earth, a low value of e (nearly circular orbit) corresponding to a less

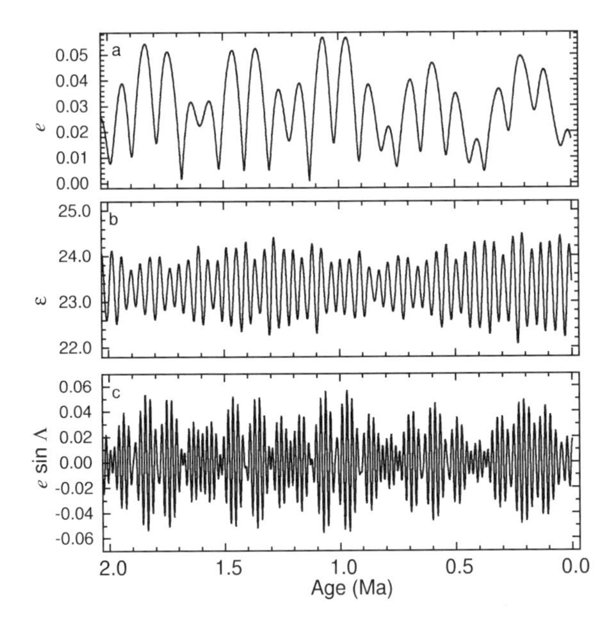

Figure 1-7 Secular variation of the orbital parameters for the past 2 My according to the expansion of the secular variations of the orbital parameters given by Berger and Loutre (1991). (a) The secular behavior of the eccentricity forcing of the solar constant, e. (b) The secular behavior of the obliquity of the Earth's orbit, ε, in degrees. (c) As above but for the precessional index $e \sin \Lambda$. Notice how the spectral characteristics do not change appreciably between the early and late Pleistocene, but, the paleoclimatic record does.

than 0.2% reduction of the solar constant from its value at a maximum value of e. Assuming the solar constant is 1370 W m^{-2}, this represents a swing of about 2 W m^{-2} over the 100-ky period. The near-41- and 20-ky periods due to variations in ε and Λ, respectively, lead to comparatively high-amplitude swings in the latitudinal distribution of incoming radiation as a function of the season (e.g., at 9 ka, when the Earth was closest to the sun in Northern Hemisphere summer, there was 7% more summer radiation and 7% less winter radiation in the Northern Hemisphere, with important consequences for the strength of the monsoons), but the overall radiation over the globe is a constant. As we shall see, the signature of the effects of these relatively high-amplitude 41- and 19/23-thousand-year cycles seems to be imprinted on the climate record of the past several million years obtained from the deep-sea cores, indicating their probable importance for climate change (Hays *et al.*, 1976).

It has been noted by Muller and MacDonald (1995) that in addition to the Milankovitch Earth-orbital changes (eccentricity, obliquity, precession), the angle of inclination of the plane of Earth's orbit is changing with a period close to 100 ky. They propose that this changing inclination can bring Earth into a zone of high cosmic dust, with possible climatic consequences at this 100-ky period, but this is regarded as highly speculative.

It has also been noted (e.g., Bills, 1998) that because the variations in Earth's orbital parameters are largely due to the oblateness of Earth's shape, any internally driven change in this shape (e.g., due to glacial mass shifts) could alter the orbital parameters. Thus, the orbital changes are not completely in the realm of "external" forcing.

1.3.1.3 Rate of the earth's rotation. There is evidence from tidal theory (MacDonald, 1964) and paleontological studies of layered fossil mollusks (Wells, 1963) that there has been a progressive decrease in the angular speed of Earth's rotation, Ω (excluding the slight variations due to angular momentum and mass redistributions that are an internal function of climate change). This leads to an increasing length of the day over geologic time that must affect the diurnal climate cycle and the global circulation through Coriolis force changes (Hunt, 1979; Jenkins, 1993) (see Fig. 1-6b).

1.3.1.4 Cosmic bolide bombardments. There is increasing evidence that Earth has been subjected episodically to impacts by cosmic fragments, some of which are believed to have led to catastrophic changes in Earth's climate and biosphere. One such event is believed to have occurred about 65 My ago, resulting in massive rapid changes in the Earth's biota, including the extinction of the dinosaurs (Alvarez *et al.*, 1984).

1.3.1.5 Other possible cosmic factors. In addition to the above factors, other long-term extraterrestrial processes have been speculated on as potential sources of external forcing (e.g., Torbett, 1989). These include possibilities such, as intersections of the Earth with interplanetary dust, as in the orbital inclination hypothesis, and changes associated with the movement of the entire solar system relative to the galaxy, in addition to the movement of galaxies relative to some fixed point.

In all of our considerations we take as constant other astronomical properties of the Earth that have a fundamental bearing on its climate—for example, the mass, dimensions, and shape of the Earth, and tide-generating forces due to other celestial bodies.

1.3.2 Tectonic Forcing

1.3.2.1 Plate tectonic crustal movements. Under the influence of massive sluggish convective motions in Earth's mantle, the face of Earth has been changing continuously on long time scales, altering the shape and distribution of continents and oceans. An estimate of the variation of the ratio of land to ocean over the age of Earth is shown in Fig. 1-8, and in Chapter 3 (Fig. 3-1) we show an estimate of the evolving geographic distribution of the continents over the past 500 My. The topographic and bathymetric variations associated with such tectonics are of further significance, as is the nature of the exposed rock that is potentially weatherable, with geochemical consequences (e.g., the carbon cycle). These changes must have first-order implications for Earth's climate, as will be discussed in Chapters 3, 7, and 13.

1.3.2.2 Geothermal heat flux. Due mainly to radioactivity in the solid Earth, a temperature gradient exists near the sublithospheric surface that is associated with a variable upward heat flux in space and time, $G^{\uparrow}$. This flux acts continuously to warm the lower portions of the atmosphere, oceans, and ice sheets, with an average value on the order of 10^{-2} W m^{-2}.

1.3.2.3 Volcanic activity. Major injections of heat, chemicals, and aerosol masses into the atmosphere and oceans, associated with both the tectonic movements and

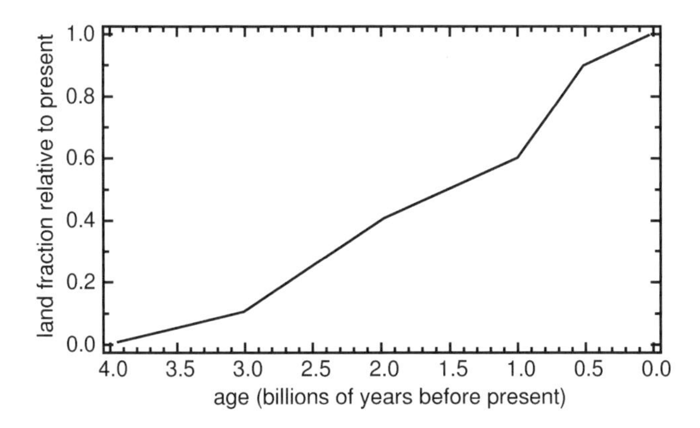

Figure 1-8 Estimate of the variation of the ratio of land to ocean over the age of the Earth. After Kuhn *et al.* (1989).

the magma-producing geothermal heating listed above, have been occurring with a varying degree of intensity throughout Earth's history.

1.3.2.4 Other terrestrial factors. Although we do not know of any direct connections to climate change, we note that the fundamental magnetic field of Earth produced by magnetohydrodynamic forces in Earth's core has varied and even reversed its polarity over geologic time. At the least, these reversals, as imprinted in rocks, have important application in dating stratigraphic climatic evidence (see Section 2.6).

The mass and composition of the planetary fluid envelope, both of which are important properties of the climate system, are influenced to some extent by the externally imposed fluxes mentioned above, as well as by factors dependent on the internal climatic variables (e.g., temperature-dependent CO_2 fluxes between the atmosphere and oceans). We shall assume here, that for terrestrial climatic variations on all but the longest time scales involving the "creation" of the fluid envelope, we can take the total mass of atmosphere, and of water in the combined forms of vapor, cloud, surficial liquid, and surficial ice, as externally given.

Moreover, we shall assume we can consider Earth's atmospheric composition to consist of an externally fixed mixture of major gases, and a smaller variable part that we have already included as internal to the climatic system consisting of water vapor and trace constituents, such as carbon dioxide, ozone, ionized species, and particulates. We should recognize, however, that in a more complete climatic theory we would have to consider the total mass and composition to be fully "internal" (i.e., dependent on the other climatic variables), thereby allowing the possibility for huge changes in the climatic system (e.g., "runaway greenhouse" processes, or oxygen generation due to biospheric evolution).

Although human activities are often considered external, we should also recognize that a good deal of human activities are a direct or indirect response to the environmental challenge posed by the climatic system (e.g., heating for warmth). Included as such "anthropogenic" influences are factors such as heat and mass (e.g., CO_2) fluxes into the atmosphere and oceans, frequently described as "pollution," and altered land surface states arising from industrial, agricultural, and urban activities.

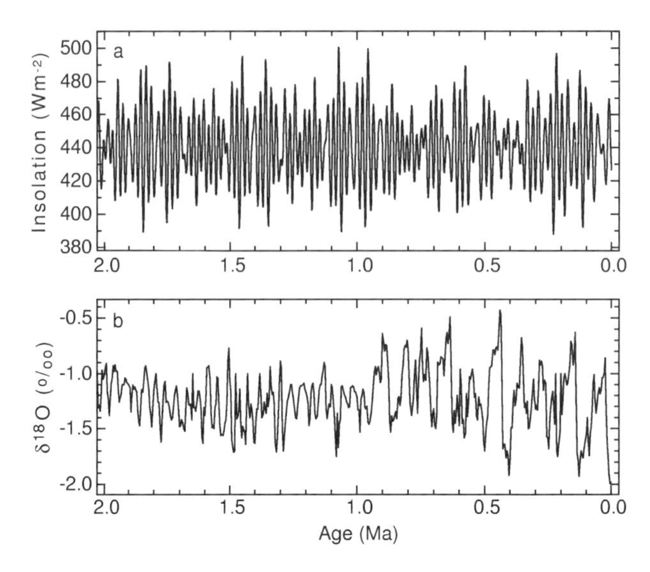

Figure 1-9 (a) Fluctuations of incoming solar radiation in summer at 65°N due to Earth-orbital variations over the past 2 My. After Berger and Loutre (1991). (b) Fluctuations of global ice mass derived from $\delta^{18}O$ variations recorded at western equatorial Pacific site ODP 806 (Berger *et al.*, 1993).

1.4 THE ICE-AGE PROBLEM

In Fig. 1-9, along with another view of the ice variations shown in Fig. 1-4, in this case over the past 2 My, we show a relevant aspect of the main external forcing of the climate system over this period—i.e., the variations due to the Earth-orbital changes of the incoming solar radiation at high Northern Hemisphere latitudes in summer (Berger and Loutre, 1991), believed to exercise a control on the survival of winter snowfall and hence glacial growth (e.g., Milankovitch, 1930). In Fig. 1-10 we compare the spectra of the ice variations and the external radiative changes for the past 1 My, showing the dominance of the near-100-ky ice cycle in spite of imperceptible orbital forcing at this period. It can be surmised, particularly from Fig. 1-9, but also from Fig. 1-10, that the main ice variations bear little resemblance to this known orbitally induced radiative forcing, except for a notable imprint of the near-41-ky obliquity forcing and a relatively lower amplitude imprint of the near-20-ky forcing due to Earth-orbital precession (Hays *et al.*, 1976). However, the 41-ky obliquity forcing was used to help determine the chronology of the $\delta^{18}O$-ice record, thereby enhancing this component in the $\delta^{18}O$ spectrum. Note also, in Fig. 1-9, the lack of agreement near 400 ka, when very weak orbital forcing exists but the $\delta^{18}O$ response is very large, and near 900 ka, when a marked transition occurs in spite of fairly uniform orbitally induced radiative changes over the whole 2-Ma interval.

In spite of the large magnitude of the changes in the climatic state of the Earth represented by all these variations, particularly the major transitions near 2.5 Ma, and near 0.9 Ma when the high-amplitude 100-ky-period oscillations began, no widely ac-

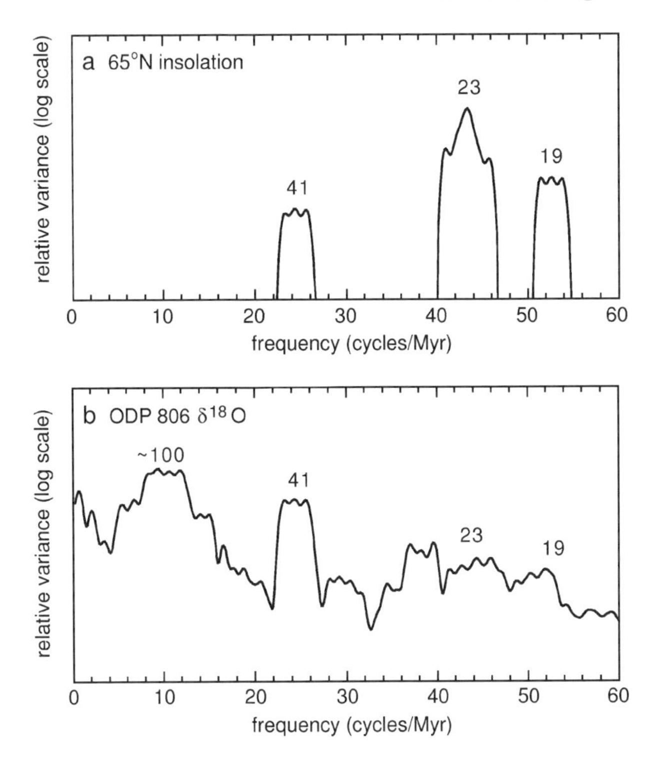

Figure 1-10 Comparison of (a) the spectrum of incoming summer radiation at 65°N due to the Earth-orbital changes shown in Fig. 1-9a with (b) the spectrum of ice mass variations shown in Fig. 1-9b over the past 1 My. The numbers at the top of the spectral peaks are the approximate periods in 10^4-year units (ky).

cepted explanation for these changes presently exists. From an inspection of Figs. 1-9 and 1-10 the following conclusions can be made:

1. The main variations of planetary ice mass do not represent a linear response to the known orbitally induced radiative forcing, having a temporal spectrum that is much different than that of the forcing.
2. These variations are not in quasi-static "equilibrium" with the known forcing, and therefore cannot be determined without taking the nonsteady rates of change of ice into account.
3. Although it is possible that the orbital external forcing may be a necessary condition for the observed ice variations, they cannot be a sufficient condition.

The implication of these facts is that we are necessarily dealing with a nonlinear, nonequilibrium system, involving mechanisms, and variables other than ice, that are internal to the system. Thus, it seems likely that the observed climatic variability represents a complex mix of both external forced and internal free effects. We know, for example, that externally imposed diurnal and seasonal radiative variations lead to significant observed responses, and there is now strong evidence that even the much weaker near-20- and 41-ky radiative variations due to precession and obliquity changes lead to significant climatic responses. On the other hand, we have just shown that the largest variations in ice volume in the late Pleistocene have a broad spectral peak centered at a period of about 100 ky, near which one can find only what appears to be an extremely weak forcing due to Earth eccentricity variations. Such a broad peak is usually indicative of an internally driven fluctuation due to instability, rather than the response to an externally driven forcing like that at the 41-ky period. More generally, we find in the full climate record a broad spectrum of variability, only a very small part of which is directly related to known forcing frequencies. These observations are indicative of strong free variability within the system, at least a part of which can perhaps best be described as "climatic turbulence."

Because any real physical system must be subject to some frictional and diffusive processes, we are led to the realization (Lorenz, 1963) that we are dealing with an extremely complex, forced, dissipative system that undoubtedly contains a rich assortment of linear and nonlinear, positive and negative, feedbacks. The theory for the evolution of such a system will inevitably require all the tools of applied mathematics usually included under the broad title of "dynamical systems analysis." This aspect is discussed in Chapter 4, along with a broader discussion of the physical considerations and difficulties that must be treated, outlining a general approach by which we can reduce the problem to a manageable level. Before embarking on this fuller discussion of the theoretical issues involved, we first review in the next chapter the methods of analysis that enable the reconstruction of past climatic evolution, including the pictures shown in Figs. 1-2 and 1-3. This is followed in Chapter 3 by a review of the inferred climatic evolution over all time scales, with emphasis on the late Cenozoic ice ages, supplementing the material already discussed in this introductory chapter.

2

TECHNIQUES FOR CLIMATE RECONSTRUCTION

The nature and reliability of climatic information available for any earlier time period are highly dependent on the techniques available for climatic reconstruction. In this chapter we present a brief survey of these techniques, ranging from those applicable on the shortest time scales, when quantitative and qualitative human inferences are possible, to the longest scales, which depend heavily on isotopic analysis of core material obtained from stratified sedimentary rocks and ice sheets. We conclude with a discussion of the critical problem of attaching a temporal chronology to these records. Fuller discussions of all these techniques can be found, for example, in Hecht (1985), Lowe and Walker (1997), Crowley and North (1991), Williams *et al.* (1998), and Bradley (1999).

2.1 HISTORICAL METHODS

2.1.1 Direct Quantitative Measurements

The most direct and desirable method for assessing climatic change is to compare statistics of the kind described in most climatology texts for present times with similar statistics developed for past times. Unfortunately, the global network of stations maintained by the national weather services, on which such statistics are based, came into being only in the mid-1800s for surface observations and not until the mid-1900s for the present more extensive surface and upper air networks. Thus we are limited to about 50 years of systematic synoptic upper air data and about 150 years of synoptic surface air measurements that can be portrayed in the form of maps. Even for the present there are serious gaps in the atmospheric network (primarily over the oceans) and there is no permanent international network at all for the synoptic observation of the ocean (though the regular use of satellite observations provides important new coverage for ocean surface temperature as well as for global cloud structure and radiation flux).

Before the formation of the national weather services many individuals and observatories began compiling local records of the weather variables. Such records, which go back as far as the mid-seventeenth century, when most of the basic surface meteorological instruments were introduced (e.g., the barometer, thermometer, hygrometer, and anemometer), are a continuing source of nonsystematic information concerning climatic change on the scale of the past few hundred years (Lamb, 1977; Bradley, 1999).

2.1.2 Descriptive Accounts of General Environmental Conditions

Before the advent of scientific meteorological instrumentation there were many qualitative accounts of the prevailing environmental conditions, in many diverse forms of historiography, literature, and art. Thus, as examples, biblical stories reveal a climatic picture of the middle east region that is probably relevant for several thousand years before the present, and primitive cave paintings can give some clues concerning the climatic conditions that were necessary for the animals portrayed to have existed near the cave locations. Of a somewhat less qualitative nature are the many examples of sequentially recorded climatic data in the form of personal diaries and records of lake levels and times of freezing, spring plantings, harvests, frosts, and other agriculturally significant events (e.g., Ladurie, 1971). More recent literature and art can also give valuable information concerning shorter term climatic changes, such as the relative extent of mountain glaciers during the past several centuries.

2.2 SURFICIAL BIOGEOLOGIC PROXY EVIDENCE

Only by indirect or proxy techniques based essentially on biological and geological information can we infer the climatic states that prevailed before the several thousand years of human recorded history. Indeed, these proxy methods probably provide the most reliable and extensive climatic indicators even for a large part of the period encompassed by recorded historical evidence.

In this section we discuss techniques that depend on analyses of surficial data derived from (1) biological entities that exhibit annual layering (e.g., tree rings, corals) and (2) surface geomorphic evidence (e.g., sea and lake level changes, sand formation and movement, ice and glacier signatures).

In Sections 2.3–2.6 we discuss data derived from biological, physical, and chemical analyses of the time-dependent record of sedimentation and snowfall, as revealed by stratified sedimentary rocks and ice sheets, which may be exposed at the surface or, more importantly, obtained by coring sections through rock, deep ocean beds, and ice sheets.

2.2.1 Annually Layered Life Forms

Perhaps the most useful of this class of climatic indicators are tree rings, i.e., the concentric annual deposits of tree trunk material, the thickness and density of which are proportional to the relative warmth and moisture availability (see Stockton *et al.*, 1985). Certain localized varieties of trees can live for up to 5000 years, permitting the inference of fairly long-term qualitative climatic variations for these localities. Of even greater significance, however, many trees that do not live quite as long (typically, up to 1000 years) cover extensive geographical regions, and hence can reveal broad spatial patterns of climatic change through the use of modern statistical analysis.

Other useful living biological indicators of climatic change are the annual growth layers of the corals that form the vast reefs around the globe, the thickness and chemical composition of which are indicative of the water temperature and salinity in which they form (Gagan, 2000). Likewise, layered calcium carbonate accretions of many forms of molluscan shells are also of use in assessing the changes in surficial water temperature.

As an aside, we note that some species of mollusks reveal both diurnal and annual cycles of layered accretions. Thus, when fossil forms of these species are found in ancient sediments they can indicate the number of days in a year or, equivalently, the rate of rotation of the earth, at the time of deposition of the ancient sediments in which they are found. From this evidence it has been estimated, for example, that 400 million years ago a year contained about 400 days (Wells, 1963). As noted in Section 1.3, rotational changes of this kind are evidence of variable external forcing of long-term climatic change.

2.2.2 Surface Geomorphic Evidence

There are at least four classes of surface geomorphic evidence that are of value in paleoclimatic reconstruction:

1. The structure of marine shorelines as evidence of sea level height. The varying level of the surface of the ocean relative to the continental margins is a function of the amount of water locked up as glacial ice, as well as of the three-dimensional tectonic and eustatic motions of the lithosphere. At some locations such sea level changes are deducible from an examination of the shoreline geologic history and from the history of coral reef growth. After separating the tectonic and eustatic effects, a most valuable indication of global ice volume can be deduced, going back several hundred thousand years (Dodge *et al.*, 1983).
2. Lake shorelines and beds and evidence of past regional hydrologic conditions. The changing levels of lakes over periods ranging up to about 50,000 years can be inferred by old lake terrace formations and by indications of the past presence of lakes where none exist at present. These are clearly of value in developing a qualitative picture of many aspects of the paleohydrologic conditions involving precipitation, evaporation (which is dependent on temperature and wind), and surface flow (e.g., Kutzbach, 1980).
3. Soils and sand. The weathering of rocks and decomposition of organic material into soil and sand, and their chemical and physical properties, are highly dependent on climatic conditions. As examples, "red beds" of soil are related to high temperatures and some dryness in the past, and the presence of evaporites (e.g., salt, potash, and gypsum deposits) is a strong indicator of extended periods of aridity (i.e., negative surface hydrological balance). On the other hand, kaolin-rich soils and peat bogs are

indicative of a past history of humid conditions. The sand formations found in inland desert regions are usually the result of arid climatic conditions that prevailed for many past years, and contain some frozen-in consequences of these past conditions. In particular, the dune structure is one of the few indicators available to determine the winds that prevailed many years in the past (Bagnold, 1954), and even the sand grain textures may give some information in this regard (Krinsley and Wellendorf, 1980). Soil conditions can be especially good indicators of cold, glacial conditions, a subject that is more appropriately included in the following major category of surface geomorphic evidence.

4. Ice and glacier effects. The advance and subsequent retreat of vast ice sheets over a given terrain leave a strong signature on the terrain, as exemplified by (a) scratches (i.e., glacial striations), which are due to moving ice and rocks encased in the ice and are imprinted on the underlying surface rock, indicating the direction of motion of the ice; (b) the form of the valleys through which the ice progresses, which changes from a characteristic preglacial V-shape to a postglacial U-shape; (c) basal water flow beneath and surrounding ice-sheets, and the rocks and soil carried by the ice and subsequently deposited on retreat, which leave characteristic patterns of debris and geomorphology in the forms of moraines, glacial till, rounded boulders, erratics, moulins, and eskers (cf. Andersen and Borns, 1994); the accumulations of blown dust from glacial outwash areas (loess) also constitute an important indicator of past ice-age conditions (e.g., Kukla, 1977), particularly of the aridity and the strength and direction of the winds that prevailed at the time of glacial retreat; and (d) the isostatic depression of the lithosphere under the ice, and its subsequent long-term rebound, which leave characteristic geomorphic patterns as evidence of the past ice distribution.

In addition to the surface effects of large ice sheets itemized above, similar evidence may be used to determine the terminal positions of local mountain glaciers (yielding variations on the order of 40,000 years in some cases). Moreover, in tundra, or so-called periglacial regions, exposed earth sections show long-term sequences of freezing and thawing that can serve as climatic indicators (e.g., Washburn, 1973).

2.3 CONVENTIONAL NONISOTOPIC STRATIGRAPHIC ANALYSES OF SEDIMENTARY ROCK AND ICE

Perhaps the most important means for estimating long-term climatic changes is analysis of the geological record resulting from the deposition in oceans (and lakes) of sediments that are progressively consolidated into layered sedimentary rock. This analysis involves the more conventional determinations of the physical structure of the sediments, including signatures of past surface geomorphic conditions (such as described in the previous section) acquired during deposition or during periods of uplift and surface exposure of the rocks, and the climate-dependent biological content of the layers. In addition, the isotopic composition of the biogenic material (to be discussed in Section 2.4) and nonisotopic chemical methods applied to the deep ocean and ice cores

(Section 2.5) can give indications of the conditions under which the biota, and hence sediments, were formed.

The stratified rock can either be exposed at the surface (as is dramatically illustrated by the Grand Canyon, for example), or by coring cylindrical sections of the subsurface rock on land or the rock below the ocean, where sediments are currently being deposited. Cores and sections through the ancient bottoms of lakes are similarly of interest. Especially revealing are the annually layered (i.e., "varved") lake sediments that can show seasonal variations.

Because ice sheets can develop over many thousands of years, they also contain climate change information determined primarily by the chemical and physical properties of the ice and entrapped solid and gaseous material as a function of core depth.

The determination of the climate associated with the properties of a given stratum of rock or ice is only a part of the problem—it is also necessary to determine the chronological date at which the stratum was deposited, and hence when the inferred climatic conditions prevailed. Thus, the problem of dating stratigraphic sequences (geochronometry) is a critical one that we shall discuss in Section 2.6.

We now review briefly some of the more conventional techniques for determining the climatic information in the stratigraphic sequences.

2.3.1 Physical Indicators

On land surfaces, the composition of the rock assemblages in the strata can give a first clue to the climatic condition of deposition, particularly with reference to glacial conditions, which are easily identified by layers containing glacial debris or moraine material (e.g., tillites).

As other examples, sedimentary layers containing much coarser grained detrital material are indicative of humid, high-energy water drainage conditions; layers containing residues from evaporating shallow sea waters are indicative of hot and arid conditions, and calcium carbonate (e.g., limestone) layers can be indicative of nutrient-rich surface waters supporting an abundance of marine organisms, both in the shallow (planktic) and deep (benthic) ocean. When such high bioproductivity is accompanied by weak oceanic circulation, and hence low oxygen levels at depth, easily identifiable black shale layers may form in the strata.

One of the most interesting proxy indicators in this category for ice-age considerations is the presence of ice-rafted debris (IRD) in the marine cores. This is measurable simply as the fraction of material of terrestrial origin (carried by melting icebergs to deeper portions of the ocean) relative the amount of calcareous ($CaCO_3$) and siliceous material that is normally deposited in the deeper ocean due to the growth and decay of microorganisms. The alternation of glacial IRD episodes and nonglacial episodes can be visually observed in the laboratory by noting the alternation of whitish ($CaCO_3$) layers and much darker IRD layers, but is quantifiable by measuring and recording the percentage of $CaCO_3$ at any core level. Thus, for example, it will be shown in the next chapter that a sharp increase in ice-rafted debris began appearing in high-latitude ocean cores simultaneously in both hemispheres, about 2.5 million years ago, indicating a fairly abrupt onset of a glacial epoch.

Also of value are assessments of thickness and rates of accumulation of dust, ash, and sand that were deposited both over land and over the ocean by paleowinds from the continents. The presence of volcanic ash in a layer is of significance in its own right as an indicator of volcanic activity and the consequent high turbidity levels of the atmosphere, an important external factor forcing climate.

A particularly illuminating feature of the land surface in these regards is the stratified layering of the wind-blown loess deposits (mentioned in Section 2.2.2), giving strong indications of episodic climatic variations.

2.3.2 Paleobiological Indicators (Fossil Faunal Types and Abundances)

The stratigraphic layers contain much fossil evidence of the climate-dependent life forms that existed at the time of sedimentary deposition. These indicators may be in several forms:

1. Pollen and spores, usually found in lake and ocean sediment deposits; these are indicative of the temperature, precipitation, and soil moisture conditions under which the parent plants flourished, as well as wind conditions responsible for transport (e.g., Hooghiemstra and van der Hammen, 1998).

2. Imprints of plant macrofossils or floral assemblages (e.g., leaves), or the organic residues of such assemblages in the form of peat, coal, or oil, for example. The transition of leaves from predominantly smooth-edged forms common at warmer temperatures to serrated-edged forms common at cooler temperatures has been particularly useful as an indicator of the cooling trend over the Cenozoic, as described in Section 3.2 (Wolfe, 1978). In addition, the stomatal properties of fossil leaves have emerged as a potentially important indicator of paleo-CO_2 levels (Royer, 2000; Royer *et al.*, 2000).

3. Fossil faunal assemblages mainly in the form of bone material of larger animals.

4. Of greatest importance because of their abundance and length of record chronologically, calcareous or siliceous shells of marine organisms, including planktonic plants and animals and larger molluscan and shellfish forms. Depending on the species studied, these can be highly indicative of surface or deep ocean (benthic) water conditions of temperature and salinity, and of the extent of sea ice. One foraminiferal species, for example, has the characteristic that its shell spirals to the right in warm water and to the left in cold water. A particularly important species for recording the sea surface temperature changes that accompanied the so-called Heinrich glacial surge events (to be discussed in Section 3.4 and Chapter 16) is *Neogloboquadrina pachyderma*, the abundance of which increases markedly in colder waters. A major advance in the application of this method is represented by the introduction of sophisticated statistical methods by which we can map the oceanic fields of temperature and salinity based on "transfer functions" relating the assemblages of various species to the above physical properties (e.g., Imbrie and Kipp, 1971). Of particular interest from a long-term climatic viewpoint are the comparatively rapid extinctions of species that appear in the stratigraphic record, probably indicative of major changes in the climatic state

of the Earth. In addition to the mere presence or absence of all these life forms as indicators of climate, the isotopic analysis of the material derived from marine plankton and molluscan shells, to be discussed next, has proved to be a particularly important tool for the reconstruction of past climates.

5. More recently a new method has been developed to infer the paleo-variations of surface temperature by determining the alkenone (ketone) content of the fossil cell membranes of certain cocoliths. In particular, in colder waters such organisms increase their production of these organic molecular chains, which are fairly well preserved in the sedimentary column, thereby representing an example of resistant organic molecules and compounds containing environmental information; these are now termed "biomarkers." Extensive applications of this "alkenone thermometry" (e.g., Prahl and Wakeham, 1987) are now being made. Alkenone biomarkers have also been used in conjunction with δ^{13}C measurements to infer CO_2 variations in the past (e.g., Pagani *et al.*, 1999).

2.4 ISOTOPIC METHODS

Among the most important indicators of climate change are the oxygen isotope ratio ($^{18}O/^{16}O$) and the carbon isotope ratio ($^{13}C/^{12}C$) of the calcareous material in sedimentary cores, along with the deuterium ratio (D/H) in the waters of oceans and ice sheets. These ratios are usually expressed as departures, δ, from a laboratory standard value (in parts per thousand) by the general formula,

$$\delta(X) = \left[\frac{R(X)_{\text{sample}}}{R(X)_{\text{standard}}} - 1 \right] \times 10^3 \tag{2.1}$$

where $R(X)$ is the ratio of the mass of the isotopic form X to that of its more abundant form; $R(X)_{\text{standard}}$ is a constant value either for $CaCO_3$-derived oxygen (PDB) or for "standard mean ocean water" (SMOW). Thus $R(^{18}O) \equiv {}^{18}O/{}^{16}O$, $R(D) \equiv D/H$, and $R(^{13}C) \equiv {}^{13}C/{}^{12}C$.

2.4.1 Oxygen Isotopes

Briefly stated, the oxygen isotope method applied to foraminiferal material in sedimentary cores depends on the fact that water that evaporates to form water vapor tends to be richer in ^{16}O than in ^{18}O. An additional climatologically significant ^{18}O–^{16}O separation (i.e., fractionation) process is also operational in the atmosphere. Thus, when oceans experience a large net evaporative loss (as must occur during periods of extensive ice formation) and subsequent additional fractionation of vapor during precipitation formation (Rayleigh fractionation), the water remaining in the ocean has proportionally more of the heavier ^{18}O, which is incorporated by biota in the formation of calcium carbonate shells. It follows that if a stratum contains shells with high levels of $\delta\,^{18}O$, it was formed during a period of large ice volume.

This effect is supplemented to some extent by another effect discovered by Urey (1947), verified by Epstein *et al.* (1953), and applied by Emiliani (1955) and others. In particular, the amount of ^{18}O in the carbonate shells during calcification can be shown to be partly controlled by the ocean temperature at which the shells are formed, with high values of the ratio indicating cold temperature. Thus the $\delta\,^{18}O$ of the fossil biota is some complex function of both ice volume and ocean temperature. During times when, from other lines of evidence, there was little ice on the planet, the $^{18}O/^{16}O$ ratio is almost entirely due to the temperature effect. Shackleton and Kennett (1975) applied this reasoning to a study of the $\delta\,^{18}O$ of the benthic foraminifera in a long core to infer the changes in deep-water temperature over a roughly 50-million-year period, the results of which will be given in Chapter 3. Conversely, when measured in regions for which there is little evidence of marked temperature change over an extended period (perhaps the western tropical Pacific and the benthic polar oceans), the variations of $\delta\,^{18}O$ can probably be attributed mainly to ice variations.

As noted above, rainwater is "heavier" (i.e., richer in ^{18}O) than the vapor from which it condenses. Thus vapor in transit from evaporative source regions in lower latitudes (e.g., the subtropical oceans) will preferentially lose ^{18}O relative to ^{16}O in amounts that are believed to be proportional to the amount of rain and snowfall, and also proportional, from chemical kinetic considerations, to the relative warmth of the atmosphere, which accelerates the fractionation process. Thus, the high-latitude vapor, and hence the snow from which it precipitates to form the glacial layers, is "lighter" (i.e., depleted in ^{18}O) than the originally evaporated vapor, with the degree of "lightness" of the snowfall over the ice sheets depending on the history of the vapor and more particularly on the temperature at which the snowflakes form in supersaturated clouds; the warmer the cloud air temperature, the richer the snowflakes are in ^{18}O. It follows that higher values of $\delta\,^{18}O$ of the accumulated ice denote warmer air temperatures in the snow-forming regions of the atmosphere, giving some information about the air temperature variations over the ice sheets. This is the basis of the technique used by Dansgaard *et al.* (1971) and others to infer climatic history (paleotemperatures) from a $\delta\,^{18}O$ analysis of deep cores in the Greenland and Antarctic ice sheets. Taking into account the internal movements of the glacial ice, it is also possible to infer the times of strong accumulation or strength of the hydrologic cycle. These results seem to be well correlated with the climatic inferences from other lines of evidence, including the deep ocean cores.

2.4.2 Deuterium and Beryllium in Ice Cores

Another important temperature-dependent isotope derived from ice cores is the heavy form of hydrogen, deuterium (D). Even more clearly than ^{18}O, deuterium fractionates more strongly to the liquid or solid phase from water vapor under warmer ambient conditions of condensation. From both observational and theoretical chemical kinetic considerations, a nearly linear relation between δD and near-surface air temperature can be established, thus providing a good downcore record of surface temperature changes (Jouzel *et al.*, 1987).

Some information concerning the past rates of precipitation over ice sheets can be obtained by the analysis of the radiogenic isotope, ^{10}Be (half-life, 1.5×10^6 y). This isotope of beryllium is created in the atmosphere at a nearly steady rate by cosmic ray bombardment, and is incorporated in the falling snowfall on an ice sheet roughly in an amount that is inversely proportional to the amount of precipitation. Thus, a downcore record of ^{10}Be in an ice sheet can serve as an index of paleoprecipitation (Beer *et al.*, 1985).

2.4.3 Stable Carbon Isotopes

The process of photosynthesis, by which organic matter is formed, is characterized by a preferred uptake of the more abundant (lighter) isotope of carbon, ^{12}C, relative to the heavier isotope, ^{13}C. Thus, during times of high bioproductivity and organic burial the remaining carbon pools are enriched in ^{13}C. In the oceans, this means that high organic productivity leaves ocean water with high values of $\delta\,^{13}C$, which are transmitted to the carbonate shells formed in those waters. As a result, the $\delta\,^{13}C$ of preserved calcareous material in the sediments is an indication of past organic productivity in the ocean. These facts have several consequences of value for making paleoclimatic inferences:

1. On shorter time scales the $\delta\,^{13}C$ content of water masses can serve as a tracer of ocean currents, with the $\delta\,^{13}C$ signature of surface water sources progressively eroding as the organic matter falling through the masses is oxidized, i.e., "remineralized." To the extent that these signatures can be recorded in sedimentary $CaCO_3$ they may be capable of recording paleoocean currents.

2. The $\delta\,^{13}C$ content of carbonates can also give some indication of the transfer of carbon between its reservoirs in the ocean, biosphere, and lithosphere, serving, for example, as a measure of the rates of carbon burial in the sediments. This is discussed further in Chapter 10.

3. By measuring the differences in the $\delta\,^{13}C$ content of planktic (surface) carbonate foraminifera compared to benthic foraminifera in a given core,

$$\Delta\delta\,^{13}C \equiv \left[\left(\delta\,^{13}C\right)_{\text{planktic}} - \left(\delta\,^{13}C\right)_{\text{benthic}}\right]$$

one may infer the strength of the "biological pump" for CO_2 at that site: Thus, when $\Delta\delta\,^{13}C$ is large it is suggested that organic matter was produced in large quantities near the surface, leaving the water and $CaCO_3$ formed in the water heavy in ^{13}C, whereas oxidation of the organic material in lower levels leads to relatively lower values of $\delta\,^{13}C$ in benthic foraminifera. It is implied that carbon is being removed from the surface waters, reducing the partial pressure of CO_2 there and hence lowering the equilibrium atmospheric CO_2 value (Shackleton and Pisias, 1985) (see Chapter 10). Although these inferences are speculative, it is interesting that they are quite consistent with the more direct and reliable measurements of past atmospheric CO_2 concentrations from the analysis of trapped air in ice sheet cores (to be discussed in Section 2.5.2).

4. One method based on $\delta\,^{13}C$ has been used to estimate changes in atmospheric CO_2 over the past few hundred million years (Arthur *et al.*, 1991; Freeman and Hayes,

1992; Hollander and McKenzie, 1991). This is based on a comparison of the $\delta\,^{13}C$ of the carbonates versus organic material in sedimentary rocks, from which one can estimate the carbon isotope fractionation during photosynthesis (a function of CO_2). Other applications of carbon isotope fractionation in the terrestrial biosphere have also been proposed to extract information on past levels of CO_2—for example, by examining the $\delta\,^{13}C$ of ancient soil (paleosol) carbonates, which bear the imprint of both atmospheric CO_2 ($\delta\,^{13}C$ enriched) and soil organics ($\delta\,^{13}C$ depleted); see Cerling (1991) and Mora *et al.* (1991). Cerling *et al.* (1997) also proposed that by examining the $\delta\,^{13}C$ content of the bone and tooth remains of grazing animals it is possible to estimate the type of vegetation that was consumed by the animals. In particular, low $\delta\,^{13}C$ values signify a preponderance of "C_3 plants" that utilize a three-carbon compound photosynthetic process (e.g., trees, grains) requiring high levels of atmospheric CO_2, whereas lower $\delta\,^{13}C$ values signify a preponderance of "C_4 plants" that utilize a four-carbon process (e.g., maize, sugar cane, and prairie grasses) requiring much lower CO_2 values. These relationships can be qualified to some extent, showing, for example, that a significant drop in CO_2 occurred in the Miocene about 7 My ago. An excellent review of this and other methods to estimate long-term changes in carbon dioxide is given by Royer *et al.* (2000).

The role of $\delta\,^{13}C$ in formulating a more structured dynamical model of CO_2 variations is discussed further in Chapter 10.

2.4.4 Strontium and Osmium Isotopes

Two other isotopes that have been of interest in climate studies are ^{87}Sr and ^{187}Os, which derive radiogenically from ^{87}Rb and ^{187}Re, respectively. The strontium isotope as measured by the ratio $^{87}Sr/^{86}Sr$ is relatively abundant in certain silicate rocks, and the osmium isotope as measured by the ratio $^{187}Os/^{186}Os$ is relatively abundant in organic-rich rocks (e.g., shales) and some silicate-bearing igneous (ultramafic) rocks. Thus, the amounts of these isotopes in ocean water, which comes mainly from river inputs containing weathering products of these rocks, have been used as a measure CO_2 drawdown associated with chemical weathering. This is also discussed further in Chapter 10 (Section 10.4).

2.5 NONISOTOPIC GEOCHEMICAL METHODS

Many other nonisotopic chemical properties of sedimentary rock and ice stratigraphy are potentially useful for climatic reconstruction. Here we remark on two of these that have been of special interest.

2.5.1 Cadmium Analysis

From the work of Boyle (1988a) it is known that the presence of cadmium (Cd) in ocean water is a good indication of the major nutrient, phosphorus, which is itself difficult to monitor. Thus, by examining the Cd content of benthic foraminifera it is possible to ascertain the amount and intensity organic matter decomposition in the lower levels of the ocean. This can provide useful information about patterns of deep water circulation from the surface source regions, and the vertical distribution of nutrients, because they may affect the carbon budget and level of atmospheric CO_2.

2.5.2 Greenhouse Gas Analysis of Trapped Air in Ice Cores

The most direct method for determining the past variations of CO_2, methane (CH_4), and other trace gases of the atmosphere is by the analysis of air trapped in ice sheets. The methods for extracting the samples and making the determinations were developed notably by groups in Switzerland (e.g., Berner *et al.*, 1980) and France (e.g., Delmas *et al.*, 1980). Extensive studies of cores in Greenland and, notably, the Vostok core in Antarctica (Lorius *et al.*, 1993; Petit *et al.*, 1999) have led to an increasingly accurate picture of the way not only CO_2 has varied over the past 400 ky, but also methane. As will be discussed next, by analysis of the ice a fairly continuous record of many soluble (sulfate, nitrate, sodium, chloride) and insoluble trace atmospheric substances can be determined.

2.5.3 Chemical and Biological Constituents and Dust Layers in Ice Cores

As noted, in addition to analyses of the trapped air in ice cores, much useful paleoclimatological information can be obtained by determining the record of deposition of solid particles and their chemical and biological constituents. These constituents include Ca, Fe, K, and S, and biogenic forms of sulfur that can be linked to the sources of windblown material and atmospheric loading processes. Of particular interest are the signatures of persistent extreme dryness and windblown dust during periods of glaciation, as well as shorter episodes of intense global volcanic activity. The deposition of nutrient elements also may affect biological activity in surface ocean waters. Many papers on this subject can be found in the summary volume "Greenland Summit Ice Cores" (1998) published by the American Geophysical Union.

2.6 DATING THE PROXY EVIDENCE (GEOCHRONOMETRY)

As we have noted, the above stratigraphic climate indicators require a method for dating the times at which the strata were deposited. The most useful method for establishing such chronologies for the past 40 ky is the radiocarbon method based on the unstable isotope ^{14}C (half-life, 5600 y).

Briefly, the surface of the Earth (in particular, the ocean surface) is approximately in chemical equilibrium with the atmosphere, where cosmic rays interact with stable carbon in the form ^{12}C to produce a known amount of radioactive carbon, ^{14}C. Thus, there is a nearly fixed ratio of ^{12}C to ^{14}C in the atmosphere and in the surface waters with which it is in equilibrium, assuming a steady rate of ^{14}C production and taking into account some upwelling of older ^{14}C from deeper waters. When material from the

Table 2-1 A Summary of Paleoclimatic Data Sources

Proxy data source	Some of the variables measured	Possible climatic inferences	Potential study period	Typical sampling interval
Ice cores	$\delta^{18}O$, δD, CO_2, CH_4, dust	Temperature	40 ky	1–10 y
	ice chemistry	Atmospheric circulation	100–500 ky	10–100 y
Tree rings	Ring width	Temperature		
	$\delta^{18}O$, δD	Precipitation (drought)	10 ky	1 ky
	$\delta^{13}C$, $\Delta^{14}C$	Solar variability		
Coral	$\delta^{18}O$,	SST, precip-evap,	100–1000 y	1 month
	Sr/Ca	sea level		1 y
Pollen	Percent, influx	Temperature Precipitation	10–100 ky	10–100 y
Soils	$\delta^{13}C$ Loess	CO_2, wind	1–10 My	Snapshots
Closed-basin lakes	Lake level	Precipitation-evaporation	10–100 ky	Snapshots
Lake sediments	$\delta^{18}O$ Diatoms	Temperature Salinity	10–100 ky	10–100 y
Ice sheets	Former extent, glacial rebound	Area, thickness, bedrock depression	1 My	Snapshots
Mountain glaciers	Former extent	Snowline, air temperature	10 ky	Snapshots
Marine sediments	$\delta^{18}O$, $\delta^{13}C$	Global ice mass,	10 My	1 ky
	Foraminiferal	ocean circulation,	500 ky	100 y
	assemblages	SST	50 ky	1–10 y
Raised shorelines	Evaluation	Sea level Bedrock depression	1 My	Snapshots
Laminated or varved sediments	Reflectance, magnetic properties	Precipitation Wind	10 ky	1 y

surface (e.g., calcareous foraminifera and molluscan shells) falls downward through the ocean and is buried in sediment (no longer exposed to the atmosphere), the ^{14}C in shell material continues to decay at its known physically determined rate without being replenished. By examining the ratio $^{12}C/^{14}C$ in shells (or corals), one can calculate the time at which the material was buried, giving a "^{14}C age." This must be corrected for variations in atmospheric ^{14}C production determined from other considerations (primarily the effect of magnetic field changes) to yield a "calendar age" (see e.g., Bradley, 1999).

To date the age of rock strata older than 40,000 years, similar use can be made of the uranium decay series, and for dating the oldest rock and mineral materials, methods based on the decay series of potassium and argon have proved most useful. In dating the formation of this older material account must be taken of the global tectonic movement of the Earth's crust, including the ocean beds, and the magnetic reversal chronology imprinted in the rocks has provided invaluable markers. A good introductory discussion of the biochemical oceanic climatic indicators and methods for dating is given by Turekian (1976, 1996), with further details in the work of Williams *et al.* (1998).

Table 2-1 depicts many of the techniques we have outlined in this chapter, showing the time periods over which the technique can yield climatic information. Some indication of the minimum sampling interval is also shown. In the following chapter we summarize what appears to be the salient global features of past climatic variation as revealed by these techniques. They should be regarded only as our present best estimate of the true variations, subject to continuing reassessment and refinement as these techniques are applied more extensively and new techniques are developed. It is clear that many of the existing techniques can give only qualitative indications of climatic states, and considerable uncertainty remains in establishing a definitive chronological record of the global climate.

3

A SURVEY OF GLOBAL PALEOCLIMATIC VARIATIONS

We now present an overview of the findings obtained by the use of the techniques described in the previous chapter. This summary will start with the more poorly known, long-term variations extending over about one-half billion years (the Phanerozoic Eon) and will proceed with increasing time resolution to the better known recent variations.

As one might surmise from our discussion of the proxy techniques for past climate reconstruction, only a crude estimate of global climate changes, primarily for surface temperature (T_S) and ice mass (I), can be made beyond the few hundred years of instrumental measurement. As we go back in time, the ability to reconstruct geographic patterns of the climatic variables is greatly reduced, especially for the very long periods when continental drift becomes important. Hence our focus here will be on the inferred variation of the two variables, T_S and I, both of which are probably coherent. Inferences are sometimes made concerning the distributions of the other variables (e.g., wind, pressure, storminess) that are likely to go along with these temperature or ice distributions based on the known present seasonal variations, but these are usually more in the nature of theoretical deductions than direct observational estimates.

We shall see that—based on what we can infer from the relatively short time period during which quantitative meteorological records have been compiled (about 200 years), from the somewhat longer period of qualitative historical accounts of weather conditions, and from the very much longer geological record imbedded in the rocks and sediments—there has always been so much variability of climate that it seems doubtful that we can ever speak of a single climatic norm for the Earth. Judging from the past, the climate we are experiencing today is almost certain to be transient, giving way to something different in the future.

To illustrate this variability, we now describe some estimates of the extent to which global temperature, or ice mass, has varied over five separate time scales measured before the present, ranging from the very long-term period of about 1000 Ma to one encompassing only the past few hundred years. These descriptions will naturally contain increasing detail and accuracy as we approach the most recent period.

3.1 THE PHANEROZOIC EON (PAST 600 My)

Because Earth is estimated to be 4600 My old, we are dealing here with only about 13% of Earth's lifetime. It is not possible to discuss the climatic change over this long time scale without first gaining some appreciation of the vast changes in the geography of Earth (e.g., the distribution of continents and oceans) that are believed to have occurred over this period. In Fig. 3-1 we present a sequence of global maps showing these changes, starting at about 515 Ma (Scotese, 1997). Before discussing these maps we note that at about 600 Ma the geologic record reveals a significant feature in unraveling the history of Earth—the early sign of life forms that can leave clearly visible

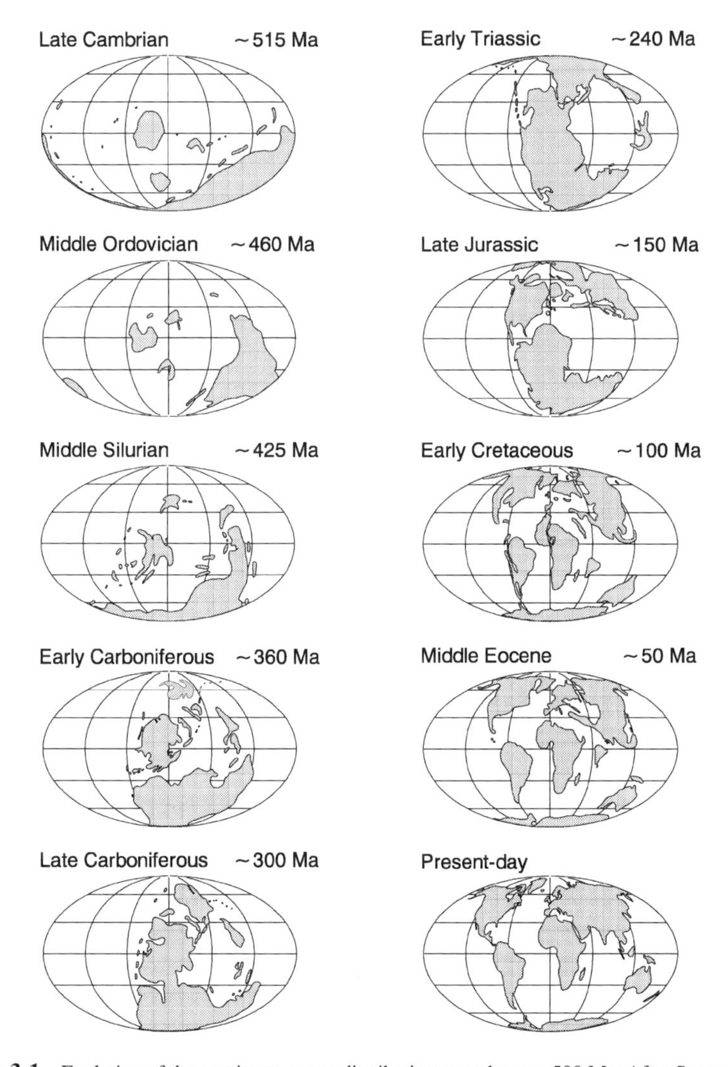

Figure 3-1 Evolution of the continent–ocean distribution over the past 500 My. After Scotese (1997).

fossil signatures in the strata. This marks the beginning of the Phanerozoic (i.e., visible life) Eon, from which time forward more definitive estimates of climatic change can be made because of the climatic dependence of the life forms. There is evidence also (e.g., Walker, 1978) that the "fixed" constituents of the present atmosphere, mainly N_2 and O_2, had achieved their current concentrations by this time. The classic subdivision of the Phanerozoic Eon into geological eras and periods is shown in Fig. 3-2. Pre-Phanerozoic time is sometimes called the Cryptozoic (hidden life) Eon, though careful recent study of this early period has begun to reveal some intriguing signatures of primitive life forms.

From Fig. 3-1 we see that in the Middle Ordovician (460 Ma) all continents existing at the time (Baltica, Laurentia, Siberia, China, Kazakhstania, and the massive Gondwana) were located near the equator, with open ocean prevailing in the polar regions. By the Middle Silurian (425 Ma) Gondwana had drifted to the South Pole, leaving

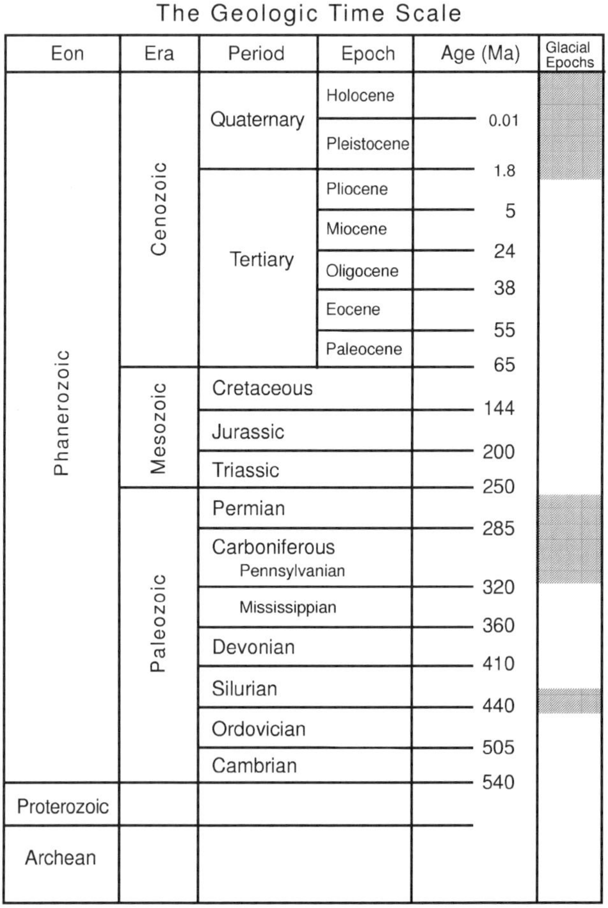

The Geologic Time Scale

Eon	Era	Period	Epoch	Age (Ma)	Glacial Epochs
Phanerozoic	Cenozoic	Quaternary	Holocene	0.01	
			Pleistocene	1.8	
		Tertiary	Pliocene	5	
			Miocene	24	
			Oligocene	38	
			Eocene	55	
			Paleocene	65	
	Mesozoic	Cretaceous		144	
		Jurassic		200	
		Triassic		250	
	Paleozoic	Permian		285	
		Carboniferous Pennsylvanian		320	
		Mississippian		360	
		Devonian		410	
		Silurian		440	
		Ordovician		505	
		Cambrian		540	
Proterozoic					
Archean					

Figure 3-2 The geologic time scale, showing intervals when extensive glaciation is believed to have prevailed.

the North Pole as open ocean. The configuration in the Early Carboniferous (360 Ma) shown in Fig. 3-1 reveals that by this time Siberia had moved to a position near the North Pole and a new continental combination of Laurentia and Baltica (Laurussia) had formed and begun its collision course toward Gondwana to the south. This collision ultimately closed the zonal oceanic passage in the tropics, thereby forming the supercontinent Pangea near the end of the Paleozoic [see the Early Triassic (240 Ma) map]. On its eastern flank this Pangean land mass formed a nearly closed ring around a large oceanic area known as the Tethys Sea, with the remaining huge oceanic expanse constituting the Panthallasic Ocean.

The past 200 My have been characterized by the breakup of Pangea into the present continental forms. In the first stage, this breakup involved the separation of a northern continental block, Laurasia, from Gondwana, which itself had begun to fragment (see Fig. 3-1, showing the Early Cretaceous map at 100 Ma). In a later stage, the breakup involved the separation of North and South America from the Eurasian–African land masses to form the Atlantic Ocean (50 Ma).

The present configuration of the continents was established over the Cenozoic, with India completing its drift to join the Asian land mass early in this period. The Early Tertiary (Paleocene) seems to have been a time of low continents and shallow seas. In the Eocene, at about 50 Ma, Australia began separating from Antarctica, opening up an oceanic passage through which the Antarctic circumpolar current seems to have been established at about 30 Ma.

Other changes have been occurring continuously through the Late Tertiary (Neogene) up to the present. Perhaps the most significant are the tectonic closing of the seaway across Panama to form the present isthmus separating the Atlantic and Pacific Oceans, estimated to have occurred between about 10 and 3.5 Ma (Berggren and Hollister, 1974; Frank *et al.*, 1999) and the similar closing of a seaway in what is now western Siberia that formerly permitted more tropical waters to flow to the Arctic Ocean. It would appear that over the past few million years Earth's lithospheric geography can be taken to be roughly in its present state, though, as we shall see, significant changes were engendered by the formation of glacial ice during this period.

Taking these changes in geography into account, crude estimates have been made of the way in which mean surface temperature or ice mass has varied over the past 500 Ma. These were summarized in Fig. 1-3. The curve shown in the figure is a modified version of that given by Frakes *et al.* (1992) and is based largely on the evidence of fossil assemblages in the sediments, corroborated by geomorphological indicators, and supplemented by the more recent evidence mainly from the isotopic indicators. We find that for the greater past of this time period the global temperature tended to be warmer than at present, with little or no ice sheets. Four noteworthy departures from this prevailing state of warmth seem to have occurred:

1. In the Late Precambrian, somewhere in the vicinity of 750 and 600 Ma, glacial ages seem to have existed.

2. Near the end of the Ordovician Period, about 460 Ma, glacial ice was present in a portion of Gondwana located near the South Pole, which is now the North African Sahara (see Fig. 3-1).

3. At about 300 Ma, from the Late Carboniferous to the Early Permian (i.e., Permo-Carboniferous), extensive ice sheets appear to have existed in the Gondwana portions of the then-forming Pangean supercontinent, now comprising parts of South America, South Africa, Australia, India, and Antarctica.

4. During the past 50 My, a gradual cooling took place and polar ice began to grow, with an especially steep drop in temperature in the mid-Miocene and culminating in periods of severe glaciation over roughly the past 2 My.

3.2 THE CENOZOIC ERA (PAST 65 My)

In Fig. 3-3 we show in more detail an estimate of the manner in which the surface ocean temperature (planktonic curve) declined during the Cenozoic. A similar temperature decline at continental sites in North America has been inferred from analyses of fossil leaf margin properties, as described in Section 2.3.2. As a part of this decline, it has been estimated that widespread local glaciers appeared in Antarctica no later than 25 Ma, and by 5 Ma the Antarctic ice sheet had grown to its present massive volume. At the same time, there is evidence (benthic curve) that the deeper waters of the ocean were also cooling down substantially from the higher values that prevailed during the Mesozoic (estimated to average about 15°C) to values much closer to the present 4°C. This benthic curve is based on an oxygen isotope analysis of foraminifera in strata laid down at a time when there was little or no ice, so that the isotopic changes can be attributed only to the Urey-temperature effect alone (see Section 2.4.1).

As we approach the beginning of the Pleistocene, at about 2 Ma, the record begins to show strong evidence for the beginning of oscillatory conditions, characterized by an alternate growth and decay of ice in the climatic system. This behavior is the hallmark of the Pleistocene. It is believed that the appearance of these fluctuations in the

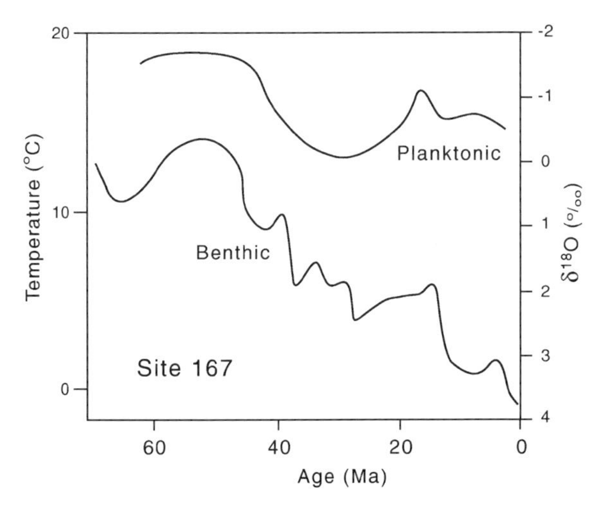

Figure 3-3 Estimate of the variation of sea surface and deep ocean temperatures over the Cenozoic (60 My) at site 167, based on planktonic and benthic $\delta^{18}O$ measurements, respectively. After Savin (1977).

climatic record represents more than the effects of better resolution and data for the more recent time period, but in fact represents a true physical change in the behavior of the system. Apparently, when some critical state is achieved the system can undergo a marked glacial–nonglacial variation. We shall discuss the possibilities for this behavior from a theoretical viewpoint in Chapters 14 and 15. In the following section we discuss the observational record of the last glacial epoch in more detail.

3.3 THE PLIO-PLEISTOCENE (PAST 5 My)

An estimate of the temperature and ice variations over the past 6 My, based on the oxygen isotope composition of benthic foraminifera in tropical eastern Pacific cores, was shown in Fig. 1-4a, and, in expanded form for the past 1 My in Fig. 1-4b. In

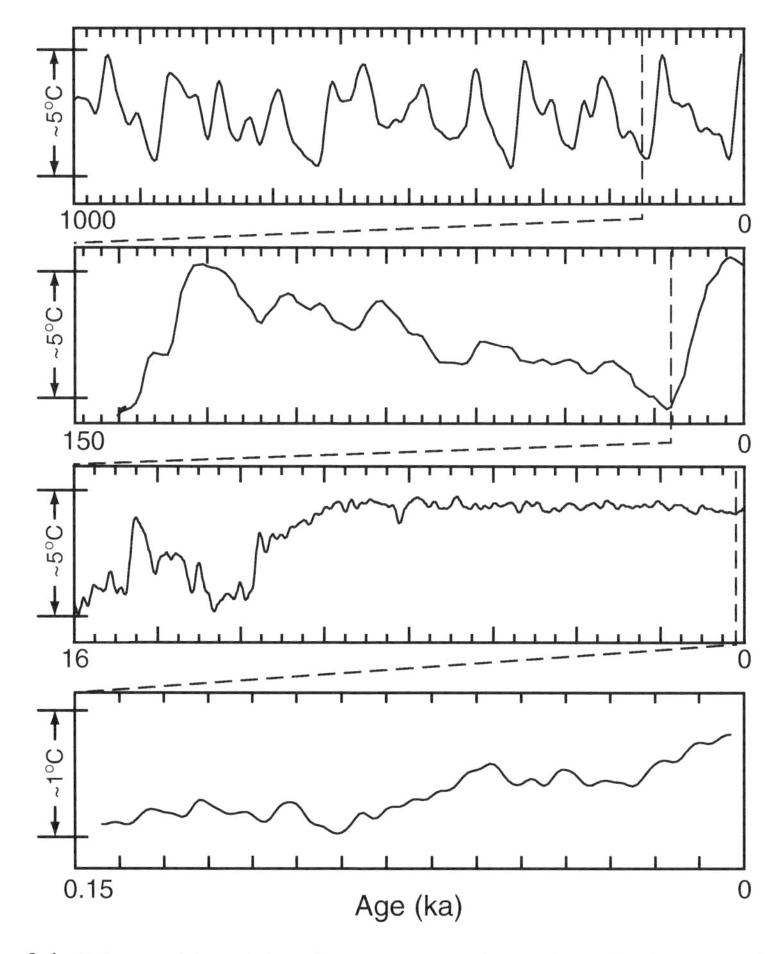

Figure 3-4 Estimates of the variation of temperature over the past 1 My, showing these variations in increasing detail up to the present based on various lines of evidence.

Fig. 1-9, we showed another record of the variations of ice over the past 2 My as represented by a western Pacific planktonic $\delta^{18}O$ record (Berger *et al.*, 1990), along with a spectral analysis of the past 1 My of this record (Fig. 1-10). Records similar to this have been obtained from many other cores distributed over the world oceans [e.g., the ~800 ky SPECMAP composite record (Imbrie *et al.*, 1984)], and are corroborated by other lines of evidence such as faunal assemblages. Particularly illuminating complementary evidence is provided by the records of ice-rafted debris in both the North Atlantic and Southern Ocean (Fig. 3-5), corroborating the onset of ice conditions at about 2.5 My and the transition to a near-100-ky period oscillation at about 900 ka in both hemispheres.

These records show a repetition of heavy glaciation (i.e., "ice ages") at roughly 100-ky intervals during the past 900 ky, with buildups of the ice tending to be somewhat slower and irregular than the more precipitous retreats. Some evidence, however, indicates that there may have been episodes of rapid buildups of ice also (e.g., Ruddiman *et al.*, 1980). The "sawtooth" appearance of some of the main glacial cycles was first noted by Broecker and van Donk (1970). As noted in Chapter 1 (see Figs. 1-7 and 1-8), spectral analysis for the past 1 My reveals variance maxima not only in the 100-ky period, but also in the region near 40, 23, and 19 ky that would appear to be associated with the obliquity and precessional variations of insolation (Hays *et al.*, 1976). It should be noted, however, that records of this kind are often tuned to the 40-ky period to obtain their chronologies.

At the same time analysis of the trapped air content of the Vostok ice core (Jouzel *et al.*, 1993; Petit *et al.*, 1999) has revealed a variation of the greenhouse gases, CO_2 and methane, over the past 420 ky similar in appearance to that of the $\delta^{18}O$ ice record over this period (see Fig. 3-6). In Chapter 8 we discuss this covariability in greater detail.

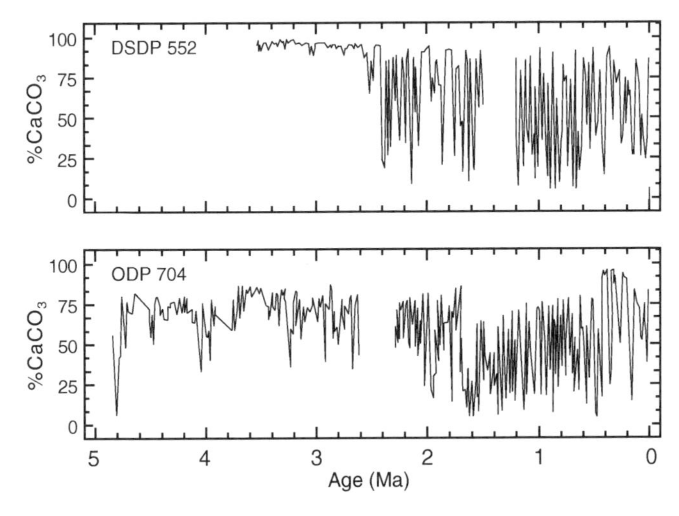

Figure 3-5 Variations of ice-rafted debris (IRD) as measured by the reduction in the ratio of $CaCO_3$ to lithic material over the past 4–5 My, as recorded in a North Atlantic core 552A (Shackleton *et al.*, 1984) and a Southern Ocean core (Hodell, 1993).

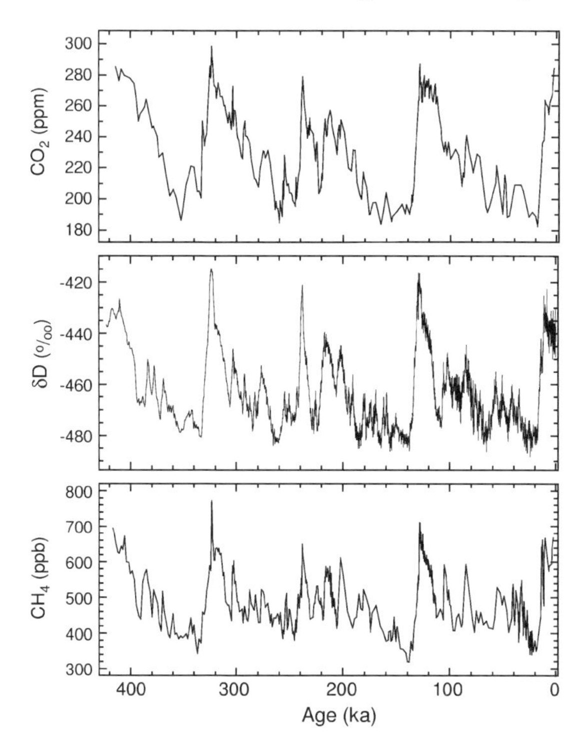

Figure 3-6 Variations of carbon dioxide, temperature, and methane as determined from analysis of the Vostok ice core over the past 420 ky. After Petit *et al.* (1999).

3.4 VARIATIONS DURING THE LAST ICE AGE: IRD EVENTS

Recent studies suggest that during the last glacial regime in the Northern Hemisphere quasi-periodic episodes occurred, during which large numbers of icebergs were discharged in the North Atlantic. As shown in Fig. 3-7a, the evidence for this is the strong signature in the sediments of ice-rafted debris of content similar to that of the basal Canadian and European terrains, alternating with the normal deposits of carbonate shells (Heinrich, 1988; Broecker *et al.*, 1992; Grousset *et al.*, 1993). Alley and MacAyeal (1994) have, in fact, estimated that the ice streams emanating from the Laurentide ice sheet could have contained enough debris to account for the IRD deposits. It is seen that these Heinrich events occurred at intervals of about 6–12 ky, between which other shorter period fluctuations are evident; these have been identified with the so-called Dansgaard–Oeschger fluctuations recorded in the Greenland ice cores (see Section 3.6).

In Fig. 3-7b, we show more detailed records of three quantities covarying with the four Heinrich IRD events (denoted by HL 1–4 at the bottom of the figure) (Bond *et al.*, 1993). These quantities are the δD and $\delta\,^{18}$O measures of temperature recorded in the Vostok (Antarctica) and GRIP (Greenland) ice cores, respectively (see Sections 2.4.1 and 2.4.2), and a measure of sea surface temperature (SST) recorded by the abundance

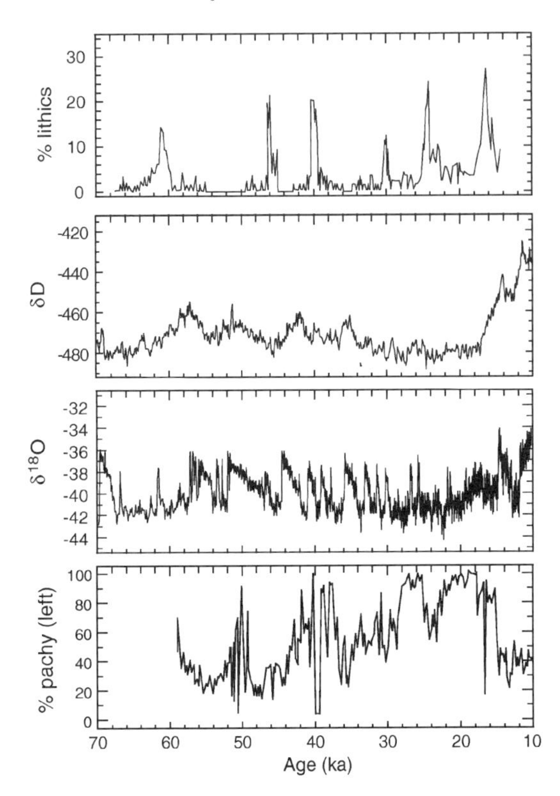

Figure 3-7 (Top) Variations of ice-rafted debris (IRD) over the past 60 ky at North Atlantic site 609, showing quasi-periodic pulses known as "Heinrich events," labeled H1–H5. Variations of temperature accompanying the Heinrich events as measured by records of δD, δ^{18}O, and *N. pachyderma* (see Chapter 2).

of *N. pachyderma* (see Section 2.3.2). We see that prior to each H-event the SST is relatively cold, and following each H-event there is a rapid increase in SST, after which there is a slower sawtooth-shaped decline until the occurrence of the next H-event (the so-called Bond cycles). Superimposed on these changes are the Dansgaard–Oeschger oscillations. These phenomena will be discussed further, from both observational and theoretical perspectives, in Chapter 16.

3.5 THE LAST GLACIAL MAXIMUM (20 ka)

The variations in the global environment that accompany the oscillations shown in Fig. 1-4 are enormous. For example, just 20 ky ago, at the time of the peak of the last great ice age (the Wurm–Wisconsin glacial), ice sheets reaching depths of over 3 km covered large portions of North America, Europe, and Asia. One estimate of this ice coverage is shown in Fig. 3-8, which is the celebrated picture of the state of Earth's surface 20 ka in a typical August, constructed by the CLIMAP group (1976), but which has since been modified, particularly with regard to a reduction in the values of the sea surface temperatures in lower latitudes. This synoptic reconstruction is based on many of the lines of evidence discussed in Chapter 2; the sea surface temperature and ice coverage, in particular, are based on the detailed analysis of faunal assemblages (and their oxygen isotope properties) in sediment cores obtained over an extensive oceanic network. This is the only glacial maximum, extending back for about 1 My, for which geological evidence is good enough to attempt such a reconstruction—but even in this case, there is much controversy regarding the thickness, volume, and location of the ice sheets (Peltier, 1994, and references therein). As an example, the presence of major ice sheets on the Siberian shelf and in Tibet is still in contention.

We can see that the main centers of ice accumulation were asymmetrically distributed over the globe. In the Southern Hemisphere we find an even more glaciated Antarctica than today, with an enlarged field of sea ice extending as much as 10° latitude equatorward of its present mean position. Glaciated regions are also to be found in mountain areas of South Africa, Australia, and South America. The most dramatic increases in glaciation, however, are to be seen in the Northern Hemisphere. Here a huge North American ice sheet (the Laurentide) appears, merging somewhat with the still-present Greenland ice sheet, as well as a Scandinavian ice sheet, engulfing the North Atlantic Ocean in a ring of glaciers and expanded sea ice. With great amounts of ocean water locked up in glacial ice, sea level is reduced by over 100 m, exposing large areas of lithosphere now below sea level (e.g., Southeast Asia and Australia, and the land bridge between Asia and North America). Other features of the inferred last glacial maximum continental state are represented by the shading patterns described in the legend of Fig. 3-8. It is also noteworthy that accompanying this cold, maximal ice state are extreme aridity and windy conditions, as evidenced by dust layers in ice cores and accumulation of windblown debris (loess); see Sections 2.2.2 and 2.5.3.

Of equal interest to this continental picture are the inferred patterns of sea surface temperature. Compared with the present distribution, with the ice-age advance of the mean sea-ice edge in both hemispheres (at which edge the temperature is fixed near 273K), there is a displacement of the isotherms toward the tropics, where there is relatively smaller cooling from the present levels. Thus we find a marked "squeezing" of the isotherms (i.e., a large temperature gradient) in middle latitudes. This is of significance in determining the position of the major ocean currents and gyres (see Chapter 11) and atmospheric storm tracks, which tend to form along "frontal" zones. Overall, the mean surface temperature of the oceans 20 ka is estimated to be about 3K cooler than at present. Further details concerning both these oceanic features and the continental features can be found in the CLIMAP Project Members (1976) report.

3.6 POSTGLACIAL CHANGES: THE PAST 20 ky

In Figs. 3-9 and 3-4c,d we show in more detail the variations of ice and temperature since the last ice-age maximum, revealing that the meltdown of the ice sheets and amelioration of climate leading to our present state have not been a uniform process. In particular, Fig. 3-9 shows that from an analysis of the sea level changes implied by coral reef growth recorded at Barbados (Fairbanks, 1989), the melting of the ice caps occurred in two main pulses near 14.5 and 11.5 ka. These pulses are separated by an interval, at about 12 ka, known as the Younger–Dryas, during which time there was a pronounced return to colder conditions in Europe and elsewhere, accompanied by some regrowth of glaciers, notably in Scotland and Scandinavia. This Younger Dryas (YD) episode, which was preceded by a warmer episode known as the Allerod–Bölling (A-B), is also reflected in the temperature proxy curves in Fig. 3-7b. Also evident in Fig. 3-4c are the smaller amplitude, higher frequency Dansgaard–Oeschger (D-O) fluctuations.

Following a period of rapid warming, the last ice sheets of North America and Eurasia disappeared about 10 ka, marking the beginning of the present Holocene interglacial epoch. A continued temperature rise during this epoch reached a peak at about 5–6 ka, a time of greater warmth than the Earth has ever again experienced. This is the so-called climatic optimum or hypsothermal. The inferred global rainfall anomalies for this period, compared to the present, shown in Fig. 3-10 (Kellogg, 1978), are of interest because they may represent the conditions at any time of extreme global warmth, such as we may soon experience if theoretical predictions of the consequences of the anthropogenic CO_2 increase discussed below are correct. Note, for example, that rainfall was more plentiful in the now-arid areas of North Africa and the Middle East.

This hypsothermal state was followed by a period of falling temperatures, which reached a minimum at about 500 BC, a time of considerable advance of mountain glaciers. Another increase in temperature followed, culminating at about 1000 AD in a "secondary" climatic optimum characterized by a warm, relatively dry, and storm-free North Atlantic Ocean. It was during this time that the Vikings were able to make their monumental expeditions to Iceland, Greenland, and Labrador, leading to farming settlements on Greenland. However, this climatic state also proved to be transient, the next swing being toward cooler temperatures, with southward advancing sea ice leading to the thermal minimum in the period 1400–1800 AD, centered near 1700 AD, that has been dubbed the "Little Ice Age" (LIA).

3.7 THE PAST 100 YEARS

This brings us to a time when meteorological instruments come into use and accurate measurements of many meteorological variables begin to be recorded. In Fig. 3-11 we show the 100-year record for global mean temperature inferred from global data, a period that is strongly influenced by anthropogenic forcing as well as natural variability (IPCC, 1996). The main features are a warming trend during the first part of the twentieth century, followed, around 1940, by a brief temperature decline, followed in turn by a continuing rise in temperature. These global temperature changes are accompanied by the changes in the extent of mountain glaciers over the past century (e.g., Ladurie, 1971), as well as a recent reduction of polar sea ice (Vinnikov *et al.*, 1999; Johannessen *et al.*, 1999; Rothrock *et al.*, 1999). The more recent years encompass

Figure 3-8 Sea surface temperature (in degrees Celsius), ice extent, and ice elevation (in meters), at the peak of the last ice age, 18,000 years ago, as reconstructed from paleoclimatic evidence. White areas denote regions covered by snow and ice. Continental outlines are based on an estimated 120-m lowering of sea level, relative to present value. After CLIMAP Project Members (1976).

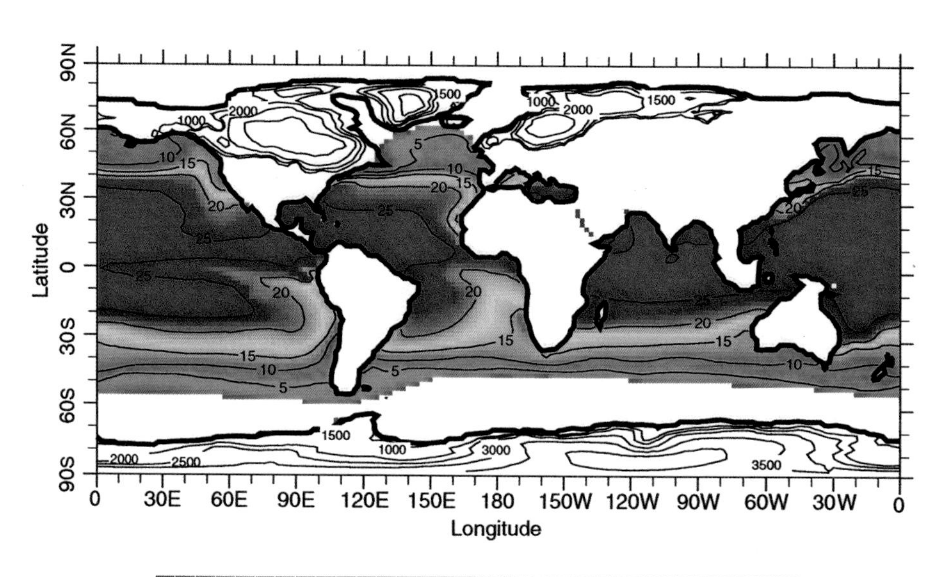
Latitude
Longitude
90N
60N
30N
0
30S
60S
90S
0
30E
60E
90E
120E
150E
180
150W
120W
90W
60W
30W
1000
1500
2000
2500
3000
3500
5
10
15
20
25

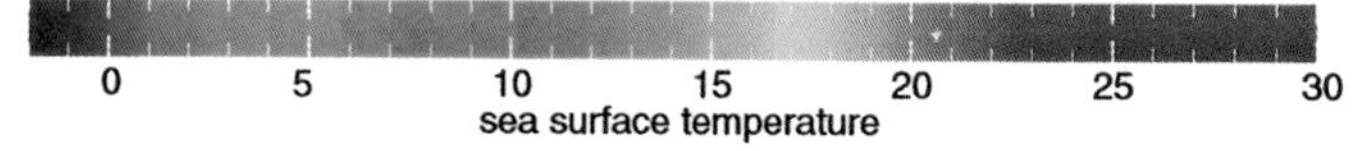
0
5
10
15
20
25
30
sea surface temperature

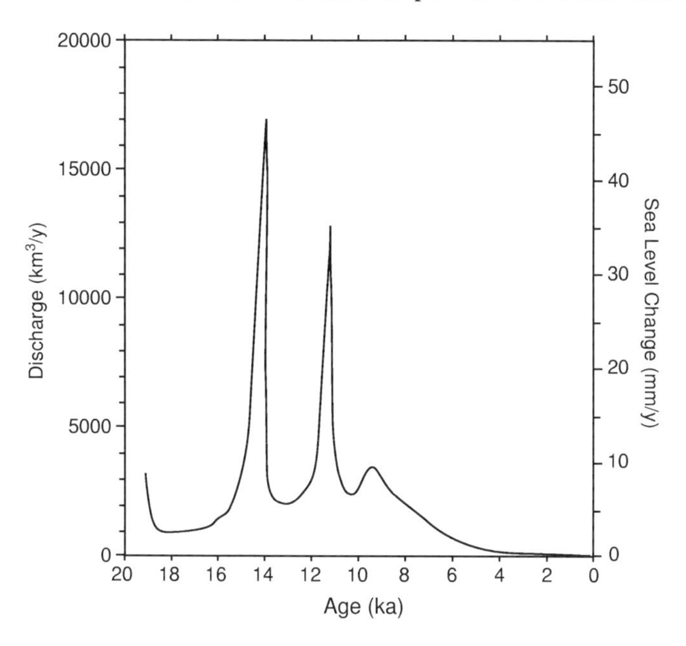

Figure 3-9 Rate of glacial melting, determined from analysis of Barbados coral reefs as an indicator of sea level. After Fairbanks (1989).

the growth of global upper air observations and it becomes possible to look not only at the temperature fluctuations, but also at the accompanying worldwide changes in atmospheric wind circulation and the other meteorological variables. All of these features are of increasing interest as the most immediate examples of "global change," the importance of which for humans is self-evident, especially if one considers the more dramatic possibilities represented by the "ice ages."

The pronounced covariation of CO_2 and temperature over this period, extending back to the Late Pleistocene (Fig. 3-6), is shown in Fig. 3-12; from other combined theoretical and observational evidence to be discussed in Chapter 12 it appears that this covariation holds over the full Phanerozoic.

3.8 THE GENERALIZED SPECTRUM OF CLIMATIC VARIANCE

Given the evidence for climatic change presented in the above sections, we are in a position to summarize the main variability in the form of a variance spectrum. Attempts at representing such a spectrum have been made by Mitchell (1976), Kutzbach (1976), and Shackleton and Imbrie (1990). In Fig. 1-2 we showed a highly idealized schematic version of this spectrum that might apply to a multimillion-year record of surface temperature at midlatitude point. We note that, above a red-noise background of increasing variance with period, there are four main maxima corresponding, respectively, to small-scale turbulence (minutes), synoptic weather variability (days), the ice-age variations (10–100 ky), and ultralong period "tectonic" fluctuations (mil-

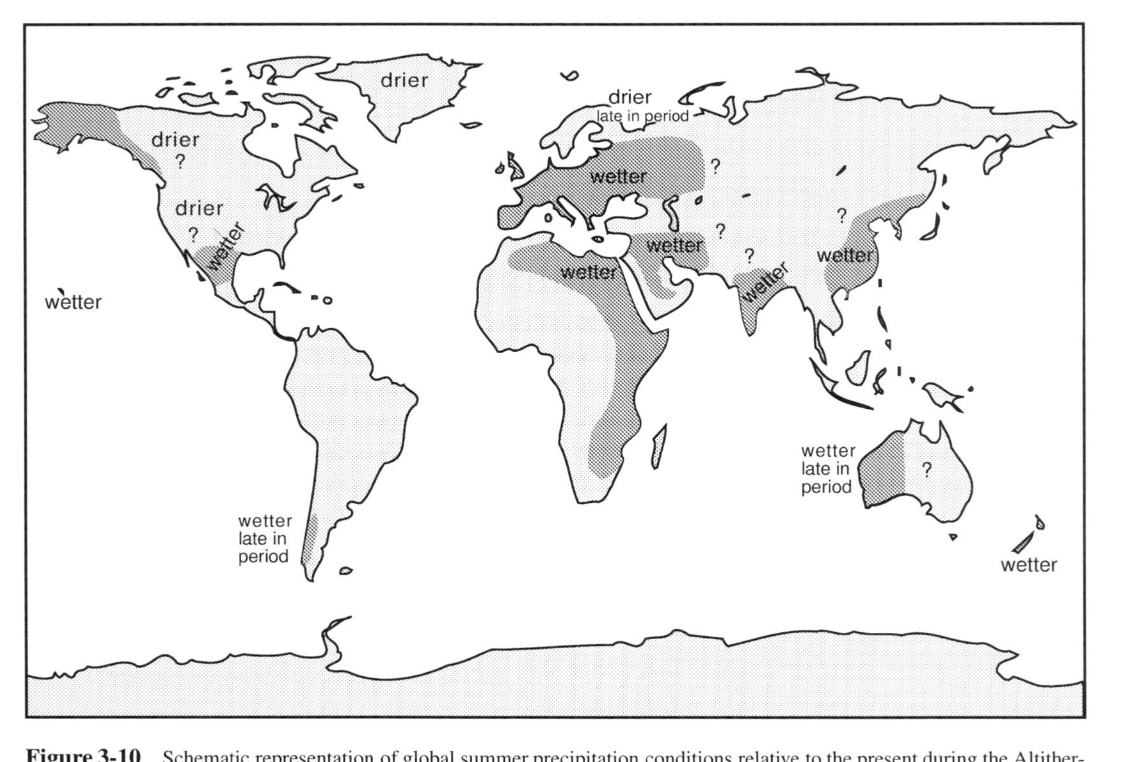

Figure 3-10 Schematic representation of global summer precipitation conditions relative to the present during the Altithermal period, 4–8 ka. After Kellogg (1978).

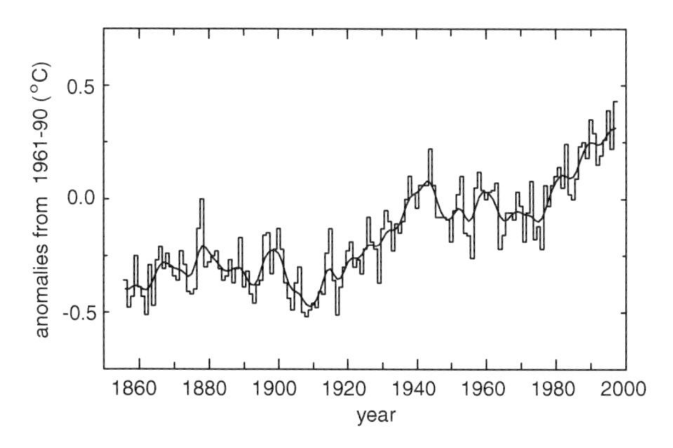

Figure 3-11 Estimate of the variation of global mean temperature since 1860.

lions of years). In the next chapter we shall make further reference to this spectrum, and to the quantities ϵ and δ shown at the bottom (Fig. 1-2), as a basis for resolving the full paleoclimatic problem into manageable components.

With the time-series and spectral descriptions described in this chapter we come closer to the main topic of this book, namely, the development of a theory of climate by means of which we hope to account for the huge changes that have occurred in the past and predict the nearer term changes. Before embarking on the more quantitative aspects of the problem, however, a few general remarks are in order regarding the possible causes of the climatic changes we have discussed.

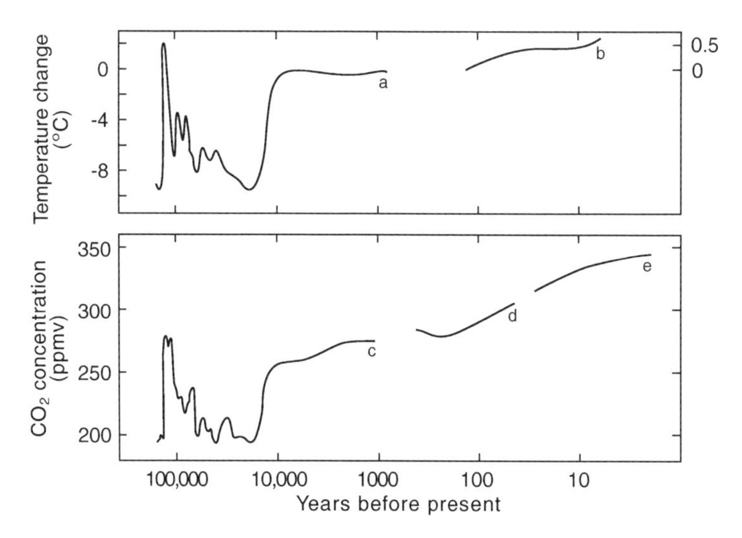

Figure 3-12 Temperature and CO_2 variations for the past 160,000 years. (a) Surface ice temperature variations in Antarctica (Barnola *et al.*, 1987). (b) Global annual mean surface temperature (Jones *et al.*, 1986). (c) CO_2 variations from ice-core analysis (Barnola *et al.*, 1987). (d) CO_2 variations from ice-core analysis (Pearman *et al.*, 1986). (e) Modern Antarctic CO_2 observations (Bacastow and Keeling, 1981; Komhyr *et al.*, 1985). After Pearman (1988).

3.9 A QUALITATIVE DISCUSSION OF CAUSES

To illustrate some of the causal mechanisms that are commonly invoked to explain climatic change, let us consider again the mean temperature changes that have taken place in recent years (Fig. 3-11). At present, the most popular explanation for the rise in world temperature during the early part of the twentieth century is the rising level of carbon dioxide (CO_2) in the atmosphere over this period due to the burning of fossil fuels such as coal and oil. Because CO_2 is a good absorber and emitter of longwave radiation, an increasing amount of this constituent leads to an augmentation of the so-called greenhouse effect whereby heat is radiatively trapped within the atmosphere. There has also been an increase of dust and particulate matter (e.g., sulfate aerosols) in the atmosphere, one cause being volcanic activity, but with industrial activity playing an increasingly important role. It is believed that the scattering and reflection of solar radiation by this global dust field is causing a reduction of the CO_2 effect. Crude quantitative estimates show that these two effects could account for the mean temperature changes shown in Fig. 3-11. Note that, in both cases, humans enter the picture not only as affected bystanders, but also as possible causal agents.

Before we accept these explanations too readily, however, it would be wise to look at the fuller possibilities for causation of climate change, on all time scales, based on our discussions. In Fig. 1-1, we categorized the various elements of the climate system bearing on the problem. The largest construct (Fig. 1-1) represents the internal climatic system, within which there is a representation of the atmosphere separated from representations of the ocean and biolithosphere by a construct representing a boundary layer, which is the main climatic environment of man. The atmospheric construct includes all the elemental variables, the statistics of which (including variances) constitutes the description of climate: the wind V, temperature T, precipitation P, cloudiness C, and chemical and particulate composition ξ_i (e.g., the two factors, CO_2 and atmospheric aerosols, which we just noted play a prominent role in the proposed explanation of the most recent temperature trends).

If we wish to discuss the way in which these atmospheric climatic variables are fluctuating in time in the free atmosphere and boundary layer, we soon find that we must bring other variables into consideration. We are thus led to consider the oceanic state and the physical and biological states of Earth's land surface (including factors such as the existing ice and snow coverage, and the kind of soil and vegetation), both of which groups are highly coupled with the atmospheric climate via fluxes through the boundary layer.

In addition to these last sets of variables, which not only influence the elemental atmospheric climatic variables but are also influenced by them (thus constituting part of the internal feedback system), there are a number of truly external forcing factors discussed in Chapter 1 that can lead to variations in all the climatic variables but which themselves are completely independent of the climate. These include two main categories of factors: those of a purely astrophysical nature involving the sun's radiative output, the Earth's relative position to the sun, and the possibilities for Earth collisions with interplanetary material; and those involving solid Earth tectonic processes such as continental drift and ocean formation, orogenesis (i.e., mountain building), and volcanic activity whereby heat and matter are injected into the atmosphere. As a special category, we might add human activities, including items such as changing the air composition and temperature, and changing the nature of Earth's surface by industrial, agricultural, and urban activities.

Almost all the proposed theories of past and future climatic change involve one or more of the variables or factors included in Fig. 1-1. Several general points can be made. From a qualitative and even semiquantitative point of view almost all factors included, and therefore every combination of factors, seem capable of causing some of the climatic changes observed. To give some examples, plausible arguments have been presented for the role of continental drift and mountain building to account for long-term climatic oscillations, and for the role of the Earth-orbital changes as a cause of the Pleistocene ice-age variations, and for the role of anthropogenic CO_2 forcing and volcanic activity as causes of the recent swings of global temperature. All of these are explanations by an essentially external forcing—that is, by the effects of processes included in the upper and lower constructs shown in Fig. 1-1.

In addition to these possibilities there is also the likelihood that oscillations of climate can occur even if there were absolutely no *variations* of external forcing of this kind, purely by internal oscillations involving feedbacks among all of the components of the Earth–air–sea system included within the main internal construct. That is, it is possible that the swings in climate are the consequence of a steady forcing of an inherently unstable system that never achieves a true equilibrium.

A simple illustration of the difference between an externally forced variation and one that results from internal instabilities can readily be found at the higher frequency end of the meteorological spectrum containing phenomena with which we are all familiar. In particular, the annual seasonal variation of the climate is a clear example of an externally forced variation imposed by the astronomical variation in the geometry of the Earth relative to the sun. As is the case in most of the proposed theories of climatic change due to external causes, a simple linear cause-and-effect relationship is involved, with the response being of exactly the same frequency as the forcing. However, we know that there are other very significant changes of the environmental variables that bear no direct relation to this fundamental annual forcing or, for that matter, to the diurnal forcing. These are the irregular fluctuations, having frequencies of several days to a week, that are associated with storm movements—i.e., with the passage of the highs and lows of the weather map. In a single day, with the passage of a strong weather "front," the temperature can change at a point as much as the mean temperature changes between summer and winter. This kind of change is the consequence of inherent instabilities in the atmosphere that lead to responses of far different frequency than the forcing. What results is a form of turbulence on a global scale. In much the same way, as noted in Chapter 1, it could be that what we are looking at in the very long-term record of temperature over millennia is the nature of "climatic turbulence." The record of temperature and ice change shown in Fig. 1-4, for example, may indeed be viewed as a response to the external forcing of the system, but it may bear little similarity in phase, frequency, or amplitude to this forcing. It should be clear, considering all of the factors involved, that the task of accounting for and predicting the climate is bound to be very difficult. What we do know is that the system we are treating is too complicated to allow us to say anything meaningful at the qualitative level of our present discussion. Words and simple thought processes alone, however strongly based on physical arguments, will not suffice to achieve the kinds of deductions (explanations and predictions) that we seek. Rather, we strive to develop quantitative theory of climate in which all relevant forcings, feedbacks, and competitive physical factors are taken into account simultaneously. That is, we consider the explanation of variations in the climatic system as a problem in mathematical physics, in which the basic conservation laws for mass, momentum, and energy are expressed in symbolic forms so that the power of mathematical deductive logic can be used to extract quantitative relationships. As we shall see, due to the indeterminacies of our statements of these conservation laws, and the special difficulties and the statistical nature of climate, we shall have to proceed with a more phenomenological approach, including "stochastic" considerations. To begin, in Chapter 4 we shall set down the relevant mathematical statements of the conservation laws, and discuss a framework for attacking the many difficulties inherent in the problem. With this as our backlog, we will be in a position, in Chapters 5 and 6, to provide the foundations for a dynamical theory, the development of which is our ultimate objective.

4

GENERAL THEORETICAL CONSIDERATIONS

As noted in Section 1.2, it is the purpose of theory to provide a predictive connection between the known external forcing described in Chapter 1 and the observed internal behavior described in the previous chapter. It is a collective belief that this connection must ultimately rest on the most basic statements of the conservation principles governing behavior in each of the domains of the full climate system. These are the classic equations of geophysical fluid dynamics that we next describe.

4.1 THE FUNDAMENTAL EQUATIONS

Individually, the atmosphere, ocean, and ice masses can be treated as continua describable by field variables representing an infinite number of degrees of freedom. The behavior of such continua is generally governed by partial differential equations that are the basic fluid dynamical equations expressing conservation of mass (continuity equations), momentum (Navier–Stokes equations of motion), and energy (first law of thermodynamics), along with diagnostic relationships for internal energy and the thermodynamic state.

Let us assume that such a set of equations governing the nearly "instantaneous" values of all the variable describing the climatic system can be written in an "exact" form. In our discussion that follows, we shall start by taking these fundamental partial differential equations governing the instantaneous, continuum, variations of an arbitrary mass of material within any domain of the climatic system as the most fundamental building block of climate theory—but it will become apparent, when we apply the necessary averaging to obtain equations governing the quantities we are really interested in, that these equations must be supplanted by new phenomenological forms that may bear little structural similarity to their generic forebears.

The following general set of conservation equations is commonly adopted for the behavior of any unit volume in the atmosphere, ocean, or ice mass:

Mass:

$$\frac{\partial \rho}{\partial t} = -\nabla \cdot \rho \mathbf{V} \tag{4.1}$$

$$\frac{\partial \rho \xi}{\partial t} = -\nabla \cdot \rho \xi \mathbf{V} + \mathcal{S}_{\xi} \tag{4.2}$$

Momentum:

$$\frac{\partial \rho \mathbf{V}}{\partial t} = -\nabla \cdot (\rho \mathbf{V})\mathbf{V} - \nabla p + \rho \nabla \Phi - 2\rho \Omega \times V + \mathbf{F} \tag{4.3}$$

Energy:

$$\frac{\partial \rho (cT)}{\partial t} = -\nabla \cdot \rho (cT)\mathbf{V} - p\nabla \cdot \mathbf{V} + \rho q \tag{4.4}$$

$$q = q_{\mathrm{SW}} + q_{\mathrm{LW}} + q_{\mathrm{C}} + q_{\mathrm{L}} + q_{\mathrm{F}} \tag{4.5}$$

In addition, there are the equations of state,

$$\rho = p/RT \qquad \text{(ideal gas atmosphere)} \tag{4.6}$$

$$\rho = \rho^*\left[1 - \mu_{\mathrm{T}}\left(T - T^*\right) + \mu_{\mathrm{P}}\left(p - p^*\right) + \mu_{\mathrm{S}}\left(S - S^*\right)\right] \qquad \text{(ocean, ice)} \tag{4.7}$$

The symbols in the above equations are defined in Table 4-1.

To a high degree of accuracy, the vertical component of the equation of motion [Eq. (4.3)] for the atmosphere, ocean, and ice masses can be taken as the hydrostatic balance equation,

$$\frac{\partial p}{\partial z} - \rho g = 0 \tag{4.8}$$

and for the oceans and ice sheets it is a good approximation to assume incompressibility,

$$\nabla \cdot \mathbf{V} = 0 \tag{4.9}$$

In order to close this system, Eqs. (4.1)–(4.7) must be supplemented by constitutive equations for friction and for all the modes of heating embraced by q, including shortwave (q_{SW}) and longwave (q_{LW}) radiation, conduction and small-scale convection (q_{C}), all water phase transformations (q_{L}), and viscosity (q_{F}). This additional set of equations forms a significant part of the full system governing climatic behavior. The radiation component, for example, must include the effects of all the constituent atmospheric gases, including important variable greenhouse constituents, water vapor, and carbon dioxide. Separate physical equations must also be formulated to describe the distribution of clouds in the atmosphere, which are not only of intrinsic interest as a climate variable but which also have an enormous effect on the flux of radiant energy in the system. Representations must also be made for the boundary fluxes of mass, momentum, and energy due to all scales of variability, providing the essential

Table 4-1 Conservation Equation Symbols

Symbol	Definition
ρ	Density
ξ	Mass mixing ratio of an arbitrary variable constituent (e.g., water vapor, cloud drops, CO_2, salinity) in nondimensional units of mass of constituents per mass of carrier medium
p	Pressure
T	Temperature
S	Salinity ($\equiv \xi_S$, usually expressed in g/kg, i.e., ‰)
$\mathbf{V}$	$\mathbf{v} + w\mathbf{k}$
$\mathbf{v}$	$u\mathbf{i} + v\mathbf{j}$
u	dx/dt (eastward speed)
v	dy/dt (northward speed)
w	dz/dt (vertically upward speed)
(dx, dy, dz)	$(a \cos\varphi \, d\lambda, \; a \, d\varphi, \; dz)$
$\mathbf{i}, \mathbf{j}, \mathbf{k}$	Unit vectors eastward, northward, and upward, respectively
λ	Longitude
φ	Latitude
z	Upward vertical distance
t	Time
a	Radius of Earth
∇	$\mathbf{i}\partial/\partial x + \mathbf{j}\partial/\partial y + \mathbf{k}\partial/\partial z$
Φ	gz
g	Acceleration of gravity
Ω	Angular velocity of Earth
$\mathbf{F}$	Frictional force per unit volume $= (\frac{\partial \tau_{x\alpha}}{\partial \alpha}\mathbf{i} + \frac{\partial \tau_{y\alpha}}{\partial \alpha}\mathbf{j} + \frac{\partial \tau_{z\alpha}}{\partial \alpha}\mathbf{k})$
$\tau_{\beta\alpha}$	Stresses acting in the β-direction on planes perpendicular to α, excluding the pressure p (normal stresses are "deviators"), $(\alpha, \beta) = (x, y, z)$
$\mathcal{S}_\xi$	Source function for variable constituent
q	Rate of heat addition per unit mass due to shortwave radiation q_{SW}, longwave radiation q_{LW}, molecular and small-scale eddy conduction q_C, water phase changes q_L, and viscosity q_F (W/kg)
c	Specific heat at constant volume (J/kg K)
R	Gas constant for air (J/kg K)
$(\chi)^*$	Standard value of (χ)
μ_T	Coefficient of volume expansion $[-(\partial\rho/\partial T)^*_{p,s}/\rho^*]$
μ_P	Coefficient of isothermal compressibility $[(\partial\rho/\partial p)^*_{T,s}/\rho^*]$
μ_S	salinity coefficient $[(\partial\rho/\partial S)^*_{p,T}/\rho^*]$
$\partial/\partial t$	$d/dt - \mathbf{V}\cdot\nabla$

interfacial coupling between the domains. The above relationships, Eqs. (4.1)–(4.7), together with these supplementary relations and boundary conditions, constitute the fundamental set of geophysical fluid dynamics (GFD) equations forming the basis for global (sometimes called "general") circulation models (GCMs) that govern the fast-response atmosphere/shallow-ocean system, and also govern the slower response parts (ice sheets, deep ocean, and geochemical inventories), that are central to paleoclimatology. In most of the classic applications of GFD, the atmosphere and oceans are treated as separate problems in fluid mechanics involving nearly instantaneous flows as modeled, for example by laboratory experiments (e.g., Gill, 1982; Pedlosky, 1987; Salmon, 1998). On the other hand, as discussed in the remainder of this chapter, the field of climate dynamics treats the time-averaged properties of the complete coupled environmental system, including the cryosphere, chemosphere,and biosphere, as essential, additional, interactive components.

Of particular interest for our paleoclimatological problem is the specialization of the mass continuity equation, Eq. (4.2), for all forms of water (vapor, liquid, and ice). Let us define $\chi_\xi (\equiv \rho\xi)$ as the concentration in kg m^{-3} of a substance whose mixing ratio is ξ, where ρ is the density of the carrier medium containing the trace substance (e.g., the atmosphere, for water vapor). If we are dealing with a pure substance (e.g., glacial ice) we simply set $\xi = 1$ and $\chi = \rho$. In particular, we define the following concentrations relevant for the hydrologic cycle: χ_{v} (atmospheric water vapor), χ_{nw} (atmospheric liquid cloud and rain drops), χ_{ni} (atmospheric ice crystals and snow), χ_{w} [all forms of surficial liquid water—e.g., ocean, lakes, rivers, groundwater subglacial (basal) water], and χ_{i} (all forms of surficial ice—e.g., snow, ground ice, sea ice, ice sheets). Then the specialized forms of Eq. (4.2) governing these hydrologic components are, respectively,

$$\frac{\partial \chi_{\mathrm{v}}}{\partial t} = -\nabla \cdot \chi_{\mathrm{v}}\mathbf{V} - \mathcal{C} \tag{4.10}$$

$$\frac{\partial \chi_{\mathrm{nw}}}{\partial t} = -\nabla \cdot \chi_{\mathrm{nw}}\mathbf{V} + \mathcal{C}_{\mathrm{w}} + \mathcal{M}_{\mathrm{n}} \tag{4.11}$$

$$\frac{\partial \chi_{\mathrm{ni}}}{\partial t} = -\nabla \cdot \chi_{\mathrm{ni}}\mathbf{V} + \mathcal{C}_{\mathrm{i}} - \mathcal{M}_{\mathrm{n}} \tag{4.12}$$

$$\frac{\partial \chi_{\mathrm{w}}}{\partial t} = -\nabla \cdot \chi_{\mathrm{w}}\mathbf{V}_{\mathrm{w}} + \mathcal{M} - \mathcal{E} \tag{4.13}$$

$$\frac{\partial \chi_{\mathrm{i}}}{\partial t} = -\nabla \cdot \chi_{\mathrm{i}}\mathbf{V}_{\mathrm{i}} - \mathcal{M} \tag{4.14}$$

where $\mathcal{C}$ $(= \mathcal{C}_{\mathrm{w}} + \mathcal{C}_{\mathrm{i}})$ is the rate of condensation to cloud drops ($\mathcal{C}_{\mathrm{w}}$) and sublimation to ice crystals ($\mathcal{C}_{\mathrm{i}}$) (negative values indicating evaporation or ablation), $\mathcal{M}_{\mathrm{n}}$ is the rate of melting of ice cloud, $\mathcal{E}$ is the rate of evaporation (including that of the liquid phase of ice ablation) from Earth's surface, $\mathcal{M}$ is the rate of melting of surficial ice (negative values of which denote freezing), and $\mathbf{V}_{\mathrm{w}}$ and $\mathbf{V}_{\mathrm{i}}$ are the three-dimensional velocities of subsurface water and ice, respectively. Equations (4.11) and (4.12) are the fundamental continuity equations for cloud mass, whereas Eqs. (4.13) and (4.14) are the fundamental equations of ground hydrology and glaciology, respectively, when applied to continental surfaces. A further specialization of these equations for ice sheet behavior is given in Chapter 9.

4.2 TIME AVERAGING AND STOCHASTIC FORCING

In attempting to apply the above fundamental equations as the basis for a general predictive theory from which we can account for phenomena of the kind represented in Fig. 1-4, we must deal with several key issues. For example, as noted above, these equations govern the space–time continuum of values of the climatic state variables. In practice, however, due to numerical and observational limitations of spatial and temporal resolution, we cannot measure and are not really interested in "instantaneous" values, but rather in an average value of the variables over somc running time period δ defined by

$$\bar{x} = \frac{1}{\delta}\int_{t-\delta/2}^{t+\delta/2} x\,dt \tag{4.15}$$

Typical averaging periods of interest are roughly an hour (defining synoptic values resolved by standard meteorological network measurements), and about 100 years (which constitutes our definition of a climatic average, the variations of which are the subject of paleoclimatology). The variations of the synoptic-average ($\delta_s = 1$ hr) state of the climatic system constitute the weather, whereas the climatic-average ($\delta_c = 100$ y) statistics constitute the climate (see Table 4-2). The averaging periods of interest are determined by the nature of the variations exhibited by the system, as pictured in the idealized spectrum shown in Fig. 1-2. An important reason for choosing $\delta_c = 100$ y is that there appears to be a relative minimum in observed natural variability of this period (i.e., this period falls within a relative spectral "gap" between more energetic long-term variations such as the ice ages and shorter term variations such as the annual cycle, mesoscale ocean eddies, and synoptic weather). Formally, we can write

$$P\left(x^\star\right) \ll \delta \ll P(\bar{x}) \tag{4.16}$$

where $P(\bar{x})$ and $P(x^\star)$ stand for the dominant periods of the fluctuations of $\bar{x}$ and departures $x^\star$, respectively (see Fig. 4-1). The x is any variable, the bar denotes the climatic average, and the asterisk denotes the departure of the instantaneous value from the climatic mean value, including the effects of all periods smaller than 100 years, i.e.,

$$x = \bar{x} + x^\star$$

This makes it likely that the 100-year running mean will vary relatively slowly and quasi-linearly within any averaging time interval, δ_c, and that there will be many higher frequency fluctuations within this averaging interval (cf. Fig. 4-1). Thus, because of Eq. (4.16), as first approximations the so-called Reynolds conditions are satisfied:

$$\bar{\bar{x}} \approx \bar{x} \tag{4.17}$$

Table 4-2 A Resolution of Climate Variability[a]

Averaging period (δ)	Spectral band	Physical phenomena	Known forcing
	0–1 hr	Micro- and mesometeorological eddies: turbulence and convection	—
1 hr		Synoptic average (δ_S)	—
	1 hr–3 mo	Diurnal cycle, cyclone waves, blocking, and index variations	Diurnal solar radiation cycle
3 mo		Seasonal average	—
	3 mo–100 y	Annual cycle, year-to-year (interannual variability)	Annual solar radiation cycle
100 y	—	Climatological average (δ_C)	—
	100 y–1 My	Historical and paleoclimatological variations, ice ages	Earth-orbital radiation cycles
10 My	—	Tectonic average (δ_t)	—
	10 My–4.5 By	Ultra-long-term climate variations influenced by global tectonics and planetary evolution	Continental drift
4.5 By	—	Age of the Earth	—

[a] Based on the identification of the main spectral gaps, δ_S, δ_C, and δ_t; see text.

$$\overline{x^\star} \approx 0 \tag{4.18}$$

$$\overline{\frac{\partial \bar{x}}{\partial t}} \approx \frac{\partial \bar{x}}{\partial t} \tag{4.19}$$

$$\overline{\frac{\partial x}{\partial t}} = \left(\overline{\frac{\partial \bar{x}}{\partial t}} + \overline{\frac{\partial x^\star}{\partial t}}\right) \approx \left(\frac{\partial \bar{x}}{\partial t} + \overline{\frac{\partial x^\star}{\partial t}}\right) \tag{4.20}$$

$$\overline{x_1 x_2} = \left(\overline{\bar{x}_1 \bar{x}_2} + \overline{x_1^\star x_2^\star} + \overline{x_1^\star \bar{x}_2} + \overline{\bar{x}_1 x_2^\star}\right) \approx \left(\bar{x}_1 \bar{x}_2 + \overline{x_1^\star x_2^\star}\right) \tag{4.21}$$

In order to obtain dynamical equations governing the evolution of such running time-average variables, the "exact" instantaneous equations [e.g., Eqs. (4.1)–(4.7)] must be averaged over the relevant time interval δ. As is well known from turbulence theory (e.g., Monin and Yaglom, 1971) there are at least two major consequences of such an averaging process: (1) eddy stress terms that are introduced must be parameterized to effect closure in terms of the averaged variable, and (2) stochastic forcing terms are introduced due to the effects of nonsystematic random departures of the instantaneous values from the mean values and the impossibility of achieving exact parameterizations.

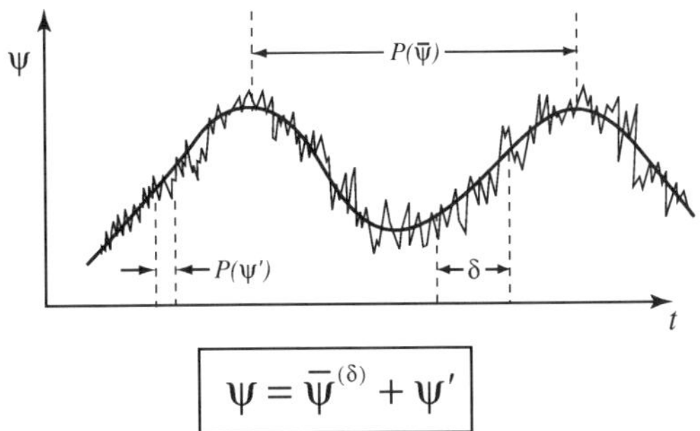

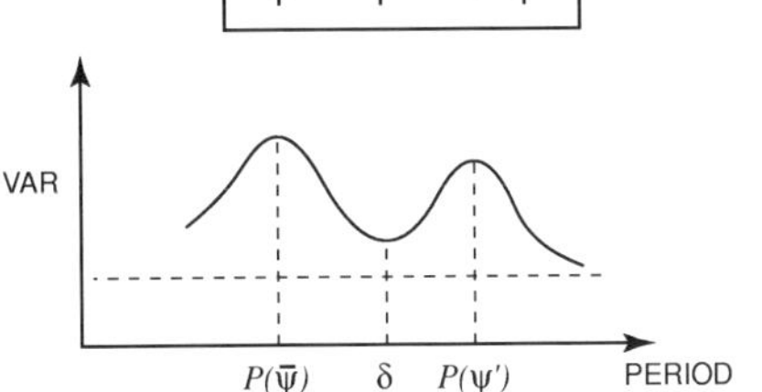

Figure 4-1 Schematic representations of the temporal change of a variable ψ containing two dominant periods of variation $P(\bar{\psi}^{(\delta)})$, and $P(\psi')$ plus some noise, where δ is the time interval defining the running average $\bar{\psi}^{(\delta)}$ having a value within the spectral gap between $P(\bar{\psi}^{(\delta)})$ and $P(\psi')$.

These consequences can be illustrated by considering a simplified form of the energy equation, Eq. (4.4), in which we take $\nabla \cdot \mathbf{V} \approx 0$ (e.g., an ocean):

$$\frac{\partial T}{\partial t} = -\nabla \cdot T\mathbf{V} + \frac{q}{c} \tag{4.22}$$

The heating function $q = q(x_j; t)$ is generally a complex set of terms that are dependent linearly and nonlinearly on many other climatic variables x_j $(j = 1, 2, 3, \ldots)$ and on t itself through external forcing that can be deterministically prescribed (such as periodic Earth-orbital radiative changes) or randomly imposed (such as atmospheric turbidity changes due to volcanic eruptions).

Now if we apply Eqs. (4.17)–(4.21) to Eq. (4.22) we obtain

$$\frac{\partial \overline{T}}{\partial t} = -\nabla \cdot \overline{T}\,\overline{\mathbf{V}} - \nabla\overline{T^{\star}\mathbf{V}^{\star}} + \frac{1}{c}\overline{q(x_j; t)} - \left(\overline{\frac{\partial T^{\star}}{\partial t}}\right) \tag{4.23}$$

At any time τ,

$$\left(\overline{\frac{\partial T^{\star}}{\partial t}}\right) = \frac{T^{\star}(\tau + \delta/2) - T^{\star}(\tau - \delta/2)}{\delta} \equiv R_1 \tag{4.24}$$

which, as can be seen from Fig. 4-1, will be essentially random if $T^{\star}$ contains an aperiodic component. The ubiquitous presence of this aperiodic component is manifest in the underlying base value of variance in the spectrum shown in Fig. 1-2, indicative of a white noise background variability. At any time step, the amplitude of this random term can be of the same order as $\partial\overline{T}/\partial t$, but will be reduced as δ is increased, thereby increasing the signal-to-noise ratio. As we have indicated in Eq. (4.24), we denote this random component by R_1.

Another source of "random forcing" must be introduced when we parameterize the stress term $\overline{\mathbf{V}^{\star}T^{\star}}$, as well as stress terms that may be included in $\overline{q(x_j;t)}$. In general, we have

$$\left(\overline{x_1^{\star}x_2^{\star}}\right) = f(\bar{x}_j) + \varepsilon \tag{4.25}$$

where $\bar{x}_j$ can be any set of climatic-mean variables, f is a deterministic formula usually obtained by some mixture of theory and empiricism (e.g., the diffusion approximation $\overline{\mathbf{V}^{\star}T^{\star}} = -k\nabla\overline{T}$, where k is an eddy diffusivity), and ε $(= \phi + R_2)$ is an error that contains a systematic part ϕ and a random part R_2. For the diffusion approximation the random part will be present if for no other reason than the variability of k when estimated over different time periods (e.g., Robock, 1978).

Additional sources of random forcing enter the problem, even at the level of the instantaneous equations. These may be due, for example, to irregular external inputs (e.g., volcanism) that constitute an unknown aperiodic forcing, which we can denote by R_3, and to the uncertainty of initial conditions. Thus, after the above parameterization, the partial differential equation, Eq. (4.22), can be written in a schematic form,

$$\frac{\partial \overline{T}}{\partial t} = f\left(\overline{T}, \bar{x}_j, \omega_T; t\right) \tag{4.26}$$

where f is a collection of terms involving partial derivatives, and ω_T $(= R_1, R_2, R_3)$ is a representation of the random forcing that must be included in equations governing variables averaged over any time period. For some averaging periods, however, the amplitude of the random component may be small enough to be neglected [as is usually done in the atmospheric general circulation models (AGCMs) applied to synoptic-average variables]. On the other hand, it seems likely that for the statistical-dynamical models (i.e., SDMs) formulated for longer term mean values (e.g., 100 y), such random forcing cannot be neglected. In the next chapter we shall formalize the definitions of the running time averages to be considered in the full paleoclimatic problem.

In order to solve these partial differential equations governing averaged values, they must be reduced either to a system of finite-difference equations governing values at a discrete space–time grid or to a finite system of ordinary differential equations governing the amplitudes of a truncated orthogonal spatial expansion of the variables. Thus, we are always forced to deal with a more approximate system that can usually be written in the form,

$$\frac{dx_i}{dt} = f_i(x_j, F_i, \omega_i) \tag{4.27}$$

where x_i now denotes the set of any time-averaged climatic-mean variable (including higher moments of the probability distribution, e.g., variances and covariances), $f_i(x_i, F_i)$ is the deterministic component that, in addition to linear and nonlinear terms dependent on x_j, can contain time-dependent deterministic forcing components F_i, and ω_i is stochastic forcing that introduces a probablistic component. Depending on whether this system is written for grid point values or orthogonal (e.g., Fourier) components, the variables x_i will be functions of (λ, ϕ, z) or wave numbers (k, l, m) in the case of a Fourier expansion.

Such a set of time-dependent ordinary differential equations governing averaged quantities and containing stochastic forcing constitutes a stochastic-dynamical system characterized by a very large but finite number of degrees of freedom. As we have just indicated, if the system governs the synoptic-average variables ($\delta = 1$ hr) of the atmosphere with the synoptic spatial resolution (thereby requiring parameterization only of subsynoptic frequencies and spatial scales), we speak of the system as an atmospheric general circulation model. In this case the solution can be iterated forward in time to an equilibrium state described by a full set of statistics (e.g., means, variances, frequencies of rarer events) that constitute the climate. This, of course, is a demanding procedure in time and resources.

For studies of long-term climatic change—that is, evolution over thousands of years—it is natural to consider a statistical-dynamical model composed of equations governing a longer term average, such as 100 years. In this case we must parameterize not only subsynoptic phenomena but all phenomena up to and including interannual and interdecadal variations, a most difficult and challenging task. Although the stochastic forcing amplitude may be relatively small in the AGCM, it will tend to be larger in the SDM. Moreover, additional equations, including the parameterization formulas, will be necessary to deduce the higher order statistics (such as the spatial and temporal variances and amplitude of the seasonal cycle) that are needed for a full description of climate. Further discussions of the AGCM and SDM governing the fast-response atmosphere and surficial layers as an essentially steady-state (equilibrium) problem are given in Chapter 7. As developed further in the next section of this chapter and in Chapter 5, the slower response variables describing the deep cryosphere and ocean, for example, will require a more dynamical, nonequilibrium, treatment.

It seems likely that to describe the macrobehavior of the climatic system, with an accuracy such that observations can reveal it, will require a much smaller number of variables (and their governing equations) than are represented by the full set described in Section 4.1 applied to all the variables, in all the domains, at a synoptic spatial grid. Thus, a major challenge in developing the theory of long-term climatic fluctuations will be to learn how to truncate (or space average) the system in ways that capture the main variables, at the same time permitting closure of the system by physically valid parameterizations. Related to this is the need to identify groupings of variables that are "coherent" or diagnostically related so that they can be represented by much reduced sets of variables (perhaps even a single variable). A formal procedure for accomplishing such reductions by successive spatial integrations starts with the very simplest one-variable model obtained by integrating over the entire climatic system and proceeds by systematically expanding the model with added variables and equations. A partial test of the success of this process is the degree to which the unexplained observed variability is of a random white noise variety, to which no further determinism can be brought to bear.

4.3 RESPONSE TIMES AND EQUILIBRIUM

We have noted that the climatic system is extremely heterogeneous, containing domains (or subsystems) that are all interactive to some extent but having vastly different properties and modes of behavior if considered alone. One important property distinguishing each of the various domains is their response time (also called the damping, relaxation, equilibration, or adjustment times, or more simply, the time constant), which is a measure of the time it would take for dissipative processes acting in the absence of continued forcing to remove departures from equilibrium (i.e., to reequilibrate after a small change in boundary conditions or forcing). (This is to be distinguished from the *time scale*, which is a measure of the typical period of the observed fluctuations.) A short response time, for example, indicates that the system (or subsystem) has the inherent capability of responding quickly to any perturbations from its equilibria. A domain with a short response time can be considered quasi-statically equilibrated to a neighboring domain that has a much longer response time. In this sense, the component with the longer response time can be considered to "carry" along the component with the more rapid response time.

Some useful statements in these regards can be made, based on the following considerations: Let $\bar{x}^{(\delta)}$ be the average of x over a time interval δ satisfying Eq. (4.16) (where x can be temperature T, or ice mass, for example), the departures of which ($\bar{x}^{(\delta)\prime}$) from equilibrium ($\bar{x}_0^{(\delta)}$) are governed by a generalized equation of the form Eq. (4.27). Separating out the main (linear) dissipative term having a time constant $\varepsilon(\delta)$ that includes the effects of all the diffusive flux modes embraced by the particular averaging period δ, we have

$$\frac{d\bar{x}^{(\delta)\prime}}{dt} = f\left(\bar{x}_0^{(\delta)}, \bar{x}^{(\delta)\prime}, F'\right) - \frac{1}{\varepsilon(\delta)}\bar{x}^{(\delta)\prime} \tag{4.28}$$

If $P(\bar{x}^{(\delta)})$ is the characteristic period of variations of $\bar{x}^{(\delta)}$, it follows from Eq. (4.16) that a characteristic is value of $d\bar{x}^{(\delta)\prime}/dt$ is

$$\frac{d\bar{x}^{(\delta)\prime}}{dt} \sim \frac{|\bar{x}^{(\delta)\prime}|}{P(\bar{x}^{(\delta)})/4} \ll \frac{|\bar{x}^{(\delta)\prime}|}{\delta} \tag{4.29}$$

Thus, if $\varepsilon(\delta) \leqslant \delta$, the dissipative term $\bar{x}^{(\delta)\prime}/\varepsilon(\delta)$ must be much larger than $d\bar{x}^{(\delta)\prime}/dt$ and can only be balanced in quasi-static equilibrium with $f(\bar{x}^{(\delta)\prime}, F')$. On the other hand, if $\varepsilon(\delta) \sim P(\bar{x}^{(\delta)})/4 \gg \delta$, then $d\bar{x}^{(\delta)}/dt$ is of the same order as the rate of dissipation and must be evaluated in a time-dependent, nonequilibrium, calculation of the right-hand side of Eq. (4.28).

In summary, the following conditions are suggested that determine the nature of the equations that must be solved to determine $\bar{x}^{(\delta)}$ (cf. Saltzman, 1978):

1. $\varepsilon_x(\delta) \leqslant \delta$ is in most cases sufficient to ensure that the equation governing $\bar{x}^{(\delta)}$ is diagnostic, i.e., that we can assume a quasi-steady state, $\partial\bar{x}^{(\delta)}/\partial t \approx 0$.

2. $\varepsilon(\delta) \sim P(\bar{x}^{(\delta)})/4 \gg \delta$ is a sufficient condition that the equation governing $\bar{x}^{(\delta)}$ is prognostic, posing a dynamical systems problem, with $d\bar{x}^{(\delta)}/dt$ of comparable magnitude to the other terms. If $\varepsilon \ll P(\bar{x}^{(\delta)})$, then, even if $\varepsilon \gg \delta$, $d\bar{x}^{(\delta)}/dt$ will be negligibly small compared to the other terms (i.e., the system is "overdamped") and $\bar{x}^{(\delta)}$ can be determined diagnostically.

These conditions must be qualified by two caveats: (1) the spectral "gaps" must be relatively deep and widely separated (which we assume is acceptably the case for the gap between synoptic weather variability and the ice-age variations), and (2) the equilibrated faster response states must be unique (i.e., there must be no unstable equilibrium that will admit rapid transitions between two or more stable equilibria, thereby requiring a time-dependent treatment that may involve a sensitive dependence on stochastic perturbations). Further analysis, highlighting some of the positive and potentially negative aspects of this averaging approach, is given by Nicolis and Nicolis (1995).

In Fig. 1-2 we presented a schematic spectrum of climatic variability, showing the location of the main averaging periods (δ_s, δ_c, δ_t) and the time constants ε_α, ε_τ, ε_θ, and ε_I for the atmosphere, surface temperature, deep ocean temperature, and ice-sheet mass, respectively. In Table 4-3 we list the main subsystems and their characteristic present dimensions and masses, thermal constants, estimates of $\varepsilon_T(\delta)$ [in most cases for both $\delta = \delta_s \sim 1$ hr (the synoptic average) and $\delta = \delta_c \sim 100$ y (the climatic average)], and rough estimates of ε_M for the ice domains. [The thermal response time is shown in Saltzman (1983) to be approximately of the form

$$\varepsilon_T = \left[b + \left(\frac{k}{D^2} \right) + \left(\frac{k_v}{D^2} \right)_\delta + \left(\frac{k_h}{L^2} \right)_\delta \right]^{-1} = \varepsilon_T(\delta) \tag{4.30}$$

where b is a cooling coefficient for longwave radiation, k is the molecular thermal diffusivity, $k_v(\delta)$ and $k_h(\delta)$ are the vertical and horizontal thermal diffusivities, respectively, due to eddies of all periods shorter than δ, and D and L are characteristic vertical and horizontal spatial scales of $\overline{T}^{(\delta)}$, respectively].

In the atmosphere, for example, it can be seen that, whereas for $\delta = \delta_s$ we have $\varepsilon_T(\delta_s) \gg \delta_s$, for $\delta = \delta_c$ we have $\varepsilon_T(\delta_c) \gg \delta_c$, implying that for the climatological mean the atmospheric thermal response is so fast that its mean can be considered as an equilibrium state. A schematic portrait showing the thermal response times for δ_c of all the domains comprising the complete climatic system was shown in Fig. 1-1. In general, as we proceed from top of this figure (i.e., the atmosphere) to the bottom (deep ocean, and ice sheets with associated bedrock) we encounter increasingly longer response times.

In forming Table 4-3 we have considered each climatic domain separately. Actually, as shown in Fig. 1-1, all adjacent domains are linked by complex physical processes involving cross-boundary fluxes of mass, momentum, and energy that constitute internal "forcing" and "feedback" in the system. As we said at the beginning of this section, domains with large ε (slow response time) will tend to "carry along" the domains with smaller ε, which, because of their fast response times, tend to adjust quasi-statically to the changing boundary conditions imposed by the slow response domain. This follows from the fact that if $\varepsilon_y \gg \varepsilon_x$ for two domains x and y, one can always adopt an averaging period $\delta = \varepsilon_x$, making the equation governing x diagnostic but leaving the domain for which $\varepsilon_y \gg \delta$ essentially prognostic. In this sense we speak of the deep ocean and ice sheets, for example, as the carriers of long-term climatic variability, the prognostic equations for these domains governing the nonequilibrium evolution of the

Table 4-3 Physical Properties of the Main Climatic Domains[a]

Variable unit Averaging period	A ($10^{12} m^2$)	D (m)	M (10^{18} kg)	c (10^3 J/kg K)	ε_T (s) δ_S	 δ_C	ε_M (s) δ_C
Climatic domain							
Atmosphere							
1. Free	510	10^4	0.5	1	10^7	10^6	—
2. Boundary layer	510	10^3	5	1	10^5	10^5	—
Ocean							
3. Mixed layer	334	10^2	34	4	10^7	10^6	—
4. Deep	362	4×10^3	1400	4	10^{11}	10^{10}	—
5. Sea-ice (pack-shelf)	30	$1–10^2$	0.5	2	$10^6–10^{10}$		$10^6–10^{10}$
Continent							
6. Lakes and rivers	2	10^2	0.2	4	10^6	10^6	—
7. Litho-biosphere	131	2	1	0.8	10^6	10^6	—
8. Snow and surf. ice layer	801	1	0.1	2	10^5	10^5	10^5
9. Mountain glaciers	1	10^2	0.1	2	10^{10}	10^{10}	10^9
10. Ice sheets	14	10^3	10	2	10^{12}	10^{12}	10^{11}

[a] Including characteristic present dimensions: A, area; D, depth; M, mass; c, thermal constants, with their response times ε_T and ε_M.

system. The nature and path of this evolution, however, can be markedly influenced by all the higher frequency effects of the fast response domain, the deterministic parts of which must be parameterized and the random parts of which must be included as stochastic forcing. By the same token, the low-frequency, long-response-time phenomena can be considered as fixed conditions in deducing the separate behavior of the domain with higher frequencies and shorter response times. As a general rule, if two domains have similar response times the behavior in both domains must be solved for simultaneously, i.e., two adjacent domains independently in equilibrium for a given δ ($\varepsilon \leqslant \delta$) must be in equilibrium with each other.

One consequence of the above remarks that we have already mentioned is that the climatic-mean ($\delta = \delta_c$) state of the atmosphere should be deducible as a steady-state problem subject to fixed boundary conditions imposed by a neighboring domain for which $\varepsilon \gg \delta$. From the values of ε given in Table 4-3 and Figs. 1-1 and 1-2, we see that the response of the surface of the Earth (e.g., the biolithosphere temperature, surface snow, and sea ice, and the mixed-layer ocean temperature) is almost fast enough to be of an equilibrium nature comparable to the atmosphere. Therefore these aspects of the surface state should be deduced simultaneously with the atmospheric state, especially insofar as there are significant feedbacks between the atmosphere and these domains (e.g., the ice-albedo feedbacks). If we arbitrarily prescribe the surface snow–ice distribution, for example, one can indeed compute the atmospheric state that is required for equilibrium with it, but this equilibrium may not be consistent with the prescribed snow–ice field (e.g., the equilibrium may require surface temperatures that are well below freezing in regions of ample moisture where no ice is prescribed).

We shall sometimes use the term "super"-general circulation model to denote a model governing this fuller atmosphere plus surface system, representing a coupled atmospheric component (the AGCM), oceanic mixed-layer component (MLM), sea-ice component (SIM), surface hydrological (e.g., soil moisture and snow cover) component (SHM), and a terrestrial biosphere component (TBM) (see Fig. 4-2). The GCM is characterized by its representation of the evolving motion fields on synoptic scales in the atmosphere and mixed layer, including their explicit effects on all the other surficial components varying on scales less than about 100 y, but excludes the much slower response parts of the climate system such as the deep ocean (governed by an OGCM), the ice sheets, and rock-geochemical inventories. It is important to recognize that even though the atmosphere and its lower boundary surface layer may be in equilibrium, judging from the geologic record of long-term climatic change, this equilibrium must entail a small but significant disequilibrium of the total ice mass field and the deep ocean temperature. In fact, assuming a response time for ice sheet mass of the order of 10^4 y, we note that the changes in ice sheet mass must be determined from an explicit, nonequilibrium, calculation. This will pose a critical difficulty that will be discussed in the next chapter along with a deeper consideration of the material in this section as applied specifically to the problem of paleoclimatic evolution.

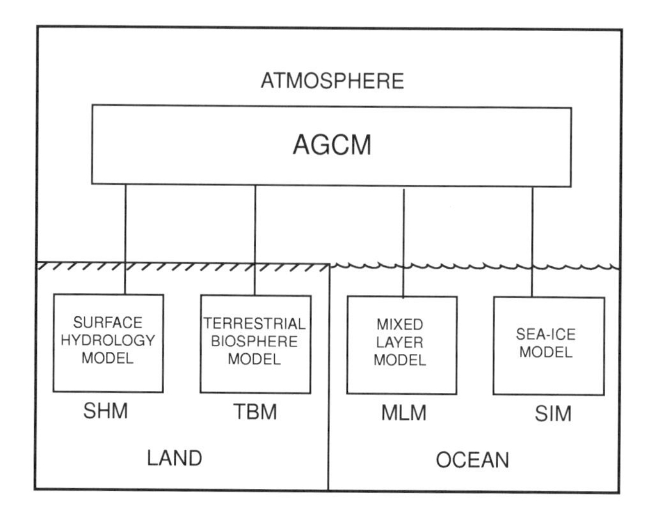

Figure 4-2 Components of a "super-GCM" in which an atmospheric component (AGCM) is coupled with models governing the terrestrial and oceanic surface boundary layer states (see text).

4.4 SPATIAL AVERAGING

The equations described above apply to the values of the climatic variables at all locations in the climatic system. We are most often interested, however, in the large spatial scale aspects of climate (i.e., the macroclimate), as distinct from details of the local variability. These larger aspects can be isolated by applying space averaging in addition to the temporal averaging discussed above. In particular, by suitably defined spatial averaging operators, it is possible to resolve formally the complete field of the climatic fields into an arbitrary number of components, each of which is identifiable with a prominent observed feature of the total macroclimatic variability.

A major resolution of this kind consists of two components: (1) the zonally averaged (i.e., axially symmetric) mean field representing the North–South climatic variations, and (2) the axially asymmetric departure of the mean field from the zonal average values, representing the East–West variations. To express this resolution in mathematical form, we first define a zonal average of any variable x,

$$\langle x \rangle \equiv \frac{1}{2\pi} \int_0^{2\pi} x \, d\lambda \tag{4.31}$$

and the departure from the zonal average,

$$x_* = x - \langle x \rangle \tag{4.32}$$

which, for the atmosphere, includes the effects of all the large-scale, weather-producing, waves and vortices. Applying Eq. (4.31) to a climatic mean variable, $\bar{x}$, we have

$$\bar{x}(\lambda, \phi, z, t) = \langle \bar{x} \rangle(\phi, z, t) + \bar{x}_*(\lambda, \phi, z, t) \tag{4.33}$$

The zonal average component $\langle \bar{x} \rangle$ isolates the primary variation of climate from equator to pole that tends to be forced by the basic axial symmetry of rotation and monthly mean solar radiation. The departure field $\bar{x}_*$, often called the stationary or standing-wave field in atmospheric studies, isolates all the nonzonal features arising from the inhomogeneity of Earth's surface (e.g., continent–ocean structure and orography); for a purely homogeneous Earth's surface this departure would vanish.

Just as it is possible to further resolve the time–eddy transients, represented by $\psi^\star$, into a spectrum giving the contributions of different frequencies to the total time variability, it is possible to further resolve ψ_* into one-dimensional Fourier components around latitude circles, or into spherical harmonics. This is especially relevant for the atmosphere, where the instantaneous departure ψ_* and mean departures $\bar{\psi}_*$ have a wavelike appearance, and where, unlike the oceans, a cyclic continuity around latitude circles prevails.

We can further average any climatic variable over all latitudes as well as longitudes to obtain a mean value over the complete global surface as a function of height only. This component is denoted here by a wavy overbar,

$$\begin{aligned} \widetilde{x} &\equiv \frac{1}{\sigma_{\mathrm{E}}} \int_{-\pi/2}^{\pi/2} \int_{0}^{2\pi} x \, d\sigma = \frac{1}{4\pi} \int_{-\pi/2}^{\pi/2} \int_{0}^{2\pi} x \cos\varphi \, d\lambda \, d\varphi \\ &= \frac{1}{2} \int_{-\pi/2}^{\pi/2} \langle x \rangle \cos\varphi \, d\varphi \end{aligned} \tag{4.34}$$

where $\sigma_{\mathrm{E}} = 4\pi a^2$ (surface area of the globe) and $d\sigma = a^2 \cos\varphi \, d\lambda \, d\varphi$ (an element of area). The departure from this area average is denoted by

$$x_* = x - \widetilde{x} \tag{4.35}$$

To complete the possibilities for spatial averaging, we define a vertical, mass-weighted, average representing a mean value of any property x through the depth of the atmosphere, ocean, ice sheet, or land surface, denoted by a wiggly overbar,

$$\check{x} \equiv \frac{1}{m} \int x \, dm \tag{4.36}$$

where dm is an element of mass per unit area given by

$$dm = \begin{cases} g^{-1}\, dp & \text{(hydrostatic atmosphere)} \\ \rho \, dz & \text{(arbitrary medium of density} \rho) \end{cases}$$

and m is the mass of the medium per unit area. The limits are $p = 0$ to p_{s} for the atmosphere. In Fig. 4-3 we show various vertical layers above an arbitrary subsurface reference level ($z = 0$) located below the deepest part of the ocean, which can serve as the limits for the subsurface vertical averages. These are as follows:

z_{SL} = height of mean sea level above the arbitrary subsurface reference level

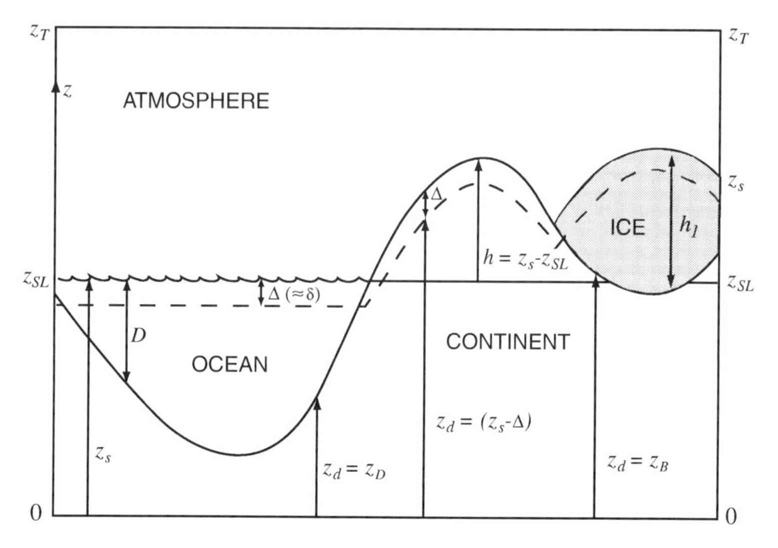

Figure 4-3 Schematic representation of the ocean–continent–ice sheet system, showing the symbols used to define various vertical dimensions described in the text.

z_s = height of Earth's surface above the reference level (z_{SL} for the ocean)

z_d = height of the lower boundary of the ocean ($z_d = z_D$), ice sheets ($z_d = z_B$), or land $[z_d = (z_s - \Delta)]$, where Δ is a shallow "active layer" below the land, ocean, or ice surface (dashed curve in Fig. 4-3) within which there is still a response to synoptic fluctuations of the surface state. For the ocean, $\Delta \approx \delta$, as defined in Chapter 11

h = height of Earth's surface (including ice and continental water bodies) above sea level ($z_s - z_{SL}$), or bathymetric ocean floor depth below sea level

D = depth of ocean ($\equiv h < 0$)

h_I = thickness of ice and/or snow.

A more detailed vertical resolution of the ice sheet and bedrock structure is given in Chapter 9.

This vertical averaging gives a representative value of the climatic variables as a function only of the horizontal coordinates, as well as providing a means of separating the barotropic dynamical processes in the atmosphere and oceans from the thermally dependent baroclinic processes (cf. Gill, 1982).

When the mass-weighted vertical average of any property x is combined with the horizontal area average, we obtain the complete global mass-weighted average over a domain (e.g., atmosphere, ocean, ice sheet), defined by

$$\breve{x} = \breve{\breve{x}} = \frac{1}{\Pi}\int_M x \, d\Pi \tag{4.37}$$

where $\Pi = \sigma_E \widetilde{m}$ is the total mass of any domain of the climate system and $d\Pi = \rho\, dV = d\sigma\, dm$, $dV = r^2 \cos\phi\, d\lambda\, d\phi\, dz$ (an element of volume of the domain).

4.5 CLIMATIC-MEAN MASS AND ENERGY BALANCE EQUATIONS

To provide integral constraints on the full three-dimensional behavior of the climate system, we now present the climatic-average, vertically integrated forms of the fundamental mass continuity and energy equations given in Section 4.1. These equations, which govern the mean global signatures of long-term paleoclimatic evolution (e.g, the water and ice balances and Earth's thermal state),will also serve in Chapter 7 as a basis for discussing simpler climate models (e.g., the energy balance model, or EBM) that have been invoked often in treatments of paleoclimatic variations.

4.5.1 The Water Mass Balance

Using the rules, Eqs. (4.17)–(4.21), we first average Eqs. (4.10)–(4.14) over the climatic-mean interval $\delta_c = 100$ y, henceforth omitting the overbar denoting this average except where it applies to a product of variables. (Such an average could be applied to the ensemble of monthly states to reveal seasonal variations.) Then, integrating from the surface (z_S) to the top of the atmosphere (z_T), we obtain equations for the rate of change of the column mass of water vapor (m_v), liquid drops (m_{nw}), and ice cloud or snow (m_{ni}) per unit area at any point,

$$(m_v, m_{nw}, m_{ni}) = \int_{z_S}^{z_T} (\chi_v, \chi_{nw}, \chi_{ni})\, dz \tag{4.38}$$

of the forms,

$$\frac{\partial m_v}{\partial t} = -\nabla \cdot \mathbf{J}_v + E - C \tag{4.39}$$

$$\frac{\partial m_{nw}}{\partial t} = -\nabla \cdot \mathbf{J}_{nw} - P_w + C_w + M_n \tag{4.40}$$

$$\frac{\partial m_{ni}}{\partial t} = -\nabla \cdot \mathbf{J}_{ni} - P_i - M_n \tag{4.41}$$

where all ice vapor transformations (e.g., sublimation) are assumed for convenience to pass through the liquid water phase, and the vertically integrated horizontal vector flux of any substance, whose mass per unit volume is denoted by χ, is

$$\mathbf{J}_\chi = \int_{z_S}^{z_T} \overline{\chi \mathbf{v}}\, dz = (J_{\chi\lambda}\mathbf{i} + J_{\chi\varphi}\mathbf{j}) \tag{4.42}$$

and, in particular, the flux vector for vapor and cloud particles are given by

$$\mathbf{J}_{(v,n)} = \int_{z_S}^{z_T} \overline{(\chi_v, \chi_n)\mathbf{v}}\, dz \tag{4.43}$$

The rate of evaporation (including ice ablation) from the surface in units of mass per unit area per unit time is given by E; C is the rate of condensation of vapor to liquid (some of which might freeze to form ice cloud particles), and M_n is the rate of melting of ice cloud in the column, given by

$$(C, M_n) = \int_{z_S}^{z_T} (\mathcal{C}, \mathcal{M}_n)\, dz \tag{4.44}$$

P $(= P_w + P_i)$ is the rate of precipitation in the form of rain (P_w) and snowfall (P_i), and F_n is the rate of snow production. Formally, $P_x = [(w_x \chi_x)_{z_S} + (v_x \chi_x \cdot \nabla z_S)]$.

Similarly, integrating Eqs. (4.13) and (4.14) through a subsurface depth from the base of the ocean (z_D) or of an ice sheet (z_B) to the surface (z_S), we obtain a pair of equations for the rates of change of the mass of the subaerial waters of the Earth (m_w) and surficial ice (m_i) per unit area at any point, defined by

$$[m_w, m_i] = \int_{z_{(D,B)}}^{z_S} \chi_{(w,i)}\, dz = \begin{cases} (\rho_w D) & \text{(ocean)} \\ (\rho_i h_I) & \text{(ice sheet)} \end{cases} \tag{4.45}$$

where D and h_I denote the thickness of a water or ice column, respectively. These equations are

$$\frac{\partial m_w}{\partial t} = -\nabla \cdot \mathbf{J}_w + P_w + M - E \tag{4.46}$$

$$\frac{\partial m_i}{\partial t} = -\nabla \cdot \mathbf{J}_i + P_i - M \tag{4.47}$$

where the climatic-mean horizontal flux vectors for liquid water and ice are given by

$$\mathbf{J}_{(w,i)} = \int_{z_{(D,B)}}^{z_S} \overline{\chi_{(w,i)} \mathbf{v}}\, dz \tag{4.48}$$

and M is the rate of melting of ice (negative values of which denote freezing of water).

If we average Eqs. (4.39), (4.40), (4.41), (4.46), and (4.47) over the total area of Earth, $\sigma_E = 4\pi a^2$, noting that $\widetilde{\nabla \cdot \mathbf{J}} = 0$, and defining the total mass of each form of water by $\Pi_x \equiv \sigma_E \widetilde{m}_x$, it is easily verified that the total amount of water in all forms is conserved, i.e.,

$$\frac{d}{dt}[\Pi_w + \Pi_i + \Pi_v + \Pi_n] = 0$$

From water inventories of the present state and estimated past planetary studies, it follows that

$$\Pi_w > \Pi_i \gg \Pi_v > \Pi_n$$

so that to first-order ice mass changes can be assumed to imply nearly opposite ocean mass changes: $d\Pi_w/dt \approx -d\Pi_i/dt$.

4.5.2 Energy Balance

As described in greater detail by Saltzman (1978), if we integrate the climatic-averaged form of the energy equation, Eq. (4.4), through the depth of the atmosphere, and through the depth of the subsurface medium (whether it is land, ocean, or ice, or some combination thereof), neglecting small terms, we obtain the following pair of equations for the rates of change of sensible heat per unit horizontal area of the atmosphere, $\Upsilon_a = \int_{z_S}^{z_T} \rho c_p T \, dz$,

$$\frac{\partial \Upsilon_a}{\partial t} = -\nabla \cdot \mathbf{J}_a + N^{\downarrow} + H_S^{(1,2,3)\uparrow} + L_v C - L_f M_n \tag{4.49}$$

and of the subsurface medium, $\Upsilon_b = \int_{z_d}^{z_S} (\rho c T) \, dz]$,

$$\frac{\partial \Upsilon_b}{\partial t} = -\nabla \cdot \mathbf{J}_b - H_S^{(1,2,3)\uparrow} - L_v E - L_f M + H_d^{\uparrow} \tag{4.50}$$

where $z_d = [z_D \text{ (ocean)}, z_B \text{ (ice)}, z_S - \Delta \text{ (land)}]$.

In these equations, the vector horizontal fluxes of energy are given by

$$\mathbf{J}_a = \int_{z_S}^{z_T} \overline{\rho (c_p T + \Phi) \mathbf{v}} \, dz = J_{a\lambda} \mathbf{i} + J_{a\varphi} \mathbf{j} \tag{4.51}$$

and

$$\mathbf{J}_b = \int_{z_d}^{z_S} \left(\overline{\rho c T \mathbf{v}} \right) dz = J_{b\lambda} \mathbf{i} + J_{b\varphi} \mathbf{j} \tag{4.52}$$

where $\mathbf{J}_b$ will be relevant for the oceans or ice sheets. $N^{\downarrow}$ $(= H_T^{(1)\downarrow} - H_T^{(2)\uparrow})$ is the net radiation at the top of the atmosphere representing the difference of the downward flux of shortwave (solar) radiation $(H_T^{(1)\downarrow})$ and outgoing longwave (terrestrial) radiation $(H_T^{(2)\uparrow})$; $H_S^{(1,2,3)\uparrow}$ $(= -H_S^{(1)\downarrow} + H_S^{(2)\uparrow} + H_S^{(3)\uparrow})$ is the upward flux of energy at the Earth's surface, composed of fluxes due to shortwave radiation $H_S^{(1)\downarrow}$, longwave radiation $H_S^{(2)\uparrow}$, and conduction plus small-scale atmospheric convection $H_S^{(3)\uparrow}$; $H_d^{\uparrow}$ is the upward conductive/convective heat flux at $z = z_d$ (including the effects of the geothermal flux $G^{\uparrow}$); Φ $(= gz)$ is the geopotential energy; the subscript b refers to the specific nature of the subsurface media; and L_v and L_f are the latent heats of vaporization and fusion, respectively.

If we consider an integral over the shallow "active layer" of depth Δ, shown in Fig. 4-3, then in the limit as $\Delta \to 0$, $\Upsilon_b \to 0$ and we obtain the "surface heat balance" condition,

$$H_S^{(1,2,3)\uparrow} - L_v E - L_f M = H_{SS}^{\uparrow} \tag{4.53}$$

where $H_{SS}^{\uparrow}$ is the upward conductive/convective heat flux just below the surface interface, which must be continuous with the upward heat flux on the atmospheric side of the surface interface represented by the left-hand side of Eq. (4.53).

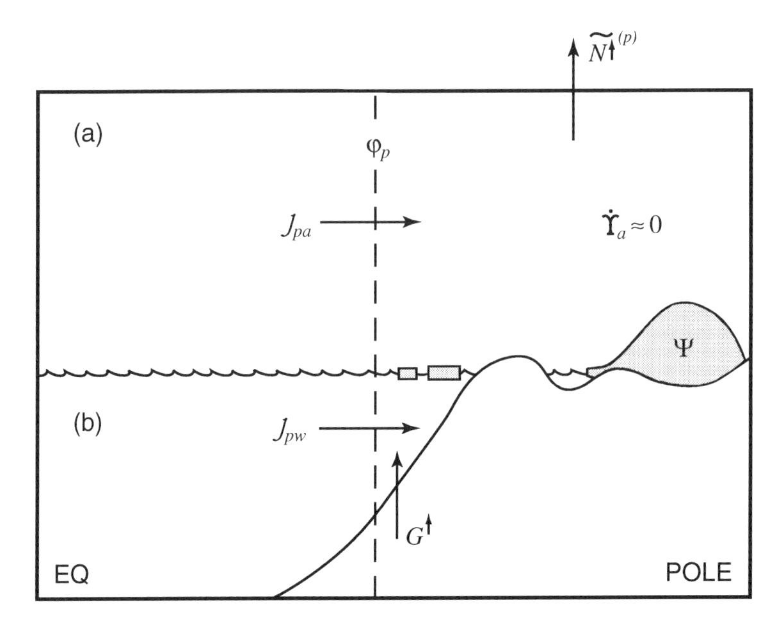

Figure 4-4 Schematic representation of the heat balance of a cap poleward of latitude φ_{P} that includes the total ice mass Ψ. See text for definitions of heat fluxes $\mathcal{J}$, $G^{\uparrow}$, and $\widetilde{N}^{\uparrow(P)}$.

Using these energy equations we can establish some relations for the rates of change of global ice mass that will be of much relevance in the further discussion of the physics of the ice ages in the next chapter. In particular, let us now integrate the ice balance equation, Eq. (4.47), over a "polar cap" bounded by an imaginary wall along a latitude (say $\varphi_{\mathrm{P}} = 50°\mathrm{N}$) that essentially includes almost all of the ice that historically formed in the Northern Hemisphere (see Fig. 4-4). Thus,

$$\frac{d\Psi}{dt} \approx \sigma_{\mathrm{P}}\left(\widetilde{P}_{\mathrm{i}}^{(P)} - \widetilde{M}^{(P)}\right) \tag{4.54}$$

where

$$\Psi = \int_0^{2\pi}\int_{\varphi_{\mathrm{P}}}^{\pi/2} m_{\mathrm{i}}\, d\sigma \tag{4.55}$$

$d\sigma = a^2 \cos\varphi\, d\lambda\, d\varphi$ is an element of the Earth's surface area, σ_{P} is the area of the polar cap north of φ_{P}, and $\widetilde{x}^{(P)}$ denotes an area average of x over the polar cap. Now, from Eqs. (4.39), (4.41), (4.49), and (4.50), assuming no significant changes in the mass of atmospheric ice clouds, or in thermal energy storage in the atmosphere Υ_{a}, and no significant flux of atmospheric ice particles across φ_{P}, we obtain

$$\frac{d\Psi}{dt} \approx \frac{1}{L_{\mathrm{f}}}\left[\left(\sigma_{\mathrm{P}}\widetilde{N}^{P\uparrow} - \sigma_{\mathrm{P}}\widetilde{G}^{P\uparrow}\right) - \mathcal{J}_{\mathrm{p}} - L_{\mathrm{v}}\mathcal{J}_{\mathrm{pv}} + \sigma_{\mathrm{P}}\frac{d\widetilde{\Upsilon_{\mathrm{b}}}}{dt}\right] \tag{4.56}$$

where $\mathcal{J}_{\mathrm{p}} = (\mathcal{J}_{\mathrm{pa}} + \mathcal{J}_{\mathrm{pw}})$ is the net poleward flux of energy across φ_{P} in both the atmosphere

$$\mathcal{J}_{\mathrm{pa}} = 2\pi a \cos\varphi_{\mathrm{P}} \left\langle \int_{z_{\mathrm{S}}}^{z_{\mathrm{T}}} \overline{\rho(c_{\mathrm{p}}T + \Phi)v\,dz} \right\rangle_{\mathrm{p}} \tag{4.57}$$

and the ocean

$$\mathcal{J}_{\mathrm{pw}} = 2\pi a \cos\varphi_{\mathrm{P}} \int_{z_{\mathrm{d}}}^{z_{\mathrm{S}}} \rho_{\mathrm{w}} c_{\mathrm{w}} \langle \overline{Tv} \rangle \, dz \tag{4.58}$$

and the net poleward flux of water vapor across φ_{P} is

$$\mathcal{J}_{\mathrm{pv}} = 2\pi a \cos\varphi_{\mathrm{p}} \int_{z_{\mathrm{S}}}^{z_{\mathrm{T}}} \langle \overline{\chi_{\mathrm{v}} v} \rangle \, dz \tag{4.59}$$

Let us now extend the integral to the whole globe ($\varphi_{\mathrm{P}} \to -\pi/2$) so that $\Psi \to I$ (total global ice mass) and $(\sigma_{\mathrm{P}}\widetilde{x}^{(P)}) \to (\sigma_{\mathrm{E}}\widetilde{x})$, where the tilde denotes a global average and σ_{E} is the area of the globe. Then, recognizing that from Eq. (4.50),

$$\frac{d\widetilde{\Upsilon_{\mathrm{b}}}}{dt} = -\widetilde{H}_{\mathrm{S}}^{(1,2,3)\uparrow} - L_{\mathrm{v}}\widetilde{E} - L_{\mathrm{f}}\widetilde{M} + \widetilde{G}^{\uparrow} \tag{4.60}$$

we obtain from Eq. (4.56),

$$\frac{dI}{dt} = \frac{\sigma_{\mathrm{E}}}{L_{\mathrm{f}}} \left(\widetilde{N}^{\uparrow} - \widetilde{H}_{\mathrm{S}}^{(1,2,3)\uparrow} - L_{\mathrm{v}}\widetilde{E} - L_{\mathrm{f}}\widetilde{M} \right) \tag{4.61}$$

or in terms of net rate of ice accumulation,

$$\frac{dI}{dt} \equiv \sigma_{\mathrm{E}} \left(\widetilde{P}_{\mathrm{i}} - \widetilde{M} \right)$$

where

$$\widetilde{P}_{\mathrm{i}} \approx L_{\mathrm{f}}^{-1} \left(\widetilde{N}^{\uparrow} - \widetilde{H}_{\mathrm{S}}^{(1,2,3)\uparrow} - L_{\mathrm{v}}\widetilde{E} \right)$$

assuming negligible storage of energy in the atmosphere.

We shall make use of this equation in the next chapter to assess the accuracy with which we must measure or calculate the energy fluxes to determine the observed values of dI/dt.

5

SPECIAL THEORETICAL CONSIDERATIONS FOR PALEOCLIMATE

Structuring a Dynamical Approach

Ideally, a theory of paleoclimate should take the form of a complete three-dimensional Earth system model comprising coupled atmosphere–hydrosphere–biosphere–chemosphere–lithosphere components. Such a model would consist of sets of equations of the forms given in Chapter 4 for each of the above domains, representing a super-GCM (including land surface, biosphere, and mixed-layer components) coupled with a deep-oceanic general circulation model (OGCM), an ice sheet model, and a geochemical (e.g., carbon) cycle model, the totality of which is often called a climate system model (CSM) (see Fig. 5-1). This CSM would then be integrated forward in time (perhaps "asynchronously" to conform with the different response times of the separate domains) over millions of years, starting with some initial conditions and subject to the changing external boundary conditions listed in Chapter 1, with the hope of accounting for all the variability revealed by the geologic evidence (Chapter 3), including the geographic distributions. However, this ultimate goal is fraught not only

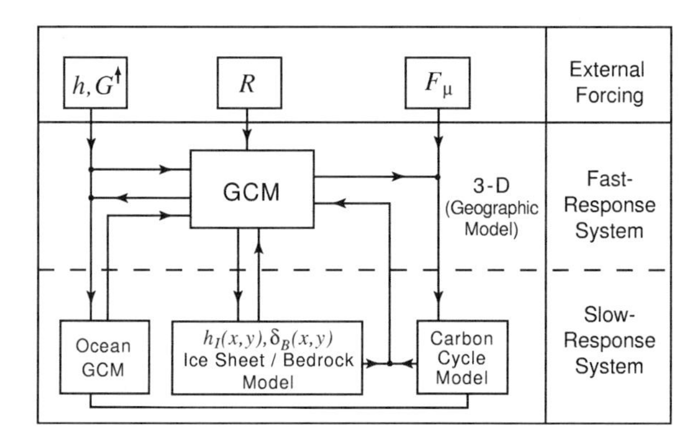

Figure 5-1 Schematic representation of the components of a climate system model. The forcing functions shown represent the global topographic height distribution (h), the geothermal flux ($G^{\uparrow}$), the incoming solar radiation ($R \equiv H_{\mathrm{T}}^{(1)\downarrow}$), and the tectonically driven outgassing of CO_2 (F_{μ}); $h_{\mathrm{I}}(x, y)$ and $\delta_{\mathrm{B}}(x, y)$ are the spatial distribution of ice sheet thickness and the underlying bedrock depression.

with the obvious practical difficulties of expense, numerics, and computer limitations, but also with the following scientific problems of a more fundamental nature.

1. **Unknown long-term forcing.** Although we may be fairly confident in variations due to earth-orbital (Milankovitch) changes in the radiation distribution over the past few million years, much weaker evidence exists for potentially significant variations in factors such as (i) volcanic inputs of heat, trace gasses (e.g., CO_2), and aerosols, (ii) the geothermal flux, (iii) the magnitude of the solar luminosity, (iv) the chronology and form of continent–ocean changes, including uplift of mountains (with ramifications for altered airflow and weathering rates) and the opening and closing of seaways and ocean sill structures, and (v) the rate of rotation of Earth.
2. **Heterogeneity and nonlinearity.** As noted, the climatic system is truly complex, being composed of many physical domains containing different working materials (atmosphere, hydrosphere, cryosphere, lithosphere, and biosphere), each having vastly different intrinsic properties, time scales, and response times (see Table 4-3). Each domain, in turn, is itself heterogeneous, describable by many variables that vary significantly over space and time scales covering many orders of magnitude. Moreover, nonlinearity is also an intrinsic property of the system, if only due to the dominance of advective fluxes in the fluid domains, but also due to the many complex feedbacks between the domains.
3. **Instability.** One consequence of this complexity is that interactions within and between the climatic system components lead to both negative and positive feedbacks that may significantly alter the response times of the variables, sometimes in counterintuitive ways. In fact, dominant positive feedbacks can lead to instabilities that drive the system to new modes of behavior that bear little resemblance to the external forcing. A good example within the atmosphere itself is baroclinic instability, which gives rise to most weather activity in middle latitudes. On a broader level, the climate system as a whole contains important constituents, such as water vapor, salinity, ice, and carbon dioxide, that can partake in self-amplifying processes. If such destabilizing processes are not properly represented, the system may not be able to display important modes of internal variability.

5.1 A BASIC PROBLEM: NONCALCULABLE LEVELS OF ENERGY AND MASS FLOW

The difficulties posed by some of the above factors might be more manageable for the paleoclimatic problem (as they are to some extent in weather forecasting) were it not for the fact that significant paleoclimatic changes (e.g., the ice age oscillations) occur on a scale (10–100 ky) that requires the consideration of nonequilibrium dynamics, wherein the rates of change of ice, deep ocean state, and geochemical inventories must be calculated explicitly because they are of the same order as the dissipative rates (see Chapter 4). However, these rates of change are too small to be either measured observationally or calculated from a physical model (e.g., GCM), given the uncertainties of the flux parameterizations involved. For example, from Figure 1-4b we find that even in periods of most rapid change (e.g., the rapid meltdown of the Laurentide and Eurasian ice sheets between 20 and 10 ka), the rate of change of global ice mass, dI/dt, is approximately 5×10^{15} kg y^{-1}, which is equivalent to a rate of sea level change that never exceeds about 5 cm y^{-1}, even for maximum ice-melting episodes during the interval (Fairbanks, 1989). It is unlikely that we will be able to compute the global fluxes of water involved in any single branch of the hydrologic cycle (e.g., evaporation, precipitation, melting/freezing) to this accuracy, let alone to compute the required small differences between them (Peixoto and Oort, 1992).

From an alternate, global energetical, viewpoint we find from Eq. (4.61) that the differences between net global outgoing radiation $\widetilde{N}^{\uparrow}$ and the net upward flux of heat at the Earth's surface $(H_S^{\uparrow} - L_v\widetilde{E} - L_f\widetilde{M})$ given by

$$\left[\widetilde{N}^{\uparrow} - \left(\widetilde{H}_S^{\uparrow} + L_v\widetilde{E} + L_f\widetilde{M}\right)\right] = \frac{L_f}{\sigma}\frac{dI}{dt} \tag{5.1}$$

(where $L_f = 3.3 \times 10^{15}$ J kg, $L_v = 2.5 \times 10^6$ J kg, and $\sigma = 5 \times 10^{14}$ m^2) would have to be calculated to an accuracy of 10^{-1} W m^{-2} (Saltzman, 1983). This is clearly beyond the capability of any GCM. The situation is not improved if one limits the area of concern to a smaller region of ice sheet growth, let us say a polar cap poleward of $\varphi_p =$ 50°N, within which area all major ice sheets are formed (see Fig. 4-4). In this case we have from Eq. (4.56) that the poleward flux of energy across 50°N $(\mathcal{J}_p + L_v\mathcal{J}_{vp})$ and the net outgoing radiation $(\sigma_P N_p^{\uparrow})$ from the top of the cap ($\sigma_P \approx 6 \times 10^{13}$ m^2) would each have to be calculated to the accuracy of 10^{14} W. However, from observational studies (Peixoto and Oort, 1992) we know that $(\sigma_P N_p^{\uparrow})$ and $(\mathcal{J}_p + L_v\mathcal{J}_{vp})$ are each of the order of 10^{15} to 10^{16} W and have observational and theoretical (e.g., GCM) uncertainties of the order of 10^{15} (Stone and Risbey, 1990).

If, as another example, we wish to calculate the net changes in mean ocean temperature that may have occurred on paleoclimatic time scales, we are again confronted with insurmountable difficulties. At various times, perhaps accompanying the Late Cenozoic ice ages, such changes were probably no greater than a few degrees per 10,000 y (i.e., on the order of less than 10^{-3} K y^{-1}). Integrating Eq. (4.50) over the world ocean we have

$$\frac{\partial}{\partial t}\int \rho c T_w\, dz = \left[\widetilde{H}_S^{\uparrow} - L_v\widetilde{E} - L_f\widetilde{M}\right]_w + G_w^{\uparrow} \tag{5.2}$$

implying that the net global heat flux across the ocean surface $[H_S^{\uparrow} - L_v\widetilde{E} - L_f\widetilde{M}]_w$ would have to be calculable to an accuracy of less than 10^{-1} W m^{-2}.

As still another example, it is similarly impossible to calculate the fluxes of carbon involved in the known changes in CO_2 that accompanied the Late Cenozoic ice ages (i.e., 10^{-2} ppm y^{-1}; see Fig. 3-6), highlighted by the fact that at present there is no acceptable explanation for the differences between observed CO_2 changes and anthropogenic fossil fuel forcing, which is at the rate of 1 ppm y^{-1}.

Thus all of the highly relevant fluxes described above are one to two orders of magnitude below the level at which we can make direct flux calculations either observationally or theoretically from models (e.g., GCMs). That is, the uncertainties in the physical parameterizations that are normally used in GCMs to achieve acceptable statistically steady atmospheric and surficial climate distributions are large enough to control the evolution of ice accumulation or ocean temperature over long time intervals. From another point of view, the "climatic drift" common to all GCM solutions is larger than the slow rates of change involved in paleoclimatic evolution. Repeating the discussion of Saltzman (1984a,b), it seems unreasonable to expect that GCMs will have the same property relative to long-term climatic change that the primitive equations have relative to weather variations (i.e., of yielding proper net quasi-geostrophic changes over many gravity wave cycles by initializing the calculations to remove gravity waves). Whereas the time derivatives in weather variation are only one order of magnitude less than the measurable magnitude of known terms representing geostrophic balance, the time derivatives of ice mass and mean ocean temperature are at least two orders of magnitude less than the measurable magnitude of the flux terms that produce them, and these flux terms involve parameterizations the forms of which are not even agreed upon.

In recognizing this state of affairs we are led to question what has generally been assumed to be a fundamental role of such models in the overall theory of long-term climatic change, namely, to provide the physically deterministic coupling between atmospheric climatic processes and models of the ice sheets and deep ocean shown in Fig. 5-1. Such a role seems implicit, for example, in Lorenz's (1970) discussion of a hypothetical CSM from which "we shall necessarily obtain changes in climate, including the great ice ages." Although this statement may be true in principle, from a practical standpoint it is difficult at present to imagine that we will be able to achieve such CSM solutions without some additional considerations.

The more general implication of the above discussion is that it will not be possible to pursue a purely deductive approach to a theory of paleoclimate and the major ice ages proceeding directly from the fundamental equations of geophysical fluid dynamics given in Chapter 4, without assigning some new free parameters. In essence, we must discover the laws and rate constants governing this system on these very slow time scales involving the complex interactions of ice, atmosphere, ocean, biolithosphere, and greenhouse gas constituents, still preserving, however, the conservation principles for mass, momentum, and energy, which take new forms. An inductive (or phenomenological) approach, involving qualitative physics, will probably be required. As will be discussed more fully in the next section, in its simplest form such qualitative physics is perhaps best expressed in the form of a stochastic–dynamical system of equations governing multiyear running-average climatic variables that are independent of the space coordinates (e.g., selected Fourier modes or global integrals). Aside from the simplicity of analysis afforded by the reduction of the order of the system, the use of global variables is also dictated by the fact that present geological evidence provides a continuous record only of global ice mass changes as revealed by the sedimentary core measurements of $\delta\,^{18}O$.

The construction of such a phenomenological theory should be judged within the usual framework of the scientific endeavor, which requires (1) that a maximum amount of the variability and covariability in the observations be deducible with the least number of free parameters and a minimum level of prescribed noise, (2) that these deductions be robust in the sense that they can survive small changes in parameters and acceptable levels of noise, and (3) that the theory not be *ad hoc*, being "at risk" in some testable way by offering predictions of variability as yet unobserved that can sharpen the search for new knowledge. In Fig. 1-5 we depicted schematically the continuing process wherein theory (i.e., sets of statements from which predictions can be made) is constantly being refined to connect externally imposed forcing with the observed world.

5.2 AN OVERALL STRATEGY

To provide some guidance in playing the game defined by the above ground rules, two basic strategies will be of value:

1. Starting, as the simplest case, with the global, spatially averaged, aspect of the full problem (aimed, for example, at accounting for the global ice mass and CO_2 variations revealed by deep-sea sedimentary cores and glacier ice cores), develop a hierarchy of models of increasing complexity that build toward a more fully geographical and physically detailed model (recall the comments made at the end of Section 4.2). The global-averaged model lends itself to treatment by a low-order dynamical systems approach that can reveal the essential dynamical structure and time evolution of the system. This will be discussed more fully in Section 5.4.
2. To make full use of the relatively highly developed theory of the atmospheric, oceanic, and surface-state climate embodied in general circulation models, a separation into the fast- and slow-response parts of the climate system will be of great value. Such a separation is not always easy to make because of the above-mentioned role of positive and negative feedbacks that alter the response times of the separate components (e.g., the ice-albedo feedback can conceivably lengthen the response time of global mean surface temperature that would otherwise be characterized as "fast"). A good start can be made, however, by drawing upon the discussion given in Section 4.3. That is, it is suggested that the natural basis for making this separation is the spectrum of known variability of the climatic system (a highly idealized schematic version of which was shown in Fig. 1-2) coupled with estimates of the response times (ε) of the main variables (x) in each domain (Figs. 1-1 and 1-2, and Table 4-2). As noted in Section 4.3 the relative "gaps" in the spectrum represent optimal averaging times (δ) to effect a separation of scales in the sense that the "Reynolds conditions" (Monin and Yaglom, 1971) can best be fulfilled, but this can only be approximate owing to the broad underlying white (or colored) spectrum due to random fluctuations. Only after comparing the response time to the averaging periods defined by these gaps can we differentiate a "fast-response variable" that equilibrates quasi-statically, from the nonequilibrium, "slow-response" (control) variable or external forcings to which the "fast variables" are "slaved" (Haken, 1983).

In particular, the spectral gaps we identified in Fig. 1-2 suggested the following choices of averaging periods (see Table 4-2):

δ_s: A synoptic averaging period of about 1 hr, which is the approximate time over which synoptic meteorological observations are representative, separating high-frequency atmospheric convective and mechanical turbulence from weather variability.

δ_c: A climatic averaging period that we have taken as 100 y (roughly at the limit of a human lifetime) in view of the relatively low power observed in fluctuations over the whole range of periods (10–1000 y), compared to the power in weather variability and the major ice-age fluctuations (even when considering sporadic fluctuations of the "little ice age" type (Kutzbach, 1976). This averaging period provides the separation between the weather maximum and the Pleistocene ice-age variance maximum in which we are most interested here.

δ_t: A tectonic averaging period of about 10 My that defines climatic states owing their variability largely to the ultralong period fluctuations in the tectonics of earth and possible long-term extraterrestrial influences such as solar luminosity changes.

As further noted in Section 4.3, if $\varepsilon < \delta$, the variable x tends to be governed by a steady-state (or diagnostic) equation relating it nearly instantaneously to slower variables or forcing (assuming a unique stable equilibrium), but if $\varepsilon \gg \delta$, and $\varepsilon \geqslant P(\bar{x}^{(\delta)})$ [period of δ-averaged value of x] the variable x tends to be governed by a nonequilibrium (dynamical) equation. In the event that multiple stable equilibria (separated by unstable states) exist for the same fixed external forcing, rapid transitions between the modes make it difficult to define δ and require the consideration of initial conditions and stochastic perturbations (i.e., consideration of the more general, nonequilibrium problem). Moreover, in this latter case a new time constant ε_s is introduced, representing the "exit-time" for stochastic transition between the equilibria. Setting aside this possibility for multiple equilibria (to be discussed more fully in the Chapter 7, Section 7.8), we can distinguish three main problems in the theory of long-term (paleo)climatic change, which can be expressed in terms of a required nested sequence of models for three different time scales (Saltzman, 1990) (see Fig. 5-2):

Problem 1 To seek equilibrium solutions for the climatic average of the fast-response (weather) variables that satisfy the condition $\varepsilon_{a,\tau} < \delta_c$ as a "slaved" function of the slow-response variables and slow external forcing.

Problem 2 To formulate and solve the closed set of nonequilibrium equations governing the evolution of the slow-response climatic variables, e.g., the ice sheets and deep-ocean state, which satisfy the condition $\varepsilon_{I,\theta} \gg \delta_c$ on ice-age time scales, as time-dependent departures from another set of equilibrium solutions, thus defining Problem 3.

Problem 3 To determine the states of ice, deep ocean, and atmospheric composition (which satisfy the condition $\varepsilon_{I,\theta} \ll \delta_T$) as a quasi-statically equilibrated response to ultraslow tectonic and astronomical variations.

A complete theory of long-term climatic change embraces all three of these problems. Our objective is to account for the slow variations of ice and the associated

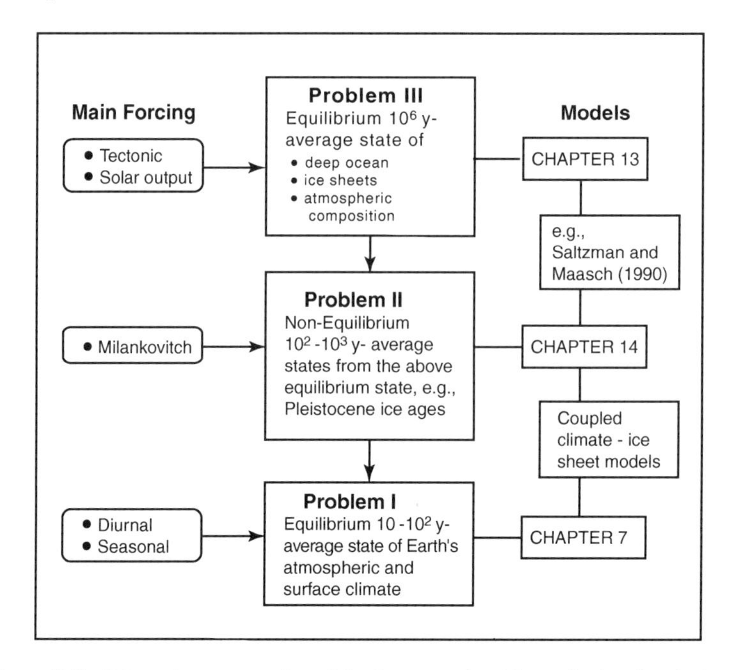

Figure 5-2 Schematic representation of the three nested problems of paleoclimatic theory.

statistical fields of atmospheric and oceanic variability. In the terminology introduced by Lorenz (1975), the steady-state Problems 1 and 3 represent "climatic prediction of the second kind," whereas Problem 2 represents "climatic prediction of the first kind" (see Section 7.5.1).

5.3 NOTATIONAL SIMPLIFICATIONS FOR RESOLVING TOTAL CLIMATE VARIABILITY

Before describing in the next section a more formal approach to implementing the preceding strategy, we first introduce some notational conventions that will be adopted in the remainder of this book.

To begin, if, as in Section 4.3, we define $\bar{x}^{(\delta_a)}$ as a running average over the interval δ_a ($a = s, c,$ or t) of any climatic variable x (which can be a grid point value, a Fourier amplitude, or a spatially averaged value), then for convenience we shall set $\bar{x}^{(\delta_a)} \equiv \bar{x}^{(a)}$, i.e., $(\bar{x}^{(\delta_s)}, \bar{x}^{(\delta_c)}, \bar{x}^{(\delta_t)}) \equiv (\bar{x}^{(s)}, \bar{x}^{(c)}, \bar{x}^{(t)})$. With this notation, we can resolve the "instantaneous" value of any variable, $x(\text{inst})$, as a nested sequence of departures from the longest term, tectonic-mean value $\bar{x}^{(t)}$ in the form

$$x(\text{inst}) = \bar{x}^{(t)} + \Delta x + x^{\star} \tag{5.3}$$

where

$$\Delta x \equiv \bar{x}^{(c)} - \bar{x}^{(t)}$$

are the paleoclimatic departures from the tectonic mean states (e.g., glacial–interglacial cycles), which, when combined with $\bar{x}^{(t)}$, defines the climatic-mean state,

$$\bar{x}^{(c)} = \bar{x}^{(t)} + \Delta x$$

The departures, $x^\star$, defined as in Section 4.2 by $x^\star \equiv [x(\text{inst}) - \bar{x}^{(c)}]$, represent all the subclimatic mean departures, which can be further resolved into two parts, $x^\star = (x^\star_{\text{S}} + x^\star_{\text{SS}})$. The first part is due to all synoptic weather fluctuations, and fast climate variations such as El Niño and decadal oscillations,

$$x^\star_{\text{S}} = \left(\bar{x}^{(s)} - \bar{x}^{(c)}\right)$$

The second part is due to all subsynoptic (e.g., mesoscale) variations and small-scale turbulence,

$$x^\star_{\text{SS}} = \left[\bar{x}(\text{inst}) - \bar{x}^{(s)}\right]$$

Because the variations of the mean climatic state $\bar{x}^{(c)} = (\bar{x}^{(t)} + \Delta x)$ are our main concern in this book (see Section 1.1), as noted in Section 4.5.1 we shall henceforth omit the overbar with the understanding that all quantities to be discussed are climatic-average quantities, i.e.,

$$x \equiv \bar{x}^{(c)} \tag{5.4a}$$

and shall further simplify our notation by defining

$$\bar{x}^{(t)} \equiv \widehat{x} \tag{5.4b}$$

Climatic averages of products or other nonlinear functions will be denoted simply by a bar, i.e.,

$$\overline{xy}^{(c)} \equiv \overline{xy} \tag{5.4c}$$

From Eqs. (5.4a) and (5.4b) it follows that

$$x = \widehat{x} + \Delta x \tag{5.5}$$

where $\widehat{x}$ is a very slowly varying component driven only by external tectonic "rock processes" and solar constant changes, representing mean conditions over roughly 10-My intervals, and Δx is the departure from this tectonic-mean state. In turn, $\widehat{x}$ can be further resolved in the form

$$\widehat{x} \equiv \widehat{x}(0) + \Delta_{\text{t}}\widehat{x}$$

where $\widehat{x}(0)$ is the present tectonic-mean value that we can identify with the Late Pleistocene state and taken as a given "initial condition" in the hindcast sense, and $\Delta_{\text{t}}\widehat{x}$ is the departure.

The departure, Δx, can also be resolved further in the form

$$\Delta x = (\Delta x)_0 + (\Delta x)' \qquad (5.6)$$

where $(\Delta x)_0$ is an *internal* equilibrium (steady state) representing a balance of fluxes between the atmosphere, ocean, and ice masses and $(\Delta x)'$ is a transient departure from $(\Delta x)_0$, representing, e.g., the glacial cycles of the Pleistocene.

Thus combining Eqs. (5.5) and (5.6) we may write

$$x = x_0 + x' \qquad (5.7)$$

where x_0 defines an equilibrium state,

$$x_0 \equiv \left[\widehat{x} + (\Delta x)_0\right] \qquad (5.8)$$

and x' defines the departure from this state,

$$x' \equiv (\Delta x)' \qquad (5.9)$$

In summary, Eq. (5.3) can be rewritten in the form

$$x(\text{inst}) = x_0 + x' + x^\star \qquad (5.10)$$

5.4 A STRUCTURED DYNAMICAL APPROACH

To put the overall strategy described in Section 5.2 in a more formal dynamical framework, we can proceed as follows (Saltzman, 1983). In accordance with Eq. (5.4a) let $x_i(\lambda, \varphi, z, t)$ $[\equiv \overline{x_i}^{(c)}]$ denote the set of roughly 100-y means of the internal variables that define "climate" (including the ensemble seasonal cycle, variances, and higher moments of their probability distributions as well as the covariances that represent fluxes of mass, momentum, and energy), and $F_i(\lambda, \varphi, z, t)$ denote the set of external forcing functions that can affect (but are not themselves affected by) these variables, where λ, φ, z and t are longitude, latitude, vertical distance, and time, respectively. Assuming appropriate parameterizations for the nonlinear effects of all phenomena of higher frequency than the 100-y climatic averaging period, including a stochastic component ω_i, the rate of change of each variable x_i is representable as a complicated, generally nonlinear, function of all the variables and forcing functions, of the form given in Eq. (4.27), i.e., $dx_i/dt = f_i(x_j, F_i, \omega_i)$.

In general, this will be an extremely "stiff" system, with many (fast-response) variables tending to equilibrate rapidly, whereas other (slow-response) variables equilibrate over very long time scales. To deal with this situation we next resolve this set of variables into a fast-response subset $X = \{x_i^{(F)}\}$, where $x_i^{(F)}$ are all variables whose dissipative response times are comparable or smaller than the climatic averaging period of about 100 y (e.g., all the "weather" variables, including water vapor content and surface temperature, and also including the ocean mixed layer, snow and sea-ice cover, and active terrestrial biosphere), and a slow-response subset $Y = \{x_{\rm i}^{(S)}\}$ comprising all the other variables having dissipative time constants much longer than the 100-y averaging period. These slow-response variables Y include the following components:

1. Cryosphere. The field's ice sheet or shelf thickness $h_{\rm I}$, bedrock depression $D_{\rm B}$, basal temperature $T_{\rm B}$ and basal water thickness $W_{\rm B}$, and thickness $h_{\rm r}$ of the deformable regolith above bedrock.
2. Oceans. The mean deep ocean temperature θ, salinity S (which may have significant positive feedbacks that can dominate rather weak dissipative processes, see Chapter 11), and biogeochemical constituents $\xi_{\rm j}$, particularly those that influence the carbon cycle. (In contrast, the deep thermohaline circulation is usually assumed to be subject to enough viscous damping to equilibrate relatively rapidly with imposed temperature and salinity gradients; see Chapter 11.)
3. Atmospheric greenhouse gases. Particularly carbon dioxide concentration, an assumed well-mixed value of which we denote by μ. The reserves of near-surface carbon in the form of peat and methane hydrates may be relevant slow-response variables in this regard.

It follows from the preceding discussion that we can write

$$\frac{dX}{dt} = f_{\rm X}(X, Y, F, \omega_{\rm i}) \approx 0 \tag{5.11}$$

This implies that the fast-response weather variables can be expressed as a quasi-equilibrated set,

$$X = X_0 \approx g_{\rm i}(Y, F, \omega_{\rm i}) \tag{5.12}$$

where $F = [F^{(F)}$ (e.g., the annual cycle of radiation) $+F^{(S)}$ (e.g., the earth orbital variations of radiation)]; i.e., from Eq. (5.7) $X' \approx 0$. This set, $X = X_0$, represents a physically realizable statistical-equilibrium climate state that can be determined asymptotically from a steady-state climate model (e.g., a GCM) as a quasi-static "slaved" response to the prescribed forcing and prescribed slow-response variables that exercise "control" of the climatic state. Included in $x_{\rm i}^{(F)}$ are covariances representing fluxes between the fast-response domains (e.g., the atmosphere) and the slow-response domains (e.g., the ice sheets). In Chapter 7 we discuss this climatic equilibrium problem, which constitutes Problem 1 in Section 5.2. Equation (5.12) therefore represents a GCM based on Eqs. (4.1)–(4.7) along with auxiliary equations representing the surficial domains (e.g., sea ice, vegetation) given prescribed values of Y and external forcing F. From one point of view, we can define fast-response variables as all those that are capable of being deduced uniquely from GCMs, given the flux resolution of the models.

On the other hand, slow-response "control" variables are the "carriers" of the long-term evolution of the climate system that must be treated more generally as a nonequilibrium set governed by dynamical equations of the form

$$\frac{dY}{dt} = f_Y\big(X, Y, F^{(s)}, \omega_i\big) \neq 0 \tag{5.13}$$

where $Y = \{x_i^{(s)}\} = Y_0 + Y'$.

We sometimes speak of the variables X as "diagnostic" in the sense that the value of any such variable at any time can be diagnosed by examining the static conditions of all the other variables and forcing prevailing at that same time, independent of any previous values. On the other hand, the variables Y are called "prognostic" in the sense that they depend on the previous conditions and can only be determined by a *dynamic* evaluation of time derivatives.

Using the equilibrium solutions given by Eq. (5.12) for X we can determine, in principle, the fluxes needed to compute the slow rates of change dY/dt or, by parameterizing these fluxes, can reduce Eq. (5.13) to a nonlinear dynamical system governing the slow evolution of these control variables (a process termed *adiabatic elimination* (Haken, 1983)),

$$\begin{aligned}\frac{dY}{dt} &= f_Y\big[g_i\big(Y, F, \omega_i^{(F)}\big), Y, F^{(S)}, \omega_i\big] \\ &= \Phi(Y, F, \omega_i)\end{aligned} \tag{5.14}$$

In the parlance of dynamical systems theory, Eq. (5.14) represents the "slow manifold" of the full climate system, to which all the fast-response variables tend to equilibrate.

In terms of the sets X and Y, the schematic portrayal of a full CSM shown in Fig. 5-1 can be redrawn in the alternate form shown in Fig. 5-3. Here we exhibit more clearly that the output of a super-GCM governing all the fast-response variables (X) provides the internal fluxes (or parameterization thereof) of mass, momentum, and energy that (along with external forcing) drives the slow-response system (Y) with some inertial delay, symbolized by the wiggled arrow in Fig. 5-3. As noted in Section 5.1, however, these fluxes are not calculable on a paleoclimatic time scale and hence must be scaled by free parameters to account for the data. At the same time, the component models governing the Y-fields provide the boundary conditions for determining the X-fields from the super-GCM.

In principle, X and Y represent huge three-dimensional, high-resolution, lattices of all the fast- and slow-response variables. For paleoclimatic purposes, where past three-dimensional fields are not well known and very long model runs are necessary, it may be more appropriate to consider lower resolution models governing both X and Y, thereby forming what are called climate (or Earth) system models of intermediate complexity (an EMIC) (Chalikov and Verbitsky, 1990; Ganopolski *et al.*, 1998; Berger *et al.*, 1998b). As will be discussed in Chapter 7, such a low-resolution model includes, for example, a statistical-dynamical atmospheric component in which all synoptic fluxes are parameterized. Even with such a reduced model containing a much more limited number of variables to be accounted for, it is still highly unlikely that

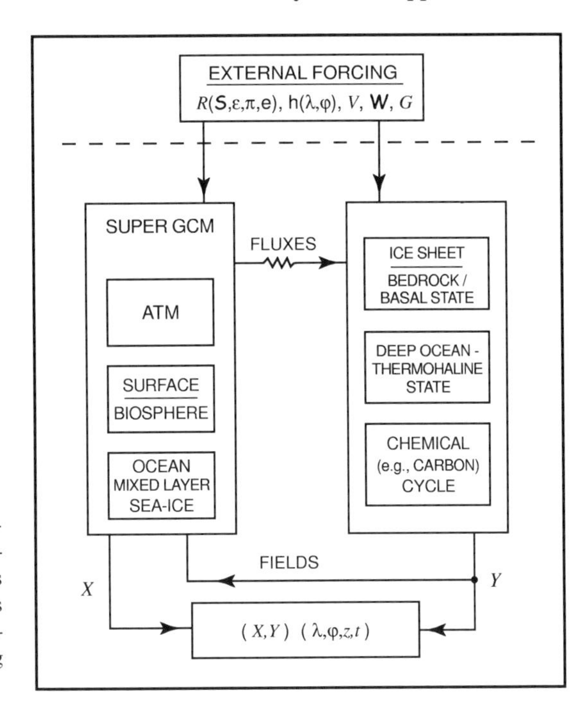

Figure 5-3 Schematic representation of the relationships between external forcing, the fast-response variables (X), and the slow-response variables (Y), showing the central role of a paleoclimatic dynamics model governing the global mean variables ($\widehat{Y}$).

one can simply set down a CSM and expect to deduce as complex a paleotime series as shown for global ice mass over 5 My in Fig. 1-4a without introducing and assigning free parameters (Saltzman and Sutera, 1984). This follows both from what we have said earlier in Section 1.4 regarding the complex mix of external forcing and free internal instability that must be be involved, and here in Section 5.1 regarding the noncalculable levels of energy and mass flux involved in the slow observed changes of Y. It then further follows that the central challenge in the pursuit of an acceptable dynamical theory of climate is to *minimize the number of such free parameters* while maximizing the amount of variability explained. More specifically, it will be necessary to discover the special mix of variable external forcing and free internal instability that can lead to this minimization of free parameters (cf. Section 1.4). Although each new free parameter admitted must surely increase the amount of variance explained in a given data set, it does so at the high cost of a loss of applicability and "significance" of the model when applied to new or improved data sets.

It is in these latter contexts that we would be well served to add a new low-order component to the full theory, as depicted in Fig. 5-4, to which we can apply the full power of dynamical systems analysis, i.e., an ultra-low-resolution, highly aggregated, model (e.g., a global average or a few-box model) capturing the coherent behavior of the full system with only a few variables and parameters. We might call this component a paleoclimate dynamics model, or PDM. Such a PDM would provide the integral constraints on the fuller system needed to explain the known time evolution with a minimum of free parameters, whereas the more detailed three-dimensional *spatial fields* of all the variables would be a major contribution of the full CSM, requiring

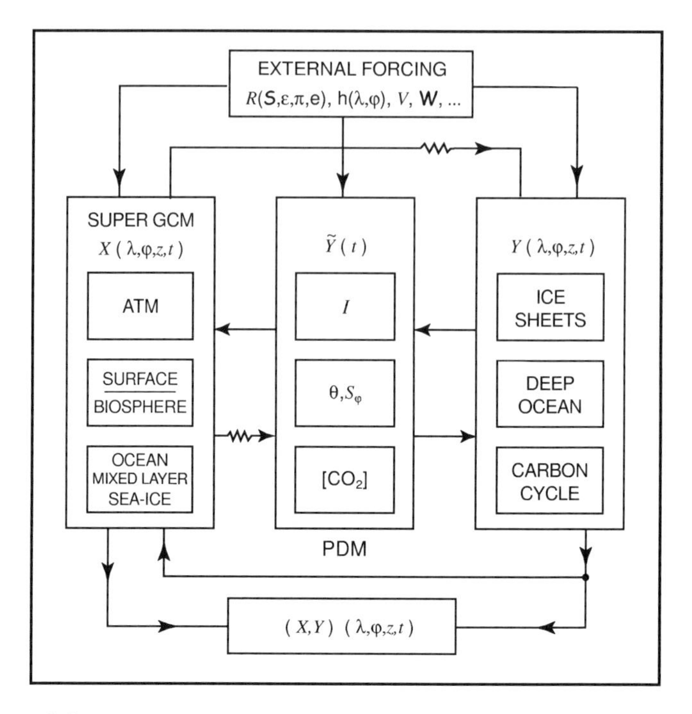

Figure 5-4 Climate system model showing the role of a paleoclimate dynamics model.

relatively few additional free parameters. It is in this latter sense that the full power of all the physics contained in a CSM comes to the fore.

Formally, we can resolve the full three-dimensional lattice of all the slow-response variables in the form

$$Y(\lambda, \varphi, z, t) = \widetilde{Y}(t) + Y_{\star}(\lambda, \varphi, z, t) \tag{5.15}$$

where, to lowest order, $\widetilde{Y}$ here stands for the global mean of the slow-response set (but this may be expanded to include some low-resolution spatial variability as in a two- or three-box model); e.g., $\widetilde{Y}$ = [global ice mass ($I \sim \delta\,^{18}$O), mean bedrock depression (D) of each of the main ice sheets comprising I, atmospheric CO_2 concentration (μ), mean oceanic temperature (θ), and meridional salinity gradient (S_φ)].

Assuming we can select a set $\widetilde{Y}$ such that a coherency condition applies to some reasonable extent, by which we can identify typical patterns of the departures $Y_{\star}(\lambda, \varphi, z)$ with the mean values of $\widetilde{Y}$, i.e.,

$$Y_{\star} \approx f_1(\widetilde{Y}, F) \tag{5.16}$$

the low-resolution PDM would take the form

$$\frac{d\widetilde{Y}}{dt} = f_2(\widetilde{Y}, F) \tag{5.17}$$

Thus, if Eq. (5.14) is designated as the "slow" manifold, then in view of Eqs. (5.12) and (5.16) the PDM, Eq. (5.17), might be termed the "center" manifold of the climate system to which all the other, both fast- and slow-response variables, are "slaved." This "center manifold," or PDM, is the lowest order set of equations that can represent the essential physics of the complete climate system. Because such a PDM cannot be deduced from the most fundamental physical equations [e.g., Eqs. (4.1)–(4.7)] for the reasons discussed in Section 5.1, it must be established in a more phenomenological manner, as will be discussed further and illustrated in Chapters 12 and 15. The central role of the PDM in a theory of paleoclimatic change is illustrated in Fig. 5-4.

Returning now to the fuller system, Eq. (5.14), because F includes all forms of long-term tectonic forcing, this system of equations embraces both Problems 2 and 3 in Section 5.2. These two problems can be separated in accord with Eq. (5.5) by resolving the variables Y into two components defined by

$$Y = \widehat{Y} + \Delta Y \tag{5.18}$$

where $\widehat{Y}$ is a tectonic-mean part identified with a roughly 10 My average state satisfying the steady-state form of Eq. (5.13), and thus posing Problem 3 described in Section 5.2, i.e.,

$$\Phi(\widehat{Y}, \widehat{F}) = 0 \tag{5.19}$$

in which $\widehat{F}$ denotes extremely slow tectonic forcing (e.g., continental drift), and ΔY is the departure, $Y - \widehat{Y}$, defining Problem 2, governed by

$$\frac{d(\Delta Y)}{dt} = \Phi\left[\Delta Y, \widehat{Y}, F, \omega_{\mathrm{i}}\right] \tag{5.20}$$

A specific application of this general formalism is made in Chapters 12 and 15, where we discuss a climatic dynamical system of the form given by Eq. (5.20), applied to a low-order PDM system governing the global mean values $\widetilde{Y}$.

In accord with Eq. (5.6) we may further separate the variables (ΔY) into their equilibrium values $(\Delta Y)_0$ and departures $(\Delta Y)'$, i.e.,

$$\Delta Y = (\Delta Y)_0 + (\Delta Y)' \tag{5.21}$$

where

$$\Phi\left[(\Delta Y)_0, F\right] = 0 \tag{5.22}$$

i.e., $(\Delta Y)_0 = f(F)$. It is important to note, however, that although the equilibria, Eq. (5.12), for the fast-response set X may be observationally realizable, the equilibria $(\Delta Y)_0$ are unlikely to be realizable (even if stable) because the departure $(\Delta Y)'$ will always be present, satisfying $d(\Delta Y)'/dt \neq 0$.

Once the evolution of the three-dimensional field of slow-response variables is determined by a solution of Eq. (5.14) [or, alternately, of the coupled system of Eqs. (5.15) and (5.16)], subject to initial conditions, all of the fast-response climatic variables corresponding to any point in time can be recovered from Eq. (5.12), i.e., from a GCM solution with prescribed slow-response variables (Y) and external forcing (F). For notational simplicity, in the remaining chapters we shall represent the combined set of slow-response variables Y and external forcing F by the symbol y_{i}, i.e.,

$$y_{\mathrm{i}} = \{y_{\mathrm{i}}\} = \left\{x_{\mathrm{i}}^{(s)}, F_{\mathrm{i}}^{(s)}\right\} = (Y, F) \tag{5.23}$$

In Fig. 5-4 we showed the relationships between external forcing, the fast-response (slaved) variables (governed, e.g., by a GCM), and the slow-response (control) variables governed by nonequilibrium (dynamical) models, in providing the essential connections to form a unified theory for the evolution of the complete climatic state $[X(t), Y(t)]$.

5.5 THE EXTERNAL FORCING FUNCTION, *F*

Before concluding this chapter a few further remarks are in order concerning the external forcing function F, a general qualitative discussion of which was given in Chapter 1 (Section 1.3). As noted there, this forcing can be divided into two main categories: (1) astronomical/cosmic forcing and (2) tectonically driven forcing. We now review the components included in each of these two categories, establishing the notation to be used for quantitative measures of these components.

5.5.1 Astronomical/Cosmic Forcing

1. $R = R(\lambda, \varphi; \mathcal{S}) \equiv H_{\mathrm{T}}^{(1)\downarrow}$ is the distribution of solar radiation (W m^{-2}) at the outer limit of Earth's atmosphere, it may vary as a function of the solar constant $\mathcal{S}$, and the Earth's orbital variations, including diurnal, seasonal, and Milankovitch periodicities (see Section 1.3). Note that $\widetilde{R} = \mathcal{S}/4$, where the wavy overbar denotes a global average.

2. Ω = angular frequency of Earth's rotation, the present value of which is $7.29 \times 10^{-5}\ \mathrm{s}^{-1}$.

3. $U^{\downarrow}$ = the rate of mass influx of cosmic material (bolides and dust) at the outer limit of Earth's atmosphere.

5.5.2 Tectonic Forcing

1. $h(\lambda, \varphi)$ = the tectonically driven geographic distribution and topography of the continents and ocean basins, measured in distance relative to prevailing sea level (i.e., $h > 0$ is continental topography, and $h < 0$ is ocean bathymetry).

2. $G^{\uparrow}$ = upward geothermal flux (W m^{-2}).

3. $V^{\uparrow} = (V_{\mathrm{a}}^{\uparrow} + V_{\mu}^{\uparrow})$ = mass flux of solid aerosol material in the form of dust and ash ($V_{\mathrm{a}}^{\uparrow}$) and gaseous material, particularly carbon dioxide ($V_{\mu}^{\uparrow}$), due to all forms of tectonic volcanism and venting.

4. $\mathcal{W}$ = a measure of the "weatherability" of carbonate and silicate rocks due to their degree of exposure to atmospheric CO_2 (see Chapter 10).

In summary, the function F represents the set of all the above variables, i.e.,

$$\begin{aligned} F &= \left\{R(\mathcal{S}), \Omega, U^{\downarrow}; h, G^{\uparrow}, V^{\uparrow}, \mathcal{W}, \ldots, F_{\mathrm{i}}\right\} \\ &= \{F_{\mathrm{i}}\} \end{aligned} \tag{5.24}$$

Before embarking on a discussion of the specific physical domains to be modeled (cf. Fig. 5-1) that are subject to the forcing F represented by Eq. (5.24), we conclude our consideration of the "foundations" of our subject by giving in the next chapter an overview of the dynamical systems ideas to which we have been referring, including some prototypical climatic applications of these ideas.

6

BASIC CONCEPTS OF DYNAMICAL SYSTEMS ANALYSIS

Prototypical Climatic Applications

It is appropriate at this point to discuss in a more general way aspects of dynamical systems theory that are relevant for analyzing a system like that given by Eq. (5.11), discussed in the previous chapter (e.g., Section 5.3). We begin by describing the formalism for determining the stability of an equilibrium of such a system.

6.1 LOCAL (OR INTERNAL) STABILITY

For generality, let us consider the stability properties of any equilibrium (steady state) of a system of equations of the form

$$\frac{dx_i}{dt} = f(x_i) \tag{6.1}$$

where x_i denotes a set of variables. Small departures, x_i', from a steady-state $x_{i(0)}$ satisfying $f(x_{i0}) = 0$ are governed by the linearized form of Eq. (6.1),

$$\frac{dx_i'}{dt} = J_0 x_i' \tag{6.2}$$

where J_0 is the Jacobian matrix, $(\partial f/\partial x_j)_0$, evaluated at $x_i^{(0)}$. Then assuming departures of the form $x' \sim \exp(\omega t)$, where ω are the eigenvalues, the local behavior near $x_{i(0)}$ is determined by the roots of the equation

$$\text{Determinant}\{\omega\delta_{ij} - J_0\} = 0 \tag{6.3}$$

where δ_{ij} is the Kronecker delta. If all the eigenvalues have negative real parts the steady state is stable, and if one or more eigenvalues has a positive real part then the steady state is unstable.

It is important to note that for the case of a single variable, i.e., $x_i \equiv z$,

$$\frac{dz}{dt} = f(z) \tag{6.4}$$

the eigenvalues can only be purely real, $\omega = (\partial f/\partial z)_0$, implying exponential decay (stability) if $\omega < 0$ and exponential growth (instability) if $\omega > 0$. However, for two or more variables, ω is determined from a higher order differential equation allowing ω to be a complex number that admits trigonometric, i.e., periodic, behavior near the equilibrium.

As an alternate way of looking at the stability properties of the single-variable equation, Eq. (6.4), it is possible to rewrite Eq. (6.4) in terms of a Lyapunov pseudo-potential, V_z, i.e.,

$$\frac{dz}{dt} = -\frac{\partial V_z}{\partial z} \tag{6.5}$$

where

$$V_z = \int f(z)\, dz$$

Thus,

$$\frac{dV_z}{dt} = \frac{\partial V_z}{dz}\frac{dz}{dt} = -\left(\frac{dz}{dt}\right)^2 \leqslant 0 \tag{6.6}$$

which means that the variable z will always decrease toward a minimum of V_z (i.e., toward a potential well that represents a stable equilibrium).

Let us return now to the more general multivariable climate case discussed in Section 5.3. If we resolve the sets X and Y in the manner of Eq. (5.6), noting that $dX_0/dt = dY_0/dt = 0$, it follows that we can write the dynamical equations for the transient departures X' and Y' in the following forms:

$$\frac{dX'}{dt} = f_x(X, Y, F, \omega) = \mathbf{A}X' + M(X, Y, F, \omega) \approx 0 \tag{6.7}$$

where $\mathbf{A}$ is the coefficient matrix for all the linear terms in X', having strongly negative real eigenvalues (i.e., the variables are rapidly damped toward a unique stable equilibrium state X_0) and M is a general nonlinear function including all the remaining terms. That is, the transient departures X' must all die away such that $X \rightarrow X_0$ as given by Eq. (5.9).

On the other hand, for the slow-response variables, we have

$$\frac{dY'}{dt} = f_Y\big(X, Y, F^{(S)}, \omega\big) = \mathbf{B}Y' + N(X, Y, F, \omega) \neq 0 \tag{6.8}$$

where $\mathbf{B}$ is the coefficient matrix for all the linear terms in Y', having weakly negative or positive real eigenvalues (indicating that the variables damp very slowly or are actually driven from the equilibrium Y_0), and N includes all the nonlinear terms. As noted in Section 5.3, by substituting for $X(Y, F, \omega)$ in Eq. (6.8) we obtain the dynamical representation of the "slow" manifold of the climate system, Eq. (5.14),

$$\frac{dY}{dt} = \Phi(Y, F, \omega)$$

6.2 THE GENERIC CUBIC NONLINEARITY

Using the above formalism, if it turns out that a system contains an unstable equilibrium (e.g., **B** has a positive real eigenvalue), it is necessary in order for a real dissipative system (e.g., the climatic system) to satisfy conservation requirements (i.e., the system not amplify exponentially to infinite values) that, for large displacements from the equilibrium, nonlinear restorative proceses represented by N dominate over the linear instability prevailing at small displacements. Moreover, it will be shown in the following discussions that it is reasonable to assume that the nonlinear damping takes a cubic form. Thus, if a single variable is involved, i.e., $Y = y_{\rm i}$ in Eq. (5.10), the following form for $N(y_{\rm i})$ is suggested:

$$N(y_{\rm i}) = \phi_0 + \phi_1 y_{\rm i} + \phi_2 y_{\rm i}^2 - \phi_3 y_{\rm i}^3 \tag{6.9}$$

where $\phi_3 > 0$.

A cubic representation of this kind arises in many areas of physics, usually called the "Landau" (1944) form when one variable in involved, but generalizes to the "Landau–Hopf" form when more than one variable is involved, thereby permitting oscillatory behavior. In the following discussion we provide a rationale for this nonlinear dissipative form as being the simplest means for satisfying the conservation requirement that an instability be contained by strong negative feedbacks leading to stability at large displacements from the unstable equilibrium.

To illustrate, for a single generalized variable $\Delta y_{\rm i} = z$ in Eq. (5.12) we have as the governing equation for this departure from a long-term average state an equation for the "amplitude" z,

$$\frac{dz}{dt} = cz + N(z) \tag{6.10}$$

where c is a control variable determined by external forcing, the collected nonlinear terms are

$$N(z) = -\left(k_2 z^2 + k_3 z^3 + \cdots\right)$$

and k_n are constants.

The even-powered terms all lead to an "asymmetric" variation of z, tending to drive the departures monotonically toward infinite positive or negative values. Thus, acting alone these even-powered terms cannot satisfy conservation requirements and hence would be physically unrealistic if they were not supplemented by the odd-powered terms. These nonlinear odd-powered terms are therefore a necessary adjunct to a destabilizing first-order effect (cz); in short, *instability must be accompanied by odd-powered nonlinearity*. The most basic truncated case is the cubic form,

$$\frac{dz}{dt} = cz - k_2 z^2 - k_3 z^3 \tag{6.11}$$

which can provide a starting point for the following more general discussion of bifurcation theory.

6.3 STRUCTURAL (OR EXTERNAL) STABILITY: ELEMENTS OF BIFURCATION THEORY

In Section 6.1 we discussed the stability of a particular equilibrium point in a system when an internal state variable is perturbed from this point, holding all external factors fixed. We now consider the changes in the number, values, and internal stability of all the equilibria of a system when an external factor (i.e., a control variable) is changed. The stability of the equilibrium structure in the face of changes in a control variable has variously been called *global* (as opposed to *local*), external, or structural stability. The critical values of the control parameters at which discontinuous changes in the equilibrium structure occur are called bifurcation points.

To illustrate we first consider the properties of the single-variable equation, Eq. (6.11), reduced to its symmetric cubic form ($k_2 = 0$),

$$\frac{dz}{dt} = cz - k_3 z^3 \tag{6.12}$$

which can be rewritten in a normal form by scaling $z = k_3^{-1/2} Z$, i.e.,

$$\frac{dZ}{dt} = cZ - Z^3 \tag{6.13}$$

This equation has three equilibria Z_0 (fixed points) for $c > 0$: $Z_0^{(0)} = 0$, $Z_0^{(1)} = c^{1/2}$, and $Z_0^{(2)} = -c^{1/2}$. These equilibria are graphed as a function of c in Fig. 6-1, which is the *bifurcation diagram* for Eq. (6.13), revealing its structural stability properties, including the "pitchfork" bifurcation at $c = 0$. At this bifurcation point the equilibrium $Z = 0$ which is stable for all $c < 0$, becomes unstable for all $c > 0$, a symmetric pair of stable equilibria emerging parabolically with increasing c. As is conventional, solid curves denote locally stable equilibria and dashed curves denote locally unstable equilibria.

In Fig. 6-2 we plot (a) $\dot{Z}$ as a function of Z and (b) the Lyapunov potential implied by Eq. (6.13), defined by $V_Z = -\int [cZ + Z^3]\, dZ$, showing the generic property of the cubic form in confining an unstable equilibrium ($Z_0^{(0)}$) between two stable equilibria $Z_0^{(1)}$ and $Z_0^{(2)}$. This common physical situation will appear in relation to the EBM

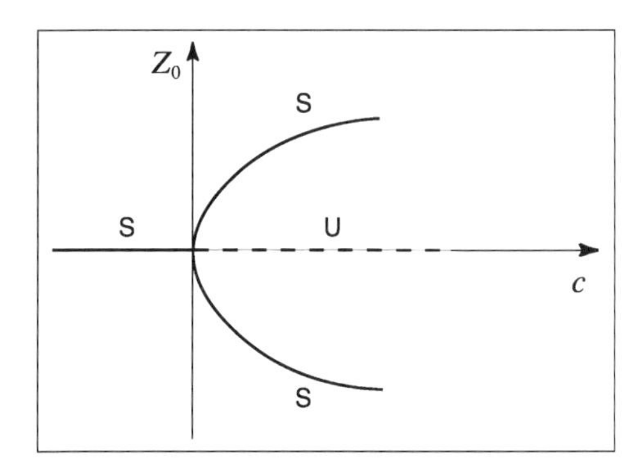

Figure 6-1 Bifurcation diagram for Eq. (6.13). Stable and unstable equilibrium branches denoted by S and U (dashed), respectively.

to be discussed in Chapter 7 and the simple two-box ocean model to be discussed in Chapter 11.

Although the pitchfork bifurcation just described provides the most elemental, physically significant representation of an unstable system, for completeness we mention two other basic "normal forms" and their bifurcation diagrams.

If, for example, instead of Eq. (6.13) we consider the form

$$\frac{dZ}{dt} = cZ + Z^3 \tag{6.14}$$

a stable equilibrium could exist only for $c < 0$ having the bifurcation diagram shown in Fig. 6-3a. This represents the "subcritical" pitchfork bifurcation, to be distinguished from the "supercritical" pitchfork bifurcation represented by Eq. (6.13) and pictured in Fig. 6-1. Although the subcritical pitchfork for $c < 0$ has no physically interesting

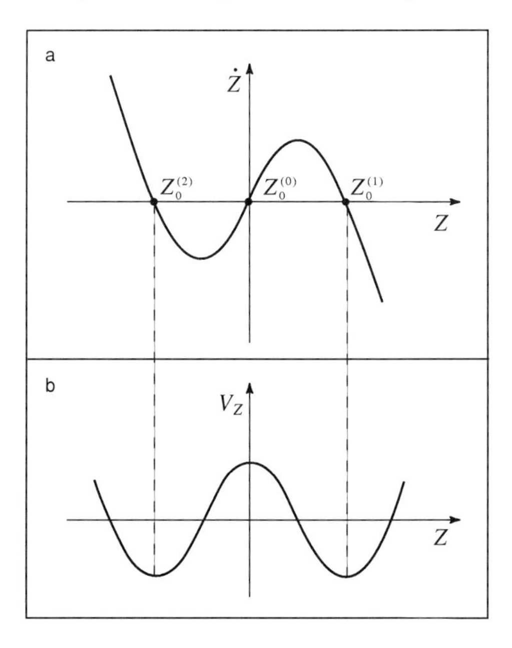

Figure 6-2 Properties of Eq. (6.13) as illustrated by plots of (a) $\dot{Z}$ as a function of Z and (b) the Lyapunov potential V_Z.

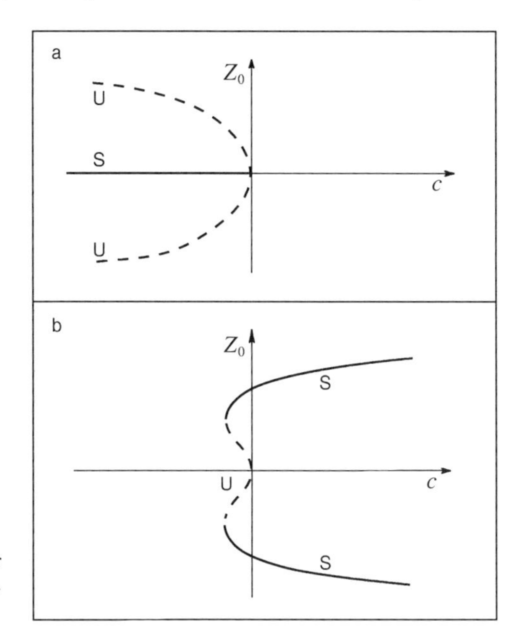

Figure 6-3 (a) Bifurcation diagram for Eq. (6.14), and (b) for Eq. (6.14) with the addition of the term $(-Z^5)$ on the right-hand side.

stable equilibrium, if $c > 0$ [thereby requiring from conservation principles that an additional higher power term, $-Z^5$, be added to Eq. (6.14) to allow stable equilibration (i.e., saturation)] we then obtain a combined supercritical–subcritical bifurcation pattern of the interesting form shown in Fig. 6-3b.

Another more general, physically significant variant of Eq. (6.13) has a normal form in which a quadratic term is also included,

$$\frac{dZ}{dt} = cZ + Z^2 - Z^3 \tag{6.15}$$

leading to an asymmetric bifurcation diagram of the kind shown in Fig. 6-4. This normal form comes closest to the representation to be discussed in Chapter 12 (Section 12.4), allowing for "symmetry-breaking" phenomena as well as bimodal behavior.

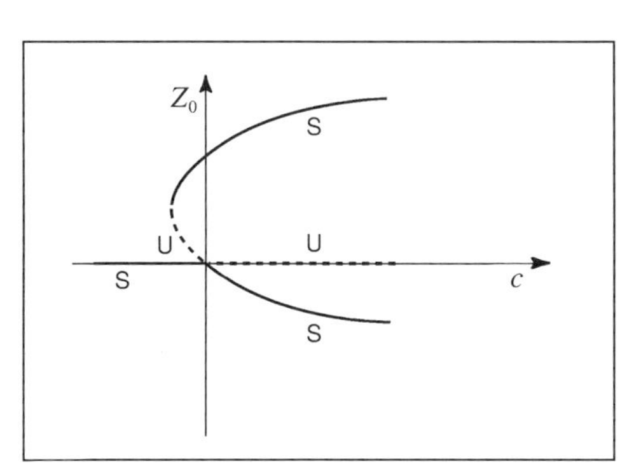

Figure 6-4 Bifurcation diagram for Eq. (6.15).

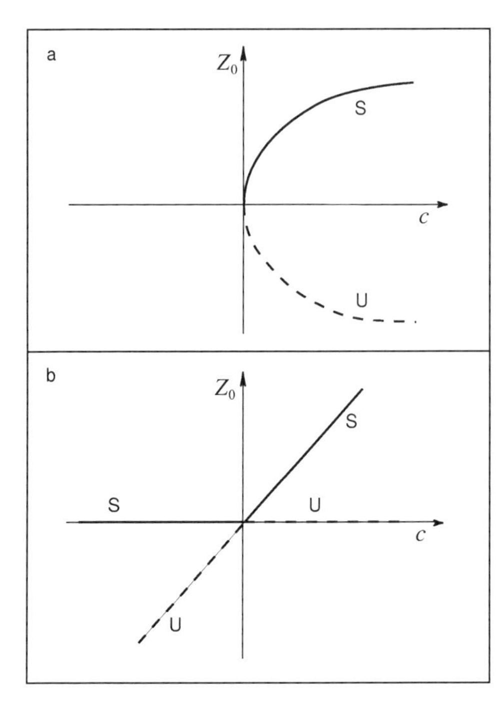

Figure 6-5 Bifurcation diagram for Eq. (6.16) illustrating a saddle node (*or limit point*) bifurcation (a), and for Eq. (6.17), illustrating a *transcritical* bifurcation (b).

Other possibilities are represented by normal forms for a single variable in which damping occurs at the quadratic rather than cubic level. These include the form

$$\frac{dZ}{dt} = c - Z^2 \tag{6.16}$$

which gives rise to the *saddle node* (or *limit point*) bifurcation shown in Fig. 6-5a, and the form

$$\frac{dZ}{dt} = cZ - Z^2 \tag{6.17}$$

which gives rise to the *transcritical bifurcation* diagram pictured in Fig. 6-5b. Note that in both cases transient departures from the equilibria will run away catastrophically from the lower unstable equilibrium, underlining the reasons why a cubic of the form given by Eq. (6.13) or (6.15) is the natural representation of the damping rate when a destabilizing first-order process is operative (i.e., $c > 0$).

To conclude our brief review of bifurcation properties for single-variable systems we consider what is called an "imperfect" form of Eq. (6.13) in which an external forcing function F is added, i.e.,

$$\frac{dZ}{dt} = cZ - Z^3 + F \tag{6.18}$$

In Fig. 6-6 we show the dependence of the equilibria on each of the two control variables F and c. The dependence on F takes the form of an S-shaped curve with two

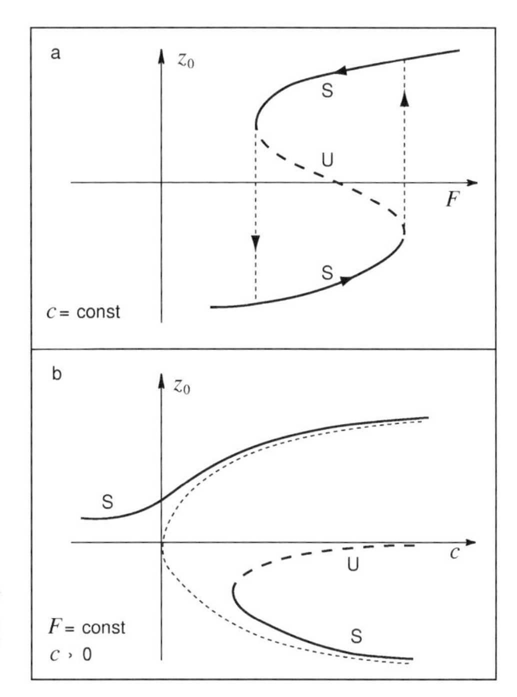

Figure 6-6 Bifurcation diagram for Eq. (6.18) (a) as a function of F with c taken as a constant and (b) as a function of c with F taken as a nonzero constant.

stable branches and one unstable branch. A periodic variation of F near the unstable branch can lead to a hysteresis loop, as shown by the arrows in Fig. 6-6a. This generic S-shaped equilibrium pattern is of the same form as the equilibrium patterns to be shown in Fig. 7-3 for the EBM and Fig. 11-5 for the Stommel two-box ocean model. In Fig. 6-6b we show the same equilibria as a function of c, holding F constant, in which case the equilibria no longer take the pure pitchfork form shown by the dotted curve; instead they are broken into two curves, one of which has an unstable branch, giving the appearance of an imperfect pitchfork.

Note that we have been discussing only the properties of a single uncoupled variable. In general, more than one variable is usually involved in a physical system. The presence of even a single additional variable permits a much richer set of possibilities, including a self-sustained nonlinear oscillation (i.e., a *limit cycle*). The point at which a control parameter change converts a stable equilibrium, to which transients damp periodically, to an unstable equilibrium, from which transients are driven toward such a sustained oscillation, is known as a *Hopf bifurcation*.

From all of the above single-variable cases we can surmise that the presence of an unstable equilibrium for a given external forcing is always accompanied by at least one stable equilibrium to which the system can be driven, thereby preventing a catastrophe. Moreover, if more than one stable equilibrium exists these equilibria must be separated by an unstable equilibrium. It follows that if an instability exists the system must be nonlinear (i.e., it must exhibit more than one equilibrium). In other terms, nonlinearity is a necessary condition for instability of a real physical system that does not "blow up," but this is not a sufficient condition because only one real stable equilibrium may exist, with others being imaginary and hence unphysical. These rules apply more generally to the multivariable system to be discussed next, with the additional provision that the stable equilibrium point may be replaced by a stable limit cycle or stable multidimensional periodic behavior.

6.4 MULTIVARIABLE SYSTEMS

A set of time-dependent variables x_i constitutes a phase space, each point of which represents a particular set of values at a particular time. As the set of values of x_i evolve in time they trace out a trajectory (orbit, or phase path) in this phase space (see Section 8.4 and Fig. 8-5). The geometry of such a trajectory represents an alternate way of describing the nature a dynamical system.

In particular, a set of ordinary differential equations of the form of Eq. (6.1), in which there is no time-dependent external forcing on the right-hand side (i.e., an autonomous dynamical system), has the important uniqueness property that the crossing of a trajectory with a portion of its previous trajectory at any point is prohibited. Thus a trajectory can either terminate at a stable equilibrium point, approach a previous portion of its trajectory asymptotically (thereafter repeating the trajectory periodically), or advance to infinity in a generally unphysical (catastrophic) way.

6.4.1 The Two-Variable Phase Plane

As the simplest multivariable case let us consider two variables forming a phase plane, each governed by an autonomous equation of the form given by Eq. (6.1), i.e.,

$$\frac{dx_1}{dt} = f_1(x_1, x_2) \tag{6.19a}$$

$$\frac{dx_2}{dt} = f_2(x_1, x_2) \tag{6.19b}$$

subject to some initial condition $[x_1(0), x_2(0)]$. We denote the equilibria (i.e., steady states or *fixed points*) satisfying

$$f_1\left(x_1^{(n)}, x_2^{(n)}\right) = f_2\left(x_1^{(n)}, x_2^{(n)}\right) = 0 \tag{6.20}$$

by $x_1^{(n)}$, where $n = 0, 1, 2 \ldots$ denotes all real (physically possible) roots of Eq. (6.20). One of these roots will clearly be $x_1^{(0)} = x_2^{(0)} = 0$ if there is no constant forcing term in Eqs. (6.19a) and (6.19b).

The local stability properties of each of these equilibria, determined by the procedure given in Section 6.1, has a much richer variety of possibilities than the single-variable system discussed earlier. In particular, as noted in Section 6.1, because the eigenvalues can have imaginary as well as real parts ($\omega = \omega_r + i\omega_i$) there can be oscillatory behavior near the equilibria. In Table 6-1 we present a classification and

Table 6-1 Classification of the Eigenvalues (ω) of the Linear System, Eq. (6.21) Determining the Local Stability of the Equilibria

Equilibrium properties	Eigenvalues	Type of equilibrium	Phase portrait near equilibrium	Symbol
$a < 0$ $a^2 > b > 0$	$\omega_+ < \omega_- < 0$	Asymptotically *stable* improper node		S1
$a < 0$ $a^2 = b$	$\omega_+ = \omega_- < 0$	Asymptotically *stable* improper node		S2
$a < 0$ $0 < a^2 < b$	$\omega_\pm = \omega_r \pm i\omega_i$ $\omega_r < 0$	Asymptotically *stable* spiral (damped oscillation)	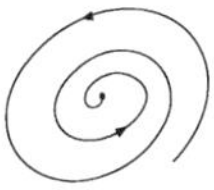	S3
$a = b$ $b > 0$	$\omega_\pm = \omega_r \pm i\omega_i$ $\omega_r = 0$	Center (undamped oscillation)	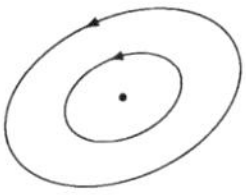	S4
$a > 0$ $a^2 > b > 0$	$\omega_- > \omega_+ > 0$	*Unstable* improper node		U1
$a > 0$ $a^2 = b$	$\omega_+ = \omega_- > 0$	*Unstable* proper node		U2
$a > 0$ $0 < a^2 < b$	$\omega_\pm = \omega_r \pm i\omega_i$ $\omega_r > 0$	*Unstable* spiral (amplifying oscillation)	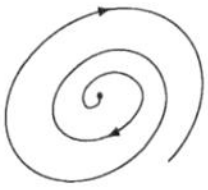	U3
$b > 0$	$\omega_+ < 0 < \omega_-$	*Unstable* saddle point	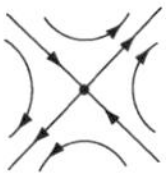	U4

pictorialization of all possible eigenvalues of Eqs. (6.19a) and (6.19b), the linearized form of which [cf. Eq. (6.2)] can be written in the matrix form,

$$\begin{bmatrix} \dot{x}_1' \\ \dot{x}_2' \end{bmatrix} = \begin{bmatrix} \left(\frac{\partial f_1}{\partial x_1}\right)_0 & \left(\frac{\partial f_1}{\partial x_2}\right)_0 \\ \left(\frac{\partial f_2}{\partial x_1}\right)_0 & \left(\frac{\partial f_2}{\partial x_2}\right)_0 \end{bmatrix} \cdot \begin{bmatrix} x_1' \\ x_2' \end{bmatrix} \tag{6.21}$$

where $(\ )_0$ denotes the value at an equilibrium point $(x_1^{(0)}, x_2^{(0)})$.

The eigenvalues of the matrix in Eq. (6.21), satisfying $(x_1', x_2') \sim e^{\omega t}$, are

$$\begin{aligned} \omega^{\pm} &= a \pm \sqrt{a^2 - b} \\ &= \omega_{\mathrm{r}}^{\pm} + i\omega_{\mathrm{i}}^{\pm} \end{aligned}$$

where

$$a = \frac{1}{2}\left(\frac{\partial f_1}{\partial x_1} + \frac{\partial f_2}{\partial x_2}\right)_0$$

and

$$b = \left(\frac{\partial f_1}{\partial x_1}\frac{\partial f_2}{\partial x_2} - \frac{\partial f_1}{\partial x_2}\frac{\partial f_2}{\partial x_1}\right)_0$$

As shown in Table 6-1, there are eight possible combinations of roots (excluding the cases $b = 0 \rightarrow \omega = 0$), four of which are stable ($\omega_{\mathrm{r}} \leqslant 0$), listed as S1–S4, and four of which are unstable ($\omega_{\mathrm{r}} > 0$), listed as U1–U4. We see that S3 and U3, in particular, represent damped and amplifying oscillatory behavior, respectively, near the equilibrium; a transition from S3 to U3 passing through a "center" S4 as a control parameter is varied leads to a stable self-sustained periodic oscillation (a *limit cycle*) if stabilizing conditions prevail at large distances from the equilibrium [e.g., a cubic damping of a form implied by Eq. (6.15)]. This special transition leading to self-sustained oscillations is called a supercritical Hopf bifurcation.

More generally, as we depart from the close vicinity of an equilibrium point the trajectories increasingly come under the influence of the "forces" represented by the nonlinear terms in Eq. (6.9) and also the possible influences of the attractor or repellor properties of other equilibria. In accord with the uniqueness theorem for trajectories in autonomous systems described above, the trajectories permitted in the presence of multiple equilibria often lead to a partitioning of the phase plane into domains, separated by special trajectories called *separatrices*, within which the set of all trajectories must be confined.

Before discussing the interesting properties of systems containing more than two variables (Section 6.6), in the next section we review the results of a prototype example of a two-variable climatic system that can illustrate the separatrix properties and many other features of multivariable dynamical systems.

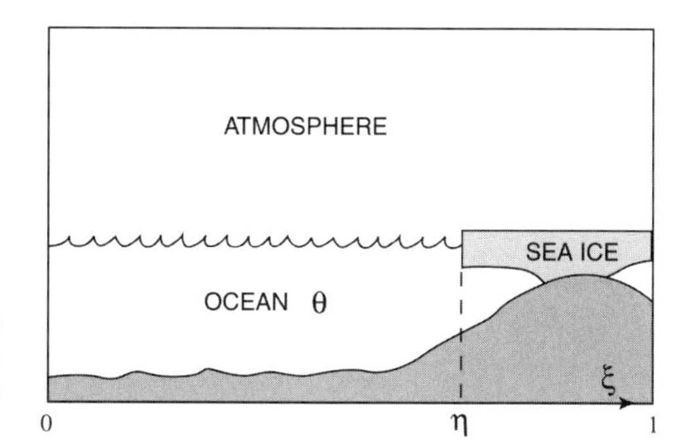

Figure 6-7 Pictorialization of a simple coupled ocean temperature (θ)/sea-ice extent (η) model represented by Eq. (6.22), where $\xi = \sin\varphi$.

6.5 A PROTOTYPE TWO-VARIABLE MODEL

As the basis for a fuller discussion of possible dynamical behavior on a phase plane we now review previous results for a coupled ocean/sea-ice model described previously in a series of articles (Saltzman, 1978, 1979, 1982; Moritz, 1979; Saltzman and Moritz, 1980; Saltzman *et al.*, 1981, 1982). It consists of a one-box ocean, with polar sea-ice coverage that can be viewed as a diagnostic for more massive marine ice sheets and shelves (see Fig. 6-7). The sea-ice coverage, which is dependent on the ocean temperature, can either prevent the loss of heat from the ocean in high latitudes by its insulating effect, thereby warming the bulk ocean (negative feedback), or cool the ocean by the ice-albedo effect (positive feedback). An additional positive feedback is possible through an assumed diagnostic dependence of the atmospheric greenhouse gas concentration (CO_2 and water vapor) on ocean temperature. Further details can be found in the above references, particularly Saltzman and Moritz (1980). Elements of this model have been incorporated into recent models of interdecadal–millennial scale oscillations (e.g., Yang and Neelin, 1993; Zhang *et al.*, 1995; Egger, 1999). The model is expressed mathematically as a coupled autonomous polynomial system of sixth degree governing two state variables, ocean temperature (θ) and the sine of the ice-edge latitude (η), having the form given by Eqs. (6.19a) and (6.19b), i.e.,

$$\frac{d\theta}{dt} = f_\theta(\eta, \theta) \qquad (6.22)$$
$$\frac{d\eta}{dt} = f_\eta(\eta, \theta)$$

Thus, we are considering a specialized form of Eq. (5.13) in which $Y = (\eta, \theta)$, $F_{\theta,\eta} = 0$, and, as anititial case, the stochastic forcing $\omega_{\theta,\eta} = 0$ (see Section 6.6 for stochastic forcing effects). Various specializations of these equations are given in the aforementioned articles, the most basic of which is a van der Pol-type system (Saltzman *et al.*, 1981), to be discussed in Section 6.5.2.

In Fig. 6-8 we show the deterministic solution for the fuller system (Saltzman, 1982), given a set of "reference" values of all the coefficients and fixed parameters that are believed to be reasonable estimates based on present observations. This solution, obtained by standard methods for autonomous systems (e.g., Boyce and DiPrima,

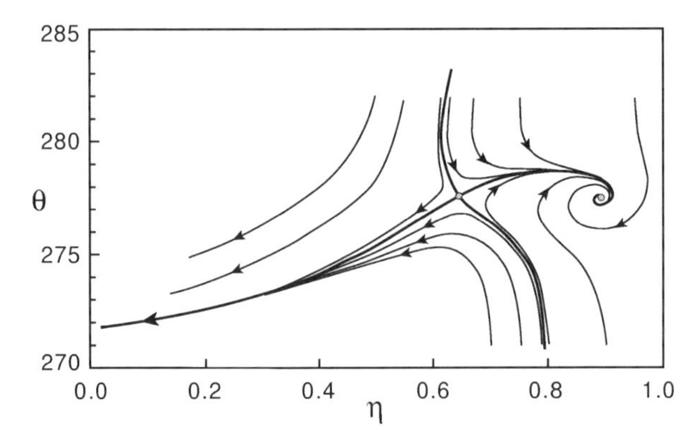

Figure 6-8 Deterministic phase-plane (θ, η) solution of the system, Eq. (6.22), for the model described in Saltzman (1982).

1977), is in the form of a $\eta - \theta$ phase-plane representation. Two physically admissable equilibria (within the bounds $0 < \eta < 1$) are found for these reference values: a stable *spiral* equilibrium ($\eta_0^{(1)} = 0.8904$, $\theta_0^{(1)} = 277.505$K) and an unstable *saddle-point* equilibrium ($\eta_0^{(2)} = 0.6469$, $\theta_0^{(2)} = 277.674$K). The time-dependent evolution of the solution for any arbitrary initial condition is indicated by the trajectories shown. The separatrices are indicated by the heavy curves is Fig. 6-8.

Figure 6-8 can also serve to illustrate another important property of dynamical systems, namely their attractor sets. These are the points toward which whole families of transient trajectories tend to flow and remain confined in a steady regime. As might be expected, more complex multivariable systems can admit more complex attractor sets than pictured here (see Section 6.7). An attractor set is ergodic (or *transitive*) if its points fill an entire physically realizable domain; if, on the other hand, the entire domain is divided into closed subdomains within which distinct attractor sets (i.e., *attractors*) are trapped, the system is intransitive. In this intransitive case the behavior of the system will depend critically on the location of the initial conditions in the phase space. A system is *almost instransitive* if it resides for long periods of time in one or another domain that is not fully closed off, so that occasional exits from one domain to another can occur. These ideas will be invoked in Section 7.8, where the transitivity properties of the atmosphere and surficial climate state are discussed.

In Fig. 6-9 we show one sample trajectory in the attractor region of the stable equilibrium in units of the departures from the stable spiral equilibrium. The initial departure is $\eta = -0.2$ and $\theta = 0$K. The *residence density* portrait in this figure gives some measure of time spent by the system in a particular point in phase space, the time interval between each dot being 60 y.

In Fig. 6-10 we show the net heat flux across the ocean sea-ice surface, $H_S^{\uparrow}$ ($\xi = \sin\varphi$), corresponding to the two equilibria. Note that

$$f_\theta(\eta_0, \theta_0)^{(1,2)} = \int H_{S0}^{\uparrow}(\eta_0, \theta_0)^{(1,2)}\, d\xi = 0$$

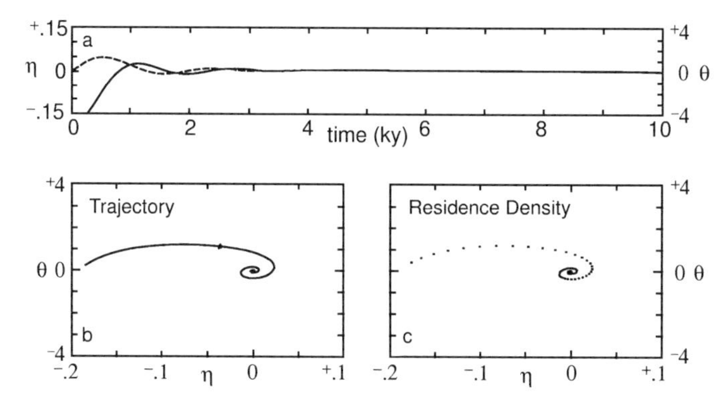

Figure 6-9 Sample deterministic solution near the stable equilibrium $(\eta_0^{(1)}, \theta_0^{(1)})$ of Fig. 6-8 in units of departure from the equilibrium (η', θ'): (a) time evolution of η' (full curve) and θ' (dashed curve); (b) trajectory; (c) trajectory speed (dots are spaced at every 60 years).

6.5.1 Sensitivity of Equilibria to Changes in Parameters: Prediction of the Second Kind

The equilibrium positions and their stability characteristics can change as a consequence of changes in any of the many parameters specified in the model. This possibility will be discussed further in more detail in Section 7.10 and applied in connection with single-variable systems in Sections 7.5 and 11.4; we now review this subject in the context of our prototype two-variable system in which more than one external forcing parameter may be involved.

If λ_i denotes the value of any particular parameter (e.g., prescribed external forcing, or a parameterization coefficient such as a diffusion constant), the change in the equilibrium position (η_0, θ_0) due to a small (say 10%) departure of λ from its reference

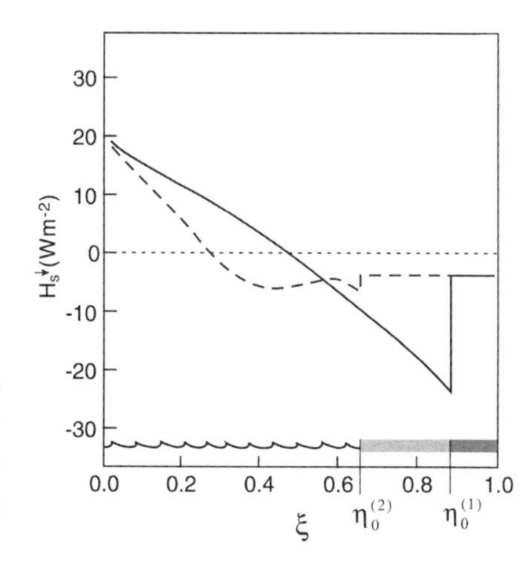

Figure 6-10 Distribution of the net downward flux of energy at the ocean surface, $H_S^{(5)\downarrow}(\xi)$, corresponding to the two reference equilibria $\eta_0^{(1)}$ (solid curve) and $\eta_0^{(2)}$ (dashed curve). After Saltzman and Moritz (1980).

value $\widehat{\lambda}$ is given by $\Delta(\eta_0, \theta_0) = 0.1\widehat{\lambda}_i[\delta(\eta_0, \theta_0)/\delta\lambda_i]$, where

$$\frac{\Delta(\eta_0, \theta_0)}{\Delta\lambda_i} = \left[\frac{\delta(\eta_0, \theta_0)}{\delta\lambda_i} + \sum_j \frac{\delta(\eta_0, \theta_0)}{\delta\lambda_j} \cdot \frac{\Delta\lambda_j}{\Delta\lambda_i}\right] \tag{6.23}$$

This represents the total sensitivity of (η_0, θ_0) to λ. If we hold fixed all parameters λ_j other than a particular parameter λ_i (that is, neglect the dependence of λ_j on λ_i), we define a *partial sensitivity* (cf. Saltzman and Pollack, 1977), $\delta(\eta_0, \theta_0)/\delta\lambda_i$. (The factor $0.1\widehat{\lambda}_i$ is often omitted in the definition of "sensitivity," a convention we shall adopt in the more formal sensitivity and feedback analysis for surface temperature given in Section 7.10.) A sensitivity analysis for all parameters of a model, based on Eq. (6.23), is important for two main reasons: (1) it provides some measure of the reliability of deduced equilibria in the face of known levels of uncertainty of the parameters, and (2) by revealing the parameters to which the equilibria are most sensitive, it points out the areas where future research efforts are most needed to establish more precise estimates of the parameters or the physical processes they represent.

In discussions of the possibilities for climatic change, we are often most interested in the shift of the equilibria in response to variations of particular parameters that can be regarded as external forcing (e.g., the solar constant $\mathcal{S}$, and CO_2 changes due to anthopogenic sources). For example, a commonly used measure of the overall sensitivity of climatic models, particularly the EBM to be discussed in Section 7.5, is the partial sensitivity corresponding to a 1% change in the solar constant,

$$\beta = 10^{-2}\widehat{\mathcal{S}}\frac{\partial \widetilde{T}_{S0}}{\partial \mathcal{S}} \tag{6.24}$$

where $\widehat{\mathcal{S}}$ denotes the present (i.e., reference) value of the solar constant and $\widetilde{T}_S$ is the global-average surface temperature. However, this latter measure is, at best, valid only very close to one of several possible equilibria that could be present. A fuller description of the sensitivity of the equilibria to variations of a particular parameter is provided by a plot of the equilibrium points as a function of the parameter, examples of which are the *equilibrium portraits* shown in Fig. 6-11 for the present model (Saltzman and Moritz, 1980). This figure shows the variations of the ice-edge equilibrium positions as a function of (a) the solar constant and (b) the reference CO_2 concentration.

The shift of an equilibrium due to the variation of a single external parameter is an important piece of information concerning the possibilities for climatic change, constituting what has been termed "predictability of the *second* kind" by Lorenz (1975). As was discussed in Section 5.2, however, such a prediction is not equivalent to a prediction of the actual trajectory that the climatic system will execute (prediction of the *first* kind). In the first place, it is likely that other external variables are also changing at the same time; in the second place, even if there were no other external changes taking place, it is highly unlikely that the complete climatic system ever actually resides at the deterministic equilibrium point in phase space. Rather, judging

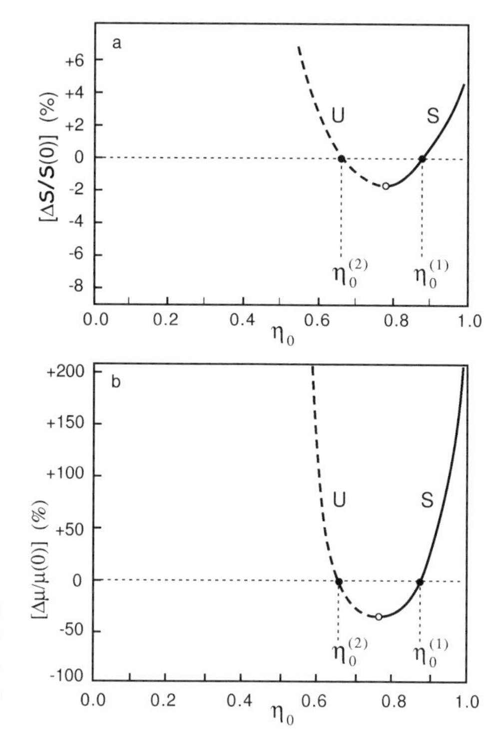

Figure 6-11 Equilibrium portrait showing the variations of η_0 in response to (a) variations only of the incoming radiation S and to (b) variations only of the CO_2 content of the atmosphere, μ. After Saltzman and Moritz (1980).

from the continuous spectrum of past climatic change, it is probable that climate always represents a transient, nonequilibrium state, the trajectory of which may be close to some stable equilibrium point or stable limit cycle but never locks firmly into one. Such departures from equilibrium must always be present, if only as a consequence of the instabilities and nonlinearities on time scales shorter than the climatic averaging period, which generate ever-present stochastic noise (e.g., "weather" fluctuations and interannual variability). As one possible scenario we can imagine that the evolution of the climatic state consists of a trajectory, under the constant influence of stochastic forcing, seeking to approach asymptotically a stable equilibrium that is drifting in response to changing external conditions. In Fig. 6-12 we show a schematic two-dimensional portrayal of such a trajectory, for η and θ, corresponding to an arbitrary displacement of the equilibrium due perhaps to a changed amount of CO_2 in the atmosphere, as depicted in Fig. 6-11. As shown in this diagram, it is entirely possible that at any time the displacement of the trajectory from the equilibrium point is greater than the displacement of the equilibrium from its initial position.

6.5.2 Structural Stability

In addition to the changes in position of the equilibrium in phase space described above, the possibility also exists that the local or internal stability of the equilibrium, discussed in Sections 6.1 and 6.3, is changed when parameters of a dynamical system are changed. The most dramatic form of such a stability change is a transition between

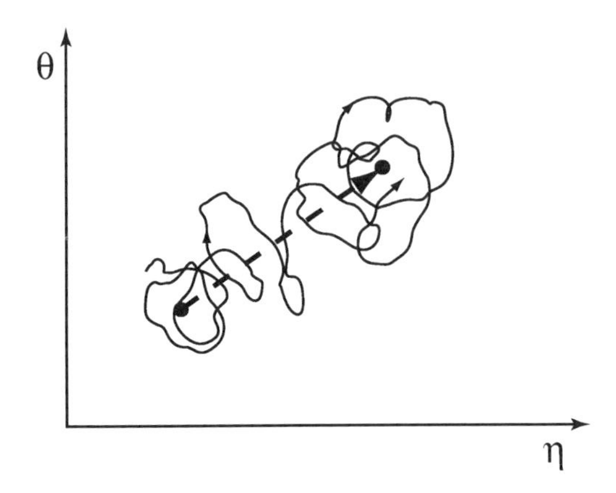

Figure 6-12 Hypothetical trajectory of the climatic system corresponding to a displacement of a stable equilibrium point of the system due to external forcing.

stability and instability (e.g., a *Hopf bifurcation*). An example of this is illustrated in Fig. 6-13 for our prototype model. By increasing the coefficient controlling the magnitude of the CO_2 positive feedback to a higher value, a bifurcation occurs in which the stable spiral equilibrium $(\eta_0^{(1)}, \theta_0^{(1)})$ becomes an unstable spiral equilibrium while retaining stability at large departures from $(\eta_0^{(1)}, \theta_0^{(1)})$. This implies the existence of the stable limit cycle shown in Fig. 6-13. Note that such a limit cycle can be viewed as a special separatrix within which all trajectories are trapped as an attractor set asymptotically approaching the limit cycle.

Along with the possibility that climatic variability occurs near a stable equilibrium point (as illustrated in Fig. 6-8), the possibililty also exists that climatic variability occurs near an unstable equilibrium point around which a limit cycle shapes the variations. These possibilities represent two major scenarios for the basic determinism of free climatic variability. Another possibility, discussed in the next section, is the existence of multiple unstable equilibria, that form a strange attractor (cf. Lorenz, 1963), or multiple stable equilibria, between which the system can flip under the influence of stochastic noise (cf. Sutera, 1980) (see Section 6.6).

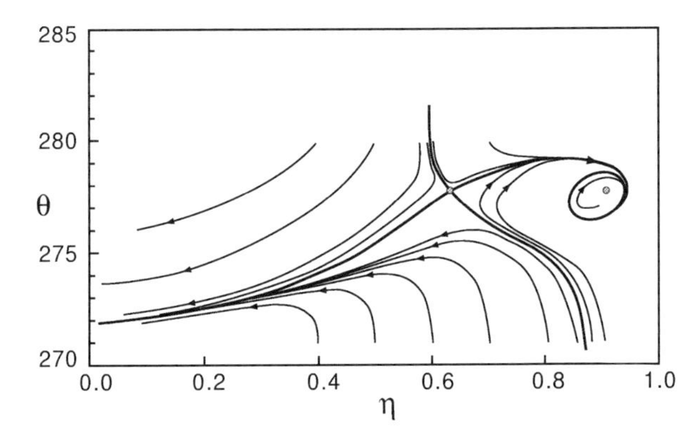

Figure 6-13 Same as Fig. 6-8 for an increased value of the CO_2 feedback parameter showing a bifurcation to a stable limit cycle around $\eta_0^{(1)}, \eta_0^{(2)}$. After Saltzman (1982).

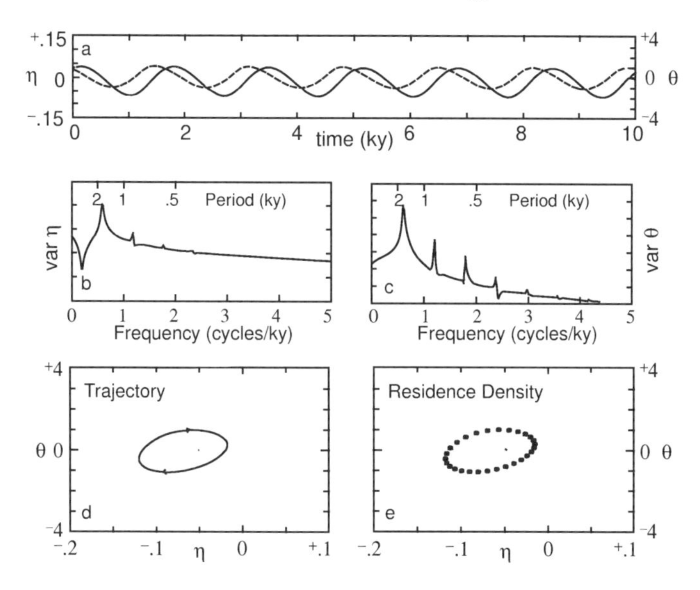

Figure 6-14 Deterministic solution near the stable limit cycle portrayed in Fig. 6-13 showing (a) the sample evolution for η (solid curve) and θ (dashed curve), (b, c) the variance spectra for η and θ, (d) the phase plane trajectory, and (e) the trajectory speed shown by dots spaced at every 60 years. After Saltzman (1982).

In Fig. 6-14 we depict the behavior of the deterministic limit cycle flow surrounding $(\eta_0^{(1)}, \theta_0^{(1)})$ in more detail, showing in parts b and c the variance spectra for η and θ, respectively. For our particular reference values a millennial-scale period of close to 2000 y is indicated.

With even greater changes in the parameters, the model can be reduced to an extremely simple form described in Saltzman (1978) in which only a single damped oscillatory stable equilibrium is admissible (i.e., there is no nearby unstable saddlepoint equilibrium). The behavior of a sample trajectory for this system is shown in Fig. 6-15. Similarly, as described in Saltzman *et al.*, (1981), it is possible to construct a model in which only a single stable limit cycle of the kind portrayed in Fig. 6-14 is present, but again with no nearby unstable saddlepoint equilibrium. The remarkable feature of this latter, van der Pol-type, relaxation oscillation is the presence of a biomodality in the trajectory speed, as was shown Fig. 6-14e (i.e., the *residence density* portrait).

In Fig. 6-16 we depict the changing physical state of the model climatic system as it executes the limit cycle, in this case without the presence of a nearby unstable equilibrium (i.e., a van der Pol-type relaxation oscillation). Here τ_0 and η_0 represent the equilibrium values of the equatorial surface temperature and ice edge, respectively, and H denotes the net flux of heat into or out of the ocean.

The Saltzman *et al.*, (1981) model is of special pedagogical interest because it illustrates in its simplest form the physical ingredients necessary for auto-oscillatory (e.g., limit cycle) behavior. In particular, for this case Eqs. (6.22a) and (6.22b) take the

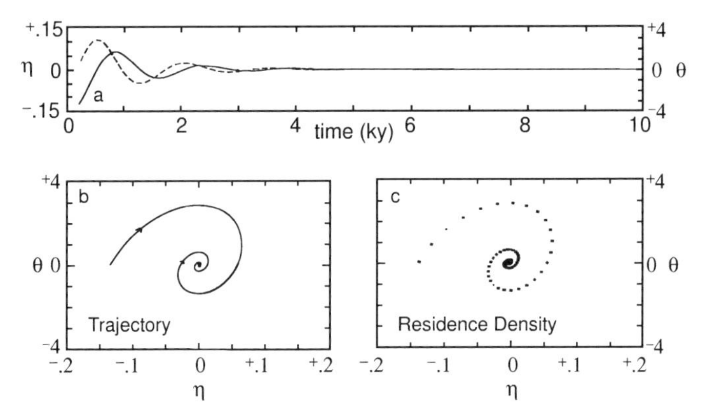

Figure 6-15 Same as Fig. 6-9, but for the simpler model described in Saltzman (1978) containing only a single stable spiral (damped oscillatory) equilibrium. From Saltzman (1982).

form

$$\frac{d\theta'}{dt} = -c_1\eta' + c_2\theta' - c_3\eta'^2\theta' \tag{6.25a}$$

$$\frac{d\eta'}{dt} = c_4\theta' - c_5\eta' \tag{6.25b}$$

where the primes represent departure from an equilibrium, and c_n are constants. The main components leading to auto-oscillatory behavior in Eq. (6.25) are as follows: (I) A basic drive toward weakly damped harmonic oscillation represented by the terms

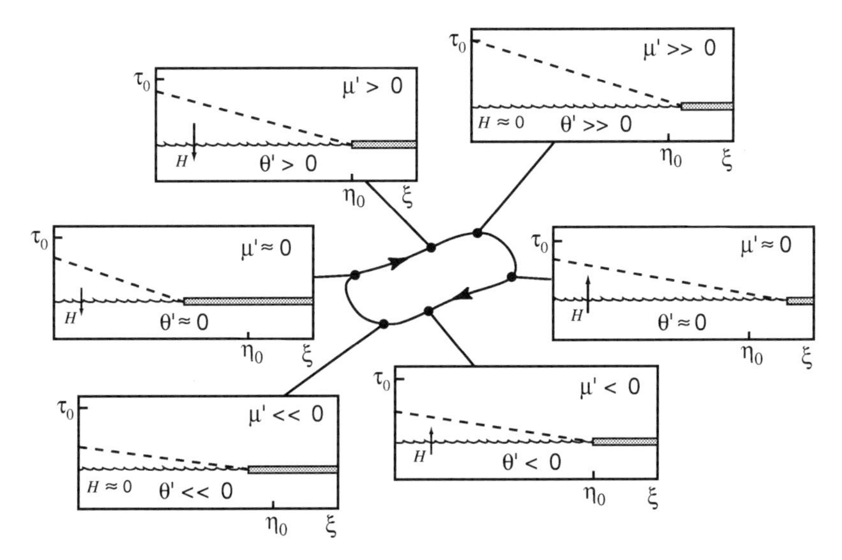

Figure 6-16 Schematic sequence of climatic states corresponding to the limit cycle solution, in this case for a van der Pol-type relaxation oscillator. After Saltzman *et al.* (1981).

$-c_1\eta'$, $+c_4\theta'$, and $-c_5\eta'$ in Eqs. (6.25a) and (6.25b), opposed by (II) a linear positive feedback that can destabilize the system ($+c_2\theta'$) dominating the weak damping ($-c_5\eta'$), and (III) a cubic nonlinear restorative mechanism represented by $-c_3\eta'^2\theta'$ that becomes dominant when the system is displaced far from its unstable equilibrium (η'_0, θ'_0).

A full mathematical analysis of this system, including additive periodic forcing, is given by Nicolis (1984, 1987).

6.6 THE PROTOTYPE TWO-VARIABLE SYSTEM AS A STOCHASTIC-DYNAMICAL SYSTEM: EFFECTS OF RANDOM FORCING

As we have said, the deterministic system just discussed is, at best, a highly speculative and simplified model of some of the feedbacks that are likely to be operating in the real climatic system. Clearly, there is much room for improvement of the parameterizations and for enrichment of the total physical content and fidelity of the model. For the reasons discussed in Section 4.2, however, it is unlikely that any amount of added physical rigor would enable us to represent the true behavior of the climatically averaged variables η_0 and θ_0 as a purely deterministic system (see also Hasselmann, 1976; Saltzman, 1982). The most we could hope for is that with such added physical rigor, the residual effects due to inadequate representation of phenomena of frequency higher than the climatic averaging period can be considered "random," i.e., representable by a white noise process of suitable amplitude. This constitutes the stochastic component of any climatic dynamical system.

In essence, this stochastic component represents the effects of determinism on scales that we cannot, or choose not to, consider, the sources of which are threefold, as noted in Section 4.2:

1. The impossibility of fulfilling the so-called Reynolds conditions for the climatic average because of aperiodic fluctuations on shorter time scales.
2. The difficulty of adequately parameterizing the nonlinear effects of the higher frequency phenomena leading to errors, at least a part of which is "random."
3. The impossibility of specifying initial conditions and boundary conditions (i.e., external forcing) without some level of uncertainty and error.

We must therefore conclude that the variations of climate derived from any model must be imprecise to some extent and hence be describable only in some probabilistic form. A schematic flowchart showing the components of a theoretical model of climatic variability, including the stochastic components, is shown in Fig. 6-17.

We now add to the completeness of the deterministic feedback system discussed in the last section by including the effects of a white noise stochastic forcing process. The most appropriate manner of writing the resulting *stochastic–dynamical* system is

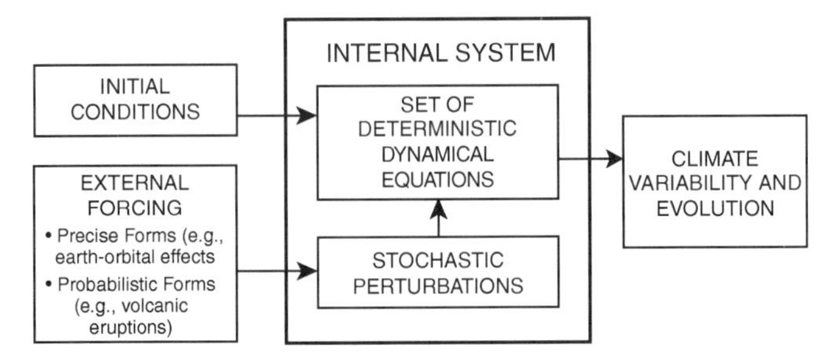

Figure 6-17 Schematic flowchart, showing the components of a theoretical model of climatic variability, including stochastic effects.

in the so-called Ito form (e.g., Schuss, 1980),

$$\begin{aligned} \delta\theta &= f_\theta(\eta, \theta)\delta t + \varepsilon_\theta^{1/2}\delta\mathbf{W} \\ \delta\eta &= f_\eta(\eta, \theta)\delta t + \varepsilon_\eta^{1/2}\delta\mathbf{W} \end{aligned} \tag{6.26}$$

where $\delta\eta$ and $\delta\theta$ are over a finite time step δt; f_θ and f_η are the deterministic functions for which solutions were obtained in the previous section; $\varepsilon^{1/2}/\delta t$ is the standard deviation (or typical amplitude) of the random fluctuations; $\delta\mathbf{W}$ is the incremental white noise function that can be identified with the output of a random number generator of variance equal to unity.

6.6.1 The Stochastic Amplitude

The first, and perhaps most critical, requirement for an analysis of the stochastic properties of the model is to estimate a reasonable reference amplitude of the noise as measured by $\varepsilon^{1/2}$. As noted above, there are several contributing factors to this noise amplitude. The simplest approximation is to assume that the past record of aperiodic variability about the running average climatic variations gives an acceptable measure of the combined effects. For an averaging interval $\delta_c = 100$ y the interannual variability of the annual mean ice edge should give a good first approximation of $\varepsilon_\eta^{1/2}$.

In Saltzman and Moritz (1980) and Saltzman (1982) we neglected any direct stochastic forcing of θ (i.e., $\varepsilon_\theta^{1/2} = 0$), and determined the interannual noise level of η to be of the order of $\varepsilon_\eta^{1/2} = 10^{-3}$. We shall take values equal or larger than this value to illustrate the possible effects.

6.6.2 Structural Stochastic Stability

A fundamental question is the degree to which the deterministic properties of a model can survive the presence of a reasonable level of stochastic noise, and to what extent the stochastic noise introduces new features when interacting with the deterministic solutions that would never appear in its absence (e.g., "tunneling" or quasi-periodic "exit" from one stable equilibrium to another). As a general rule, stochastic noise is

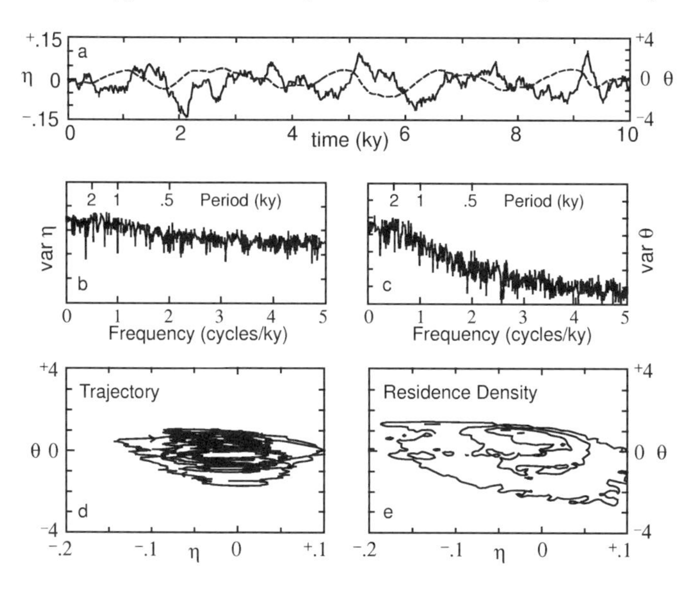

Figure 6-18 Sample stochastic solution for the deterministic system shown in Fig. 6-9, obtained by perturbing η with random forcing of an amplitude corresponding to observed interannual variations: (a) sample evolution; (b) the variance spectra for η; (c) the variance spectra for θ; (d) the sample phase plane trajectory; (e) the "residence density" of the solution. After Saltzman (1982).

a source of instability in a system, tending to prevent it from ever settling down to a fixed stable point or limit cycle.

In Figs. 6-18, 6-19, and 6-20 we show the influence of various levels of white noise on the deterministic sample trajectory solutions shown in Figs. 6-9, 6-15, and 6-14, respectively. In addition, in Figs. 6-21 and 6-22 we show the effects of two levels of stochastic forcing on the pure van der Pol-type limit cycle described by Saltzman *et al.* (1981) of the form shown in Fig. 6-15. These illustrate a few of the possible consequences of stochastic forcing, the "residence densities" in particular, giving evidence of the probabilistic nature of the response to be expected in any model purporting to represent climatic solutions realistically. For our present two-parameter stochastic system the "residence density" portrait is a mapping of the number of 1-y time steps per 10^4 y that the solution resides in an elementary rectangle formed by a 60×60 grid of the $\eta - \theta$ phase plane shown. More generally, when the dynamical system contains more than two variables, the residence density portrait should be in the form of a multidimensional "cloud" in phase space.

Fig. 6-18 illustrates how the presence of stochastic noise can lead to sustained quasi-periodic behavior near a stable spiral equilibrium. Note also the asymmetry of the residence density with respect to the equilibrium point at $(\eta', \theta') = (0, 0)$. This asymmetry results from the influence of the second, unstable, equilibrium shown in Fig. 6-8, which tends to drive the stochastic trajectories in the manner shown. We can see that in systems with multiple equilibria under the influence of stochastic perturbations, the solution not only does not permanently coincide with the stable equilibrium point but may not even have a maximum probability of being at that equilibrium point.

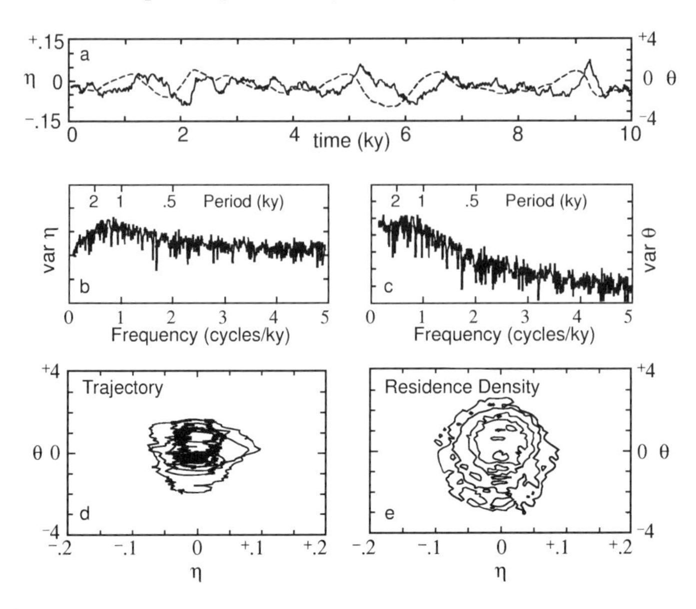

Figure 6-19 Stochastic solution near the stable limit cycle for the deterministic system portrayed in Fig. 6-15, using the same format as Fig. 6-18. After Saltzman (1982).

In the case where the system contains only a single stable spiral equilibrium, shown in Fig. 6-15, the addition of stochastic perturbations again leads to sustained periodic variations but with a symmetric residence density portrait centered on the equilibrium point (Fig. 6-19).

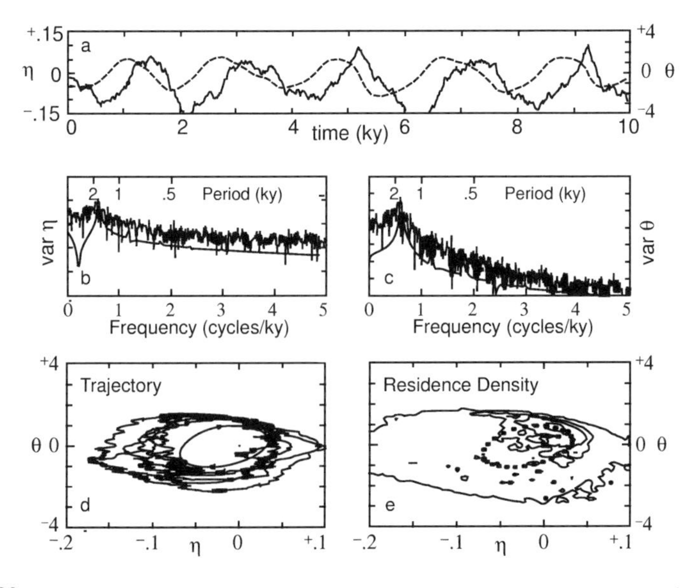

Figure 6-20 Same as Fig. 6-18, but for the model having the deterministic solution shown in Figs. 16-13 and 6-14.

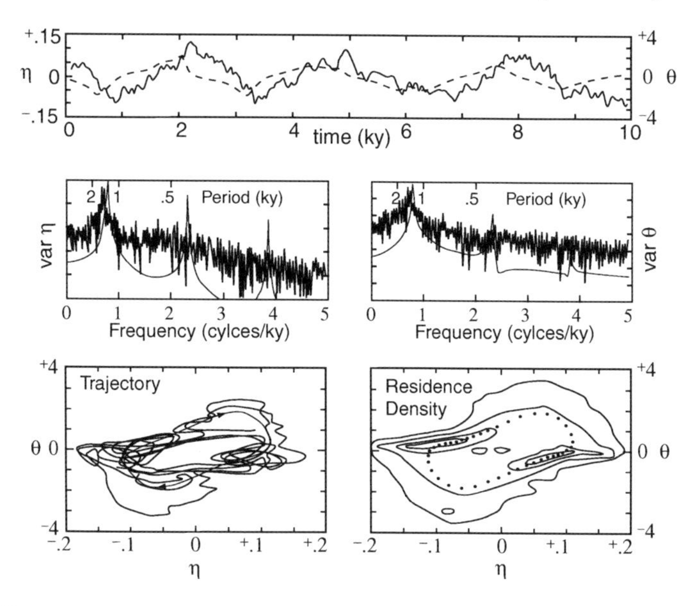

Figure 6-21 Sample stochastic solution for the idealized van der Pol-type model described in Saltzman *et al.* (1981), using a reference noise amplitude. Format is the same as in Fig. 6-18. The deterministic solution is shown by the dotted curve.

In Fig. 6-20 we show the effects of stochastic perturbations on the limit cycle deterministic flow, shown in Figs. 6-13 and 6-14, that was obtained by increasing the magnitude of the CO_2 feedback, thereby passing from a stable to an unstable spiral equilibrium (i.e., a Hopf bifurcation). It is of interest to note the similarities of the residence density portraits for the deterministic flows on either side of this bifurcation subjected to similar stochastic perturbations. The possibility for this similarity of response was pointed out by Sutera (1980) in a discussion of the Lorenz chaotic attractor (Section 6.7). Thus, similar residence density signatures obtained from observations may not uniquely reveal the nature of the nearby equilibria and the underlying deterministic flow.

The addition of stochastic perturbations to a relaxation oscillator of the general form shown in Fig. 6-14, but containing no other equilibria outside the limit cycle as in Fig. 6-16, leads to the interesting bimodal residence density portrait shown in Fig. 6-21. This represents a good example of an almost-intransitive system (cf. Lorenz, 1968) in which, because of the double maximum in the limit cycle trajectory speed, the system tends to reside for relatively long time periods in two distinct regions of phase space (see Section 7.8).

Finally, in Fig. 6-22 we show the effects on this van der Pol-type system of increasing the amplitude of the stochastic perturbations above what we believe is a "reasonable" level. This exercise illustrates how a high level of stochastic forcing in one part of a dynamical system (η in this case) can damp the response in another part of the system (θ). Thus, the possibility exists that in spite of a basic limit cycle determinism

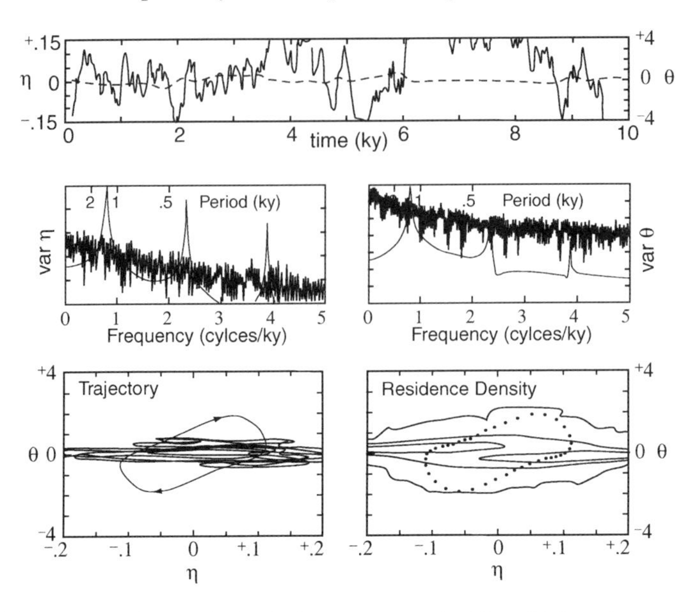

Figure 6-22 Sample stochastic solution for the system treated in Fig. 6-21, but using a noise amplitude three times larger. After Saltzman *et al.* (1981).

with significant θ variability, no signature of such variability may be observed due to the presence of stochastic noise.

6.7 MORE THAN TWO-VARIABLE SYSTEMS: DETERMINISTIC CHAOS

If the system has more than two variables, in addition to the possibility for a *point* attractor of spatial dimension 0, or a *curved* attractor of dimension 1 (e.g., limit cycle), the attractor could take the form of a smooth two-dimensional torus-shaped surface or some multidimensional "hypersurface" or "manifold" of some integral dimension greater than 2. In this case the trajectories would wind around such a surface with more than the single frequency possible for the one-dimensional curved attractor, the additional frequencies being incomensurate. In order for such periodic behavior to be realized there cannot be any stable equilibrium *points* to serve as attractors.

A significant additional possibility arises in the case of three or more variables: the attractor can be a nonsmooth manifold of noninteger dimension (i.e., a *fractal*). The trajectories on such a "strange" attractor are aperiodic or "chaotic," the pioneering example of such an attractor given by Lorenz (1963), following the discovery of such an aperiodic solution of a low-order deterministic model of Bénard convection by Saltzman (1962). Further discussions of this possibility for deterministic aperiodic behavior can be found in many books (e.g., Thompson and Stewart, 1986; Moon, 1992; Ghil and Childress, 1987; Lorenz, 1993) and an excellent discussion of the bifurcation, predictability, and stochastically forced properties of this system is given by Moritz and Sutera (1981). As it turns out, this chaotic behavior is a ubiquitous property of most complex nonlinear dynamical systems and of the observed physical phenomena they are meant to represent (such as global weather variations).

This chaotic, aperiodic, behavior is a consequence of the instability of the trajectories; that is, any small displacement from a given trajectory will result in the divergence of the trajectory from the given trajectory. It follows that aperiodicity indicates a lack of predictability, and, as a corollary, a chaotic system is strongly sensitive to initial conditions. A measure of the exponential rate of this divergence, characterizing the magnitude of the instability and lack of predictability, is called the *Lyapunov exponent*. Conversely, for an attractor whose trajectories are stable, small displacements will remain close to the trajectory, implying that when the trajectory comes close to a previous section of the trajectory it will retrace the trajectory periodically; this implies perfect predictability. Note that the spatial dimension of an attractor, whether integer or fractal, must be smaller than the number of variables in the dynamical system. (Included in the number of variables are the amplitudes of nonautonomous external forcings of different periods.) Thus, a determination of the dimension implied by a time-series record of a climatic variable can give an important clue regarding the minimum number of variables and forced frequencies that are likely to be involved in generating the time series dynamically. In addition, it would be of further value to determine whether the irregular variations generally characterizing observational time series are the result of a deterministic dynamical system generating a chaotic solution or of a nonfractal (e.g., periodic) system that is subjected to stochastic perturbations (noise). Methods have been proposed with the aim of ascertaining all of these properties of an observed record, notably by Grassberger and Procaccia (1983). However, as discussed by Maasch (1989) in his application of the method to $\delta\,^{18}$O-ice-sheet records of the type shown in Fig. 1-4 (see also Nicolis and Nicolis, 1984), very long records are needed to obtain reliable estimates; a rough estimate places the dimension of the ice-sheet records in the range between 4 and 6 (Maasch, 1989). A good discussion of the Grassberger–Procaccia method is given by Nicolis and Nicolis (1984, 1986), who, using a relatively short $\delta\,^{18}$O data set, suggested a fractal dimension of 3.1, lower than the Maasch value.

PART II

Physics of the Separate Domains

7

MODELING THE ATMOSPHERE AND SURFACE STATE AS FAST-RESPONSE COMPONENTS

In Sections 4.3 and 5.4 we suggested that the climatic-mean state of the atmosphere and upper layer of the Earth's surface (including the biosphere) are fast-response domains that can be considered to be in a unique equilibrium with slow external forcing (e.g., earth-orbital changes) and the slow variations of the ice sheets, deep ocean, and geochemical (e.g., CO_2) inventories. We shall adopt this suggested transitivity (or ergodicity) of the fast-response variables (see Section 6.5) as a working hypothesis, to be checked for consistency with the most complete models of these domains (GCMs). Up to now all such GCMs have indeed yielded unique (though, depending on the model, slightly different) asymptotic solutions that represent a dynamic equilibrium with prescribed forcing and slow-response variable distributions (e.g., ice sheets, CO_2 level). This uniqueness result is corroborated by more systematic, initial-value sensitivity studies to be described in Section 7.8.

Models governing these equilibrium states are of intrinsic interest in accounting for (i.e., "simulating") the present climate, thereby establishing the degree of their credibility, and in reconstructing the past atmospheric paleoclimates that accompanied altered geography, atmospheric composition, radiative forcing, and distributions of the slow-response variables (e.g., ice sheets). Because in any theory of climatic variability we are usually most interested in the changing state of the atmospheric and surface boundary layers where human activities are concentrated, all models of climate must ultimately include such equilibrium solutions either explicitly or implicitly.

In addition to their above "simulation" role, the equilibrium models are of special theoretical interest as the basis for establishing "sensitivity" functions of the form necessary for closure of the theory for the evolution of the slow-response variables. More fundamentally, as was portrayed in Fig. 5-3, it would be desirable to use the equilibrium models in a coupled mode to determine directly the small net fluxes of mass (e.g., water, carbon) and energy between the atmosphere, ocean, and ice sheets that accompany the equilibrium state. For this reason the ultimate model of long term paleoclimatic variations will inevitably require an interactive calculation involving time-dependent changes in ocean and ice states in quasi-static equilibrium with the atmosphere. As discussed in Section 5.1, however, on the "ice-age" time scale the fluxes are several orders of magnitude below the resolution of these models (or, for that matter, below our ability even to measure them observationally). From another point of view, the variations of the slow-response variables are well within the "drift" error of the models. In the end, we must resort to parameterizations based on relationships of the form given by Eq. (5.12), requiring the assignment of some free parameters.

Numerous models have been formulated and solved to deduce the quasi-equilibrium climatic state of the atmospheric and surficial variables with fairly good, but far from perfect, fidelity to the observations. These range from the above-mentioned general circulation models, which are the most physically complete models, to the more heavily parameterized time-averaged or statistical-dynamical models that include as the simplest special case the purely thermodynamical energy balance models.

7.1 THE GENERAL CIRCULATION MODEL

General circulation models (also called *global circulation* or *explicit-dynamical models*) are based on equations of the form given by Eqs. (4.1)–(4.7) governing the synoptic-average variables $\bar{x}^{(s)}$ [$= (\bar{x}^{(t)} + x_1^\star)$; see Section 5.3] with a geographic spatial resolution comparable to the global synoptic weather network. The subsynoptic departures $x_2^\star$ must be represented in a parameterized form as part of the functions $\mathcal{S}_\xi$, F, and q. As noted, Eqs. (4.1)–(4.7) are *prognostic* equations. The time-dependent solutions of these equations represent a sequence of weather maps that are allowed to evolve numerically until the statistically steady state is achieved. The output is then averaged to yield the model "climate" in much the same manner as observational weather records are processed to determine climatic norms. It is assumed that although the individual details of the evolving numerically generated maps cannot be accurate by the standards of "weather prediction," their ensemble climatic statistics can be fairly accurate. Because these models treat most of the phenomena included in the high-energy frequency band between 1 hr and 1 week (see Fig. 1-2) by *explicit* deterministic physics, this approach represents the most rigorous and complete theory of the equilibrium of the atmosphere and surface layers with the quasi-steady boundary conditions imposed by the deep ocean and ice sheets. In essence, GCMs represent a culmination and synthesis of our present level of knowledge of the dynamics of the atmosphere, mixed-layer ocean, and surface state of the Earth, including the biosphere. We can expect these models to give the best estimates obtainable to establish the relations needed for closure of the form given by Eq. (5.5). There are now many examples of such models and solutions for present boundary conditions, reviews of which, and discussions of the basic physics included, are given, for example, by Washington and Parkinson (1986), McGuffie and Henderson-Sellers (1997), Schlesinger (1988), and Trenberth (1992). In Sections 7.9 and 7.11 we will describe some of the applications of the GCM to paleoclimate sensitivity and simulation studies, respectively.

It is widely recognized, however, that many questions arise concerning the quality of the physics included in the models, and that, in spite of all the detail with which synoptic variations are treated explicitly, there is still a great deal of variability on smaller scales ($x_2^\star$) that must still be parameterized. Errors in these parameterizations as well as in the representation of the synoptic processes can be amplified in the iterative numerical process to such a degree that the final output may be no more accurate than that of much simpler models in which the synoptic scale phenomena are also parameterized. Moreover, because of the great number of phenomena included simultaneously in a highly nonlinear system, a "cause and effect" understanding of the processes involved is difficult to diagnose.

From another viewpoint, given the enormous amount of computer time required for a single experimental run to an asymptotic equilibrium for a high-resolution GCM, it is prohibitive to do very much experimentation and long-term integration with these models. The solutions obtained give only the "snapshot" equilibrium pictures for a single time in the past. To simulate long-term climate change, for which the carrier time-dependent equations are those governing the deep ocean and ice masses, it would be necessary to calculate such equilibria at some regular interval, say every few hundred years, at each stage determining the net flux of heat into or out of the oceans (which must be balanced by release or consumption of the latent heat of fusion and/or by a net radiative imbalance at the top of the atmosphere), and the net rate of accumulation or melting and ablation of ice mass. As we have noted, it is not possible to compute these fluxes to the required accuracy.

Finally, we must recognize that the GCM typically contains a huge number of degrees of freedom, as represented, for example, by the thousands of different equations for each dependent variable at each grid point in the three-dimensional spatial lattice. It seems reasonable to expect, however, that the macroturbulent behavior represented by these many equations is not altogether "free" but possesses some coherent organizations to effect the bulk transports of mass, momentum, and energy required by the conservation principle applied to the climatic-average state. All these considerations point to the desirability of advancing the development of statistical-dynamical models in which the task of generating the synoptic eddy statistics by continuous hour-by-hour integration of a huge dynamical system is supplanted by the development of physically based, deterministic representations of the parameterized effects of these eddies on the climatic-mean state.

7.2 LOWER RESOLUTION MODELS: STATISTICAL-DYNAMICAL MODELS AND THE ENERGY BALANCE MODEL

Statistical-dynamical models are based on equations governing the climatic-mean variables, $\bar{x}^{(c)}$, which, as stated in Section 5.3, we henceforth simply call x, omitting the overbar. These equations, are essentially diagnostic when applied only to the atmosphere and underlying surface layer, taking exactly the same form as Eqs. (4.1)–(4.7) but with the trace constituent source function $\mathcal{S}_\xi$, friction function F, and heating function q now modified to include the flux convergences due to all sub-climatic mean departures $x^\star \equiv (x_1^\star + x_2^\star)$, i.e.,

$$\mathcal{S}_\xi \rightarrow \left[\mathcal{S}_\xi - \nabla \cdot \rho \overline{\xi^\star \mathbf{V}^\star}^{(c)}\right] \tag{7.1}$$

$$F \rightarrow \left[F - \nabla \cdot \overline{(\rho v^{\star})\mathbf{V}^{\star}}^{(c)}\right]$$

$$q \rightarrow \left[q - \nabla \cdot \overline{\rho} c \overline{T^{\star}\mathbf{V}^{\star}}^{(c)}\right]$$

where we have invoked the Boussinesq approximation, $\rho = \overline{\rho}$. Thus, SDMs must include parameterizations not only of the effects of fluctuations of frequencies less than about an hour ($x_2^{\star}$), as in the GCMs, but also the effects of all frequencies up to the 100-y climatic averaging period ($x_1^{\star}$). This can include the diurnal and seasonal cycles, mesoscale phenomena, synoptic weather waves, and even interannual and decadal variations. In view of the complexity and high energy level of all this subclimatic variability, it is clear that the development of physically sound parameterizations is a major challenge. It is also likely that the stochastic terms in the governing SDM equations will be significant so that one could not expect to observe that the climatic system resides at any deterministic equilibrium, even for fairly good parameterizations. Despite the difficulties of parameterization, SDMs based on rather simple approximations can account for a good deal of the large-scale spatial variability of the climatic-mean state of the atmosphere and its underlying boundary. A thorough review of the general foundations of SDMs is given by Saltzman (1978) and further general reviews of the hierarchy of SDMs that have been formulated by successive spatial averaging of the basic three-dimensional time-averaged equations have been given by Schneider and Dickinson (1974), Saltzman (1985), MacCracken and Ghan (1988), and Taylor (1994).

As a brief review here, we recall the expansions (Section 4.4)

$$x = \langle x \rangle + x_{*} \tag{7.2}$$

where

$$\langle x \rangle = \frac{1}{2\pi} \int_0^{2\pi} x \, d\lambda \tag{7.3}$$

is the zonal average and

$$x_{*} = x - \langle x \rangle \tag{7.4}$$

are the departures from the zonal average representing the "standing" or "stationary" waves associated with the continent–ocean and topographic distribution. This expansion forms the basis for resolving SDMs into two basic components: (1) two-dimensional "meridional plane" or "axially symmetric" models governing $\langle x_i \rangle(\varphi, z)$, and (2) three-dimensional axially asymmetric (standing-wave) models governing $x_{i*}(\lambda, \varphi, z)$ (see Saltzman, 1968).

7.2.1 A Zonal-Average SDM

Using the simplest vertical resolution capable of representing mean poloidal motions (e.g., Hadley and Ferrel circulations) and the hydrologic cycle, both of which are fundamental to a description of the atmospheric climatic state, Saltzman and Vernekar

(1971) constructed the first SDM, from which most of the presently observed zonal-mean, winter and summer, climatic statistics can be deduced to roughly the same accuracy as can be deduced from a GCM. This model is based on a two-layer representation of the atmospheric wind field, superimposed on the vertically averaged fields.

The deduced dependent variables of the model include the climatic means and variances of the temperature and wind, the humidity, and all components of the surface and atmospheric heat, momentum, and water balances (e.g., precipitation and evaporation, horizontal and vertical fluxes of sensible and latent heat and momentum due to eddies and mean poloidal motions). The vertical fluxes are determined within the atmosphere and at the interface between the atmosphere and subsurface media (ocean, land, ice), by means of a full set of parameterizations of shortwave and longwave radiation, convection of sensible heat, and latent heat processes. The temperature at the base of the seasonal thermocline, T_{d}, is prescribed as a lower boundary condition, and the average winter and summer radiation is prescribed as an upper boundary condition. To effect closure of the system, baroclinic and barotropic wave theory is used in a simplified manner to parameterize the horizontal transports of heat, momentum, and water vapor in the atmosphere. More detailed and comprehensive reviews of the model specifications and results have been given in Saltzman (1978).

In Fig. 7-1 we show some sample comparative results of applying this model to present lower boundary conditions, and to the conditions that prevailed 20 ka, the ice-age maximum (Saltzman and Verneker, 1975).

7.2.2 Axially Asymmetric SDMs

It is clear from any map of surface climatic fields (e.g., temperature) that the departures from zonal symmetry associated with the continent–ocean distribution are of the same order as the meridional variations of the zonally averaged climatic fields. Moreover, when one examines the present pattern of glaciation, and the patterns of glaciation that prevailed during the last glacial maximum about 20 ka, we see a marked asymmetry around latitude circles (e.g., Fig. 3-8). Over shorter, secular, time scales, we note that for regional changes in climate over the past 10 to 100 y observational evidence points to the importance of shifts in the "standing-wave" patterns that are probably associated with changes in mean asymmetric heat source and sink distributions (e.g., changes in snow cover and sea surface temperatures).

For all these reasons it follows that a theory for the axially asymmetric climatic state is at least as important as the theory of the axially symmetric climatic state. Although, as described in Section 7.3, reasonably good deductions of asymmetric surface temperatures are possible with purely thermodynamic models (EBMs), a fuller theory by means of which mean winds, the hydrologic cycle, and the upper air climate can also be deduced requires also a consideration of the equations of motion (i.e., momentum models). This is because all these features are intimately related to the three-dimensional circulations in latitudinal as well as meridional planes (e.g., the "monsoons") and to the associated standing horizontal wave motions.

Reviews of linear statistical-dynamical models of the mean asymmetric flow, in which the symmetric mean flow is prescribed, have been presented by Saltzman (1968)

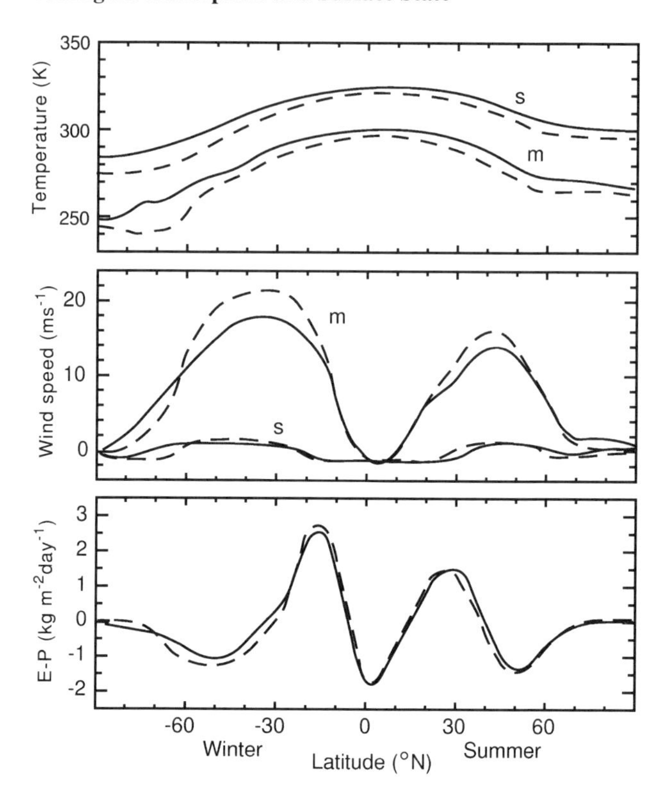

Figure 7-1 The distribution of zonally averaged potential temperature (top panel), zonal wind (middle panel), and evaporation–precipitation difference (bottom panel) as simulated by an atmospheric SDM for modern conditions (full line) and for 20-ka glacial conditions (dashed line). In the top and middle panels, s and m refer to surface values and mean atmospheric values, respectively. After Saltzman and Vernekar (1975).

and Held (1983). Note that most of the studies on this topic have been based (1) on prescribed diabatic heat sources (or prescribed surface temperatures from which heat sources were obtained from a "Newtonian"-type parameterization), and (2) on prescribed asymmetric transient eddy sources of heat and momentum. As such, these models are not full-fledged, internally "self-determining," climate models. Moreover, in contrast to most of the zonally averaged climate studies discussed thus far, these asymmetric studies have usually been concerned with deducing the wind field rather than the temperature field (i.e., the emphasis has been on the dynamics rather than the thermodynamics).

In spite of their shortcomings, the above studies have been of great value as diagnostic guides to a more valid equilibrium climate theory for the atmosphere. For example, it is clear from these studies that the effects of mountains and heating in generating the asymmetric climate patterns are of comparable magnitude. One implication is that the topographic blocking effect of ice sheets (such as the Laurentide at the LGM) on the prevailing atmospheric global climate may have been as large as the thermal effects associated with the increased ice coverage. Also, it is possible that quasi-resonant horizontal modes can become excited or damped, depending on the distribution of continent and ocean. This will be of interest in considering the terrestrial climates that prevailed millions of years ago when land–sea distributions were much different.

Early attempts, within the framework of a statistical-dynamical model, to deduce the asymmetric mean surface temperature field along with the free atmospheric wave structure, forced by an internally determined (not a prescribed) heating field, have been made by Ashe (1979), and by Vernekar and Chang (1978), who use a vertical resolution similar to that employed in the zonal-averaged model of Saltzman and Vernekar (1971). In treating the heating as a parameterized function of the other climatic variables, these studies open up for consideration a wide variety of possible feedbacks that have hitherto been ignored. Of special interest will be the feedbacks between atmospheric temperature or snow cover mentioned above, though there are many other interesting possibilities (e.g., Oerlemans, 1979).

7.2.3 The Complete Time-Average State

In the last analysis, we are not interested in the separate axially symmetric (i.e., zonally averaged) climatic component or the axially asymmetric component, but in the sum of these two representing the complete climate (i.e., the temporal means and variances) prevailing at all geographic locations, as obtained in GCM solutions. A statistical-dynamical theory for this complete time-average state can be achieved by coupling models for the two separate components, e.g., in the manner suggested by Saltzman (1968), but as yet no attempts have been made to effect such a combination. It may in fact turn out to be more convenient and plausible to bypass this scheme entirely in favor of solving the complete time-average equations by numerical process for a global grid network, or by proceeding directly to the GCM. It should be recognized however, that the information gained from the separate problems will be invaluable in formulating and interpreting such a complete climatic-mean model or GCM.

7.3 THERMODYNAMIC MODELS

By further spatial averaging, SDMs simpler than those described above can be formed. As a consequence of this averaging, all explicit references to momentum flux processes governed by the equations of motion are removed, leaving an essentially thermodynamic model. Thus, no calculations can be made, for example, of the mean poloidal circulations (e.g., Hadley–Ferrel circulations), which are of great importance in a full description of climate (particularly the cloud and precipitation zonation).

7.3.1 Radiative–Convective Models

Averaging globally in the horizontal reduces the time-averaged equations to a one-dimensional model governing the basic vertical profile of temperature, $\widetilde{T}(z)$, where, as defined in Section 4.4,

$$\widetilde{x}(z) \equiv \frac{1}{\sigma_E} \iint x\, dx\, dy \tag{7.5}$$

defines the horizontal average of any variable x over the area of the globe, σ_E. Such one-dimensional models are called "vertical column" or "radiative–convective" models (RCMs) because they incorporate detailed representations of vertical radiative fluxes, usually supplemented by a "convective adjustment" scheme to simulate the effects of vertical sensible heat fluxes associated with high lapse rates that may tend to form radiatively. Excluded, however, are any explicit considerations of the fluxes and dynamics associated with the field of motion.

Aside from providing a first-order representation of the vertical structure (e.g., temperature lapse rate) of the atmosphere, these models have proved to be of great value in perfecting the radiation and convection algorithms used in GCMs and for preliminary testing of the sensitivity of the climatic state to changes in solar radiation, atmospheric composition, and cloud structure. As a particularly useful application for paleoclimatic considerations, these models are the basis for estimating the climatic effect of inferred paleocompositions of Earth's earlier atmosphere (e.g., O_2, CO_2, methane) (e.g., Kasting, 1989). Good general reviews of these models are given by Ramanathan and Coakley (1978) and Karol and Rozanov (1982).

7.3.2 Vertically Averaged Models (the EBM)

The most widely applied group of thermodynamic models are based on vertical averaging of the time-averaged equations. These include models governing the following categories:

1. Two-dimensional, vertically averaged horizontal fields $\check{x}(\lambda, \varphi)$, where the vertical average is defined by

$$\check{x} \equiv \frac{1}{p_S} \int_0^{p_S} x\, dp \tag{7.6}$$

where p_S is surface pressure. Thus, these models are governed by the equations given in Section 4.5.

2. One-dimensional, vertically and zonally averaged atmospheric fields, e.g., $\langle \check{T} \rangle(\varphi)$, often coupled with, or expressed in terms of, the surface fields $\langle T_S \rangle(\varphi)$.

3. A zero-dimensional, vertically and horizontally averaged value, e.g., the mean planetary temperature $\widetilde{T}$.

In spite of their much greater simplicity, compared to either the GCM or the SDM "momentum models" described in Section 7.2, extensive paleoclimatic applications continue to be made with vertically averaged thermodynamic systems that fall in the above three categories. We shall discuss these models in more detail in the next two subsections.

7.4 THE BASIC ENERGY BALANCE MODEL

Because we are dealing in this chapter only with the atmosphere and the surficial layer, to the exclusion of the presumably slower response state of the deep ocean and ice masses, let us consider a column at any point extending from the top of the atmosphere (z_T) to the base of the surficial layer (z_δ), which, for the ocean, is taken as the base of a "mixed layer" of about 50 m depth. We now apply the equations in Section 4.5 [i.e., Eqs. (4.49) and (4.50)] to such a column, with the following critical approximations that are implicit in all EBMs:

1. Assume no vertical heat exchange at the base of the mixed layer (z_δ), eliminating from explicit consideration all interactions with the deep ocean. Thus, $z_d \to z_\delta$ with $H_\delta^\uparrow = 0$.
2. Assume no latent heat processes, i.e., water phase transformations, eliminating from consideration all branches of the hydrologic cycle, i.e., $C = F_n = E = M = 0$.
3. Assume the total thermal energy (sensible heat) in a column is locally proportional to the surface temperature, T_S, i.e.,

$$(\Upsilon_a + \Upsilon_b) \equiv \int_{z_\delta}^{z_T} \rho c T \, dz = \Lambda(\lambda, \varphi) T_S \tag{7.7}$$

where $\Lambda(\lambda, \varphi)$ is a heat capacity factor in $\mathrm{J\,m^{-2}\,K^{-1}}$ that depends on the nature of the underlying surface (e.g., land, snow, sea ice, oceanic mixed layer) and may vary in time.

4. Assume no horizontal heat exchange due to mean poloidal motions or standing waves, implying that the total flux is due to transient eddies (e.g., atmospheric cyclone waves).
5. Assume that this horizontal heat flux can be represented in terms of surface temperature only, either as (i) a diffusive process (Sellers, 1969; North, 1975a,b).

$$(\mathbf{J}_a + \mathbf{J}_b) \approx \int_{z_\delta}^{z_T} \rho c \overline{T^\star \mathbf{v}^\star}^{(c)} \, dz \approx -D \nabla T_S \tag{7.8}$$

where D is an eddy conduction coefficient in $\mathrm{W\,K^{-1}}$ that may be taken as an empirical function of latitude $D = D(\varphi)$ to allow for some of the explicitly excluded heat flux processes (item 4 above), or (ii) a nearly equivalent Newtonian cooling process (Budyko, 1974), i.e.,

$$\nabla \cdot (\mathbf{J}_a + \mathbf{J}_b) = \gamma \left[T_S(\lambda, \varphi) - \widetilde{T}_{S0} \right] \tag{7.9}$$

where $\widetilde{T}_{S0}$ is a global mean value to which the system would tend to be restored under the influence of heat transport, and γ is a constant.

6. Following Budyko (1969) and Sellers (1969), assume the incoming solar radiation at the top of the atmosphere $H_{\mathrm{T}}^{(1)\downarrow}$ can be represented by

$$H_{\mathrm{T}}^{(1)\downarrow} = R(\varphi, t)\big[1 - \alpha(\lambda, \varphi)\big] \tag{7.10}$$

where $R(\varphi, t) = (\mathcal{S}/4) f(\varphi, t)$; $\mathcal{S}$ is the solar constant, $f(\varphi, t)$ represents the distribution of the solar radiation in space and time (including the effects of seasonal and earth-orbital variations) satisfying $\tilde{f}(\varphi, t) = 1$, and α is the albedo, which is a function of angle of incidence and surface state (e.g., ice coverage), sometimes taken simply as a function of the temperature.

7. Assume the outgoing longwave radiation, $H_{\mathrm{T}}^{(2)\uparrow}$, can be expressed in the form

$$H_{\mathrm{T}}^{(2)\uparrow} = A + BT_{\mathrm{S}}(\lambda, \varphi) \tag{7.11}$$

where A and B are empirical constants that implicitly represent the effects of the present gaseous components of the atmosphere, but may be made explicit functions of the atmospheric composition (e.g., CO_2).

With these assumptions, assuming the diffusive heat transport, the basic two-dimensional governing equation becomes

$$\Lambda(\lambda, \varphi)\frac{dT_{\mathrm{S}}}{dt} = \nabla \cdot \big[D(\varphi)\nabla T_{\mathrm{S}}\big] + R(\varphi, t)\big[1 - \alpha(\lambda, \varphi)\big] - A - BT_{\mathrm{S}} + \omega(T_{\mathrm{S}}) \tag{7.12a}$$

where, in accordance with the discussion in Chapter 4, we have added a term, $\omega(T_{\mathrm{S}})$, representing random (or stochastic) forcing.

More rigorously, if the additive stochastic forcing is modeled in the form of Gaussian white noise, the changes in temperature T_{S} should be calculated from the "Ito form" of the Langevin type, Eq. (7.12a), i.e.,

$$\begin{aligned} dT_{\mathrm{S}} = {} & \frac{1}{\Lambda(\lambda, \varphi)}\big\{\nabla \cdot \big[D(\varphi)\nabla T_{\mathrm{S}}\big] + R(\varphi, t)\big[1 - \alpha(\lambda, \varphi)\big] - A - BT_{\mathrm{S}}\big\}\, dt \\ & + \mu_{\mathrm{T}}^{1/2}\, dW \end{aligned} \tag{7.12b}$$

where the term in braces is the deterministic part and the stochastic part is represented as the product of the white noise standard deviation $\mu_{\mathrm{T}}^{1/2}$ and dW, the differential increment of a Wiener process (e.g., Sutera, 1981).

If the time average is taken to be for a roughly 100-y ensemble of monthly mean states, this equation can be assumed to govern the seasonal cycle as well as the annual mean state. In spite of the simplicity of this model, North *et al.* (1983) have shown that a good deal of the seasonal geographic distribution of surface temperature can be accounted for, and, moreover, Crowley *et al.* (1986) have shown that the sensitivity of the equilibria of Eq. (7.12) to orbitally induced changes in the solar radiation distribution closely approximates the results a more physically complete GCM. For these reasons, the EBM has been used as the fast-response component of (1) ice sheet models by DeBlonde and Peltier (1991a,b) and DeBlonde *et al.* (1992), wherein the net ice accumulation is parameterized in terms of T_S (see Chapter 9), and (2) deep ocean thermohaline circulation models (e.g., Sakai and Peltier, 1997).

7.5 EQUILIBRIA AND DYNAMICAL PROPERTIES OF THE ZERO-DIMENSIONAL (GLOBAL AVERAGE) EBM

As noted in Section 7.3.2, even simpler versions of the EBM are obtained by further spatial averaging. Thus, taking the zonal average of Eq. (7.12) we obtain the one-dimensional EBM,

$$\Lambda(\varphi)\frac{\langle dT_S\rangle}{dt} = \frac{1}{a^2\cos\varphi}\frac{2}{\partial\varphi}\left[D(\varphi)\frac{\partial\langle T_S\rangle}{\partial\varphi}\cos\varphi\right] + R(\varphi,t)\big[1-\alpha(\varphi)\big] - A - B\langle T_S\rangle + \omega(\langle T_S\rangle) \tag{7.13}$$

and by also averaging over latitude we obtain the following zero-dimensional equation governing the global mean thermal state $\widetilde{T}_S$:

$$\Lambda\,\frac{d\widetilde{T}_S}{dt} = \frac{\mathcal{S}}{4}\langle\langle f(\varphi)\big[1-\alpha(\varphi)\big]\rangle\rangle - A - B\widetilde{T}_S + \omega\big(\widetilde{T}_S\big) \tag{7.14a}$$

where for notational simplicity we set $(\widetilde{\ }) \equiv \langle\langle(\ )\rangle\rangle$.

Recognizing that $\widetilde{f(\varphi)} = 1$, and defining a "planetary albedo," $\alpha_P = \langle\langle f(\varphi)\alpha(\varphi)\rangle\rangle$, Eq. (7.14a) can be written in the alternate form,

$$\Lambda\,\frac{d\widetilde{T}_S}{dt} = \frac{\mathcal{S}}{4}\,(1-\alpha_P) - \big(A + B\widetilde{T}_S\big) + \omega\big(\widetilde{T}_S\big) \tag{7.14b}$$

As the simplest approximation for the dependence of α_P on the planetary ice coverage it can be assumed that

$$\alpha_P = \alpha_P\big(\widetilde{T}_S\big) = \alpha_{P0} + \left(\frac{\partial\alpha_P}{\partial\widetilde{T}_S}\right)_0\big(\widetilde{T}_S - \widetilde{T}_{S0}\big) \tag{7.15}$$

where $\widetilde{T}_{S0}$ denotes an equilibrium value corresponding to a constant planetary albedo α_{P0}, i.e.,

$$\widetilde{T}_{S0} = \frac{\mathcal{S}(1-\alpha_{P0})/4 - A}{B} \tag{7.16}$$

Given the value $\mathcal{S} = 1368\ \mathrm{W\,m^{-2}}$, and Budyko's (1969) empirical values for present atmospheric conditions, $A = 203.3\ \mathrm{W\,m^{-2}}$, $B = 2.09\ \mathrm{W\,m^{-2}\,{}^\circ C^{-1}}$, and $\alpha_{P0} = 0.31$, we obtain $\widetilde{T}_{S0} \approx 15^\circ\mathrm{C} = 288\mathrm{K}$, a value close to the present mean surface temperature of Earth. On the other hand, had we assumed that the outgoing longwave radiation is given by the basic Stefan–Boltzmann law,

$$\widetilde{H}_{\mathrm{T}}^{(2)\uparrow} = \sigma \widetilde{T}_{\mathrm{Se}}^{4} \tag{7.17}$$

where σ is the Stefan–Boltzmann constant ($5.71 \times 10^{-8}\ \mathrm{W\,m^{-2}\,K^{4}}$), the "effective planetary" radiative equilibrium temperature, $\widetilde{T}_{\mathrm{Se}}$, would be given by

$$\widetilde{T}_{\mathrm{Se}} = \left[\frac{(1 - \alpha_{\mathrm{P0}})\mathcal{S}}{4\sigma} \right]^{1/4} \tag{7.18}$$

For the same value of α_{P0}, this mean effective radiating temperature is $\widetilde{T}_{\mathrm{Se}} = 250\mathrm{K}$ (a value representative of the midtroposphere), indicating that a large part of the outgoing radiation emanates from Earth's atmosphere, not its surface, which has a mean temperature of 288K. This illustrates the critical role of the atmospheric "greenhouse effect" in maintaining a warmer, more hospitable surface temperature, which, in its absence, would be close to $\widetilde{T}_{\mathrm{Se}}$ (assuming the same value of α_{P0}).

From Eqs. (7.14b) and (7.15) the transient departures, $\widetilde{T}_{\mathrm{S}}' = \widetilde{T}_{\mathrm{S}} - \widetilde{T}_{\mathrm{S0}}$, are governed by a linear equation of the form,

$$\frac{d\widetilde{T}_{\mathrm{S}}'}{dt} = -\frac{1}{\varepsilon_{\mathrm{T}}} \widetilde{T}_{\mathrm{S}}' + \omega(\widetilde{T}_{\mathrm{S}}') \tag{7.19}$$

where the time constant ε_{T} is given by

$$\varepsilon_{\mathrm{T}} = \frac{\Lambda}{\left[B + \left(\frac{\partial \alpha_{\mathrm{P}}}{\partial \widetilde{T}_{\mathrm{S}}}\right)_0 \frac{\mathcal{S}}{4}\right]}$$

Thus, for $(\partial\alpha_{\mathrm{P}}/\partial\widetilde{T}_{\mathrm{S}})_0 > -(4B/\mathcal{S})$ [e.g., for α_{P} = a constant, i.e., $(\partial\alpha_{\mathrm{P}}/\partial\widetilde{T}_{\mathrm{S}})_0 = 0$], $\varepsilon_{\mathrm{T}} > 0$ and all transients damp exponentially to a single stable equilibrium value $\widetilde{T}_{\mathrm{S0}}$. In this case, under the influence of additive stochastic forcing $\omega(\widetilde{T}_{\mathrm{S}})$, assumed to be in the form of white noise, the time-dependent behavior of the system can be written in the more appropriate Ito form,

$$d\widetilde{T}_{\mathrm{S}}' = -\frac{1}{\varepsilon_{\mathrm{T}}} \widetilde{T}_{\mathrm{S}}'\,dt + \mu_{\mathrm{T}}^{1/2}\,dW \tag{7.20}$$

A detailed examination of the properties of this stable dissipative system for the climate problem was given by Hasselmann (1976) and Lemke (1977), showing how relatively short-term random white noise fluctuations in the atmosphere and oceans can induce a "red noise" variability in the more heavily averaged climatic state represented by $\widetilde{T}_{\mathrm{S}}$, including at lower frequencies longer term paleoclimatic variability. However, the paleoclimatic fluctuations in which we are mainly interested are the departures from the red noise spectrum (e.g., the 100-ky ice-age cycles of the Late

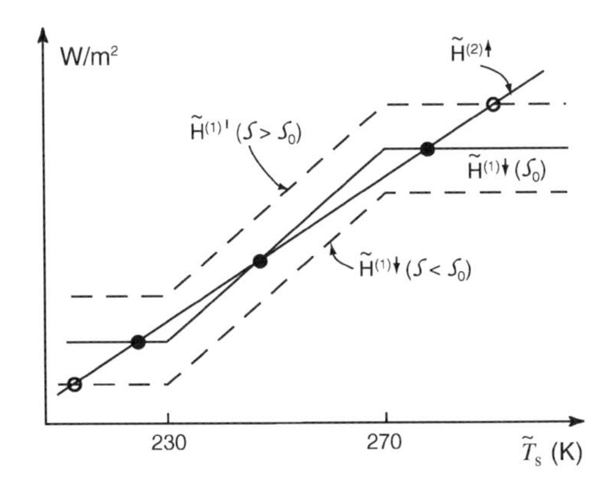

Figure 7-2 Plot of $\widetilde{H}_{\mathrm{T}}^{(1)\downarrow}$ $(= \mathcal{S}[1 - \alpha_{\mathrm{P}}(\widetilde{T}_{\mathrm{S}})])$ vs. $\widetilde{H}_{\mathrm{T}}^{(2)\uparrow}$ $[= A + B(\widetilde{T}_{\mathrm{S}})]$, showing possible multiple equilibria when $\widetilde{H}_{\mathrm{T}}^{(1)\downarrow} = \widetilde{H}_{\mathrm{T}}^{(2)\uparrow}$ for $\mathcal{S} = \mathcal{S}_0$ (black dots), and single equilibria for $\mathcal{S} > \mathcal{S}_0$ and $\mathcal{S} < \mathcal{S}_0$ (open dots).

Pleistocene). Such departures can be induced by white noise only if the system is nonlinear; such nonlinearity can be readily obtained, for example, when we consider the more general case in which the planetary albedo decreases strongly with increasing temperature (i.e., decreasing ice coverage).

Thus, as $(\partial\alpha_{\mathrm{P}}/\partial\widetilde{T}_{\mathrm{S}})$ approaches $-4B/\mathcal{S}$, the time constant ε_{T} becomes extremely long, approaching infinity. When $(\partial\alpha_{\mathrm{P}}/\partial\widetilde{T}_{\mathrm{S}}) < -4B/\mathcal{S}$, ε_{T} becomes negative, indicating that the equilibrium value $\widetilde{T}_{\mathrm{S0}}$ is unstable, with small departures growing exponentially. In this latter case, conservation of energy requires that a nonlinear mode of dissipation must be present in the system that damps the departures when they become very large. These possibilities bring to the fore the fundamental question regarding the appropriateness of considering surface temperature as a "fast-response" variable, which was raised at the beginning of this chapter. More generally, if positive feedbacks (such as that illustrated here due to the ice-albedo dependence on temperature) are large enough, the time constant can be lengthened significantly to the point where a variable may no longer be considered to have a "fast" response (cf. Section 4.3).

As it turns out, the most commonly invoked assumptions regarding $\alpha(T_{\mathrm{S}})$ (e.g., Budyko, 1969; Sellers 1969) introduce a nonlinearity that leads to multiple equilibria, at least one of which is unstable. To illustrate, let us adopt the representation of Sellers (1969):

$$\alpha_{\mathrm{P}}(\widetilde{T}_{\mathrm{S}}) = \begin{cases} \alpha_1 & (\widetilde{T}_{\mathrm{S}} > 270\mathrm{K}) \\ \left[\alpha_2 - (\alpha_2 - \alpha_1)(270 - \widetilde{T}_{\mathrm{S}})/40\right] & (230\mathrm{K} \leqslant \widetilde{T}_{\mathrm{S}} \leqslant 270\mathrm{K}) \\ \alpha_2 & (\widetilde{T}_{\mathrm{S}} < 230\mathrm{K}) \end{cases} \tag{7.21a}$$

The empirically derived albedos for warm ($\widetilde{T}_{\mathrm{S}} > 270\mathrm{K}$) and cold ($\widetilde{T}_{\mathrm{S}} < 230\mathrm{K}$) states, respectively, are $\alpha_1 = 0.3$ and $\alpha_2 = 0.7$. An analytically continuous form of this representation is given by Matteucci (1989) in the form

$$\alpha_{\mathrm{P}}(\widetilde{T}_{\mathrm{S}}) = d_1 - d_2 \tanh\left[d_3(\widetilde{T}_{\mathrm{S}} - \widetilde{T}_{\mathrm{S0}})\right] \tag{7.21b}$$

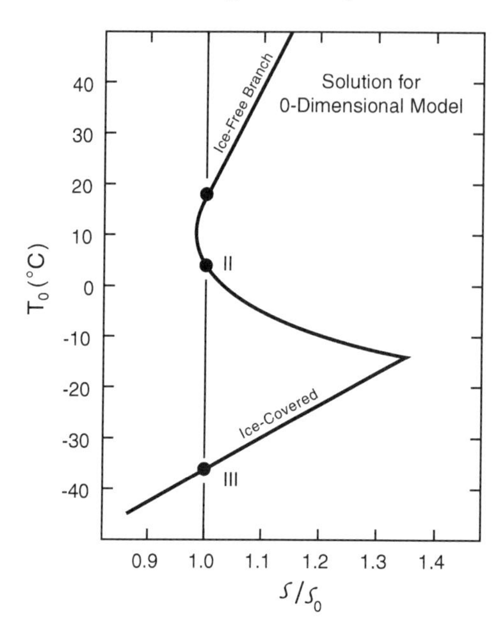

Figure 7-3 Steady-state temperatures corresponding to the climate solutions for a zero-dimensional climate model with variable ice cap as a function of solar constant in units of its present value. The roots (I, II, and III) correspond to the solutions for the present level of solar forcing. After North *et al.* (1981).

where d_1, d_2, and d_3 are empirical constants, which essentially introduces a cubic nonlinearity (see Section 6.2) because $\tanh x = (x - x^3/3 + \cdots)$. Then, from Eqs. (7.13b) and (7.21a) we have at equilibrium a balance of the incoming shortwave radiation $\widetilde{H}_T^{(1)\downarrow} = \mathcal{S}[1 - \alpha_P(\widetilde{T}_S)]/4$ and outgoing longwave radiation $\widetilde{H}_T^{(2)\uparrow} = (A + B\widetilde{T}_S)$, i.e., $\widetilde{H}_T^{(1)\downarrow} = \widetilde{H}_T^{(2)\uparrow}$, each of which is plotted in Fig. 7-2 as a function of $\widetilde{T}_S$, for the present value of the solar constant $\mathcal{S} = \mathcal{S}_0 = 1368\ \mathrm{W\,m^{-2}}$, and for $\mathcal{S} < \mathcal{S}_0$ and for $\mathcal{S} > \mathcal{S}_0$ (dashed curves). The equilibria $\widetilde{T}_{S0}$ are located at the intersections of these two curves; for the present values of $\mathcal{S}$ there are three equilibria (black circles), the central one being unstable, but for $\mathcal{S}$ being sufficiently smaller or greater than $\mathcal{S}_0$, we see that there may be only a single stable equilibrium (open circles).

In accordance with the discussion in Chapter 6, a fuller portrayal of the dependence of the equilibria on $\mathcal{S}$ is shown in Fig. 7-3, variously called the *equilibrium portrait, operating curve, or bifurcation diagram.* As described in Chapter 6, this latter designation derives from the fact that as external forcing (in this case $\mathcal{S}$) is varied as a "control parameter," the number and stability of the equilibria will undergo transitions at critical values known as bifurcation points. Thus we have distinguished between the stability at a given equilibrium point for small *internal* perturbations ("internal" or "local" stability) and the change in the number, values, and local stability of the equilibria for small changes in an *external* parameter ("external" or "global" or "structural" stability). A good discussion of this subject in the context of climate theory is given by Ghil and Childress (1987).

It should be noted that the equilibrium portrait gives the most complete picture of the "sensitivity" of the equilibria of a system to a change in a given external parameter. As discussed in Section 6.5.1, the slope of stable branches of the equilibrium curve (e.g., $\partial T_S/\partial \mathcal{S}$) is often taken as a measure of this sensitivity. Thus, at a bifurcation

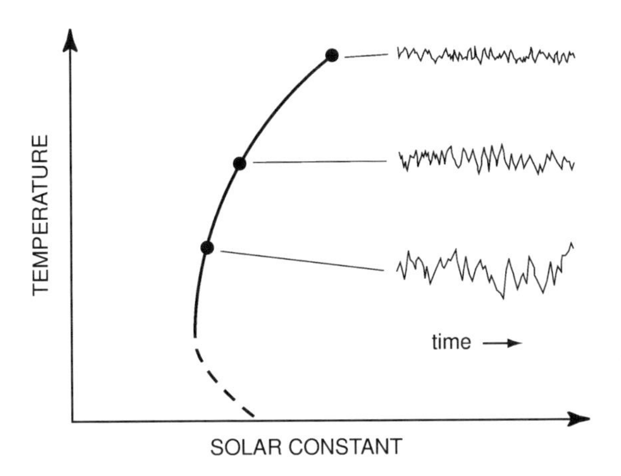

Figure 7-4 Schematic depiction of the time series of temperature fluctuations about three different equilibrium climates near a bifurcation point. Equilibrium curve is for an energy balance model of global average temperature. Different time series represent solutions to stochastic forcing of a linearized version of the model. Time and temperature scales are arbitrary, but the relative magnitude of different temperature amplitudes represents actual model output. After Crowley and North (1988).

point the system approaches infinite sensitivity to changes in external forcing (which may include stochastic forcing). In the unstable branches [negative slope in our case—e.g., Cahalan and North (1979)] the equilibria are supersensitive, both locally and globally, and cannot be realized.

In the example shown in Fig. 7-3 by North *et al.* (1981) there are two stable branches (ice free and ice covered) separated by an unstable branch. For the present value of the solar constant [$\mathcal{S}(0) \equiv \mathcal{S}_0$] there are three equilibria, the upper stable one presumably representing the present nonglacial climate. In Fig. 7-4 we show Crowley and North's (1988) portrayal of the increasing amplification of imposed noise as the bifurcation point is approached.

7.6 STOCHASTIC RESONANCE

As described above, when the single stable state equilibrium associated with a constant value of the planetary albedo α_{P0} is subjected to white noise forcing in accordance with Eq. (7.20), a red noise response is obtained representing localized variability near the single equilibrium. A much more interesting possibility arises when stochastic forcing (e.g., weather noise) is applied to a model exhibiting instability and multiple equilibria, as in the variable albedo EBM discussed at the end of the last section. In this case, the noise can provide the mechanism for transitioning (i.e., "tunneling") between the two stable equilibria. Discussions of the stochastic properties of this zero-dimensional EBM are given by Fraedrich (1978, 1979) and Sutera (1981). An illuminating way to examine the problem is by rewriting the deterministic part of Eq. (7.14b) in terms of a Lyapunov "pseudopotential" V (Ghil, 1976; North *et al.*, 1979). In particular, if we

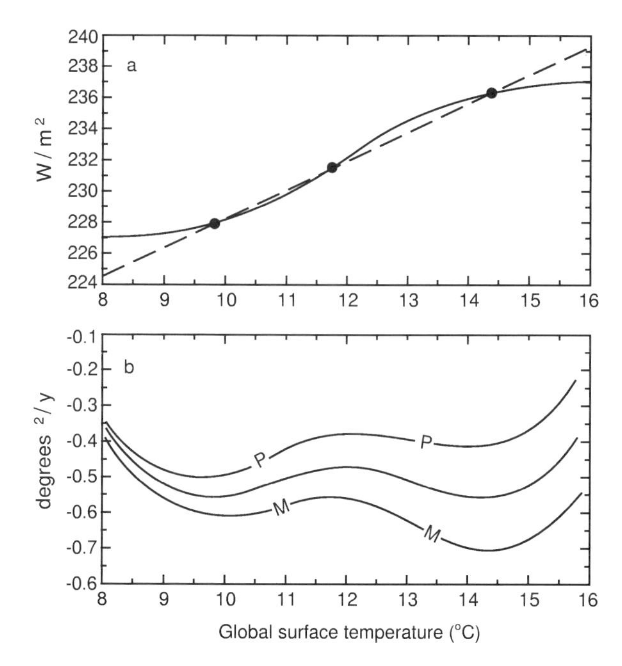

Figure 7-5 (a) Plot of the total solar radiation absorbed (solid curve) $\mathcal{S}(1-\alpha)$, and the total outgoing infrared radiation (dashed curve), $A+BT$, as functions of the global surface temperature $\widetilde{T}_{\mathrm{S}}$, in °C. The intersections of these curves correspond to steady-state climates. (b) The behavior of the pseudopotential V for different values of Q. The solid curve corresponds to $\mathcal{S}=\mathcal{S}_0$ and the curves labeled P and correspond, respectively, to $\mathcal{S}=\mathcal{S}_0(1+\beta)$ and $\mathcal{S}=\mathcal{S}_0(1-\beta)$. The minima of V represent stable equilibria; the local maximum corresponds to an unstable steady state. After Matteucci (1989).

define

$$V=-\frac{1}{\Lambda}\int\left\{\frac{\mathcal{S}}{4}[1-\alpha(\widetilde{T})]-A-B\widetilde{T}_{\mathrm{S}}\right\}dT_{\mathrm{S}} \tag{7.22}$$

then Eq. (7.14b) can be written in the form

$$\frac{d\widetilde{T}_{\mathrm{S}}}{dt}=-\left(\frac{dV}{d\widetilde{T}_{\mathrm{S}}}\right)+\omega_{\mathrm{T}} \tag{7.23}$$

where minima (potential "wells") and maxima (potential "barriers") of V correspond to the stable and unstable steady states (equilibria) of Eq. (7.14b), respectively. Thus, following Matteucci (1989), if we adopt the representation of $\alpha(\widetilde{T}_{\mathrm{S}})$ with the parameter values $d_1=0.323$, $d_2=0.015$, $d_3=0.5°\mathrm{C}^{-1}$, $\widetilde{T}_{\mathrm{S}0}=12°\mathrm{C}$, $A=210\ \mathrm{W\,m^{-2}}$, $B=1.85\ \mathrm{W\,m^{-2}\,°C^{-1}}$, $\Lambda=3\times10^{8}\ \mathrm{J\,M^{-2}\,°C^{-1}}$, and $\mathcal{S}_0=1368\ \mathrm{W\,m^{-2}}$, we obtain the curves for $\widetilde{H}_{\mathrm{T}}^{(1)\downarrow}$ and $\widetilde{H}_{\mathrm{T}}^{(2)\uparrow}$ shown in Fig. 7-5a. As before, the intersections represent the equilibria of the system occuring at $\widetilde{T}_{\mathrm{S}}=9.75°\mathrm{C}^{-1}$ (stable cold mode), 12°C (unstable), and 14.25°C (stable warm mode). The pseudopotential V is shown in Fig. 7-5b for $\mathcal{S}=\mathcal{S}_0$ (central curve), $\mathcal{S}<\mathcal{S}_0$ (upper curve), and $\mathcal{S}>\mathcal{S}_0$ (lower curve). Thus, it can be seen that if $\mathcal{S}$ oscillates between two values $\mathcal{S}\pm\beta$, the relative depths of the stable potential wells and the unstable barrier vary such that the cold-mode well is deeper (i.e., favored as an attractor) for low $\mathcal{S}$ and the warm-mode well is favored for high $\mathcal{S}$. In the absence of perturbations due to either external causes or internal stochastic forcing, the system would remain at either of these stable steady states, depending on the initial conditions.

In the presence of noise the fluctuations of $\widetilde{T}_S$, governed by the Ito form of the Langevin-type equation, Eq. (7.14b),

$$d T_S = \left\{ \frac{\mathcal{S}}{4}[1 - \alpha(\widetilde{T})] - A - B\widetilde{T}_S \right\} \frac{dt}{\Lambda} + \mu_T^{1/2}\, dW \tag{7.24}$$

can flip back and forth (i.e., "tunnel" the barrier) between one stable equilibrium and the other with a certain "mean exit time" determined by the strength of the noise and the parameters of the problem. Significantly, for the special case when the period of the external forcing [e.g., $\mathcal{S}(t)$] is the same as the mean exit time, a resonance can occur that can amplify even a weak periodic forcing. This phenomenon was discovered by Benzi *et al.* (1982) (see also Nicolis, 1982) as a proposed explanation for the dominance of the 100-ky period oscillation in the climate record of the Late Pleistocene in spite of relatively weak near-100-ky period forcing due to orbital eccentricity variations, which can be represented by $\mathcal{S} = \mathcal{S}_0[1 + \beta \cos(2\pi t/100 + 290)]$, where $\beta \approx 10^{-3}$ (Berger, 1978a,b,c).

The most extensive application of this "stochastic resonance" mechanism as an explanation of the ice-age oscillations has been made by Matteucci (1989, 1991). Although the idea can account for some aspects of the observations, it rests on the fundamental assumptions inherent in the EBM that imply the multiple equilibria of $\widetilde{T}_S$. As we shall see in Section 7.8, the existence of these multiple equilibria does not appear to be sustained by the more complete representation of climate dynamics in a GCM. Nonetheless, "stochastic resonance" turns out to be highly relevant and important in many other physical problems (e.g., Moss, 1991; Gammaitoni *et al.*, 1998).

7.7 THE ONE-DIMENSIONAL (LATITUDE-DEPENDENT) EBM

In order to study the effects of horizontal heat transport on T_S in the simplest possible manner, we return to the one-dimensional, latitude-dependent EBM governed by Eq. (7.13). Following Budyko (1969), in this case a more explicit representation of the "ice line" can be made through the following prescription of the albedo for a hemisphere:

$$\alpha(\varphi, \varphi_i) = \begin{cases} \alpha_1(\varphi < \varphi_i) \\ \alpha(\varphi_i) \\ \alpha_2(\varphi > \varphi_i) \end{cases} \tag{7.25}$$

where φ_i is the ice-line latitude, poleward of which ice or snow cover is assumed to prevail with a uniformly high albedo $\alpha_1 = 0.62$. This ice-line latitude is further assumed to coincide with $\langle T_S \rangle = -10°C$ (i.e., $\varphi_i = \varphi(\langle T \rangle) = -10°C$).

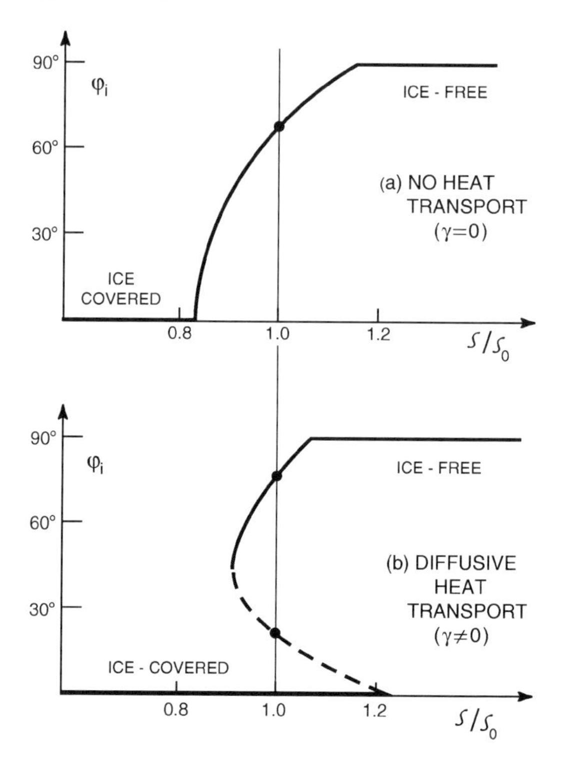

Figure 7-6 Equilibrium curves [Eq. (7.27)] for (a) the case of no heat transport ($\gamma = 0$) and (b) the case of a diffusive heat transport ($\gamma \neq 0$).

Adopting also Budyko's (1969) Newtonian cooling representation of the effects of horizontal heat flux, Eq. (7.9), the governing deterministic equation becomes

$$\frac{d\langle T_S\rangle}{dt} = \frac{\mathcal{S}}{4} f(\varphi)\big[1 - \alpha(\varphi, \varphi_i)\big] - A - B\langle T_S\rangle - \gamma\big(\langle T_S\rangle - \widetilde{T}_S\big) \tag{7.26}$$

At equilibrium

$$\frac{\mathcal{S}}{4} f(\varphi)\big[1 - \alpha(\varphi, \varphi_i)\big] - A - B\langle T_S\rangle - \gamma\big(\langle T_S\rangle_0 - \widetilde{T}_S\big)_0 = 0 \tag{7.27}$$

where, from Eq. (7.16), $\widetilde{T}_{S0} = 1/B[(\mathcal{S}_0/4)(1 - \alpha_{P0}) - A]$.

Applying Eq. (7.27) at the ice line ($\varphi = \varphi_i$) we can solve for the value of the solar constant $\mathcal{S}$ needed to maintain this ice line at any given latitude φ_i:

$$\mathcal{S}(\varphi_i) = \frac{A + (B + \gamma)\langle T_S\rangle_0(\varphi_i) + A(\gamma/B)}{f(\varphi_i)[1 - \alpha(\varphi_i)] + [1 - \alpha(\varphi_i)](\gamma/B)} \tag{7.28}$$

For the case of no heat transport ($\gamma = 0$) the functional relation between φ_i and $\mathcal{S}$ is plotted in Fig. 7-6a, showing only stable equilibria, with an ice-free or ice-covered

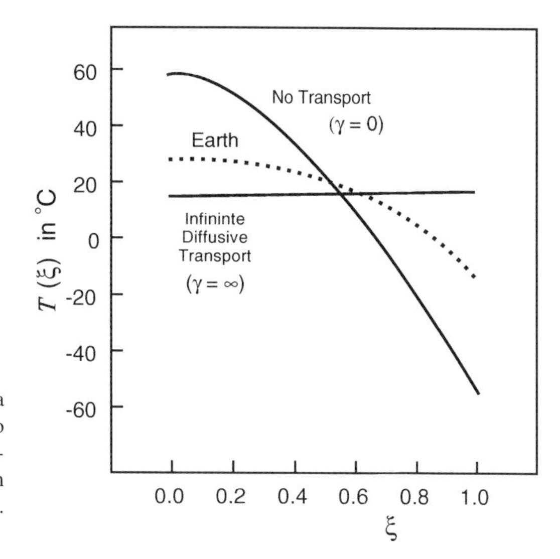

Figure 7-7 Temperature variation as a function of ξ ($= \sin\varphi$) for the cases of no heat transport ($\gamma = 0$) and infinite diffusive transport ($\gamma = \infty$), compared with the present variation. After North *et al.* (1981).

Earth possible for certain ranges of $\mathcal{S}$. When the heat transport is included ($\gamma > 0$) the equilibrium relation between φ_{i} and $\mathcal{S}$ for the same parameter values is shown in Fig. 7-6b, revealing an increased sensitivity that is manifested by an unstable branch and multiple equilibria. Any reduction of the solar constant below the bifurcation point near $\mathcal{S} = 0.9\mathcal{S}_0$ can admit only a single equilibrium representing an ice-covered Earth. This increased sensitivity is due to the effect of poleward heat transport in drawing energy from, and hence cooling, latitudes just equatorward of the ice edge. Thus an *ice-baroclinicity* (i.e., temperature gradient) feedback now can amplify the *ice-albedo* feedback that creates the baroclinicity. In this regard, a useful discussion of the difference between the zero and one-dimensional EBM is given by Watts (1981) [see also the discussions by North *et al.* (1981) and Hartmann (1994)].

The latitude dependencies of $\langle T_{\mathrm{S}} \rangle$ for the no-transport ($\gamma = 0$), infinite-transport ($\gamma = \infty$), and an intermediate Earth-like case are shown in Fig. 7-7. For this Earth-like case a comparison of the observed and calculated $\langle T_{\mathrm{S}} \rangle$ profile for the globe, using a diffusive heat flux, Eq. (6.8), and empirical values of parameters for the Northern Hemisphere winter, is shown in Fig. 7-8 (North and Coakley, 1979). This shows that a respectable amount of agreement can be achieved with a simple EBM, though some of the deficiencies of the model are already apparent—e.g., the flatness of the observed curve in low latitudes has not been represented due to the exclusion of the nondiffusive effects of the Hadley poloidal cell. If, as in the above North and Coakley (1979) study, the diffusive model of the heat flux (Sellers, 1969) is employed, results similar to those obtained by the Newtonian cooling model (Budyko, 1969) are obtained, but some differences arise—e.g., North *et al.* (1981) note the occurrence of a small additional region of instability near the pole (the *small ice-cap instability*), which plays a role in some ice-age theories.

A direct application of the one-dimensional EBM to ice-age theories can be made by considering the effects of Earth-orbital radiative forcing on the variations of the ice

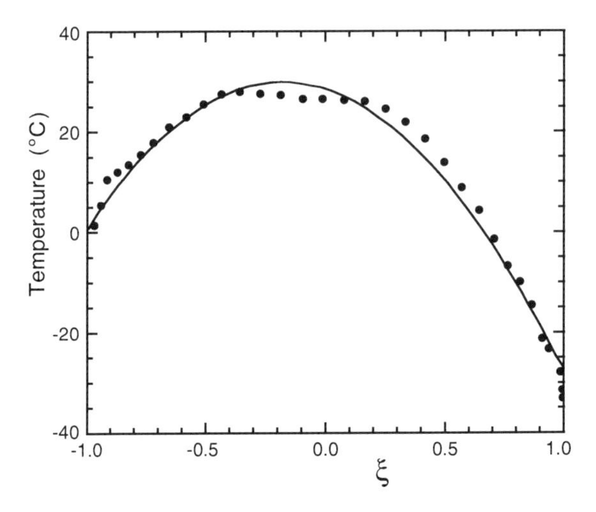

Figure 7-8 Zonal-average mean surface temperature for December, January, and February, deduced from an energy balance model (solid curve) compared with observed values (dots), plotted against the sine of the latitude, ξ. After North and Coakley (1979).

line (e.g., Suarez and Held, 1976). This aspect will be discussed further in Chapter 14 in which we review all models of Late Cenozoic climate change and glaciation.

For present purposes, in this chapter we are concerned mainly with the EBM as a theory of the surface temperature and ice as a thin surficial cover having special albedo properties. In this context, the sensitivity and multiple-equilibrium results, if valid, certainly place in question the notion that the climatic mean surface temperature is a fast-response variable that always damps relatively rapidly to a stable equilibrium. Whether or not these are valid results, the possibility exhibited by the EBMs for extreme sensitivity of the ice line to variations to external forcing (e.g., $\mathcal{S}$) and for a "runaway" ice coverage that is self-maintaining due to the ice-albedo effect has served to alert the climate research community to the potential sensitivity of climate to all kinds of external changes, both natural and human induced.

The question remains, however, whether the EBM results are just the properties of an oversimplified model (note all the approximations listed in Section 7.4), or whether a more complete model such as a GCM will contain enough negative feedback to invalidate these results as properties of the real climate system. In the next section we discuss this issue in the broader context of the transitivity properties of dynamical systems (see Section 6.5).

7.8 TRANSITIVITY PROPERTIES OF THE ATMOSPHERIC AND SURFACE CLIMATIC STATE: INFERENCES FROM A GCM

A fundamental issue in climate dynamics concerns the degree to which the climate depends on initial conditions. Using the terminology of Lorenz (1968) we can distinguish between a *transitive* system in which the dynamics imply a unique, stable, climate (de-

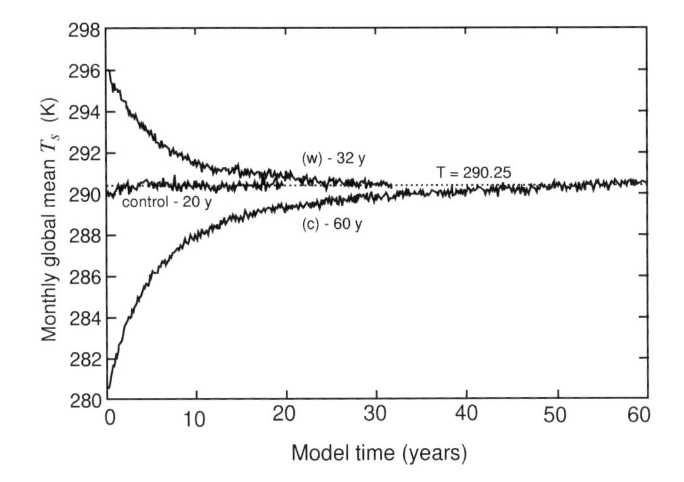

Figure 7-9 Approach to the same equilibrium of the global-mean surface temperature (dotted line) computed from a nonseasonally forced GCM (the NCAR CCM1) in which Earth's eccentricity is zero, the obliquity is essentially zero (0.1), and the carbon dioxide concentration is 330 ppm, starting from warm (W), cold (C), and present control initial conditions. After Saltzman *et al.* (1997).

fined by an infinitely long set of statistics) that is independent of initial conditions, and an *intransitive* system in which more than one permanent stable climate can exist, depending on the initial conditions. Such intransitivity may seem implausible for Earth's climate, if only due to changes in both external forcing and stochastic perturbations that could drive the system from one stable climate regime to another over an infinite time period. This situation, in which the system may reside for extended periods near each of the stable climatic states, with intermittent transitions between them, is a more likely possibility and has been termed "almost-intransitive" by Lorenz.

According to the EBM results, one would conclude that intransitivity and almost-intransitivity are indeed features of the climate system. However, as noted, these models are grossly oversimplified. Although the GCM is also a simplification of the true dynamics of the atmosphere, it is certainly more physically complete, and its transitivity properties should provide a first-order check on the EBM results. Note that at this point we are considering only the atmosphere and its "veneer" lower surface boundary climatic state, given the prescribed states of the large ice sheets, deep ocean (e.g., thermohaline circulation), and geochemical constituents (e.g., CO_2).

In this regard, it is relevant that up to this time there have been no examples of atmospheric GCM solutions in which anything but a single equilibrium climate has been obtained. Of more special relevance is the study by Saltzman *et al.* (1997) designed specifically to test the sensitivity of one GCM, the NCAR CCM1, to an extremely warm, high water vapor, low ice coverage initial state versus an extremely cold, low water vapor, high ice coverage initial state. In both cases, the geographic and spatially averaged distributions of all of the climatic variables converged to the same state. An example of this convergence for the global mean surface temperature $\widetilde{T}_S$ for fixed equinoctial mean conditions (that exclude seasonal variations) is shown in Fig. 7-9.

Note that this figure implies a response time of about 50 y. Further tests with the CCM1 corroborate these transitivity properties (Oglesby *et al.*, 1997).

Returning to the question posed at the beginning of this chapter, we conclude that, in spite of the EBM studies, it is still a valid working hypothesis to assume that the atmosphere and surficial state (e.g., snow and sea-ice coverage) are in the category of *fast-response* variables possessing the property of transitivity. That is, the behavior in these domains are "slaved" to the truly slow-response variables such as glacial ice mass, deep ocean state, and greenhouse gas concentrations (e.g., CO_2), whose response times are greater than 100 y, as well as to slow external forcing (e.g., Earth-orbital variations).

In Sections 7.9 and 7.10 we discuss the use of the GCM with two goals; (1) to establish by systematic experiments the nature of this "slaved" relationship and (2) to simulate, as "snapshots," the atmospheric and surficial climate that accompanied inferred paleoboundary conditions.

7.9 CLOSURE RELATIONSHIPS BASED ON GCM SENSITIVITY EXPERIMENTS

On the most fundamental level it would be desirable to use the climatic equilibrium models in an asynchronously coupled manner to determine the net fluxes of mass (e.g., water, ice, carbon) and energy that drive the slow evolution of the deep ocean, ice sheets, and geochemical inventories. However, because these fluxes cannot be calculated to the required accuracy it is desirable to approach the problem initially from a more "global" viewpoint by using the GCM to establish relationships of the form given by Eq. (5.12). Thus, for example, the fluxes involved in the slow net rates of ice accumulation may be representable (with the assignment of at least one free parameter) in terms of a mean surface temperature T_S, which, in turn, is a function of the distribution of external forcing and all of the relevant slow-response variables (to be discussed below).

As the simplest approximation, fast-response variables such as T_S can be described in terms of the departures of the slow-response variables from a reference state that represents an ultra-long-term mean condition, i.e., the tectonic state denoted by $\widehat{(\;)}$ (Section 5.3). Thus, with reference to Section 5.3,

$$\begin{aligned}(X, Y, F) &= \left(\widehat{X}, \widehat{Y}, \widehat{F}\right) + \Delta(X, Y, Z) \\ &= \left(\widehat{X}(0), \widehat{Y}(0), \widehat{F}(0)\right) + \Delta_t(X, Y, Z) + \Delta(X, Y, Z)\end{aligned} \tag{7.29}$$

where as defined in Section 5.4, $X\ (= \{x_j^{(F)}\})$, $Y\ (= \{x_j^{(S)}\})$, and F are the sets of fast-response variables, slow-response variables, and forcing functions, respectively, and $[\widehat{X}(0), \widehat{Y}(0), \widehat{F}(0)]$ are present tectonic-mean values assumed to be given from Late Pleistocene data as an "initial condition." In particular, from Eqs. (5.5) and (5.12),

$$\Delta X = \left[\{g_i(y)\} - \{g_i(\widehat{y})\}\right] \tag{7.30}$$

where $y = (Y, F)$.

If g_i is a linear function of Y or F, then

$$\Delta X \approx \sum_j k_j^{(i)} (y_j - \widehat{y}_j) \tag{7.31}$$

where $k_j^{(i)} = [\partial\{g_i(y)\}/\partial y_j]$ are linear "sensitivity functions" to be determined from systematic GCM experiments.

The sensitivity functions, $k_j^{(i)}$, are usually taken as constants determined with respect to the present (control) climatic state, but, more generally, will be a function of the tectonic-mean state itself if $x_j^{(F)}$ is a nonlinear function of $y = (Y, F)$ (as in the case of the temperature response to CO_2 discussed below). Assuming that unique response functions do indeed exist, as argued in the last section, once established they can, in principle, replace the need for making further *ad hoc* GCM runs for simulations of the fast-response climate. Thus, Eqs. (7.31) and (7.32) can provide the means for introducing the "fast-response physics" embodied in equilibrium climate models (e.g., GCMs) into a low-order, global theory of the nonequilibrium slow-response paleoclimatic changes (Saltzman, 1987a,b; Saltzman and Maasch, 1988, 1990).

Only a few systematic studies based on GCMs have yet been made to determine the variations of the geographic fields of the climatic variables $X(\lambda, \phi, z)$ as a function of the change in any slow-response variable or external forcing over a broad range of possible values. The more typical attempts in this regard result in maps, or zonal averages, of climatic states calculated from GCMs to correspond to (1) doubled or quadrupled atmospheric CO_2 concentration compared with those for the present concentration [see review by Schlesinger and Mitchell (1987)], (2) the ice mass conditions that prevailed at the last glacial maximum (Williams *et al.*, 1974; Saltzman and Verneker, 1975; Gates, 1976; Manabe and Broccoli, 1985; Broccoli and Manabe, 1987), and (3) altered values of incoming solar radiation due to Earth-orbital (Milankovitch) variations (e.g., North *et al.*, 1983; Kutzbach and Guetter, 1986). In more relevant studies, however, the response of all the weather variables governed by a GCM have been determined for systematic changes embracing the broad range experienced in climatic history for atmospheric CO_2 (Oglesby and Saltzman, 1990a, 1992; Kothavala *et al.*, 1999), solar radiation (Marshall *et al.*, 1994; Felzer *et al.*, 1995), Earth-orbital changes (e.g., Gallimore and Kutzbach, 1995; Phillips and Held, 1994), Northern Hemisphere ice-sheet coverage (Felzer *et al.*, 1996), and the Earth's rotation rate (Hunt, 1979; Jenkins, 1996). General reviews are given by Felzer *et al.* (1998, 1999). It is of interest that in many of these studies the changes that occur are strongly related to the induced changes in cloud cover, which is one of the least accurate components of the GCM.

7.9.1 Surface Temperature Sensitivity

As noted, the variations of the surface temperature of Earth are of particular interest for the evolution of global climate and ice mass. By asymptotically equilibrated numerical experiments with GCMs, subject to systematic changes in both the slow-response variables listed in Section 5.4 and prescribed external forcing, we can, in principle, establish a diagnostic equation of the form given by Eq. (5.12) for mean surface temperature (T_S) for any season or geographical region. Although all of the slow-response variables that form the set Y have three-dimensional distributions, for simplicity in this section and to form a foundation for the low-order dynamical systems approach to be developed in later chapters, we shall emphasize global measures of these slow-response variables. These are the ice-sheet mass $I = \sum \Psi_j$ [where Ψ_j $(= \iint \rho_i h_{Ij}(\lambda, \varphi) a^2 \cos\varphi \, d\lambda \, d\varphi)$ is the mass of the j, the individual ice sheet, and $h_{Ij}(\lambda, \varphi)$ is its geographical thickness distribution], the mean bedrock depression beneath the individual ice sheets (D), the greenhouse gas composition of the atmosphere (particularly the carbon dioxide concentration (μ)), and the mean temperature of the whole ocean (θ) related strongly to the depth and extent of the main thermocline and to the thermohaline circulation. Other possibly important slow-response variables related to these are regolith amount in ice sheet areas, and carbon reserves near the surface.

As noted in Sections 1.3 and 5.4.1, there are several sources of external forcing on the time scale of the major variability, exhibited in Fig. 1-3. The two main categories of such forcing are (1) astronomical/cosmic variations associated with known radiative changes $R(\varphi, t)$ $(\equiv H_T^{(1)\downarrow})$ due to both Earth-orbital variations and possible changes in solar luminosity $\mathcal{S}$, possible changes in the rate of rotation of Earth Ω, and changes in the mass influx of cosmic material at the outer limit of Earth's atmosphere $U^\downarrow$, and (2) tectonically induced geological changes in the distribution and topography of the continents and ocean basins $h(\lambda, \varphi)$, in the mass influx of solid and gaseous material to the atmosphere due to all forms of volcanism and venting $V^\uparrow$, in weatherability of rocks $\mathcal{W}$, and in the magnitude of the upward geothermal flux $G^\uparrow$. Along with the direct effect on surface temperature of changed continent–ocean distribution $h(\lambda, \varphi)$, some of these tectonic forcings have significant indirect effects due to dust load and greenhouse gas changes, which can significantly affect the radiation budget.

In line with Eqs. (7.29) and (7.30), it follows that, to first order, the climatic-mean temperature at any location can be expressed as a function of global measures of all these slowly varying quantities $y = (Y, F) = [I, D, \mu, \theta;\, R(\mathcal{S}), \Omega, U^\downarrow, h, G^\uparrow, V^\uparrow, \mathcal{W}]$, expressible in the form

$$\begin{aligned} T(\lambda, \varphi, z) &= \widehat{T}(\lambda, \varphi, z) + \Delta T(\lambda, \varphi, z; Y, F) \\ &= \widehat{T}(0) + \Delta_t \widehat{T} + \Delta T \end{aligned} \tag{7.32}$$

where $Y = \{x_i^{(s)}\}$, and $\Delta T = \sum_y k_y(\lambda, \varphi, z)\Delta y_i$. If $\Delta T(y)$ is a linear function, then $k_y(\lambda, \varphi, z) = (\partial \widehat{T}/\partial y) =$ a constant.

In view of the uncertainties in the chronology and magnitude of tectonic forcing (including volcanic aerosol loading of the atmosphere) on the near-100-ky scale of the main ice-age cycles, we shall here neglect $\Delta(h, V^\uparrow)$, as well as $\Delta G^\uparrow$, and $\Delta\Omega$, but shall formally include the longer term effects of all external forcings on the tectonic mean state.

As a first important result of GCM experiments it is found that for values of $\widehat{y} = \widehat{\mu}, \widehat{I}, \widehat{D}, \widehat{\theta}, \widehat{R}, h$, and Ω appropriate for the present tectonic mean state i.e., $\widehat{y}(0)$ (which from geologic evidence is apparently not too different from the state that prevailed before the onset of the Plio-Pleistocene ice epoch about 2.5 Ma), the values of $T_S(\varphi)$ obtained range from a maximum of about 300K in the tropics to about 250K near the poles. As noted in the introduction, these values embrace the possibility for water to exist in all its three forms near the surface and admit the presence of suspended liquid and solid states in the form of atmospheric clouds. It is not unexpected that a system so poised can exhibit wide fluctuations in the ratio of ice to liquid water (ocean), the variations in the amount of vapor being small due to its steady control by the processes of saturation, condensation, and precipitation imposed by the above temperatures.

The same type of GCM experiments that give the above results for present values of the slow forcing and slow-response variables can be performed for systematic changes in these variables to give estimates of the sensitivity (or response) functions $k_y = \partial T/\partial y$. At this point we shall assume that for all variables except CO_2 the response is linear (k_y = a constant). The thermal response function for carbon dioxide has been found to be logarithmic rather than linear (Oglesby and Saltzman, 1990a), i.e.,

$$\Delta T(\mu) = \mathcal{B} \ln\left(\frac{\mu}{\widehat{\mu}}\right) \tag{7.33}$$

where $\mathcal{B}$ = a constant. For the linear approximation valid in the neighborhood of $\widehat{\mu}$,

$$\Delta T(\mu) \approx k_\mu(\mu - \widehat{\mu})$$

where

$$k_\mu = \left(\widehat{\frac{\partial T_S}{\partial \mu}}\right) = \frac{\mathcal{B}}{\widehat{\mu}}$$

An example of the CO_2 results for the zonal average value of surface temperature at 51°N in summer is shown in Fig. 7-10. The logarithmic shape of this curve is similar

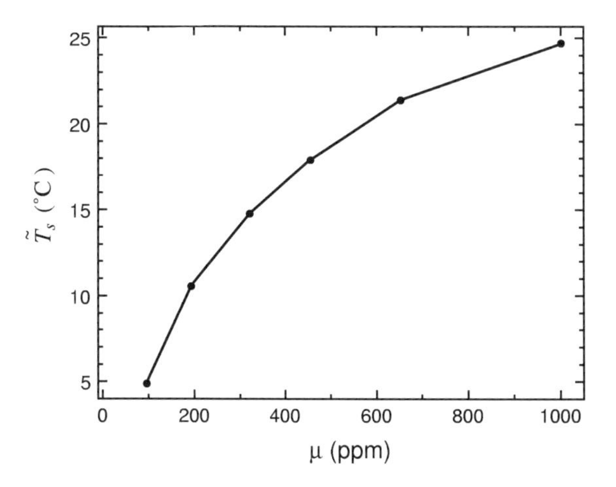

Figure 7-10 Dependence of the equilibrium Northern Hemisphere, high-latitude, summer surface temperature ($\widetilde{T}_S$) on the atmospheric CO_2 concentration (μ) for the NCAR CCM1 general circulation model. After Oglesby and Saltzman (1990a).

to that derived for an idealized Earth–ocean system by Manabe and Bryan (1985), to estimates given by Budyko (1982), as well as to the earliest estimate of this sensitivity curve by Callendar (1938). It is significant for paleoclimatic studies that this nonlinear curve shows higher sensitivity of the climate system to CO_2 changes for low values of CO_2 than for higher values. The geographic distribution of the sensitivity of T_S to CO_2 for summer and winter is shown in Fig. 7-11. Although the sensitivities are lower in the tropics, the weather variability is also much lower there, leading to a higher signal-to-noise ratio in the low latitudes than in high latitudes (Syktus *et al.*, 1997).

On the other hand, the sensitivity function for insolation appears to be more linear (Marshall *et al.*, 1994), implying a near constant value of k_R (see Fig. 7-12). If we assume the responses of T_S to I, and θ, are also linear to first order, we can specialize Eq. (7.32) in the more explicit form

$$T = \widehat{T} + \mathcal{B}\ln(\mu/\widehat{\mu}) + k_I\Delta I + k_\theta\Delta\theta + k_R\Delta R + k_U\Delta U + k_h\Delta h + k_G\Delta G^{\uparrow} + k_V\Delta V + k_{\mathcal{W}}\Delta\mathcal{W} \tag{7.34}$$

Typical values for some of these global sensitivity coefficients are $\mathcal{B} \approx 6°C$ for summer (Oglesby and Saltzman, 1990a), $k_I \approx -10^{-19}$°C/kg (Broccoli and Manabe, 1987; Felzer *et al.*, 1996, 1999), $k_\theta \approx 0.5$ (Oglesby and Saltzman, 1990a,b), and $k_R \approx 0.1$°C/W m^{-2} for winter (Marshall *et al.*, 1994) (see also Syktus *et al.*, 1994). These sensitivities should be even larger if the positive feedbacks between vegetative cover and temperature are taken into account along with the water vapor, snow, and sea-ice feedbacks that are included in the GCMs on which these calculations are based. Given the range of values exhibited by μ, θ, I, and R over the past few million years, all of

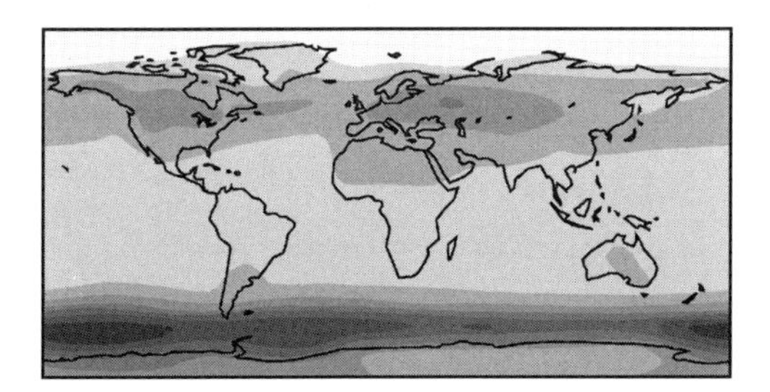

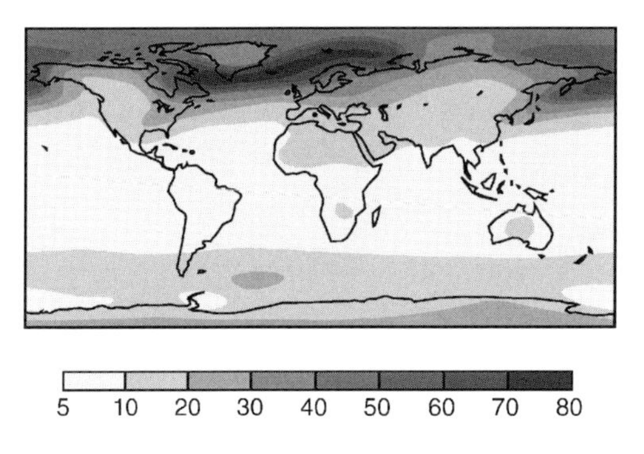

Figure 7-11 Global maps of the log sensitivity coefficient ($d\widetilde{T}_S/d\log\mu$) for the response of surface temperature to CO_2 forcing expressed as the temperature rise (°C) for a rise of CO_2 from 100 to 3000 ppm for two seasons (June–July–August, top; December–January–February, bottom) for CCM1 with a static mixed-layer ocean. After Syktus *et al.* (1997).

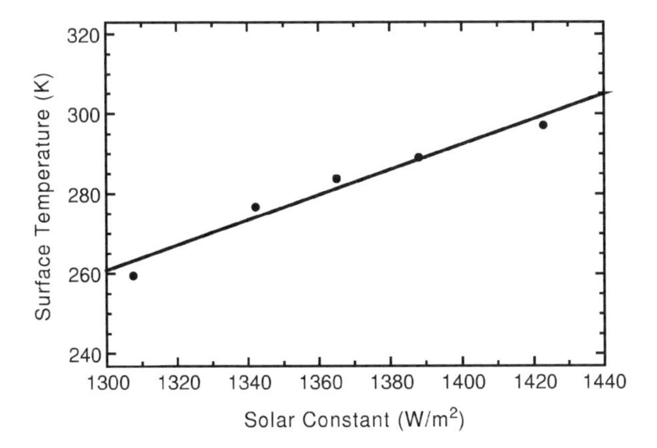

Figure 7-12 Equilibrium globally averaged surface temperature as a function of the solar constant, as determined from the CCM1 with a linear fit yielding a correlation of 0.960. After Marshall *et al.* (1994).

these slow-response variables should be of equal importance in influencing the thermal state of Earth's surface and hence the amount of ice. The observed covariability of CO_2 and temperature over this period was illustrated in Figs. 3-5 and 3-12.

We note here that in this chapter we have focused mainly on the atmospheric side of Earth's surface. In Chapter 11, we will discuss in a little greater detail the dynamics of the surface layer of the ocean, which can also be considered part of the *fast-response* system that can be compatibly coupled with the atmosphere on roughly the same time scale.

7.10 FORMAL FEEDBACK ANALYSIS OF THE FAST-RESPONSE EQUILIBRIUM STATE

The sensitivity analysis discussed in Section 6.5.1 that was applied to the prototype two-variable system can be generalized to apply to all the fast-response variables including the surface temperature T_S. Such an analysis can reveal not only the sensitivity to prescribed external forcing (e.g., insolation) and slow-response states (e.g., CO_2, and ice-sheet distribution), but can also provide the formalism for assessing what part of the response is due to the pure effect of these prescribed distributions and what part is due to the various internal feedbacks resulting from concomitant changes in other fast-response variables (e.g., water vapor, clouds, surficial ice albedo).

In general, we are interested in the change of an atmospheric or surficial climatic mean variable $x^{(F)}$ (e.g., T_S) near a particular reference value that we can identify with the tectonic-mean value, $\widehat{x}_i^{(F)}$, resulting from a change in any of the following prescribed parameters:

1. External forcing, F_i (e.g., the solar constant)
2. A slow-response variable, $x_i^{(S)}$ (e.g., μ, I, θ)
3. A parameterization coefficient, α_i (e.g., a diffusion coefficient)

Denoting any of the above parameters by $\lambda_i = \{F_i, x_i^{(S)}, \alpha_i\}$, and for notational simplicity setting $x_i^{(F)} \equiv \psi_i$, we can represent the total departure of ψ_i from a reference value $\widehat{x_i}$, due to all possible changes in λ_i, by the expansion

$$\Delta\psi_i \equiv (\psi_i - \widehat{\psi_i}) = \sum_j \left(\widehat{\frac{\delta\psi_i}{\delta\lambda_j}}\right)\Delta\lambda_j + [\text{higher order terms}] \tag{7.35}$$

where the use of δ as a symbol for a differential denotes changes calculated by taking all of the principal feedbacks internal to a GCM into account.

Separating for particular attention a specific parameter λ_i and neglecting the higher order terms, we have

$$\Delta\psi_i \approx \left(\widehat{\frac{\delta\psi_i}{\delta\lambda_i}}\right)\Delta\lambda_i + \sum_{j\neq i}\left(\widehat{\frac{\delta\psi_i}{\delta\lambda_j}}\right)\Delta\lambda_j \tag{7.36}$$

or, dividing by $\Delta\lambda_i$,

$$\left(\widehat{\frac{\Delta\psi_i}{\Delta\lambda_i}}\right) \approx \left(\widehat{\frac{\delta\psi_i}{\delta\lambda_i}}\right) + \sum_{j\neq i}\left(\widehat{\frac{\delta\psi_i}{\delta\lambda_j}}\right)\frac{\Delta\lambda_j}{\Delta\lambda_i} \tag{7.37}$$

which is of the same form as Eq. (6.23). As was discussed in Section 6.5.1, the left-hand side represents what we have called the *total sensitivity* of ψ_i to λ_i near the reference equilibrium state $\widehat{\psi_i}$; the term $(\delta\psi_i/\delta\lambda_i)$ is the partial sensitivity of ψ_i to λ_i holding all other λ_j constant (i.e., assuming λ_i is independent of all other λ_j), and the last term represents the effects of the interdependence of all λ_j. If, as is often done, we neglect this last term, we isolate for consideration only the partial sensitivity of ψ_i to a particular λ_j.

Let us now particularize the analysis by considering the partial sensitivity of the equilibrium surface temperature ($\psi_i = T_S$) to different prescribed values of carbon dioxide (i.e., $\lambda_i = \mu$), keeping the ice-sheets (I) and deep ocean state (θ) and other slow-response variables fixed, as was discussed in Section 7.9.1 (see, e.g., any of the reference points shown in Fig. 7-10). Our concern now is to separate the part of $(\delta T_S/\delta\mu)$ resulting from the direct effect of μ on T_S that would occur if there were no internal feedbacks due to the other fast-response variables (e.g., ψ_j representing water vapor, clouds, surface ice-albedo effects) from that part due to the individual feedbacks arising from all such variables that are internal to a GCM.

Thus, with the understanding that all derivatives are evaluated at a given reference point on the curve shown in Fig. 7-10, we now further expand the partial sensitivity $(\delta T_S/\delta\mu)$ in the form

$$\frac{\delta T_S}{\delta\mu} = \left(\frac{\partial T_S}{\partial\mu}\right) + \sum_j\left(\frac{\partial T_S}{\partial\psi_j}\right)\cdot\frac{\delta\psi_j}{\delta\mu} + [\text{higher order terms}] \tag{7.38}$$

where the partial derivatives denote changes holding all other variables constant, therefore requiring a much more specialized model than a GCM that actively includes variations of these other variables. The sensitivity, $\delta T_S/\delta\mu$, is the slope of the curve shown in Fig. 7-10 as determined from GCM experiments. The first term on the right-hand side $(\partial T_S/\partial\mu)$ is the sensitivity T_S would have to μ if there were no internal feedbacks due to simultaneous changes in other ψ_j; in principle, this pure response to CO_2 can be estimated from a radiation model (e.g., the radiation module of a full GCM). The factors $(\partial T_S/\partial\psi_j)$ in the second term must also be determined from specialized models (e.g., a radiation model in which factors such as water vapor, clouds, lapse rate, and surface albedo are varied independently). On the other hand the factors $(\delta\psi_j/\delta\mu)$ representing the full sensitivities of these other internal variables ψ_j to changes in CO_2 can be evaluated from the complete set of physical interactions embodied in a GCM. These latter quantities $(\delta\psi_j/\partial\mu)$ can each be further expanded in the form

$$\frac{\delta\psi_j}{\delta\mu} = \frac{\delta\psi_j}{\delta T_S}\frac{\delta T_S}{\delta\mu} + \sum_{k\neq j}\frac{\delta\psi_j}{\delta\psi_k}\frac{\delta\psi_k}{\delta\mu} \tag{7.39}$$

where the last term represents the effects of the interdependence of all the fast-response variables being considered. Making the usual assumption (Schlesinger, 1988) that the individual feedback variables are independent i.e., neglecting the last term in Eq. (7.39), and then substituting, Eq. (7.39) into Eq. (7.38), again neglecting the higher order terms, we obtain

$$\frac{\delta T_S}{\delta\mu} = \left(\frac{\partial T_S}{\partial\mu}\right)\left(\frac{1}{1-\sum_j f_j}\right) \tag{7.40}$$

where $f_j = (\partial T_S/\partial\psi_j \cdot \delta\psi_j/\delta T_S)$ are the so-called *feedback factors* (Schlesinger, 1988) and

$$G \equiv \frac{1}{(1-\sum_j f_j)}$$

is the "gain" representing the factor by which the pure radiative response $(\partial T_S/\partial\mu)$ without feedback is amplified (or reduced if $G < 1$) due to the internal feedbacks.

From analyses of this type it has been estimated that the largest feedback is due to water vapor with a magnitude $f(\text{water vapor}) \approx 0.4$, with lesser though still positive values for cloud and ice cover albedo of magnitude $f(\text{cloud}) \approx 0.2$ and $f(\text{albedo}) \approx 0.1$ (Houghton *et al.*, 1990). It follows from Eq. (7.40) that the purely radiative response of surface temperature to CO_2 is amplified by a factor of about three. A pictorialization of this feedback system, similar to one given by Peixoto and Oort (1992), is given in Fig. 7-13. Owing to the deficiencies of GCMs, there is much debate, however, concerning the true values of these feedback factors, even concerning the signs, pointing to the continuing need for improvement in the representation of these feedback mechanisms in GCMs.

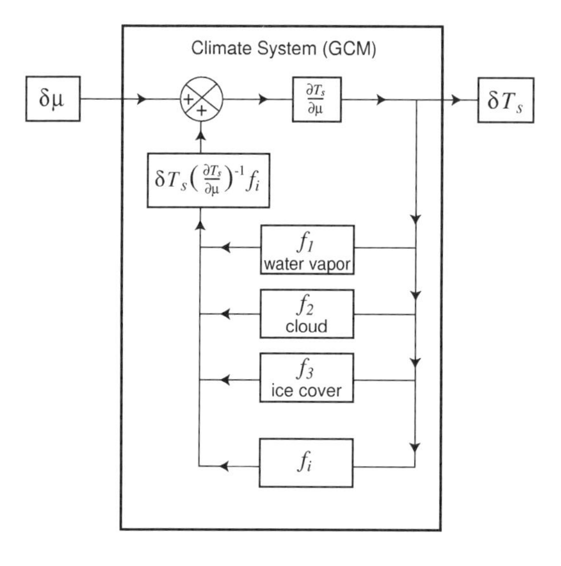

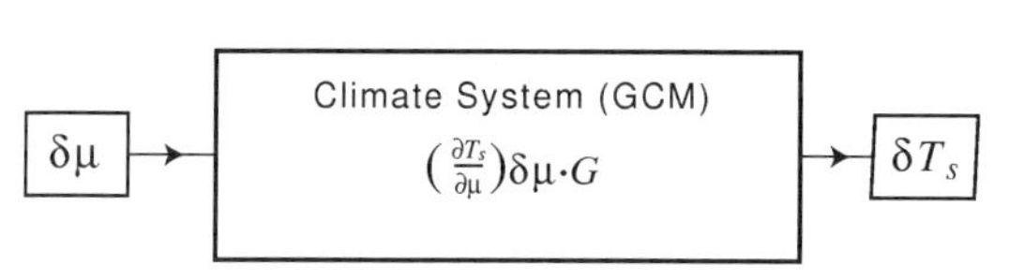

Figure 7-13 Feedback (input–output) loop representation of the response of surface temperature to a change of CO_2 ($\delta\mu$) due to interactions within the internal climate system involving water vapor, cloud, and ice cover. See text for definitions of symbols.

As a more idealized example, in the case of the EBM represented by Eqs. (7.14b) and (7.15), neglecting stochastic forcing $[\omega(T_S) = 0]$, the equilibrium is given by

$$\widetilde{H}_T^{(1)\downarrow} = H_T^{(2)\uparrow}$$

where $\widetilde{H}_T^{(1)\downarrow} = (\mathcal{S}/4)[1 - \alpha_P(T_S)]$ and $H_T^{(2)\uparrow} = A + BT_S$. The implied sensitivity of T_S to the value of the solar constant, $\mathcal{S}$, assuming α_P, B, and $\mathcal{S}$ are independent, is given by [cf. Eq. (7.24)]

$$\frac{\delta T_S}{\delta \mathcal{S}} = \frac{(1 - \alpha_P)}{B + (\mathcal{S}/4)(\delta\alpha_P/\delta T_S)} \tag{7.41}$$

For the prescribed dependence of α_P on T_S given by Eq. (7.22) this sensitivity is simply the slope at any point of the equilibrium curve shown in Fig. 7-3. From Eq. (7.41) it is clear that the sensitivity is increased for a reduced dependence of $H_T^{(2)\uparrow}$ on T_S (B small) and for an increased negative dependence of the albedo α_P on T_S (note that $\delta\alpha_P/\delta T_S < 0$). Using the value given in Section 7.5, for $B = 2.09\ \mathrm{W\,m^{-2}\,°C^{-1}}$ (an empirical value that takes into account some of the postive feedback greenhouse properties of the present atmosphere) and a fixed value of $\alpha_P = 0.31$, the sensitivity function, Eq. (6.24), is given by $\beta = 10^{-2}\mathcal{S}$, $\delta T_S/\delta\mathcal{S} \approx 1.12\mathrm{K}$, whereas if one assumed "no atmosphere" and hence no positive feedback, as in the pure

Stefan–Boltzman calculation based on Eq. (7.18), the sensitivity function would be $\beta \approx 0.63$K.

7.11 PALEOCLIMATIC SIMULATIONS

As highly approximated forms of the conservation laws governing the continuum behavior of the atmosphere, oceans, and land surface, equilibrium climate models for the fast-response variables (e.g., GCMs, SDMs, EBMs) are, in essence, hypotheses to be tested continually for their simulation capabilities compared to observations. As noted, these observations include not only the present climatic state, but also past paleoclimatic states as inferred from geologic evidence, which represent a significant part of the ground truth against which to test the models.

A list of early examples of simulation studies aimed at determining the atmospheric and surface states corresponding to the various altered boundary conditions believed to have prevailed in the past was given by Saltzman (1990). Summaries of more recent GCM studies are given by Crowley (1994), Barron and Fawcett (1995), Sloan and Rea (1995), Otto-Bliesner (1996), and Broccoli and Marciniak (1996). These studies illustrate the manner in which the full climatic distributions can be deduced once the slow-response features (e.g., CO_2) and slowly changing boundary conditions (e.g., continent–ocean structure) are specified. At the same time, because carbon dioxide levels before the Late Pleistocene are not known with much confidence, these studies are often used in an inverse way to determine the CO_2 levels needed to account for the probably better known temperature and ice-cover states.

Snapshot conditions at several significant times in Earth history (see Figs. 1-3, 1-4, 3-1, 3-2, and 3-3) have received special attention. These include the three major cold, glacial episodes, i.e., during the following times:

1. The Ordovician (~440 Ma), when ice sheets appeared on the south polar continent, Gondwana (e.g., Crowley and Baum, 1995).
2. The Permo-Carboniferous (~300 Ma), when ice sheets appeared on the supercontinent, Pangea (Crowley and Baum, 1992; Crowley, 1994; Kutzbach, 1994; Barron and Fawcett, 1995; Otto-Bliesner, 1996).
3. The Late Cenozoic, for which a particularly well-documented set of boundary conditions exists for the last glacial maximum at about 20 ka (e.g., Gates, 1976).

Also receiving special attention are four major warm, minimally-glaciated, episodes, i.e., during the following times:

1. The Late Cretaceous (~65 Ma) (e.g., Bush and Philander, 1997).
2. The Eocene (55 Ma), representing the warmest time period during the Cenozoic (e.g., Barron, 1987; Sloan and Rea, 1995; Sloan *et al.*, 1995; Stott, 1992).
3. The Early Pliocene (5–3 Ma) (e.g., Crowley, 1991).
4. The Holocene period (~6 ka), following the last glacial maximum (e.g., Liao *et al.*, 1994; Tempo, 1996; Hall and Valdes, 1997; Hewitt and Mitchell, 1997).

The Eocene results are of interest because they reveal a discrepancy between the GCM solutions and geologic evidence of the surface conditions. In particular, even for highly elevated prescribed values of CO_2, the temperatures deduced from a GCM are too cold in high latitudes and too warm in low latitudes, leading to a stronger meridional temperature gradient than is inferred from proxy data (Sloan and Rea, 1995). This discrepancy raises the possibility either for (1) a much different mode of operation of the fast climate system than is represented by the physics presently included in GCMs, (2) some unknown external forcing of the system, or (3) some glaring misinterpretations of proxy climate data. As one possible explanation it may be that the extremely warm deep ocean temperatures θ at these times (see Fig. 3-3) may be the critical determinant of T_S (Oglesby and Saltzman, 1990b), placing the burden on an explanation of this warm deep ocean temperature as a dynamical part of the slow-response system (see Chapter 11). Further discussion of these ultra-long-term, tectonically driven changes in the atmosphere and surface states is given in Chapter 13.

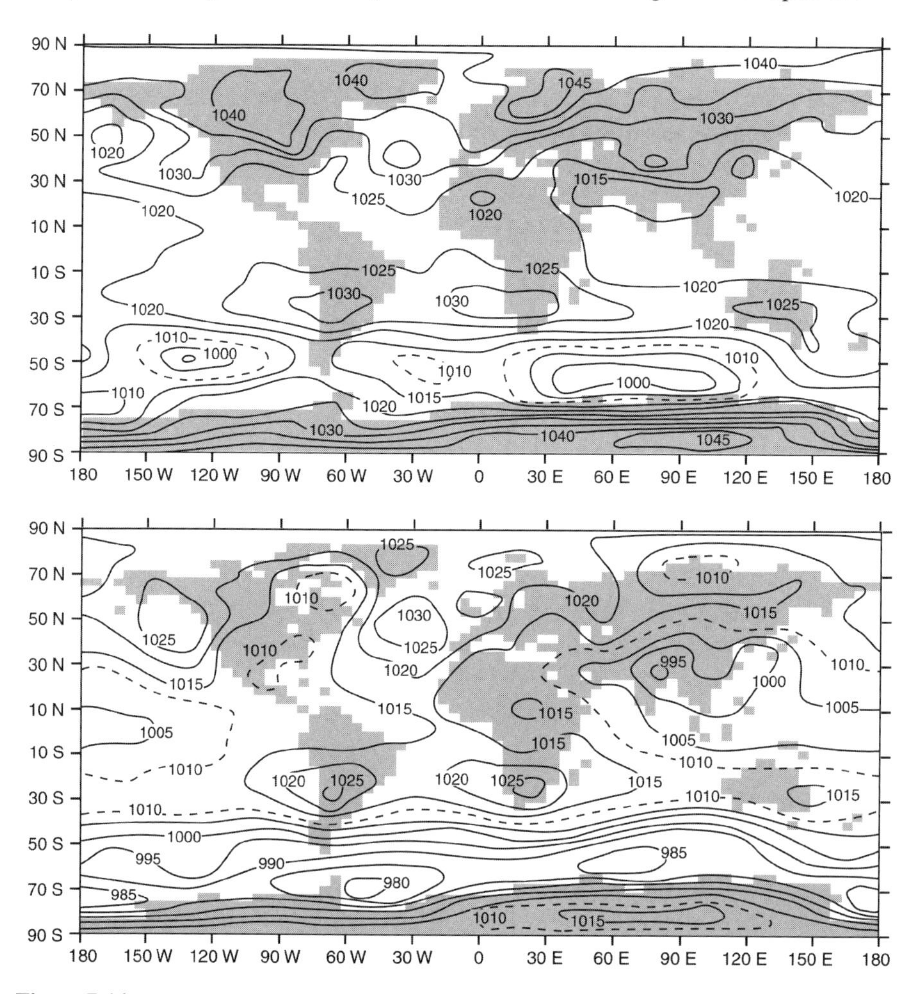

Figure 7-14 July sea level pressure (mb) deduced from a GCM for surface boundary conditions prevailing at 18 ka (top) compared with present boundary conditions (bottom). After Gates (1976).

As noted, the best documented set of CO_2, ice coverage, and geographic boundary conditions for a glacial state is the CLIMAP and Vostok reconstructions of the prevailing state at 20 ka (see Figs. 3-5 and 3-7). Many studies have been made to determine the atmospheric response to the sea surface temperatures and ice conditions that prevailed about 20 ka in comparison with corresponding solutions for the present boundary conditions, which serve as a control (Williams *et al.*, 1974; Gates, 1976; Manabe and Hahn, 1977). The most general results obtained in all these studies are the expected ice-age reductions in atmospheric temperature and in water vapor content, and in the general intensity of the hydrologic cycle, including precipitation. In Fig. 7-14, we have reproduced a sample solution obtained by Gates (1976) for one of many climatic variations, i.e., for the global sea-level pressure distribution. Similar distributions are obtained by Manabe and Hahn (1977) based on the same CLIMAP July boundary conditions. The agreement of the results using the two different GCMs is generally good, but there are many differences in the details.

Another interesting early example of the application of a GCM to paleoclimatic modeling is the study of Kutzbach (1981). Using the low-resolution model of Otto-Bliesner *et al.* (1982), it is demonstrated that an enhanced monsoonal circulation forced by the unusually large seasonal radiation variations that prevailed 9000 BP, due to Earth-orbital changes, can explain the observed paleoclimatic evidence for stronger rainfall in Africa, Arabia, and India at that time. A more general example of a GCM simulation for the mid-Holocene was made by Hall and Valdes (1997).

8

THE SLOW-RESPONSE "CONTROL" VARIABLES

An Overview

Whereas studies of the equilibrium response of the combined atmosphere and surface boundary layer to both external forcing and *prescribed* slow-response variables might be able to give an adequate account of the climatic state of the fast-response domains, they cannot account for the evolution of the slow-response variables themselves. To a large degree it is these slowly varying domains that provide the main signals of the longer term climatic change of our planet. As noted, because of their relatively slow response times we must deal with a time-dependent, or *prognostic*, system to determine their variations. In turn, this prognostic system must be coupled with a diagnostic system of the type described in the previous section to account for the atmosphere and surface boundary layers.

In Fig. 8-1 we portray schematically, in the form of a cross-section along a high Northern Hemisphere latitude, most of the global mean slow-response variables that have been considered up to now in attempts to model long-term climatic change. Included in this figure is the atmospheric variable μ representing the mean CO_2 concentration; the cryospheric variables representing the mean ice sheet mass Ψ, ice thickness h_I, basal temperature T_B and liquid water depth W_B, bedrock depression δ_B, and basal soft sediment (regolith) thickness h_R, and the mean equilibrium height of a continent above sea level when the continent is not depressed by an ice load Z; and the

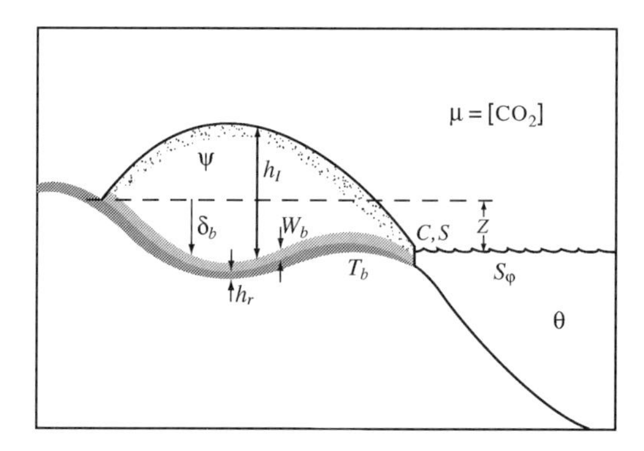

Figure 8-1 Schematic pictorialization of the main components of the slow-response climatic system.

oceanic variables θ representing the mean ocean temperature (which is assumed to be closely related to the main thermocline structure) and poleward salinity gradient S_φ, both of which are diagnostically related to the thermohaline circulation. We can consider the deformable regolith material (e.g., soil, peat) above the bedrock to be another slow-response variable, the form and composition of which vary slowly over time, especially in conjunction with ice-sheet growth and decay affecting the ice and carbon budgets (e.g., Clark and Pollard, 1998; Franzén, 1994). In the following subsections we define and describe briefly each of these slow-response variables, and present some further proxy evidence regarding their variability and covariability, supplementing that given in Chapters 1–3. We shall defer until Chapters 9, 10, and 11 the discussion of dynamical equations for these variables.

8.1 THE ICE SHEETS

8.1.1 Key Variables

At any point on Earth's surface we define the local ice thickness to be $h_I(\lambda, \varphi, t)$ implying that the local ice mass per unit area is $(\rho_I h_I)$, where ρ_I is the density of ice ($917\,\mathrm{kg\,m^{-3}}$). The mass of the jth ice sheet, Ψ_j, is therefore given by

$$\Psi_j = \iint \rho_I h_I(\lambda, \varphi) a^2 \cos\varphi \, d\lambda \, d\varphi \tag{8.1}$$

where the surface integral is taken over the total area occupied by the ice mass, A_{Ij}, extending to its edge or coastal grounding line. Beyond this grounding line marine ice shelves may exist; for simplicity here we shall assume their relatively smaller masses are strongly related to that of their adjacent ice sheets and can be absorbed in the definition of Ψ_j. The mean (scale) thickness of the ice sheet is given by

$$H_j = (A_{Ij}\rho_I)^{-1}\Psi_j \approx c_1 \Psi_j^{1/5} \tag{8.2}$$

where c_1 is a constant determined from glaciological theory (see Section 9.2). The total ice sheet mass of the globe at any time is given by

$$I = \sum_j \Psi_j \tag{8.3}$$

at present, $I(0) = (\Psi_A + \Psi_G)$, where Ψ_A and Ψ_G are the ice masses of Antarctica and Greenland, respectively. From a Late Pleistocene perspective, if we assume as a first approximation that Ψ_A and Ψ_G were relatively constant over the past 2.5 My, we would have

$$I \approx I(0) + \sum_j \Psi_j$$

where Ψ_j now represents the mass of the additional ice sheets that varied episodically over the glacial–interglacial cycles. Thus, the variations in ice mass shown in Fig. 1-4 would be mainly due to the new ice sheets that developed over North America and Eurasia (see Fig. 3-8). Over much longer periods (e.g., the Phanerozoic) significant departures occurred in the amount and location of ice sheets on new continental areas. The mass of any ice sheet is limited by its continental area, A_c a simple constraint being $\Psi \leqslant \gamma \rho_I A_c^{5/4}$, where $\gamma = 0.9\ m^{1/2}$ (Verbitsky, 1992). The response time for an ice sheet is typically taken to be about 10 ky. Being thinner and more vulnerable, the marine ice shelves probably have response times closer to 1 ky.

Basal conditions and surging. During the lifetime of an ice sheet, slow temperature variations occur at its base due to the combined effects of advective heat transfer from the upper surface, frictional warming, and the geothermal flux from below. If the basal temperature, T_B, attains the pressure melting point of ice (Shumskiy, 1955),

$$T_M = 273.16 - 7.52 \times 10^{-8} \rho_I g h_I \quad (K) \tag{8.4}$$

melting will occur that can create a lubricating layer of water of varying depth W. This water layer and the associated saturation of deformable subglacial till and sediments (regolith) make possible surging and rapid wasting of an ice sheet, which introduces another potentially important process capable of introducing unstable behavior, namely, the *surge flux*, $\mathcal{S}_I$, defined as the rate of lateral-mass discharge of ice into an ocean.

8.1.1.1 Bedrock depression. Under the weight of the ice sheets the crustal lithosphere is isostatically depressed, along with the viscous asthenosphere below it, to a local depth $\delta_B(\lambda, \varphi, t)$. For present purposes, however, we shall consider an average depression beneath each ice sheet of mass Ψ_j, denoted by D_{Bj}. It has been estimated that a reasonable value for the response time of the bedrock is about 5 ky (Peltier and Jiang, 1996), though earlier studies have assumed values ranging up to 50 ky.

The potential role of this variable was discussed early by Ramsay (1925) and developed in more physical-mathematical forms by Birchfield *et al.* (1981), Pollard (1982), Ghil and Le Treut (1981), and Peltier (1982), for example. The net effect of bedrock depression on the normal growth of ice sheets is uncertain because of the near balance between the effects of increased temperature and increased snowfall at lowered glacier elevation; we adopt the view that bedrock depression tends to increase the inertia of ice sheets by their anchoring effect and also by allowing more mass to accumulate at the depressed elevation than would otherwise be the case. However, there also exists the possibility for a major negative feedback in the form of a catastrophic collapse of the ice sheet due to the formation of proglacial lakes or ocean inundation when the ice within the depressed region is reduced (say, by Milankovitch forcing) below a critical size (Pollard, 1983a).

8.1.2 Observations

In Fig. 1-4 we showed some estimates of the global variations of ice mass, I, over the past 5 My based on the assumption that the oceanic sedimentary core record of $\delta^{18}O$ is largely due to ice variations rather than temperature changes (Shackleton and Opdyke, 1973). Questions still remain concerning the exact chronology of these quasi-cyclic variations. If, as noted above, one further assumed that the changes in Antarctica and Greenland were relatively small (a subject of some debate), then most of the ice mass variations were likely to be in the form of massive ice sheets over North America and Eurasia. In Fig. 3-8 we showed one idealized representation of the ice sheet configuration at the peak of the last glacial maximum about 20 ka. This is the only glacial maximum of the cycles extending back for about 1 million years for which geologic evidence is good enough to attempt such a reconstruction; but even in this case there is much controversy regarding the thickness, volume, and location of the ice sheets (Peltier, 1994). As examples, the presence of major ice sheets on the Siberian shelf and in Tibet are still in contention.

8.2 GREENHOUSE GASES: CARBON DIOXIDE

The main naturally varying greenhouse gases that can influence surface temperature and hence ice amount are water vapor (H_2O), carbon dioxide (CO_2), and methane (CH_4), assuming that variations of nitrous oxide (N_2O) and chlorofluorocarbons (CFCs) are more closely related to anthropogenic rather than natural sources. Whereas the variations of water vapor, the primary atmospheric greenhouse gas, are treated as an intrinsic fast-response part of the GCM and its thermal effects are implicit in GCM solutions, the much slower response natural variations of CO_2 are too poorly understood to be included rigorously as an internal variable in a GCM. However, the response to a wide range of prescribed constant values of CO_2 have been calculated with GCMs, leading, for example, to the result that $\Delta T(\mu) = \mathcal{B}\ln(\mu/\widehat{\mu})$, described in Section 7.9.1.

For our purposes we shall deal mainly with carbon dioxide, cognizant that "equivalent CO_2" (Houghton *et al.*, 1990), defined as the concentration of CO_2 that would result in the same longwave radiative forcing (or temperature change) as the combined change due to both CO_2 and CH_4, would be a better measure of the total greenhouse gas concentration. From all indications, however, these two gases vary nearly in phase (Jouzel *et al.*, 1993), probably because both are similarly dependent on processes related to surface temperature; the suggestion has been made that methane may be considered a source for CO_2 through oxidation (Loehle, 1993). Thus, the effects of CO_2 alone may be considered a lower bound on the full effects of greenhouse gases, as measured more rigorously by "equivalent CO_2."

Although there is some global variability of CO_2, horizontally and vertically as well as seasonally, to a good first approximation we can consider CO_2 to be uniformly mixed in the atmosphere. We use the symbol μ to denote the global mean, climatic-mean ($\delta_c = 100$ y) value of the CO_2 volume mixing ratio defined as the fraction of CO_2 molecules in a given volume of air molecules, measured in parts per million (ppmv). It has been estimated from measurements of the disposition of known anthropogenic inputs of CO_2 that the synoptic-mean value has a damping response

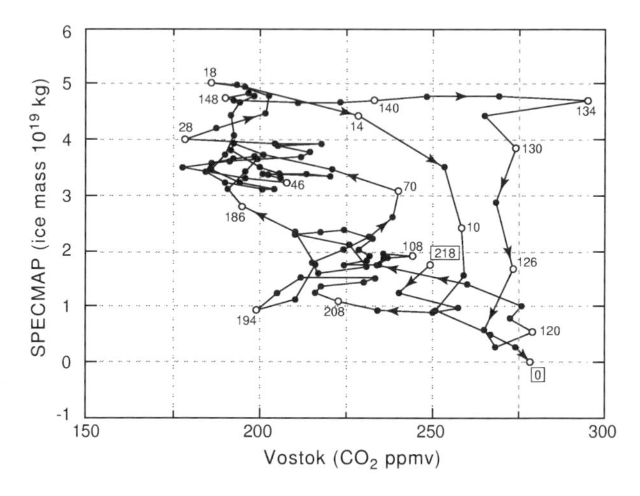

Figure 8-2 Evolution of the global climate system in the phase plane of global ice mass as deduced from the SPECMAP reconstruction (Imbrie *et al.*, 1984) and carbon dioxide concentration as deduced from the Vostok ice core (Jouzel *et al.*, 1993). Numerals mark time (ka), beginning at 218 ka and ending at the present, $t = 0$. Each dot represents a 2-ky time step. Arrows point in the direction of decreasing age, toward the present. After Saltzman and Verbitsky (1994a).

time near 100 y. For our climatic-mean values, however, we shall assume that positive feedbacks, involving the longer term behavior of the oceans, for example, may significantly lengthen the response time, possibly preventing CO_2 from achieving a long-term stable equilibrium. It is this consideration that casts atmospheric CO_2 into the category of a slow-response rather than a fast-response variable. In this connection, it is likely that, just as it is necessary to consider other slow-response variables in conjunction with ice-sheet variations (e.g., bedrock depression), it will also be necessary to consider slow-response variables such as subsurface carbon storage (e.g., peat) and ocean alkalinity and nutrient concentrations (e.g., phosphorus) in conjunction with CO_2 variations (e.g., Shaffer, 1989, 1990; Bacastow and Maier-Reimer, 1990).

In Fig. 3-6 we showed the variations in atmospheric CO_2 over the past 218 ky as determined from analyses of air trapped in the Vostok (Antarctica) ice core (Jouzel *et al.*, 1993) indicating that atmospheric CO_2 does indeed vary freely in conjunction with ice variations of the kind shown in Fig. 1-4. These results are consistent with estimates of CO_2 variations obtained from $\delta\,^{13}C$ measurements (Shackleton and Pisias, 1985) (see Section 2.4.3). To demonstrate the nature of the covariability of ice and CO_2 more clearly, we show the trajectory in the phase plane of these two variables (Fig. 8-2). In this representation, each point is separated by a 2-ky time step, starting at 218 ky and proceeding in the direction of the arrows to the present, 0 ky. As noted by Saltzman and Verbitsky (1994a), there exist two major 100-ky-period, clockwise, loops, indicative of a "sawtooth" oscillation in which high (or low) CO_2 is associated with decreasing (or increasing) ice mass; this oscillation is likely to be driven by an internal instability because external (eccentricity) forcing on this 100-ky time scale is minimal (see Fig. 1-8). One possible source of such an instability lies in positive feedbacks in Earth's carbon cycle (e.g., Plass, 1956), to be elaborated on in Chapter 10, that may augment other possible sources of instability due, for example, to positive feedback involving the ice coverage or to the behavior of the deep ocean circulation. On the other hand, the minor loops forming a denser set of points on the main 100-ky-period loop (particularly at lower CO_2 values) are of a more random direction, suggestive of a stable linear response to the near-20-ky and 40-ky components of Earth-orbital forcing, as noted by Imbrie *et al.* (1992), independent of any relationship between CO_2 and ice on these scales.

To account for the increase of CO_2 during periods of glacial maxima and the decrease of CO_2 during glacial minima probably requires that consideration be given to the changing ocean and terrestrial biospheric states engendered by the glacial variations, which must play a large role in regulating atmospheric carbon dioxide, as has been discussed at length in many studies (e.g., Taylor, 1992). In the next section we discuss the ocean state variables.

On an even longer time scale it is believed that variable tectonic factors such as volcanic outgassing of CO_2 produced by metamorphic processes, and weathering of exposed silicate rocks, have resulted in large changes of CO_2 over geologic time (e.g., Walker *et al.*, 1981; Berner, 1994).

8.3 THE THERMOHALINE OCEAN STATE

As was suggested at the beginning of the Prologue, there can be little doubt that a key to understanding long-term climatic behavior lies in the deeper ocean. Aside from intrinsic interest in its state, the deep ocean is the main "capacitor" of the climate system, representing a slow-response reservoir of heat and carbon for the atmosphere as well as other chemical constituents. Moreover, the deep ocean thermal state is a major determinant of surface temperature, and hence ice mass change as represented in Eq. (7.34).

The full climatic state of the oceans involves a three-dimensional description of the fields of motion, temperature, salt concentration (salinity), and all biogeochemical trace constituents that are relevant for climate (e.g., nutrients and carbon). A good review of our present knowledge about the state of the oceans is given by Niiler (1992). Here we shall emphasize globally averaged aspects of this full state, and, in particular, we assume that the mean deep ocean temperature is the most relevant property of the oceans for paleoclimate, on which all other relevant ocean properties might be projected, and which might serve as the key slow-response variable in a low-order representation of paleoclimatic evolution. Formally, we define the deep ocean temperature as

$$\theta = \frac{1}{V} \iiint T \, d\sigma \, dz \tag{8.5}$$

where $dV = d\sigma \, dz$ is an element of volume, and the integrals are taken over the volume of the deep world ocean below an extended mixed-layer depth δ.

The variations of θ are closely related to the ocean circulation, the thermohaline component of which, in turn, is driven by horizontal density gradients (see Chapter 11 for a fuller discussion of all the modes of oceanic circulation). These density gradients are influenced not only by thermal gradients ∇T (which respond relatively rapidly to forcing) but also by salinity gradients ∇S (which, as will be discussed in Chapter 11, may have a slow enough response time to be classified as a slow-repsonse variable). The response time for the large-scale circulation itself is a matter of debate; a typically assumed value is about 100 y, which would make it a relatively fast-response variable compared to the ice sheets, the mean ocean temperature (θ), and even the salinity gradients. This supposition is embodied in many models of the thermohaline circulation, wherein the equations of motion are reduced to a diagnostic balance between the pressure gradient, Coriolis, and frictional forces, while the salinity gradient is considered to be governed by a prognostic nonequilibrium equation (see Chapter 11).

In general, strong deepwater production in high latitudes, both by bouyancy-driven deep local convective cells and by the thermally direct part of the main thermohaline circulation, tends to fill the world ocean with cold water that would raise (i.e., shallow) the level of the thermocline and move the polar oceanic front equatorward. This cooling is constantly being opposed by smaller scale mixing processes that slowly diffuse heat downward by a complex "fingering" process from the warmer ocean surface, particularly in low and middle latitudes where surface heating and temperatures are high. This diffusion process might be intensified somewhat in the thermocline itself, which is an "internal boundary layer" characterized by a strong thermal gradient in the vertical. Nonetheless, if the deep circulation were "turned off" the bulk ocean temperature would probably take several thousand years to warm up to the mean surface values (even considering the upward geothermal flux at the ocean bottom), giving θ a longer response time than the circulation. In Fig. 3-3 we showed the dramatic way that θ is believed to have varied over the very long period of 60 My based on deep sea isotopic evidence. It would appear from paleoceanographic evidence (e.g., Chappell and Shackleton, 1986) that θ varied by no more than about 2°C in the most rapid interval of change between glacial maximum to minimum conditions (i.e., $|d\theta/dt| \approx 2°\mathrm{C}/10^4$ y).

Following Imbrie *et al.* (1992) and the many other contributions based on observation, theory, and modeling (e.g., Kellogg, 1987; Ghil *et al.*, 1987; Maier-Reimer *et al.*, 1993), an emerging view is that the Atlantic Ocean circulation fluctuated over the Late Cenozoic between the two idealized end-member states shown schematically in Fig. 8-3, representing, respectively, a relatively warm, interglacial mode similar to that prevailing at present and a cold, glacial mode. These different patterns, corresponding to different values of the single variable θ, are an example of a "coherency condition," as discussed in Section 5.4. Note the similarity of this pictorialization with that given in Kellogg (1987). In reality, for the present ocean (particularly the Atlantic) there is a marked asymmetry in the circulation characterized by a cross-equatorial flow, with the rising branch occurring in the Southern Ocean rather than in the tropics. More generally, the idealized circulation shown in this figure can only be viewed as a significant component of the full deep ocean circulation that also includes components arising from geographic and wind-driven effects that make the full circulation very much more complex than shown.

Specifically, the warm mode (high θ) is characterized by a relatively deep and latitudinally extensive thermocline, and perhaps also by a more salinity-influenced zonal-

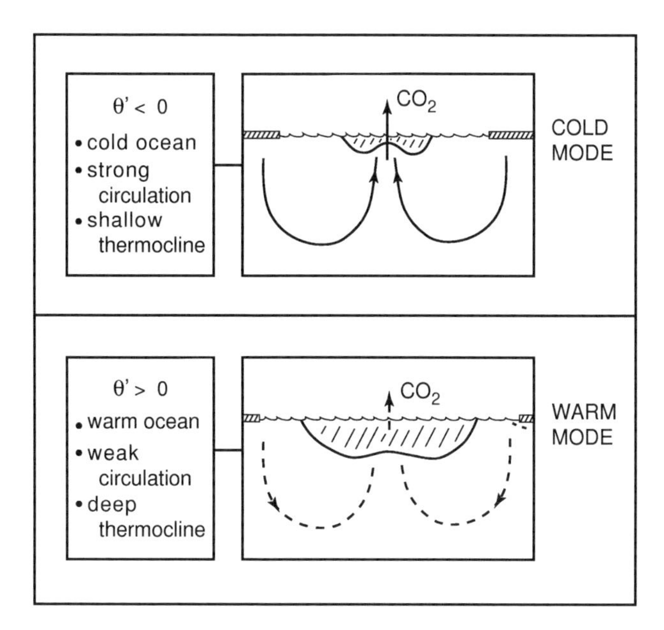

Figure 8-3 Schematic pictorialization of the thermocline structure and thermohaline circulation for a warm interglacial mode (θ large) and a cold glacial mode (θ small). Note the difference in the depth and extent of the "warm pool" indicated by the shaded area.

mean, relatively weak, symmetric thermohaline (TH) circulation (Fig. 8-3, dashed lines) that may give way to the more asymmetric circulation discussed above, forming the North Atlantic deep water (NADW) "conveyor." This conveyor helps transport heat to high latitudes in the North Atlantic, favoring warming and melting of ice; at the same time it may also lead to increased winter snowfall in colder interior regions. There are indications that in this warm ocean state, the upward flux of CO_2 from the deep ocean is inhibited, both in higher latitudes (Takahashi *et al.*, 1993) and in the main low-latitude source regions where the "thermocline barrier" is strong. The El Niño effects on CO_2 may be an analog of this inhibiting effect. On the other hand, it has been inferred from paleoceanographic observations (e.g., Kellogg, 1987) and modeling studies (e.g., Fichefet *et al.*, 1994) that the cold, glacial, model (low θ) is characterized by a shallower, less extensive thermocline, and by a more symmetric and energetic thermally driven, zonal-mean, thermohaline circulation, as pictured in Fig. 8-3. In this state, the low-latitude thermocline barrier would be weakest, favoring the upward flux of CO_2 to the atmosphere. From the inferred warmth of the ocean during the Phanerozoic (Figs. 1-3 and 3-3) we suppose that the salinity-driven warm mode was dominant over most of this longer period.

As a possible proxy for the variations of θ we draw on the analysis of Imbrie *et al.* (1992, 1993) by considering the record of sea surface temperature in the North Atlantic revealed by core K708-1 (see Fig. 8-4), a sensitive indicator of activity in the thermocline outcrop region that, according to Imbrie *et al.* (1992), is also highly coherent with other indices of the fuller oceanic state and circulation [e.g., the $\Delta\delta\,^{13}$C

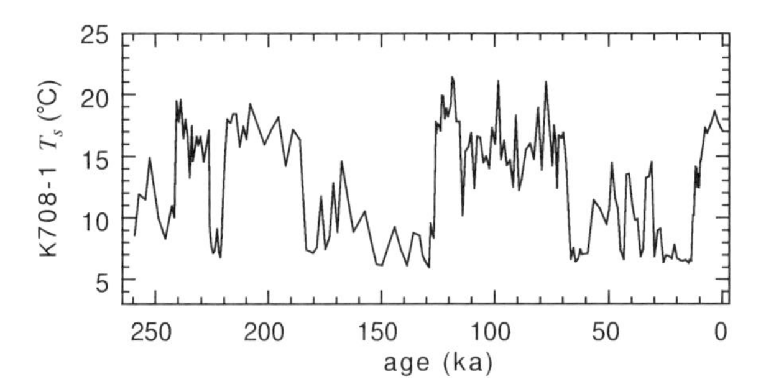

Figure 8-4 Variations of sea surface temperature recorded at core site K708-1 in the North Atlantic over the past 250 ky. From data originally presented by Ruddiman and McIntyre (1984) (see Saltzman and Verbitsky, 1994a).

"NADW ventilation index," and the Boyle and Keigwin (1982) Ca/Cd index]. As also noted by Imbrie *et al.* (1992) the frontal activity is clearly related to the wind stresses that control the gyre and ocean current structure (the variation of which are associated at a time lag with variations of the ice sheets) (see Ruddiman and McIntyre, 1977; Keffer *et al.*, 1988), and is probably also related on longer time scales to the variations in the thermohaline circulation, which can modulate the poleward transport of heat in upper oceanic levels (e.g., Broecker and Denton, 1989) as well as the breadth and depth of the thermocline. It is certainly reasonable to expect that marked changes in the oceanic frontal zone and the thermocline structure accompanying the ice variations also have important consequences for the oceanic role in regulating atmospheric CO_2 (e.g., Sarmiento and Toggweiller, 1984).

The coherence of the K708-1 sea surface temperature (SST) record with other measures of the fuller oceanic state mentioned above, coupled with the fact that very similar records can be found in the Southern Ocean [e.g., MD88-770 (46°S, 96°E), studied by Sowers *et al.* (1993)], lends some support to the possibility that this record might be taken as a broader measure of the state of the ocean, giving some indication of the variations between these two end-member states of the ocean. Whether or not this broad interpretation is valid, we can still consider the variability of K708-1 SST in its own right as a measure of the North Atlantic state, and take as a further hypothesis that the variable θ and T_S (K708-1) are linearly related

$$\theta = a + bT_S \quad \text{(K708-1)} \tag{8.6}$$

where a and b are constants.

8.4 A THREE-DIMENSIONAL PHASE-SPACE TRAJECTORY

When the above ocean record [T_S (K708-1)] for the past 200 ky is combined with the phase-plane trajectory of ice (Ψ) and CO_2 (μ) shown in Fig. 8-2, we obtain the three-dimensional trajectory shown in Fig. 8-5. The points shown are again separated by 2 ky. This trajectory is seen to take the form of two main near-100-ky-period loops

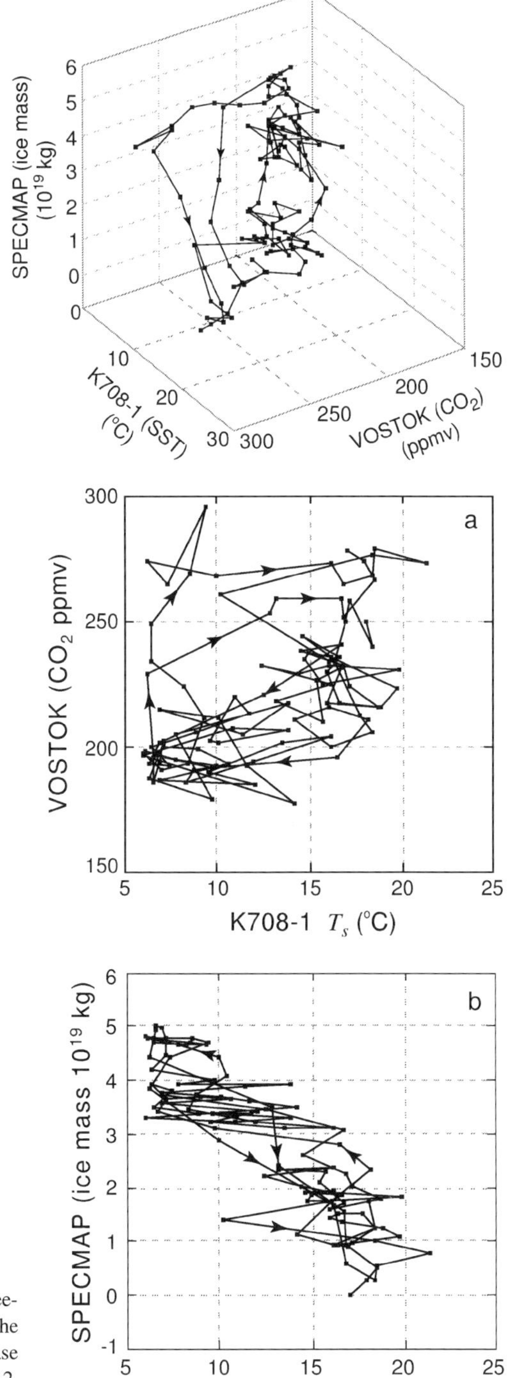

Figure 8-5 Three-dimensional trajectory of global ice mass, CO_2, and the ocean state, as inferred from SPECMAP $\delta^{18}O$, the Vostok ice core, and North Atlantic core K708-1, respectively. Dots represent 2-ky time steps. After Saltzman and Verbitsky (1994a).

Figure 8-6 Projection of the three-dimensional trajectory shown in Fig. 8-5 on the (μ, θ) phase plane (a), and on the (Ψ, θ) phase plane (b), using the same format as in Fig. 8-2. After Saltzman and Verbitsky (1994a).

in the direction shown by the arrows, with relatively rapid deglaciations occurring at higher values of CO_2, and exhibiting quasi-periodic variations in the form of the smaller, near-20- and 40-ky-period, loops that are probably forced by precession and obliquity orbital variations. The points along the main loop, particularly the regions of high density, are indicative of the stable regions of the (Ψ, μ, θ) phase space where the climate system prefers to reside, constituting the *slow climatic attractor* (Saltzman, 1988). These points surround a "hole" in the main loop from which region the system tends to be driven, indicative of a region in which an unstable equilibrium might be located.

These and other features of this climatic attractor are more clearly exposed on the two-dimensional projections of this trajectory on the separate (Ψ, μ), (Ψ, θ), and (μ, θ), phase planes, the first of which we have already shown (Fig. 8-2). In Fig. 8-6 we now show the projections on the remaining two phase planes, respectively.

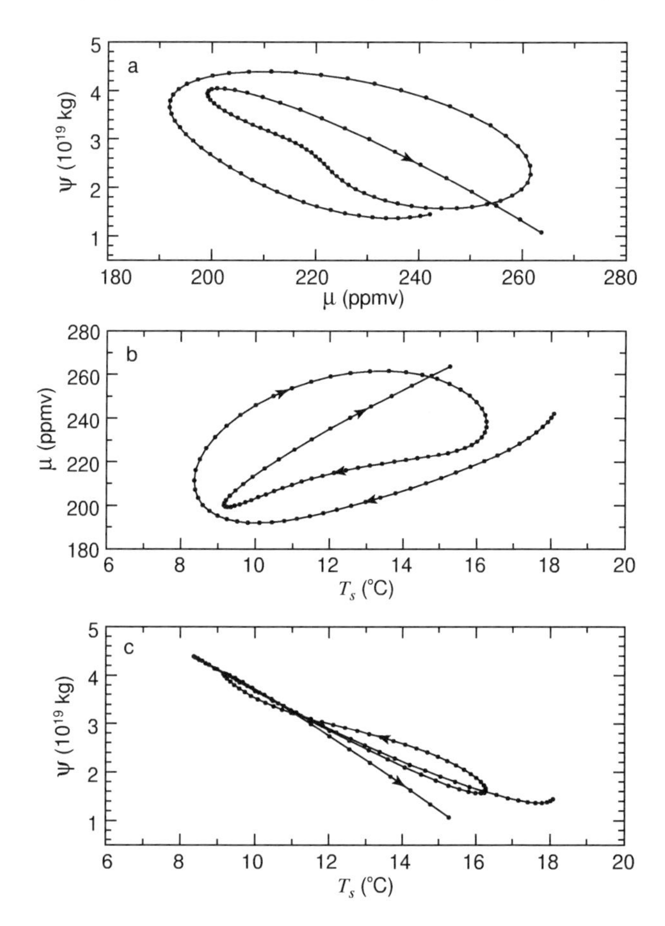

Figure 8-7 Smoothed trajectories in the phase planes of (a) (Ψ, μ), (b) (μ, θ), and (c) (Ψ, θ) that were shown in Figs. 8-2 and 8-6, revealing more clearly the near-100-ky period of oscillation. After Saltzman and Verbitsky (1994a).

From Fig. 8-6a we see that the state of the ocean, as represented by T_S (708-1), forms a near-100-ky-period loop, with CO_2 changes directed such that low (high) values of T_S (708-1) correspond to increasing (decreasing) CO_2. This conforms with the supposition that the cold-mode (shallow thermocline) ocean tends to favor a flux of CO_2 to the atmosphere and a warm-mode (deep and extensive thermocline) ocean tends to inhibit such a flux.

From Fig. 8-6b we see that ice mass and ocean state are much more closely in phase, but still show a small tendency for a counterclockwise trajectory loop in which, for example, warmer ocean states favor increasing ice mass (Ruddiman and McIntyre, 1981, 1984). To demonstrate more directly the behavior of the high-amplitude near-100-ky-period cycle, we show (Fig. 8-7) the results of a 40-ky running average filter of the three phase-plane trajectories.

9

GLOBAL DYNAMICS OF THE ICE SHEETS

In trying to build a theory of paleoclimatic variation (e.g., the ice ages) we are inexorably led, as a central concern, to a consideration of the cryosphere, particularly the behavior of the ice sheets and their marine extensions, the ice shelves. In this chapter we develop the sets of glaciological relationships that have been established and used as the basis for modeling this important component of the climate system.

9.1 BASIC EQUATIONS AND BOUNDARY CONDITIONS

Unlike the atmosphere and oceans, the glacial motions are characterized by a near balance between pressure gradient and viscous forces, and hydrostatic equilibrium, very nearly obeying the "shallow ice sheet" approximation of Grigoryan *et al.* (1976) and Hutter (1983), in which ice flow is considered to be quasi-horizontal "creep." Thus, the general equations given in Chapter 4 are specialized to the set:

$$-\nabla_2 p + \mathbf{F}_2 = 0 \tag{9.1}$$

$$-\frac{\partial p}{\partial z} - \rho_i g = 0 \tag{9.2}$$

$$\nabla \cdot \mathbf{V} = 0 \tag{9.3}$$

$$\frac{dT}{dt} = \left(\frac{\partial T}{\partial t} + \mathbf{V} \cdot \nabla T\right) = k\nabla^2 T + \frac{q_F}{c_i} \tag{9.4}$$

where ∇_2 and $\mathbf{F}_2$ denote the horizontal (x, y) components, and c_i and k are the thermal capacity and diffusivity of ice, respectively. The frictional heating per unit mass (q_F) is most conveniently expressed in tensor form (Paterson, 1994):

$$\rho_i q_F \approx 2\tau_{\alpha\beta}\dot{\varepsilon}_{\alpha\beta}. \tag{9.5}$$

The directional indices α or β are (x, y, z) [or (λ, φ, z)], and $\tau_{\alpha\beta}$ and $\dot{\varepsilon}_{\alpha\beta}$ are, respectively, the "stress deviator" and strain rate tensor components related by the generalized Glen's rheological law,

$$\dot{\varepsilon}_{\alpha\beta} = \frac{1}{2}\left(\frac{\partial V_\alpha}{\partial x_\beta} + \frac{\partial V_\beta}{\partial x_\alpha}\right) = K(n)\tau^{n-1}\tau_{\alpha\beta} \tag{9.6}$$

where $V_\alpha = (u, v, w)$; $\tau = [(\tau_{\lambda\lambda}^2 + \tau_{\varphi\varphi}^2 + \tau_{zz}^2)/2 + (\tau_{\lambda\varphi}^2 + \tau_{\varphi z}^2 + \tau_{z\lambda}^2)]^{1/2}$, the "effective" stress; n is a power usually taken to be in the range of 1 to 4, and $K(n)$ $(= A^{-n})$, where A is a function of temperature and the crystalization grain fabric of the ice (Hughes, 1998) [see Eqs. (4.16) and (4.17)].

These glaciological equations form the basis for the most complete models of ice sheets and shelves, from which solutions for the coupled motion and temperature fields can be obtained, in principle, at a three-dimensional lattice (Hutter, 1983). Examples of partial attempts to apply such models to the present Antarctic and Greenland ice sheets are provided by Budd and Jenssen (1989), Huybrechts (1990), and Greve and Hutter (1995). The many assumptions required in such models regarding the full three-dimensional stress fields and internal properties of ice (e.g., viscosity and thermal conductivity), as well as regarding basal and surficial (e.g., snow accumulation) processes, introduce enough uncertainties that equally relevant deductions for evolutionary paleoclimatic purposes can be obtained from more simplified models.

As the foundation for these more simplified models (which are to be discussed in Sections 9.2 and 9.3), we now present approximate, scalar, forms of the fundamental equations, Eqs. (9.1)–(9.6), expressed in a Cartesian geometry (x, y, z) and applied to a shallow, hydrostatic ice sheet in which viscous stresses in vertical planes and longitudinal deviatoric stresses are assumed to be negligible compared to stresses due to shears across horizontal planes (e.g., Mahaffy, 1976; Jenssen, 1977). These equations are as follows:

$$\frac{-\partial p}{\partial x} + \frac{\partial \tau_{xz}}{\partial z} = 0 \tag{9.7}$$

$$\frac{-\partial p}{\partial y} + \frac{\partial \tau_{yz}}{\partial z} = 0 \tag{9.8}$$

$$\frac{-\partial p}{\partial z} - \rho_{\mathrm{i}} g = 0 \tag{9.9}$$

$$\frac{\partial u}{\partial x} + \frac{\partial v}{\partial y} + \frac{\partial w}{\partial z} = 0 \tag{9.10}$$

$$\frac{\partial T}{\partial t} + u\frac{\partial T}{\partial x} + v\frac{\partial T}{\partial y} + w\frac{\partial T}{\partial z} = k\frac{\partial^2 T}{\partial z^2} + \frac{1}{c_{\mathrm{i}}} q_{\mathrm{F}} \tag{9.11}$$

From Eq. (9.5) and Glen's rheological (creep) law, Eq. (9.6), the stresses are related to the shear by the following relations:

$$\frac{\partial u}{\partial z} = \mu_{\mathrm{i}}^{-1}\tau^{n-1} \cdot \tau_{xz} \tag{9.12}$$

$$\frac{\partial v}{\partial z} = \mu_{\mathrm{i}}^{-1}\tau^{n-1} \cdot \tau_{yz}$$

and the heat added by internal friction is

$$q_F = \frac{1}{\rho_i}\left(\tau_{xz}\frac{\partial u}{\partial z} + \tau_{yz}\frac{\partial v}{\partial z}\right) = \mu_i^{-1}\tau^{n+1} \tag{9.13}$$

where $\tau = (\tau_{xz}^2 + \tau_{yz}^2)^{1/2}$, and the ice viscosity, μ_i, is defined by

$$\mu_i = \frac{1}{2K(T)} \tag{9.14}$$

For simplicity we shall take μ_i as a constant, independent of temperature. Although this is considered an acceptable approximation for some purposes, as was noted with regard to Eq. (9.6), in more complete three-dimensional models the rheological coefficient K is taken as a function of internal temperature and ice-grain fabric (thermomechanical coupling) or, more simply, of internal temperature alone, e.g., Shumskiy's (1975) representation for $n = 3$,

$$K = K_0 \exp\{-\kappa[T_M/T - 1]\} \tag{9.15}$$

where T_M is given by Eq. (7.4), $K_0 \approx 1.5 \times 10^{-16}$ $[(\text{Pa})^3\ \text{y}]^{-1}$, and $\kappa \approx 20$.

Note that for the linear approximation, $n = 1$, the system given by Eqs. (9.7)–(9.11) reduces to a familiar form used in most atmospheric and oceanic studies. Although the most commonly prescribed value is $n = 3$, due to the experimental uncertainties and great variability of ice properties within glaciers, calculations for $n = 1$ can give acceptable first-order results.

Boundary conditions. On the upper (free) ice sheet surface (i.e., at $z = h$), where h is the height of the top of the ice sheet above prevailing sea-level (see Fig. 9-1).

$$\begin{aligned} &p \approx 1 \quad \text{atm} \\ &\underline{\tau}_h = 0 \\ &T = T_h(x, y, t) = T_S - \gamma h \end{aligned} \tag{9.16}$$

and the kinematic condition defining the "surface" is

$$\frac{dh}{dt} = \frac{\partial h}{\partial t} + \left(u_h\frac{\partial h}{\partial x} + v_h\frac{\partial h}{\partial y}\right) = w_h + \rho_i^{-1}(P_i - M) \tag{9.17}$$

where the subscript h refers to $z = h$; $\underline{\tau}_h$ is the three-dimensional vector stress tangent to the ice surface at any point, the slope of which is usually assumed to be nearly horizontal (except possibly in a narrow zone near the ice edge) so that $\underline{\tau}_h \approx \underline{\tau} \equiv (\tau_{xz}\mathbf{i} + \tau_{yz}\mathbf{j})$; T_S is the sea level air temperature near the edge of the ice sheet [given by Eq. (7.34b)]; γ is the atmospheric lapse rate; $(\mathbf{v}_h \cdot \nabla h)$ is the ice accumulation at a point due to creep and drifting or blowing snow at the surface; and, in accord with the notation introduced in Section 4.5.1, $\mathcal{A} \equiv \rho_i^{-1}(P_i - M)$ is the net accumulation rate

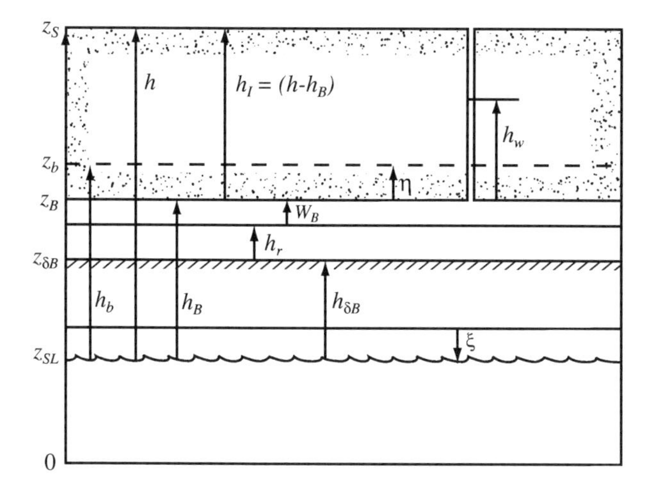

Figure 9-1 Notation for vertical thicknesses relevant for a description of an ice sheet and its lower boundary layers. Compare Fig. 8-1.

of ice in meters/year. The rate of ice accumulation due to snowfall (P_i/ρ_i) dominates over the net rate of melting, ablation, and calving (M/ρ_i) over most of the surface of an ice sheet like Antarctica, except near the periphery; however, this may not be true for Greenland. Included in M are peripheral processes such as calving and ice shelf melting.

On the bottom of the ice sheet ($z = h_B$) we set $(u, v) = (u_B, v_B)$. It is useful to distinguish between the case in which the basal temperature T_B is below the pressure melting-point temperature T_M given by Eq. (8.4), and the case in which T_B is at the melting point, forming a layer of liquid water that promotes sliding. Thus, for $T_B < T_M$, u_B and v_B will represent a relatively slow "sheet flow" (Hughes, 1998) and the vertical component of velocity and heat balance are given by

$$w_B = 0 \tag{9.18}$$

and

$$-\lambda \frac{\partial T}{\partial z} = G^{\uparrow} \tag{9.19}$$

where $G^{\uparrow}$ is the upward geothermal heat flux, and $\lambda = \rho_i c_i k$ (conduction coefficient for ice in units of W/m K)).

For $T_B = T_M$ (melting), u_B and v_B are greatly enhanced, representing "stream flow," which, if it terminates in the ocean, can become "shelf flow." In this stream flow case,

$$w_B = \rho_i^{-1} M_B \tag{9.20}$$

and

$$-\lambda \frac{\partial T}{\partial z} = \left[G^{\uparrow} + \mathcal{D} - L_f M_B\right] \tag{9.21}$$

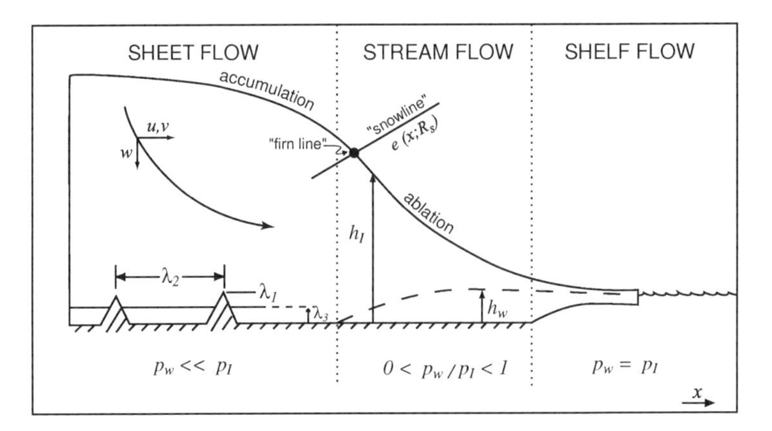

Figure 9-2 Idealized vertical cross-section of an ice sheet showing the three main regions of ice flow. See text for definitions of symbols. After Hughes (1998).

where M_B is the net rate of excess basal melting over refreezing at any point, in mass per unit area, per unit time, and

$$\mathcal{D} = \mathbf{v}_B \cdot \underline{\tau}_B = (u_B \tau_{Bxz} + v_B \tau_{Byz})$$

is the rate of heat addition per unit area due to the friction of basal sliding.

The basal sliding velocity, $\mathbf{v}_B = (u_B\mathbf{i} + v_B\mathbf{j})$ is a complicated function not only of the basal liquid water content, but also of the subglacial topography and deformable regolith. These complexities lead to great uncertainties in any representation of $\mathbf{v}_B$, the most general relation being due to Hughes (1998). This relation has the scalar form

$$|\mathbf{v}_B| = \left(k_B |\underline{\tau}_B|\right)^m \tag{9.22}$$

where $k_B = [B(1 - p_W/p_I)^2]^{-1}$, $B = B_0[(\lambda_1 - \lambda_3)/\lambda_2]^2$, B_0 is a measure of ice hardness, λ_1 and λ_2 are the mean height and lateral spacing of the bedrock roughness features, respectively, and λ_3 is the height of deformable regolith material covering the bedrock (see schematic representation in Fig. 9-2), $p_I = \rho_I g h_I$ (basal ice pressure), $h_I = h - h_B$ is the ice-sheet thickness, $p_W = \rho_W g h_W$ (basal water pressure, equivalent to the hydrostatic pressure of a borehole water column of height h_W), and $\underline{\tau}_B$ is the basal shear stress. If $p_W \ll p_I$ (i.e., $k_B \approx B^{-1}$) and $m = (n+1)/2$, this formula reduces to the simple sliding law proposed by Weertman (1957), which is most appropriate for regions where $T_B < T_M$ (sheet flow). More generally, however, when $0 < (p_W/p_I) < 1$, basal water can partially support the weight of the ice above and stream flow can ensue. When $p_W = p_I$ we have the condition for ice shelf flow (Hughes, 1998) (see Fig. 9-2).

A special lateral boundary condition is needed at the point where an ice sheet impinges on the ocean and is no longer grounded on bedrock. Beyond this grounding line the ice discharged from the ice sheet becomes a floating ice shelf, which requires a separate treatment (Oerlemans and Van der Veen, 1984; Paterson, 1994). As the simplest approximation it is common to define the grounding line junction of sheet and shelf as the point beyond which the ice thickness is small enough relative to the water depth to float; this is expressed formally as the "flotation condition"

$$\begin{aligned}\rho_i h_I &= \rho_w(-h_B) \qquad (h_B < 0)\\ &= \rho_w\left(\xi + \delta_B - \widehat{h}_B\right)\end{aligned} \tag{9.23}$$

where h_B is the prevailing height of the bedrock above prevailing sea level (negative if bedrock is below sea level), ξ is the change in sea level from its undisturbed value $(\widehat{z}_{SL})$ due to the water locked in the ice sheet, δ_B is the bedrock depression below the undisturbed bedrock level (always positive), and $\widehat{h}_B$ is the undisturbed bedrock height above the undisturbed sea level. If $h_I < (\rho_w/\rho_i)(-h_B)$, then h_I may be set to zero assuming no calving ice walls (Hughes, 1998), thereby neglecting a potential ice shelf. To consider such a potential ice shelf new physical considerations must be introduced (see Section 9.7). We see from Eq. (9.23) that the grounding line will advance toward the ocean if bedrock is rebounding, sea level is dropping, or ice thickness is increasing, and retreat landward if the reverse changes are occurring.

The equations and boundary conditions formulated in this section provide the basis for the scaling relationships discussed next (Section 9.2), which provide a good first-order representation of the equilibrium properties of ice sheets, and also for the vertically integrated, plan-form model governing the horizontal thickness profile ice sheets, $h_I(\lambda, \varphi) = [(h - h_B)$, see Fig. 9-1]. This latter model, which is the most extensively used in paleoclimatic studies, is discussed in Section 9.3.

9.2 A SCALE ANALYSIS

Let us now adopt the notation listed in Table 9-1 for characteristic (scale) values of key ice-sheet variables (see Fig. 9-1). Note that for an ice sheet in equilibrium with a characteristic snowfall rate, a, we find from Eqs. (9.17) and (9.18) the scaling approximation $W \sim a$. A typical trajectory of the internal ice flow is shown in Fig. 9-2. In the following discussion we proceed along the lines discussed by Verbitsky (1992).

To begin, we have from the continuity equation, Eq. (9.10), the relation

$$U \approx \frac{WL}{H} \approx \frac{aL}{H} \tag{9.24}$$

As shown by Verbitsky (1992), the equations of motion, Eqs. (9.7) and (9.8), together with Eqs. (9.9) and (9.12), imply that the scale of the shear stress τ_{xz} is $(\rho_i g H^2/L)$. It then follows from the power creep law, Eqs. (9.12) and (9.13), that the scale height of an ice sheet, H, is related to its lateral extent L, area A $(= L^2)$, or volume V $(= AH)$ by the respective relationships,

$$\begin{aligned}H &\approx \gamma^{\frac{4}{2(n+1)}} L^{1/2} \qquad (9.25)\\ &\approx \gamma^{\frac{4}{2(n+1)}} A_I^{1/4}\\ &\approx \gamma^{\frac{8}{5(n+1)}} V^{1/5}\end{aligned}$$

Table 9-1 Scaling Notation for Ice-Sheet Variables

Variable	Symbol	Characteristic scale
Ice thickness	h_{I}	H
Horizontal distance (summit to edge)		L
Area	A_{I}	$(\sim L^2)$
Volume		$V(\sim SH)$
Horizontal motion	u, v	U
Vertical motion	w	$W \sim a$
Basal pressure	p_{B}	$\rho_{\mathrm{i}} g H$
Top surficial temperature	T_h	$\widetilde{T}_h$
Vertical and horizontal temperature difference		ΔT
Snowfall rate (m s^{-1})	$\rho_{\mathrm{i}}^{-1} P_{\mathrm{i}}$	a
Advective time scale (surface to base)		$t^* \sim H/a$

where $\gamma = [\mu_{\mathrm{i}} a/(\rho g)^n]^{1/4}$. Thus, for example, if $n = 1$, $H = (\mu_{\mathrm{i}} a/\rho g)^{1/4} \cdot A_{\mathrm{I}}^{1/4}$. It also follows from Eqs. (9.6) and (9.7) that $U = (\rho_{\mathrm{i}} g)^n H^{2n+1}/\mu_{\mathrm{i}} L^n$.

To discuss the scaling of thermal processes in an ice sheet, particularly at its base where melting may lead to liquid water formation and surging, we show in Fig. 9-3 the specialization of Eq. (9.11) for three main regions of an ice sheet (Verbitsky and Saltzman, 1994).

1. The summit ($x < L_{\mathrm{c}}$), where L_{c} represents the characteristic distance from the center beyond which the horizontal flow becomes large enough to control the advective heat flux and friction, i.e., $\partial T/\partial t + w\,\partial T/\partial z = k\,\partial^2 T/\partial z^2$.
2. The region above the bottom boundary layer of the ice sheet ($z > h_{\mathrm{B}}$), where advection dominates, conveying the temperature at the upper surface (T_{H}) to the basal layer with an advective time-delay $t^* \approx H(t)/a$, i.e., $T_{\mathrm{b}} = T_{\mathrm{H}}(t - t^*)$, i.e., $dT/dz = 0$.
3. The bottom boundary layer of the ice sheet ($z \leqslant h_{\mathrm{b}}$), where upward vertical diffusion balances heat added due to the internal friction and geothermal heat flux, i.e., $\lambda_{\mathrm{i}}\,\partial^2 T/\partial z^2 + \mu_1^{1/n}(\partial u/\partial z)^{1+1/n} = 0$.

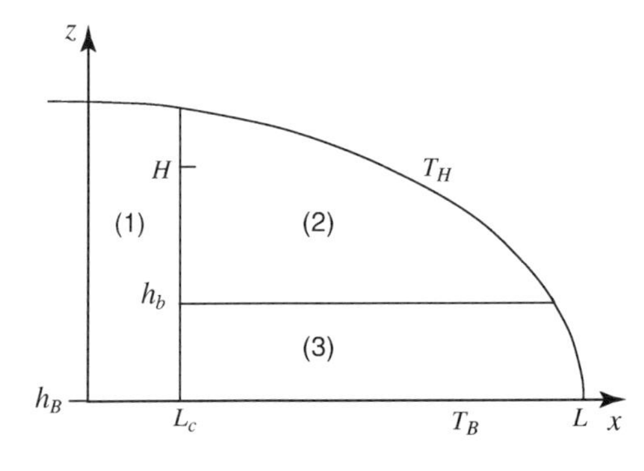

Figure 9-3 Idealized cross-section of an ice sheet showing thermal balances in three main domains of an ice sheet: (1) the central (summit) region ($L < L_{\mathrm{c}}$), where $\partial T/\partial t + w\,\partial T/\partial z = k\,\partial^2 T/\partial z^2$, (2) the main domain above the bottom boundary layer, where thermal advection dominates, i.e., $dT/dz = 0$, and (3) the basal boundary layer ($z < h_{\mathrm{b}}$), where $\lambda_{\mathrm{i}}\,\partial^2 T/\partial z^2 + \mu_{\mathrm{i}}^{1/n}\,(\partial u/\partial z)^{(1+n)/n} = 0$. After Saltzman and Verbitsky (1996).

Table 9-2 Physical Parameters of Ice Sheets

Quantity	Value
ρ	$917\ \mathrm{kg\,m^{-3}}$
μ_i $(n = 1)$	$3.4 \times 10^{12}\ \mathrm{Pa\,s}$
μ_i $(n = 3)$	$4.8 \times 10^{20}\ \mathrm{Pa^3\,s}$
g	$9.8\ \mathrm{m\,s^{-2}}$
c	$2.0 \times 10^{3}\ \mathrm{J\,kg^{-1}\,{}^\circ C^{-1}}$
k	$1.0 \times 10^{-6}\ \mathrm{m^2\,s^{-1}}$
$G\uparrow$	$0.04\ \mathrm{W\,m^{-2}}$
γ	$6.5 \times 10^{-3}\ \mathrm{{}^\circ C\,m^{-1}}$

This partitioning of thermodynamic balances is determined from the consideration that for the bulk of an ice sheet the advective heat flux F_1 (e.g., $w\partial T/\partial z$) is of the scale $a\Delta T/H$, whereas the conductive heat flux F_2 above a shallow boundary layer is of the scale $k\Delta T/H^2$. It follows that the ratio $F_2/F_1 \ll 1$, meaning that, except for the boundary layer, advection has a dominant influence on temperature. Thus temperature can be considered to be carried along the flow trajectories, i.e., delivered from the central part of the ice sheet surface to the top of the boundary layer, h_b (see Fig. 9-2). This means that $T_\mathrm{H} \sim T_\mathrm{b}$, and $\Delta T = (T_\mathrm{H} - T_\mathrm{B}) \approx (T_\mathrm{b} - T_\mathrm{B})$.

The scale thickness of the boundary layer, $\eta = h_\mathrm{b} - h_\mathrm{B}$, is obtained from the condition $F_1 = F_2(\eta) = k\Delta T/\eta^2$; i.e., $\eta = (kH/a)^{1/2}$. To estimate ΔT in terms of the snowfall rate a and the size of the ice sheet we can invoke the thermodynamic balance in the boundary layer for the $x - z$ plane shown in Fig. 9-3,

$$\lambda\,\frac{\partial^2 T}{\partial z^2} + \mu_\mathrm{i}^{1/n}\left(\frac{\partial u}{\partial z}\right)^{\frac{1+n}{n}} = 0 \tag{9.26}$$

Using the above scalings in the simplest linear rheology case $(n = 1)$ we have (Verbitsky, 1992)

$$\Delta T \approx \frac{\mu_\mathrm{i}}{\lambda}\,a^2\left(\frac{L}{H}\right)^2 \tag{9.27}$$

In Table 9-2 we list the commonly adopted values of the physical parameters of ice sheets, and in Table 9-3 we list estimates of the areas and volumes of present (P) and past ice sheets at the last glacial maximum (G). The quantities K_f and K_g listed in the last three columns of Table 9-3 will be discussed in Chapter 16 (Section 16.1.2).

Assuming reasonable observational values of a and A_I for Antarctica and Greenland, Verbitsky (1992) calculated the implied values of the other scaled variables from the above formulas with $n = 1$ (linear approximation). The results demonstrate good agreement with independent evidence of these properties. When the same calculations are made with the more conventional power value $(n = 3)$ results not too different

Table 9-3 Estimates of Nondimensional Ratios K_f and K_g for Ice Sheets of the Present and the Last Glacial Maximum

Ice sheet	Period[a]	Area A (10^6 km^2)	Volume V (10^6 km^3)	K_f $n = 1$	K_f $n = 3$	K_g
Antarctic	P	12.53	23.45	1.83	0.64	0.63
	G	13.81	26.00	1.71	0.60	0.68
Greenland	P	1.73	2.60	4.43	0.86	0.26
	G	2.3	3.50	3.55	0.80	0.33
Laurentide	G	13.39	29.46	3.85	0.82	0.30
Cordilleran	G	2.37	3.55	3.18	0.77	0.36
Scandinavian	G	6.66	13.32	4.80	0.88	0.24

[a] P, Present day; G, last glacial maximum.

from the linear results are obtained, suggesting that ice sheet behavior is not too sensitive to this factor (cf. Alley, 1992).

9.3 THE VERTICALLY INTEGRATED ICE-SHEET MODEL

The fundamental continuity equation for the rate of change of ice thickness at any point, $h_{\mathrm{I}}(\lambda, \varphi) = \rho_{\mathrm{i}}^{-1} m_{\mathrm{i}}$, is Eq. (4.47), also derivable from Eq. (9.10) and the upper boundary condition, Eq. (9.17). Assuming ρ_{i} is a constant this equation can be written in the form

$$\frac{\partial h_{\mathrm{I}}}{\partial t} = -\nabla \cdot \int_{h_{\mathrm{B}}}^{h} \mathbf{v}(z)\, dz - \nabla \cdot h_{\mathrm{I}} \mathbf{v}_{\mathrm{B}} + \rho_{\mathrm{i}}^{-1}(P_{\mathrm{I}} - M - M_{\mathrm{B}}) \tag{9.28}$$

where $\mathbf{v}(z)$ denotes a slow ice-creep component of the *total* velocity $\mathbf{v}_{\mathrm{I}}(z)$ at height z, and $\mathbf{v}_{\mathrm{B}}$ denotes a basal sliding or "surge" component assumed to be uniform with height, to be determined by the subglacial boundary conditions (see Fig. 9-4) i.e.,

$$\mathbf{v}_{\mathrm{I}}(z) = \mathbf{v}(z) + \mathbf{v}_{\mathrm{B}} \tag{9.29}$$

Following Paterson (1994), $\mathbf{v}_{\mathrm{I}}$ is determined as a function of the ice thickness h_{I} $(= h - h_{\mathrm{B}})$ from the equations of motion, Eqs. (9.7), (9.8), and (9.9), and the flow law, Eq. (9.12), as follows: First, integrating these equations from the top of the ice sheet $(z = h)$, where $p = 0$ to any level z,

$$\tau_{xz}(z) = -\rho_{\mathrm{i}} g (h - z) \frac{\partial h}{\partial x} \tag{9.30}$$

$$\tau_{yz}(z) = -\rho_{\mathrm{i}} g (h - z) \frac{\partial h}{\partial y} \tag{9.31}$$

$$p(z) = \rho_{\mathrm{i}} g (h - z) \tag{9.32}$$

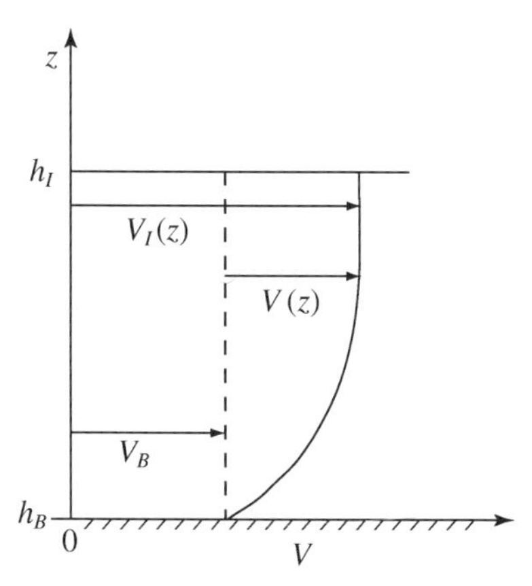

Figure 9-4 Idealized horizontal velocity components in an ice sheet, expressing the total velocity $V_{\mathrm{I}}(z)$ as the sum of an creep velocity $V(z)$ and a basal sliding velocity V_{B}.

Then, substituting Eqs. (9.30), (9.31), and (9.32) into Eq. (9.12), and using Eq. (9.14), we obtain

$$\frac{\partial(u, v)}{\partial z} = \Xi_{(x,y)} \cdot (h - z)^n \tag{9.33}$$

where

$$\Xi_{(x,y)} = -2K \cdot (\rho_{\mathrm{i}} g)^n \left[\left(\frac{\partial h}{\partial x}\right)^2 + \left(\frac{\partial h}{\partial y}\right)^2\right]^{\frac{n-1}{2}} \frac{\partial h}{\partial(x, y)}$$

Integrating Eq. (9.33) from the base ($z = h_{\mathrm{B}}$) to any level z,

$$(u, v) = \frac{\Xi_{(x,y)}}{n+1}\left[-(h - z)^{n+1} + h_{\mathrm{I}}^{n+1}\right] + (u_{\mathrm{B}}, v_{\mathrm{B}}) \tag{9.34}$$

Finally, substituting Eq. (9.34) into Eq. (9.28), we obtain, in vector form, the fundamental equation governing the thickness of a column-averaged ice sheet at any point,

$$\begin{aligned}\frac{\partial h_{\mathrm{I}}}{\partial t} &= \frac{2(\rho_{\mathrm{i}} g)^n}{n+2} \nabla \cdot \left(K h_{\mathrm{I}}^{n+2} |\nabla h|^{\frac{n-1}{2}}\right) \nabla h - \nabla \cdot h_{\mathrm{I}} \mathbf{v}_{\mathrm{B}} \\ &\quad + \rho_{\mathrm{i}}^{-1}(P_{\mathrm{i}} - M - M_{\mathrm{B}})\end{aligned} \tag{9.35}$$

This equation is the basis of many models of ice-sheet variation, some paleoclimatic applications of which are discussed in Section 9.10. The first term represents the thickness change due to the slow steady creep of ice under its own weight, and the second term represents the potentially rapid effect of the sliding velocity $\mathbf{v}_{\mathrm{B}}$ that may include ice stream flow [cf., Eq. (9.22)].

In view of the uncertainties of basal conditions, using Eqs. (9.30) and (9.31) coupled with Eq. (9.22) in its simplest form (Weertman, 1957), with $n = 1$, Greve and MacAyeal (1996) and Verbitsky and Saltzman (1997) have approximated $\mathbf{v}_{\mathrm{B}}$ for

$T_B = T_M$ as

$$\mathbf{v}_B = k_B \underline{\tau}_B = (\rho_i g h_B) h_I \nabla h \tag{9.36}$$

where k_B is treated as a free parameter.

It might also be possible, as suggested by Clark and Pollard (1998), to artificially represent the effects of altered states of basal water and deformable till by changing the creep coefficients n and K. A more rigorous approach than either this method or by the use of Eq. (9.36) is to deal more explicitly with the physics of subglacial water and deformable sediment (Boulton and Dobbie, 1993) starting from the governing continuity equations (see Sections 9.5 and 9.6).

9.4 THE SURFACE MASS BALANCE

Ideally, the rates of ice accumulation (in m s^{-1}) $[\rho_i^{-1} P_i(\lambda, \varphi, t)]$ and the normal melting/ablation/calving processes $[\rho_i^{-1} M(\lambda, \varphi, t)]$ should be determined from a GCM coupled to an ice-sheet model. Some first attempts to do this have been made, for example, by Verbitsky and Oglesby (1995). A fundamental problem of applying this procedure, especially for paleoclimate applications, is that, although the individual components, $\rho_i^{-1} P_i$ and $\rho_i^{-1} M$, might be calculable with a reasonable degree of accuracy (Ohmura *et al.*, 1996), the very small differences between them on paleoclimatological time scales is of the order of the errors in each of these components (see Section 5.1). Even for present conditions on Greenland and Antarctica it is not possible to use a GCM to determine the sign of the net mass balances.

Thus, it has been a common practice to adopt relatively crude representations of the net ice accumulation $\mathcal{A} = \rho_i^{-1}(P_i - M)$ that contain some adjustable parameters tuned to fit present observations and proxy paleoglaciological variations. For example, Weertman (1964, 1976), followed by Birchfield (1977a,b) and Oerlemans (1980b, 1981a), introduced a representation of $\mathcal{A}$ as a function of the difference between the ice-sheet height, h_I, and the height of a "snowline" (or "equilibrium line altitude;" ELA), e, separating an accumulation region and an ablation region. This representation, which has been widely used [e.g., the above contributions, and also by Pollard (1982) and Deblonde and Peltier (1991a,b)] is of the form

$$\mathcal{A} = \begin{cases} a(h-e) - b(h-e)^2 & (h - e \leqslant 1.5 \text{ km}) \\ c \quad (\text{e.g., } 5.6 \text{ m y}^{-1}) & (h - e > 1.5 \text{ km}) \end{cases} \tag{9.37}$$

where a, b, and c are tunable constants. The height of the snowline, e, is assumed to increase linearly with equatorward distance x, and increase uniformly over the ice sheet as a function of the summer insolation due to Earth-orbital changes ΔR_S (Berger and Loutre, 1991), i.e.,

$$e = e_0 + k_1 \Delta x + k_2 \Delta R_S \tag{9.38}$$

where e_0 is a constant. The particular equilibrium point at which $h = e$ defines the *firn line* (see Fig. 9-2).

Other representations that are more directly a function of surface temperature, providing a stronger link to the atmospheric climate as determined by a GCM or EBM, have also been proposed. For example, Pollard (1980, 1983a) suggests the following separate forms for P_i and M:

$$P_\mathrm{i} = \begin{cases} P(\varphi) & (T \leqslant 0^\circ\mathrm{C}) \\ 0 & (T > 0^\circ\mathrm{C}) \end{cases} \tag{9.39}$$

where $P(\varphi)$ is the presently observed latitudinal variation of the zonal and annual mean precipitation, and

$$M = \max[0; aT_h + bR + c] \tag{9.40}$$

where a, b, and c are constants, R is the zonal mean insolation as a function of the orbital parameters, and the temperature at the top of the ice sheet is given by $T_h = T_\mathrm{S}(\varphi) - \gamma h$ (γ is the lapse rate). This representation has been used recently by Peltier and Marshall (1995) in a coupled ice-sheet/EBM system.

Temperature-dependent representations have also been proposed by Huybrechts (1994) for accumulation, in the quadratic form

$$P_\mathrm{i} = a + bT + cT^2 \tag{9.41}$$

and by Khodakov (1965) for ablation, in a cubic form

$$M = \max\left[0; a(T_\Psi + b)^3\right] \tag{9.42}$$

where a, b, and c are again empirical constants (different for each formula), and T_Ψ is the summer mean surface temperature.

In all these temperature-dependent formulas the effects of factors such as Earth-orbital and CO_2 changes are implicitly included through their influence on temperature, as represented, for example, by Eq. (7.34b).

9.5 BASAL TEMPERATURE AND MELTING

If the basal temperature attains the pressure melting point, given by Eq. (8.4), liquid water forms and a new sliding mode of ice-sheet behavior can ensue. Following Verbitsky and Saltzman (1997), we now develop a simplified representation of the basal temperature compatible with the vertically integrated model, which yields results comparable to that obtained in more detailed three-dimensional thermodynamic calculations (Huybrechts, 1992).

Within the basal boundary layer of scale thickness $\eta \equiv h_\mathrm{b} - h_\mathrm{B} = (kH/a)^{1/2}$, a near balance between vertical diffusion and heating due to internal viscous stresses can be assumed to prevail (see Section 9.2), i.e.,

$$\lambda \frac{\partial^2 T}{\partial z^2} + \rho_\mathrm{i} q_\mathrm{F} = 0 \tag{9.43}$$

where q_F is given by Eq. (9.13) in which τ_{xz} and τ_{yz} are given by Eqs. (9.30) and (9.31). In accordance with the scale analysis of Section 9.2, it can be assumed that the temperature at the top of the basal boundary layer ($x = h_b$) is advected without significant change from the top of the ice sheet with a time lag $t^* = H/a$, i.e.,

$$\frac{\partial T_b}{\partial t} \approx \frac{T_h - T_b}{H/a} \tag{9.44}$$

At $z = z_B$ the basal boundary conditions, Eqs. (9.18) and (9.19), apply if $T_B < T_M$, and Eqs. (9.20) and (9.21) apply if $T_B = T_M$. Formally, the solution of Eq. (9.43) with the appropriate boundary conditions is

$$T_B = T_b + \frac{1}{\lambda}\left\{\int_{h_B}^{h_b}\int_{h_B}^{z} q_F\, dz'\, dz + \eta\left[G^{\uparrow} + (\mathcal{D} - L_f M_B)_{T_B=T_M}\right]\right\} \tag{9.45}$$

where, in accord with Eq. (9.36), if $T_B = T_M$

$$\begin{aligned}\mathcal{D} \equiv \mathbf{v}_B \cdot \underline{\tau}_B &= \frac{1}{k_B}|\mathbf{v}_B|^2 \\ &= (\rho_i g)^2 k_B (h_I \nabla h)^2\end{aligned} \tag{9.46}$$

Setting $\mathcal{D} - L_f M_B = 0$ *ab initio*, if it turns out that $T_B < T_M$ then "frozen-to-base" (no melting or sliding) conditions prevail, and the conditions $\mathcal{D} = M_B = 0$ are satisfied *a posteriori*. If, on the other hand, it turns out that $T_B \geqslant T_M$ (*temperate* glacier conditions) then basal melting can be assumed to be occurring at a rate necessary to keep the basal temperature T_B at pressure melting-point T_M. From the lower boundary condition, Eq. (9.21), this melting rate is given in terms of the rate of water thickness production, $\mathcal{M}_B = \rho_i^{-1} M_B$, by

$$\mathcal{M}_B = \frac{1}{\rho_i L_f}\left\{\lambda\left(\frac{\partial T}{\partial z}\right)_{h_B} + G^{\uparrow} + \mathcal{D}\right\} \qquad (T_B = T_M) \tag{9.47}$$

where $(\partial T/\partial z)_{h_B}$ is determined as the first integral of Eq. (9.43) (see Saltzman and Verbitsky, 1996):

$$\left(\frac{\partial T}{\partial z}\right)_{h_B} = \frac{T_b - T_M}{\eta} + \frac{q_F}{2\lambda}\eta$$

Corrections due to the effects of M_B and $\mathcal{D}$ must then be iteratively incorporated in Eq. (9.45), with the prospect of converging on the values of T_B and T_M satisfying all constraints. These values are time varying on a longer time scale associated with the evolutionary changes in h_I, for example.

The rate of variation of the basal water thickness, $W_B = m_{wB}/\rho_w$, due to melting is given by the continuity equation, Eq. (4.36), which we can write in the form

$$\frac{dW_B}{dt} = -\nabla \cdot W_B \mathbf{v}_B + \mathcal{M}_B - \mathcal{S}_w^{\downarrow} \tag{9.48}$$

where $\mathcal{M}_B$ is the net rate of water thickness production due to excess basal melting over freezing and $\mathcal{S}_w^{\downarrow}$ is the rate of water thickness loss due to downward fluxes into the underlying regolith and bedrock. Further discussion of this formalism is given by Verbitsky and Saltzman (1997) and an alternate development is given, for example, by Van der Veen and Oerlemans (1984).

9.6 DEFORMABLE BASAL REGOLITH

As noted above the disposition of basal water is influenced by the nature of the underlying lithospheric properties, particularly by the presence of more loosely aggregated, soil, till, and sediment material (i.e., regolith) that can absorb and channel water. At the same time the deformable material can have a major influence by initiating or amplifying sliding of portions of the ice sheet. This subject has been receiving increasing attention for paleoclimatic considerations, in which case the regolith thickness itself becomes an important long-term variable (Clark, 1994).

A detailed treatment of this subject requires a rigorous continuum mechanics approach applied to a poorly observed nonhomogeneous medium having complex boundary conditions. Here we simply set down the most fundamental conservation statement expressed in the form of the continuity equation, Eq. (4.2), specialized for regolith with a local density (or mass concentration) of χ_r.

$$\frac{\partial \chi_r}{\partial t} = -\nabla \cdot \overline{\chi_r \mathbf{V}_r} \tag{9.49}$$

where $\mathbf{V}_r$ is the regolith velocity to be determined from the equations of motion applied to the nonhomogeneous continuum. The total mass of regolith per unit area at any point, m_r is obtained by integrating Eq. (9.49) between the bedrock $z = h_B$ and the base of the ice sheet ($z = h_B + h_r$), i.e.,

$$\frac{\partial m_r}{\partial t} = -\nabla \cdot \mathbf{J}_r + \mathcal{S}_r \tag{9.50}$$

where $m_r = \rho_r h_r$ (the mass of regolith of mean density ρ_r and thickness h_r), $\mathbf{J}_r$ $(= \int \overline{\mathbf{v}_r \chi_r}\, dz)$ is the net horizontal flux of sediment, and $\mathcal{S}_r$ is a source function representing the production of regolith by glacial till deposition from above or bottom erosion of bedrock.

Thus, as ice sheets advance and retreat, the thickness of the regolith may vary geographically over time, influencing the behavior of subsequent ice sheets. For example, it has been suggested by Clark and Pollard (1998) and Clark *et al.* (1999) that the presence of regolith enhances sliding and peripheral ablation, leading to thinner, less massive, ice sheets. However, after repeated removal of regolith by the action of the ice sheets, the subsequent ice sheets may have more of a bedrock base that can sustain taller and more massive ice sheets. There is emerging evidence, however, that rather than acting as the deformable viscous material implied by this hypothesis (Boulton and Hindmarsh, 1987; Alley *et al.*, 1987a,b), the unconsolidated regolith material behaves more like a plastic material over which the ice can slide without significant modification when water lubricated (Engelhardt and Kamb, 1997, 1998; Hooke *et al.*, 1997; Iverson *et al.*, 1994, 1995).

9.7 ICE STREAMS AND ICE SHELVES

The detailed treatment of the ice streams that terminate as ice "lobes" over land, and as ice "shelves" in the ocean, poses difficult mechanical and thermodynamic questions that may be important in determining the behavior of the ice sheets, especially in assessing the possibility for sudden massive surges. General discussions of these questions are given, for example, by Thomas (1979), Paterson (1994), Oerlemans (1982a,b), Muszynski and Birchfield (1987), Hindmarsh (1993), and Hughes (1998).

In most models dealing with paleoclimatic variations the ice shelves have either been neglected or crudely parameterized as a function of the behavior of the much more massive ice sheet to which it is connected. This would be in keeping with the suggestion (Hindmarsh, 1993) that the mechanics of ice sheets is not likely to depend strongly on the stresses at the sheet–shelf junction. However, other studies (see Paterson, 1994) do suggest that the *pinning* of ice shelves at raised topographic points in the ocean and at the sides of embayments may play a key role in *buttressing* large ice sheets through the *backforce* exerted (Thomas, 1973a,b). In addition, it seems likely that thermal effects at the ocean shelf/sheet interface play a significant role in the ablation/calving downdraw processes governing the behavior of the whole ice sheet. Here we take the simplest view inherent in many paleoclimatic studies, by assuming that to the first order an ice shelf simply represents the outflow discharge flux of a marine-bounded ice sheet, the details of which are not critical. Using the theory of Hughes (1996, 1998), however, coupled with the thermodynamic considerations of MacAyeal (1993) and Verbitsky and Saltzman (1994), it should be possible to model numerically some significant aspects of both ice streams and their extensions as ice lobes and marine ice shelves, as suggested by Hughes (1996).

9.8 BEDROCK DEPRESSION

The importance of bedrock depression for ice-sheet modeling was noted in Section 8.1. In the most complete treatments of this topic (Peltier, 1982) careful attention is paid to geologic considerations, notably the stratified nature of the upper layers of the earth. It is generally assumed that an elastic crust and lithosphere overlie a deformable viscous asthenosphere, with the consequence that an ice sheet essentially "floats" on a depressed elastic plate. When considering details such as proglacial lake formation, or local sea level changes associated with peripheral bulges of the lithosphere due to asthenospheric motions, these geologic considerations are of great relevance. For present purposes, however, we present here the conventional simplified representation that is compatible with the level of detail embodied in vertically integrated ice-sheet models.

Briefly stated, it is usually assumed that, at any point, the bedrock depression (δ_B) below its ice-free equilibrium height will tend to relax back to a state of isostatic equilibrium with the overlying ice sheet. That is,

$$\frac{d\delta_B}{dt} = \frac{(\rho_i/\rho_b)h_I - \delta_B}{\varepsilon_\delta} \tag{9.51}$$

where ρ_b is the bedrock density and ε_δ is a relaxation time "constant," which actually depends on the ice mass distribution but is usually taken to be about 3–5 ky (Peltier and Jiang, 1996). At isostatic equilibrium, $\delta_{B0} = (\rho_i/\rho_b)h_{I0}$, where $\rho_i/\rho_b \approx 1/4$ (Hughes, 1998). This means that an equilibrium ice sheet will experience a significant drop in its height due to the isostatic adjustment with its bedrock. For example, the top of a 4-km-thick ice sheet will be about 7°C warmer due to the 1-km depression, given a typical polar atmospheric lapse rate.

In the following section we discuss briefly a potentially significant consequence of bedrock depression, namely, its effect on the calving rate in conjunction with ocean incursions.

9.9 SEA LEVEL CHANGE AND THE ICE SHEETS: THE DEPRESSION-CALVING HYPOTHESIS

The largest rates of change of sea level on paleoclimatic time scales are undoubtedly due to the growth and decay of ice sheets. At the same time, sea level changes have a significant effect on the behavior of the ice sheets and their adjacent ice shelves, as well as on the overall behavior of the oceans and atmosphere. For example, a rise in sea level will not only cause the grounding line to retreat inland, as implied by Eq. (9.23), but as a consequence of the increased flotation the ice shelf might rise above a *pinning point*, thereby weakening its buttressing effect on the main ice sheet.

More generally, the distance (h) of any topographic point above sea level can change as a consequence either of a global (i.e., *eustatic*) change in mean sea level ($\Delta_E z_{SL}$) measured relative to the center of Earth (with allowance for the geoidal mass distribution), or of a local uplift or depression of the topography (δh) due to tectonic or isostatic causes, i.e.,

$$\Delta h = -\Delta_E z_{SL} + \delta h \tag{9.52}$$

At a coastal point ($h = 0$) this change can be measured, in principle, by long tide-gauge records after averaging out higher frequency variations, usually expressed in terms of the "relative sea-level change," defined by

$$\Delta_R z_{SL} \equiv -\Delta h = \Delta_E z_{SL} - \delta h$$

Thus, for example, the local relative sea level change might be negative at a point due to a near-coastal glacioisostatic rebound, in spite of an overall glacioeustatic global increase in mean sea level.

The causes of worldwide eustatic change in sea level are resolvable into two main categories:

1. Change in the volume of ocean water, which can result from (a) the above-mentioned glacioeustatic processes involving net mass changes due to ice–water phase transformations, (b) transfers of water mass to and from the continents, e.g., lakes, groundwater, juvenile waters (it is assumed that the total mass of Earth waters in all forms has remained constant), and (c) steric processes that can occur without any ocean mass change, due to changes in temperature and salinity, in accordance with the equation of state, Eq. (4.7).
2. Change in the worldwide shape (depth profile) of the ocean basins, which results from both tectonic (e.g., sea floor spreading) and isostatic (e.g., glacial and oceanic loading) processes.

On longer Phanerozoic time scales tectonoeustatic changes of the shape of the ocean basins (e.g., sea floor spreading) are probably the main determinants of sea level, except for the intervals of pronounced glaciation. An estimate of these variations based on geologic evidence is shown in Chapter 13 (Fig. 13-3); note the similarity of this curve with the temperature record shown in Fig. 1-3.

On shorter paleoclimatic time scales, e.g., the Pleistocene, glacioeustatic/glacio-isostatic processes are undoubtedly of central importance. For this Pleistocene period, assuming a constant mean ocean area (A_W), and assuming that the steric effects of salinity changes are small compared to thermal effects, the eustatic change in sea level can be expressed to first order in the form

$$\Delta z_{SL} \approx -J_I \Delta I + J_\theta \Delta\theta \tag{9.53}$$

where $J_I = (\widehat{\rho}_i A_W)^{-1}$, $J_\theta = M_W \widehat{\mu}_T / \widehat{\rho}_W A_W$, $\widehat{\rho}_W$ is the mean density of seawater, M_W is the mean mass of the oceans over the Pleistocene, $\widehat{\mu}_T$ is the coefficient of thermal expansion weighted toward a value representive of the upper levels of the ocean, which probably provide the main contribution to the change in mean temperature of the global ocean, $\Delta\theta$.

Taking $\widehat{\rho}_W = 10^3$ kg/m^3, $A_W = 3.6 \times 10^{14}$ m^2, $M_W = 1.4 \times 10^{21}$ kg (the present value), and $\widehat{\mu}_T = 2.00 \times 10^{-4}\text{K}^{-1}$, it follows that $J_I = 2.8 \times 10^{-18}$ m/kg and $J_\theta \approx 0.8$ m/K. This latter value is probably near an upper bound because $\widehat{\mu}_T$ decreases significantly with depth and the main temperature changes probably occur in only a fraction of the total mass of the ocean. Given the plausible Pleistocene ranges, $|\Delta I| \approx$ 4–5 $\times 10^{19}$ kg and $|\Delta\theta| \approx 2$K, it follows that $\Delta_I z_{SL} \equiv J_I|\Delta I| = 112$ to 140 m and $\Delta_\theta z_{SL} \equiv J_\theta|\Delta\theta| = 1.6$ m. Thus the thermal steric effects are small compared to the glacioeustatic effects. Although there is still some argument about the magnitude of ΔI during the Pleistocene, independent evidence from sea level stands on the shores of Barbados and New Guinea seem to confirm sea level changes of more than 100 m.

At least two related, dynamical feedbacks of such sea level changes on the ice sheets have been considered:

1. The effect on the ice-sheet/shelf junction, as expressed by the flotation condition [Eq. (9.23)], that will influence the slow movements of the grounding line.
2. A potentially more catastrophic consequence due to the flooding of a depressed ice-sheet basin by rising seawater or proglacial lakes, a suggestion first made by Ramsay (1925). Such a possible instability has been postulated by Pollard (1982) as the cause of the rapid deglaciation phase of sawtooth-shaped 100-ky-period cycles of the Late Pleistocene (see Fig. 1-4b). According to this scenario, if the ice-sheet mass is reduced strongly (perhaps by an increase in solar radiation due to Earth-orbital changes) in a region already depressed due to the ice load, the rising sea level and proglacial lakes formed in the basins surrounding the ice sheet will induce rapid calving and disintegration of the remaining ice sheet. In addition to Pollard's (1982) discussion several more detailed treatments of the processes involved have been made, for example by Fastook (1983) and Van der Veen (1985).

A much simpler scaling representation of the rate of ice mass loss of a single ice sheet of mass Ψ due to this "calving catastrophe" ($\mathcal{C}_{\rm I}$) is given by Saltzman and Verbitsky (1992) in the form

$$\mathcal{C}_{\rm I} = \alpha_{\rm B}(\Psi/H) \quad \text{if} \quad D_{\rm B} > Z, \quad \text{and} \quad D_{\rm B} > D_{\rm B0} = (\rho_{\rm i}/\rho_{\rm b})H \tag{9.54}$$

and $\mathcal{C}_{\rm I} = 0$ otherwise, where $\alpha_{\rm B}$ is a constant (a suggested value of which is $20\ {\rm m\,y^{-1}}$ (Pollard, 1983); $\Psi/H \approx \rho_{\rm i} A_{\rm I}$ is the ratio of the ice-sheet mass to its scale thickness H, $A_{\rm I}$ is the scale of the ice-sheet area, $D_{\rm B}$ and Z are the scales of the depressions of bedrock ($\delta_{\rm B}$) and sea level ($z_{\rm SL}$), respectively, below the undisturbed (no ice sheet) equilibrium value of the continental elevation, $z_{\rm s}$ (see Fig. 4-3). If Z^* is the equilibrium scale value of ξ corresponding to the tectonic-mean continental elevation in the absence of any ice sheets, then

$$Z = Z^* + J_{\rm I}\Psi$$

The scale value of the equilibrium bedrock depression is denoted by $D_{\rm B0} = \rho_{\rm i}\rho_{\rm b}^{-1}H$ (see Fig. 9-5). Thus, the calving instability depends on the combined effects of bedrock depression, sea level change, and external influences that cause ice-sheet mass changes.

Other effects of sea level change on paleoclimatic evolution include the changes of (1) total landmass (see Fig. 3-8) that, e.g., can affect global biological/chemical balances as well as atmospheric behavior, (2) shorelines and sills that can alter ocean currents, (3) the moment of inertia of Earth that probably alters the orbital constants (Bills, 1994; Jiang and Peltier, 1996), and (4) the weight of the ocean that might affect large-scale tectonics. Further discussion of possible effects of changing sea level over the Phanerozoic is given in Chapter 13 (Section 13.2).

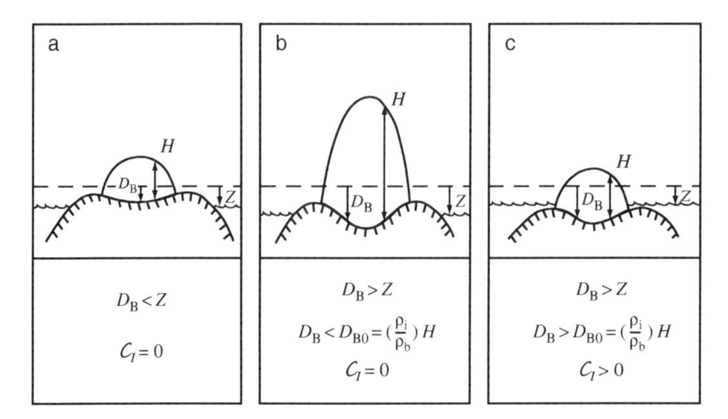

Figure 9-5 Schematic representation of the ice-sheet/bedrock/asthenosphere model described in the text. (a) Subcritical state ($\mathcal{C}_I = 0$); (c) supercritical state ($\mathcal{C}_I > 0$). After Saltzman and Verbitsky (1992).

9.10 PALEOCLIMATIC APPLICATIONS OF THE VERTICALLY INTEGRATED MODEL

Following Mahaffy's (1976) study of the Barnes Ice Cap, many applications of the simplified model described in Section 9.3 have been made to account for present-day ice sheets, notably Greenland and Antarctica. In these studies the vertically integrated ice-sheet models have been augmented to various extents by the inclusion of representations of bedrock depression, basal melting, and ice shelf effects. Some examples are Oerlemans and Van der Veen (1984), Herterich (1988), and Huybrechts and Oerlemans (1988).

These studies have shown that an acceptable first-order account of the ice thickness distributions can be obtained as a quasi-steady-state solution. In another study, using the representation of basal thermodynamics discussed in Section 9.5 and the flotation condition, Eq. (9.23), Verbitsky and Saltzman (1997) showed that the main features of Antarctica, including details of the ice streams, could be deduced to the same accuracy as the more detailed three-dimensional results of Huybrechts (1992).

Our emphasis here, however, is on paleoclimatic applications. Notable contributions in this regard have been made, for example, by Weertman (1976), Oerlemans (1980b), Birchfield *et al.* (1981), Pollard (1978, 1982, 1983a,b), Ghil and Le Treut (1981), Birchfield and Grumbine (1985), Chalikov and Verbitsky (1990), DeBlonde and Peltier (1991a,b), and Gallée *et al.* (1992), all based on an idealized one-directional (north–south) ice sheet, usually including bedrock depression, and forced by a net accumulation function in the form of an orbitally controlled snow line (Section 9.4). Two-directional (λ, φ) versions have been studied, for example, by Budd and Smith (1981), Neeman *et al.* (1988), Marsiat (1994), and Peltier and Marshall (1995).

With some necessary tuning, all of these studies can readily reproduce the orbitally forced near-20-ky- and 40-ky-period fluctuations observed in $\delta\,^{18}$O records, but cannot attain any success in reproducing the main near-100-ky-period variations of the Late Pleistocene unless additional factors are included. For example, Oerlemans (1980b) includes basal melting and surging, Pollard (1982) and Chalikov and Verbitsky (1990) include the calving mechanism, Pollard (1983a) and Deblonde and Peltier (1991a) invoke a meltwater discharge feedback, Peltier and Marshall (1995) and Tarasov and Peltier (1997a) prescribe near-100-ky-period forcing of dust loads and CO_2, Gallée *et al.* (1992) invoke a representation of albedo change as a function of snow-aging, and Clark and Pollard (1998) suggest the importance of variable subglacial sediment behavior.

As a further review, in Chapter 14 we shall place all of these models in the context of the broader suite of models proposed to account for the full Late Cenozoic ice-age variations.

9.11 A GLOBAL DYNAMICAL EQUATION FOR ICE MASS

We now develop a system of dynamical equations governing the time-dependent evolution of the mass Ψ_j of ice sheet j, the sum of such ice sheets ($I = \sum \Psi_j$) representing the total planetary ice sheet mass (the variations of which are measured by the proxy indicator, δ^{18}O. This zero-dimensional system is obtained by multiplying Eq. (9.28) by ρ_{i} and integrating over the ice-sheet area A_{i}. Thus, dropping the subscript j in this subsection for notational simplicity, with the understanding that we are dealing with an individual ice sheet, we obtain

$$\frac{d\Psi}{dt} = \oint_{\Gamma} \int_{h_{\mathrm{B}}}^{h} \rho_{\mathrm{i}} v_{\Gamma}\, dz\, d\gamma + \iint (P_{\mathrm{i}} - M)\, d\sigma + \iint M_{\mathrm{B}}\, d\sigma \tag{9.55}$$

where $d\sigma$ is an element of surface area and $d\gamma$ is an element of the total distance Γ around the edge of the ice sheet at which the normal outward ice velocity is v_{Γ}, representing a wasting discharge of ice. As discussed above, this discharge velocity can be decomposed into a normal creep component v_{n}, and fast discharge components associated either with a calving instability resulting from depression of the bedrock underlying the ice sheet (Pollard, 1983a), denoted by v_{c}, or with a sliding surge independent of bedrock depression resulting from basal melting mainly in the outer portion of an ice sheet (Paterson, 1994) denoted by v_{s}:

$$v_{\Gamma} = v_{\mathrm{n}} + v_{\mathrm{c}} + v_{\mathrm{s}} \tag{9.56}$$

Thus, we can write Eq. (9.55) in the form

$$\frac{d\Psi}{dt} = N - \mathcal{S}_{\mathrm{I}} - \mathcal{C}_{\mathrm{I}} - \mathcal{M}_{\mathrm{B}} \tag{9.57}$$

where the normal processes of slow net accumulation minus perimeter discharge of ice associated with slowly creeping flow are given by

$$N = \iint (P_{\mathrm{i}} - M)\, d\sigma - \oint\!\!\int \rho_{\mathrm{i}} v_{\mathrm{n}}\, dz\, d\gamma \tag{9.58}$$

$\mathcal{M}_B = \iint \rho_i M_B \, d\sigma$ is the total basal meltwater production rate, generally negligible compared to observed values of $d\Psi/dt$ but important nonetheless because the thin layer of water produced can facilitate a more significant rapid loss of ice due to the *surge flux* (Oerlemans and Van der Veen, 1984).

$$\mathcal{S}_I = \oint\!\!\int \rho_i v_s \, dz \, d\gamma. \tag{9.59}$$

Another related, potentially rapid, ice discharge is due to a possible bedrock/calving flux,

$$\mathcal{C}_I = \oint\!\!\int \rho_i v_c \, dz \, d\gamma. \tag{9.60}$$

Because, as noted in Section 5.1, it is not possible to calculate the net mass flux (N) representing the difference between precipitation, melting, ablation, and normal calving to the required order of magnitude of observed Pleistocene variations, we shall simply adopt the physically plausible hypothesis of Milankovitch (1930) that the summer surface temperature in high latitudes near the ice sheet (which we denote by T_Ψ) determines the survival of winter snowfall and the magnitude of the normal perimeter melting/calving, thereby decisively influencing the growth and decay of ice sheets.

The growth and decay of an ice sheet obviously depend on more than the summer mean temperature T_Ψ and its ice shelf structure. In relating $d\Psi/dt$ to T_Ψ we are essentially assuming that the full panoply of variables that describe the relevant thermohydrological processes can be projected onto a single variable, the summer high-latitude surface temperature. This has been a common assumption in many low-order models of ice variations, as was described in Section 9.4.

Formally, we set

$$N = \begin{cases} a_0 - a_1 T_\Psi - \nu_1 \Psi & (\Psi \geqslant 0) \\ 0 & (\Psi = 0,\ T_\Psi > a_0/a_1) \end{cases} \tag{9.61}$$

where T_Ψ is a fast-response variable related to slow-response variables by an equation of the form given by Eq. (7.34b),

$$T_\Psi = \widehat{T}_\Psi + B^{(\Psi)} \ln(\mu/\widehat{\mu}) + \sum_y k_y^{(\Psi)} \cdot \Delta Y \tag{9.62}$$

where $\Delta y = y - \widehat{y}$ and $y = (I, \theta, F)$; a_0 is the net rate of accumulation when $T_\Psi = 0°\mathrm{C}$ and $\Psi = 0$, taken as a constant to be determined from an initial condition; a_1 is a free parameter, and ν_I^{-1} is a dissipative time constant for the rate of collapse of an ice sheet by its own weight due internal deformation in the absence of snowfall and ablation, which in its simplest form can be taken to be a constant representative of mean ice sheet conditions. Defining $a = (\rho_i A_I)^{-1}(a_0 - a_1 T_\Psi)$ as the rate of ice accumulation at the surface of the ice sheet, this dissipative time constant is of the order $\nu_I^{-1} = H/a$.

The bedrock/calving catastrophe function, $\mathcal{C}_I$, for any ice sheet, is formulated by Saltzman and Verbitsky (1992) as an adaptation of Pollard's (1983a) development in the form given by Eq. (9.54). The simplest equation governing the bedrock depression is of the form given by Eq. (9.51), applied now to the mean (scale) bedrock depression D_B, i.e.,

$$\frac{dD_B}{dt} = eH - \nu_D D_B \tag{9.63}$$

where $e = (\rho_i/\rho_b)/\varepsilon_D$, and $\nu_D^{-1} = \varepsilon_D$ is an intrinsic response time for the bedrock (order of 10^3 to 10^4 y).

We note that in some models the bedrock depression plays a more important role in determining ice-sheet changes than simply influencing the calving process ($\mathcal{C}_I$). In particular, a depressed bedrock reduces the height of the overlying ice sheet above sea level, thereby increasing the availability of moisture for snowfall that can increase the accumulation if the temperatures are still below freezing, as in central Antarctica today. On the other hand the increased temperatures due to a reduced height might increase ablation and melting. Because it is difficult to assess the sign of the balance of these two competing processes, we shall neglect these possible effects of bedrock depression.

Concerning the surge flux, $\mathcal{S}_I$, let us assume that the rate of sliding discharge of ice mass will vary in proportion to the mean depth of basal water, W_B (cf. Oerlemans, 1982b). Moreover, $\mathcal{S}_I$ will be further accelerated or restrained in a possibly nonlinear manner as a function of the surging intensity itself. Thus we postulate a relation of the form

$$\frac{d\mathcal{S}_I}{dt} = \kappa_W \widetilde{W}_B - \nu_B\left(\mathcal{S}_I, \Psi, \widetilde{W}_B\right) \cdot \mathcal{S}_I \tag{9.64}$$

where κ_W is a free parameter, $\widetilde{W}_B$ is the area-mean basal water thickness, and ν_B is a damping coefficient that may be a nonlinear function of $\mathcal{S}_I$, Ψ, and $\widetilde{W}_B$. In accordance with Eq. (9.48) the variations of mean basal water depth, $\widetilde{W}_B$, are in turn expressable in terms of the area-mean rate of net excess of basal melting over refreezing $\widetilde{M}_B$ ($= \mathcal{M}_B/\rho_i a$), and if the water discharge [which can take the form either as a direct flux to the periphery through channels (Walder and Fowler, 1994) or as a downward subsurface flux to the underlying strata (Boulton and Dobbie, 1993)] is taken to be proportional to $\widetilde{W}_B$, then

$$\frac{d\widetilde{W}_B}{dt} = \widetilde{\mathcal{M}}_B - \nu_W \cdot \widetilde{W}_B \tag{9.65}$$

The net melting rate $\mathcal{M}_B$ is given by Eq. (9.47). Saltzman and Verbitsky (1996) show that if the basal temperature $\widetilde{T}_B$ reaches the pressure melting point, T_M [given by Eq. (8.4)],

$$\widetilde{\mathcal{M}}_B = \frac{1}{\rho_i L_f}\left\{G^{\uparrow} + \rho c_i a\left[T_H(t - H/a) - T_M\right] + 1.5\rho g a H\left(\frac{aH}{k}\right)^{1/n}\right\} \tag{9.66}$$

where L_f is the latent heat of fusion, ρ is the density of water, $G^{\uparrow}$ is the geothermal flux, c_i and k are the heat capacity and thermal diffusivity of ice, respectively, a is the mean accumulation rate at the top of an ice sheet of mean height H, n is the integer power of Glen's rheological law, and the temperature at the top of the ice sheet corresponding to a sea-level surface temperature T_Ψ and a lapse rate γ is

$$T_H = (T_\Psi - \gamma H) \tag{9.67}$$

The mean basal temperature, $\widetilde{T}_B$, given by Eq. (9.45), is expressible in the form

$$T_B = T_H(t - H/a) + (\rho_i c_i a)^{-1} G^{\uparrow} + \frac{g}{2c_i}\left(\frac{a}{k}\right)^{1/n} H^{1+1/n} \tag{9.68}$$

Thus, in summary, our basic evolution equation for a given ice sheet is

$$\frac{d\Psi}{dt} = a_0 - a_1 T_\Psi - \nu_I \Psi - \mathcal{C}_I(\Psi, D_B) - \mathcal{S}_I + \omega_\Psi \qquad (\Psi \geqslant 0) \tag{9.69}$$

where $\mathcal{C}_I$, D_B, and $\mathcal{S}_I$, are given by Eqs. (9.54), (9.63), and (9.64)–(9.67), respectively, T_Ψ is given by Eq. (9.62), and ω_Ψ represents stochastic forcing.

As noted above, in Eq. (9.61) a_0 is a coefficient that, in principle, can be determined from an initial condition, and, in its simplest form, ν_I^{-1} is a damping time constant that can be estimated to be of the order of 10^3 to 10^4 y from glaciological studies (ν_I^{-1} roughly represents the time it would take for an ice sheet to collapse of its own weight if the forcing representated by $a_0 - aT_\Psi$ were removed). Thus, if we omit consideration of $\mathcal{C}_I$ and $\mathcal{S}_I$ (which introduce some unconstrained parameters in any model) we remain only with one free parameter, a_1 $(= \partial\Psi/\partial T_\Phi)$. In view of the impossibility of calculating the difference between snow accumulation and ablation or melting of an ice sheet to the required accuracy of about 1 cm per year in equivalent sea level change (Section 5.1), we conclude that, insofar as the rate of change of the total global ice mass on geological time scales (Fig. 1-4) is concerned, it would be difficult to model the temporal variability of global ice-sheet mass more rigorously than is given by Eq. (9.69).

10

DYNAMICS OF ATMOSPHERIC CO_2

The carbon content of the atmosphere is almost completely in the form of CO_2, with only small amounts residing in other molecules such as methane. The fundamental equation governing the variations of this carbon is the continuity equation [Eq. (4.2)], which, using Eq. (4.1) can be written in the form

$$\frac{\partial \chi_C}{\partial t} = -\nabla \cdot \chi_C \mathbf{V} + \mathcal{S}^{(C)} \tag{10.1}$$

where χ_C $(= \rho\xi_C)$ is the mass of carbon per unit volume of any carrier medium (air or water) whose density is ρ, and ξ_C is the mass mixing ratio of carbon (kg C/kg atm).

Given the atomic or molecular weights (g/mol) of carbon (12), CO_2 (44), and air (29), and assuming all of the atmospheric carbon is in the form of CO_2 [i.e., $\chi_C = (12/44)\chi_{CO_2}$], we can convert χ_C to the more often used *volume mixing ratio* (or *mole fraction*) measure of CO_2 (here denoted by μ), (i.e.,

$$\mu = \frac{29}{12\rho}\chi_C$$

This volume mixing ratio is the fraction of the total number of molecules in a volume of air that are CO_2 molecules (i.e., moles of CO_2 per mole of air in a given volume), usually expressed in units of parts per million by volume (ppmv). Integrating Eq. (10.1) over the depth of the atmosphere at any point, as in Section 4.4,

$$\frac{\partial m_C}{\partial t} = -\nabla \cdot \mathbf{J}^{(C)} + q^{\uparrow} + w_G^{\uparrow} + v_{\mu}^{\uparrow} - w^{\downarrow} + w_{GA}^{\uparrow} \tag{10.2}$$

where m_C $(= \int_{z_S}^{z_T} \chi_C \, dz)$ is the mass of carbon in an atmospheric column per unit area, $\mathbf{J}^{(C)}$ $(= \int_{z_S}^{z_T} \chi_C \mathbf{v} \, dz)$ is the net horizontal flux of carbon for the column, and the vertical fluxes of carbon mass across any unit area of the surface per unit time are due to upward sea–air exchange $q^{\uparrow}$ (including the effects of biospheric uptake and surface carbonate processes in upper layer waters), net terrestrial biospheric release/uptake of

CO_2 via photosynthesis and respiration plus net organic weathering release of CO_2 to the atmosphere $w_G^{\uparrow}$ (difference between oxidative weathering of old organic matter on land and organic burial), outgassing $v_\mu^{\uparrow}$ (e.g., due to terrestrial volcanism, metamorphism, and diagenesis), rock weathering downdraw ($w^{\downarrow}$), and anthropogenic inputs due to the (rapid!) oxidation of buried organic carbon (i.e., fossil fuel) $w_{GA}^{\uparrow}$. The rock weathering downdraw ($w^{\downarrow}$) can be further partitioned into weathering of carbonates ($w_c^{\downarrow}$) and silicates ($w_S^{\downarrow}$), to be discussed below in Section 10.4. Note that a net organic *burial* at any point is given by $w_G^{\downarrow}$ ($= -w_G^{\uparrow}$), assuming no change in the carbon content of the active biosphere.

It is found, for example, that atmospheric CO_2 decreases in the spring and summer hemisphere due to uptake by the active terrestrial biosphere ($w_G^{\downarrow}$) and increases during the fall and winter. Differences also occur between the anthropogenic source regions (e.g., the Northern Hemisphere continents) and the predominantly oceanic tropical and Southern Hemisphere regions, and between local source-sink regions in the atmospheric boundary layer and the free atmosphere. For our present paleoclimatic considerations we shall ultimately neglect anthropogenic sources ($w_{GA}^{\uparrow} = 0$), and, more generally, shall consider the atmosphere to be well mixed, having a very nearly uniform annual-mean value of the volume mixing ratio μ. This uniform value of μ can be related to the total mass of atmospheric carbon ($C_a = \iint m_C \, d\sigma$, where $d\sigma$ is an element of the area of the earth) and the total mass of atmosphere ($M_A = 5.14 \times 10^{18}$ kg) by the relation, μ (ppmv) $= \kappa_\mu C_a$, where $\kappa_\mu = (29/12\ M_A) = 0.47$ ppmv/GtC (1 GtC $= 10^{12}$ kg of carbon).

Thus, retaining the anthropogenic source for completeness at this point, if we further integrate Eq. (10.2) over the total area of Earth, we have

$$\frac{dC_a}{dt} = Q^{\uparrow} + W_G^{\uparrow} + V_\mu^{\uparrow} - W^{\downarrow} + A^{\uparrow} \tag{10.3}$$

where $C_a = \kappa_\mu^{-1}\mu$, and the net vertical CO_2 fluxes across the surface of Earth (which can be expressed in units of Gt/y) are given by

$$\left(Q^{\uparrow}, W_G^{\uparrow}, V_\mu^{\uparrow}, W^{\downarrow}, A^{\uparrow}\right) = \iint \left(q^{\uparrow}, w_G^{\uparrow}, v_\mu^{\uparrow}, w^{\downarrow}, w_{GA}^{\uparrow}\right) d\sigma \tag{10.4}$$

A schematic portrayal of these fluxes is shown in Fig. 10-1, and in Fig. 10-2 we show an estimate of the present-day fluxes $Q^{\uparrow}$, $W_G^{\uparrow}$, and $A^{\uparrow}$ and of the most significant reservoir amounts of carbon. All of these fluxes are taking place continuously and varying over all time scales. It will be useful to make a separation of these fluxes into slowly varying components [denoted by the circumflex overbar, $\widehat{(\ )}$] that remain after long-term averaging (~10 My) associated with tectonic changes, and more rapidly varying components [denoted in accordance with Eq. (5.5) by $\Delta(\ )$] associated with internal and forced climate changes on shorter than purely tectonic time scales. Thus

$$\left(Q^{\uparrow}, W_G^{\uparrow}, V_\mu^{\uparrow}, W^{\downarrow}\right) = \left(\widehat{Q}^{\uparrow}, \widehat{W}_G^{\uparrow}, \widehat{V}_\mu^{\uparrow}, \widehat{W}^{\downarrow}\right) + \Delta\left(Q^{\uparrow}, W_G^{\uparrow}, V_\mu^{\uparrow}, W^{\downarrow}\right) \tag{10.5}$$

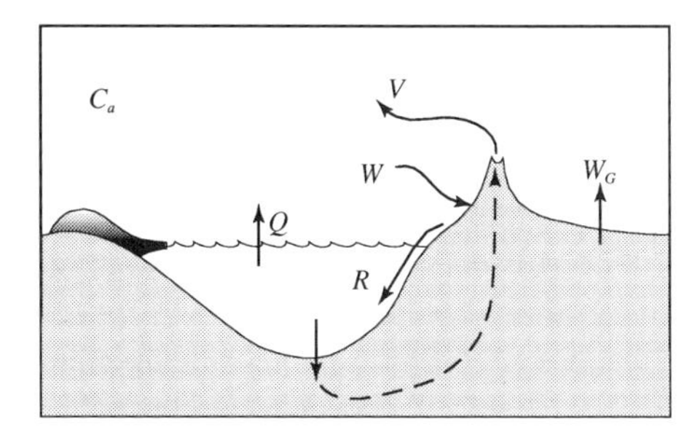

Figure 10-1 Schematic portrayal of the main fluxes comprising the carbon cycle.

10.1 THE AIR–SEA FLUX, $Q^{\uparrow}$

It is usually assumed (e.g., Broecker and Peng, 1982) that the flux of carbon in the form of CO_2 at the air–sea interface is proportional to the difference between the mole concentration (mol/m^3) of CO_2 in a thin diffusive layer of water just below the interface, denoted by $[CO_2]_S$, and the concentration in the mixed layer below this thin diffusive layer, $[CO_2]_{ml}$. The proportionality constant $\lambda(|\mathbf{V}|)$ is a function of the magnitude of mechanical mixing in the water near the surface, which is largely dependent on the magnitude of the surface wind speed $|\mathbf{V}|$. Thus, at any location

$$q^{\uparrow} = -\lambda\big(|\mathbf{V}|\big)\big([CO_2]_S - [CO_2]_{ml}\big) \tag{10.6}$$

where $q^{\uparrow}$ is the mass flux of carbon in the form of CO_2, and a square-bracketed quantity $[x]$ denotes the mole concentration of x in units of mol/m^3.

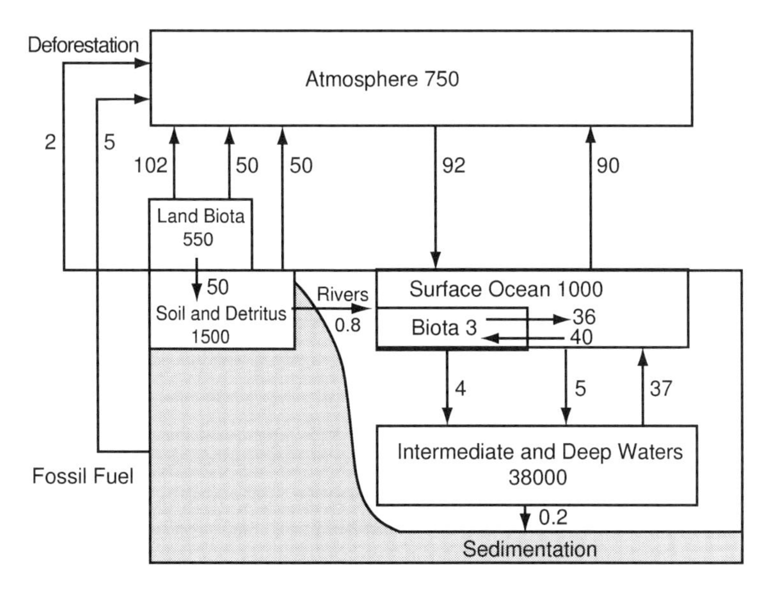

Figure 10-2 Schematic portrayal of the main carbon reservoirs and annual mean fluxes of carbon for the present day. Units are GtC for reservoirs, and GtC y^{-1} for fluxes. After Watson *et al.* (1990).

The surface water layer concentration $[CO_2]_S$ is determined by the partial pressure of CO_2 in the atmosphere at the interface, $(pCO_2)_{atm}$, and the solubility of $CO_2(\alpha)$ in accordance with Henry's law expressing the proportionality of the partial pressure of a solute to the concentration, i.e.,

$$[CO_2]_S = \alpha(T, S)(pCO_2)_{atm} \tag{10.7}$$

in which $\alpha(T, S)$ is a tabulated experimental function of temperature T and salinity S (Weiss, 1974), decreasing with both T and S. Assuming the partial pressures on either side of the air–sea interface are equal (i.e., in equilibrium), with volume mixing ratios of CO_2 near this interface at the well-mixed atmospheric value μ, we can set $(pCO_2)_{atm} \approx \mu p_{atm}$, where p_{atm} is the surface atmospheric pressure. This partial pressure $(pCO_2)_{atm}$ is sometimes called the atmospheric CO_2 *piston pressure*, continually tending to drive CO_2 into the ocean against an opposing partial pressure of CO_2 in surface seawater, $(pCO_2)_{ml}$, determined by the carbon content of the mixed layer.

Thus, using Eq. (10.7), Eq. (10.6) for $q^{\uparrow}$ can be rewritten in terms of $(pCO_2)_{atm}$ in the alternate form

$$q^{\uparrow} = -K_{\mu}\big(|\mathbf{V}|, T, S\big)\big[(pCO_2)_{atm} - (pCO_2)_{ml}\big] \tag{10.8}$$

where $K_{\mu}(|\mathbf{V}|, T, S) = [\lambda|\mathbf{V}| \cdot \alpha(T, S)]$ is the *bulk gas-exchange coefficient* for CO_2 (e.g., Liss and Merlivat, 1986), and $(pCO_2)_{ml} = \{[(CO_2)_{ml}/\alpha(T, S)]\}$ is the partial pressure of CO_2 in surface seawater in solubility-equilibrium with the atmosphere.

The concentration of CO_2 in this mixed layer, $[CO_2]_{ml}$, is determined by complex processes involving the circulation, biology, and chemistry of the ocean. Formally, $[CO_2]_{ml}$ is determined by the *law of mass action* applied to the main carbon species present in the mixed layer; these species are carbonate ions whose mole concentration is $[CO_3^{2-}]$, and bicarbonate ions of concentration $[HCO_2^-]$, i.e.,

$$\begin{aligned} \big[H^+\big]\big[HCO_3^-\big] &= K_1[CO_2]_{ml} \\ \big[H^+\big]\big[CO_3^{2-}\big] &= K_2\big[HCO_3^-\big] \end{aligned}$$

where K_1 and K_2 are the equilibrium constants that are functions of temperature and salinity. Solving for $[CO_2]_{ml}$,

$$[CO_2]_{ml} = \frac{K_2[HCO_3^-]^2}{K_1[CO_3^{2-}]} \tag{10.9}$$

In practice this formula is not useful for determining $[CO_2]_{ml}$ in the ocean because neither $[HCO_3^-]^2$ nor $[CO_3^{2-}]$ is directly measureable. Instead it is desirable to rewrite Eq. (10.9) in terms of two measureable quantities, the values of which are the end result of complex physical/chemical/biological processes in the ocean:

1. The "dissolved inorganic carbon" (usually denoted by ΣCO_2, or simply DIC),

$$\Sigma CO_2 = \big[HCO_3^-\big] + \big[CO_3^{2-}\big] + [CO_2] \tag{10.10}$$

2. The "carbonate alkalinity,"

$$A_c = \left[HCO_3^-\right] + 2\left[CO_3^{2-}\right] \tag{10.11}$$

The other components comprising the total alkalinity, A, are either small or independently measureable (e.g., the borate component).

Neglecting $[CO_2]$ in Eq. (10.10), which is generally less than 1% of the ΣCO_2, we can write $(pCO_2)_{ml}$ in terms of ΣCO_2 and A_c as follows:

$$(pCO_2)_{ml} = \frac{[CO_2]_{ml}}{\alpha(T,S)} = \frac{1}{k^*(T)}\alpha(T,S)\,\frac{(2\Sigma CO_2 - A_c)^2}{(A_c - \Sigma CO_2)} \tag{10.12}$$

where $k^*(T) = K_1/K_2 \approx 1780 - 20T$ (Broecker and Peng, 1982).

Thus, as a first-order thermodynamical equation of state for CO_2 in the mixed layer, we can write

$$(pCO_2)_{ml} = (pCO_2)^*_{ml} + \sum_x \left[\frac{\partial(pCO_2)_{ml}}{\partial x}\right]^* \delta x \tag{10.13}$$

where $x = (T, \Sigma CO_2, A_c, S)$, $(pCO_2)^*_{ml}$ is a constant appropriate for standard values x^*, and $\delta x = (x - x^*)$. As discussed more fully by Takahashi *et al.* (1993), for global mean values (with the second equation being the Revelle buffering factor),

$$\frac{[\partial(pCO_2)_{ml}/\partial T]^*}{(pCO_2)^*} \approx 0.423 \quad {}^\circ C^{-1}$$

$$\left[(pCO_2)_{ml}/\partial(\Sigma CO_2)\right]^* \cdot \frac{(\Sigma CO_2)^*}{(pCO_2)^*} \approx 10.0$$

$$\left[\partial(pCO_2)_{ml}/\partial A\right]^* \cdot \frac{A_0}{(pCO_2)^*_{ml}} \approx -9.4$$

$$\left[\partial(pCO_2)_{ml}/\partial S\right]^* \cdot \frac{S^*}{(pCO_2)^*_{ml0}} \approx 0.94$$

10.1.1 Qualitative Analysis of the Factors Affecting $Q^{\uparrow}$

It can be inferred from Eq. (10.8) that $q^{\uparrow}$ will increase with the following conditions:

1. Lowered values of atmospheric CO_2 partial pressure, which is proportional to the well-mixed atmospheric CO_2 concentration μ ("piston pressure" effect).
2. Increased surface water temperature through the solubility effects embodied in $\alpha(T)$ and $k^*(T)$.
3. Increased values of the dissolved inorganic carbon, ΣCO_2, the magnitude of which can be reduced due to consumption of carbon by soft-tissue organic productivity or increased due to upwelling of carbon-rich waters from the huge reservoir of ΣCO_2 in the deep ocean (a major part of which originates from the oxidation of soft-tissue organic matter at deeper levels) (see Fig. 10-2). This organic productivity can

be enhanced by the presence of nutrient elements such as phosphorus, nitrogen, and iron. The transfer of carbon to the deep ocean by falling organic matter is known as the "organic biological pump." A simplified form of the chemical reaction involved in this organic pump is

$$CO_2 \rightleftharpoons C(org) + O_2 \tag{10.14}$$

The upper (forward) arrow represents oxidation, respiration, and remineralization; the lower (reverse) arrow represents production and photosynthesis.

4. Decreased values of the carbonate alkalinity, A_c, that would result from increased calcium carbonate productivity in surface waters (e.g., as hard-tissue organic shells or coral reefs) according to the reaction $Ca^{2+} + CO_3 \rightleftharpoons CaCO_3$, or

$$Ca^{2+} + 2HCO_3^- \rightleftharpoons CaCO_3 + \underbrace{CO_2 + H_2O}_{H_2CO_3} \tag{10.15}$$

Note from Eqs. (10.10) and (10.11) that the reduction of A_c due to $CaCO_3$ production is twice that of ΣCO_2. The $CaCO_3$ shells that rain to the deeper ocean are either deposited in the sediments or dissolve by the reverse reaction [Eq. (10.15)], thereby transferring alkalinity to the deeper ocean (constituting an *alkalinity pump*). On the other hand, upwelling waters will bring alkalinity back to the surface along with ΣCO_2, but from Eq. (10.12) the ΣCO_2 increase will have the dominant effect on $(pCO_2)_{ml}$ and hence on $q^{\uparrow}$.

In addition, alkalinity will tend to increase due to organic productivity, which requires that nutrients (e.g., nitrate cations $[NO_3^-]$) be consumed. This nitrate consumption requires a simultaneous decrease in $[H^+]$ ions. Thus, in accordance with Eq. (10.12), the alkalinity effect will diminish (but not reverse) the effect of the ΣCO_2 decrease on pCO_2 arising from the organic biological pump. A good supplementary discussion of the biogeochemical factors affecting $Q^{\uparrow}$ is given by Najjar (1992).

From all the above considerations it is clear that the net global flux of CO_2 at the atmosphere–ocean interface, $Q^{\uparrow}$, represents a delicate balance of upward and downward fluxes in different regions of the ocean surface that depend on the spatial and temporal variations of wind, sea surface temperature, atmospheric CO_2 concentration, rates of upwelling and downwelling, and local production of soft-tissue organisms and hard-tissue ($CaCO_3$) organisms. The following broad generalizations regarding the role of the oceanic state on the net CO_2 flux are suggested.

10.1.1.1 Ocean temperature. A colder ocean, particularly at the surface, can dissolve more atmospheric CO_2. Thus, the net flux of CO_2 to the atmosphere should vary as a function of global surface temperature, and also deeper ocean temperature, i.e.,

$$\Delta Q^{\uparrow} \sim \Delta T_S \sim \Delta \mu \tag{10.16}$$

representing a positive feedback.

At the same time, colder surface waters favor the growth of siliceous planktonic organisms (e.g., diatoms and radiolaria) over calcareous ($CaCO_3$) organisms. The net effect would be a reduction in the production and flux of CO_2 to the atmosphere [see

Eq. (10.15)] in cold (low CO_2) intervals and an increased flux to the atmosphere in warm (high CO_2) intervals. This represents another, independent, global mechanism whereby a positive feedback is introduced in the carbon cycle, satisfying Eq. (10.16).

10.1.1.2 Ocean circulation. The vertical circulation of the oceans takes two main forms (to be discussed in more detail in Chapter 11):

1. The global-scale thermohaline circulation resulting from density-driven downwelling in high latitudes, where heavy cold water is produced at the surface (aided at present in the North Atlantic by the horizontal transport of salt by the gyre circulation and Mediterranean outflow). This sinking water is compensated by upwelling in lower latitudes (particularly near the equator, where the convergent trade winds induce upward Ekman pumping of water) that ultimately flows poleward. In the Atlantic a large component of this upwelling branch actually occurs in the Southern Ocean, apparently due to the inhibiting effect of a deep stable thermocline in tropical latitudes and more intense wind-driven Ekman pumping in the Southern Ocean.
2. Smaller scale overturnings mainly in higher latitudes in the form of localized deep convection cells and vertical circulations in baroclinic eddies that form in regions of strong horizontal density gradients (e.g., near the Gulf Stream).

Because the carbon content of the deeper ocean is larger than near the surface, where CO_2 values are lowered due to organic productivity (see Fig. 10-3), all of these circulations tend to produce a net upward flux of carbon and hence an increase in the flux of CO_2 to the atmosphere. This effect is most pronounced in the tropical upwelling branch of the TH and Ekman circulations, where relatively cold, carbon-rich, waters are brought to the surface and radiatively warmed, releasing large amounts of CO_2 to the atmosphere (see Fig. 10-4) (Takahashi, 1989). When such upwellings are reduced, as in El Niño episodes, there is a measureable decrease in the CO_2 efflux (Feely *et al.*,

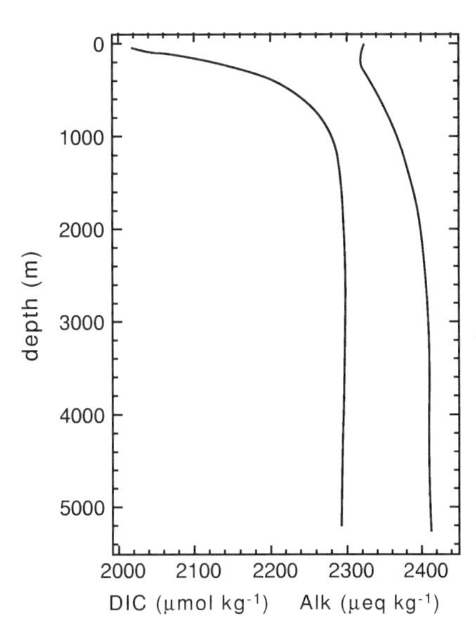

Figure 10-3 Average vertical profiles of DIC and Alk, derived from GEOSECS and TTO cruise data. After Najjar (1992).

1997, 1999; Siegenthaler, 1990). On the other hand, the vertical circulations not only transport carbon but also nutrients (notably nitrogen, phosphorous, and iron) that can increase productivity. This will tend to consume carbon in surface waters, reducing the effect of the upward carbon transport.

The uptake of CO_2 from the atmosphere in colder polar waters and subsequent *in situ* transport to deeper levels in the downward branch of the TH circulation [sometimes described as a *solubility pump* (Volk and Hoffert, 1985)] are reduced by the upward transfer of carbon in winter months due to the local smaller scale deep overturnings. It appears that, on balance, the combined "circulation-solubility" effects over the world ocean tend to pump CO_2 upward to the atmosphere.

The strength of these vertical circulations is strongly related to the depth and extent of the thermocline, which acts as a stabilizing lid or "barrier" to vertical motions, including those that might be caused by Ekman wind stresses. This is especially true in tropical latitudes, where the density stratification is more fully determined by temperature than in higher latitudes, where salinity plays a more important role (Pond and Pickard, 1983). At the same time, an extensive thermocline also implies that warm surface waters extend poleward, tending to intensify the frontal zone, storminess, and rainfall, in higher latitudes, thereby freshening the surface waters there and further inhibiting convection. The stabilizing effect of density stratifications on the vertical CO_2 flux in higher latitudes has been discussed by Francois *et al.* (1997).

As was discussed in Chapter 8, and will be discussed more fully in the next chapter, this thermocline state is in turn proportional to the mean temperature of the deep ocean, which we have denoted by θ (see Section 8.3). Thus,

$$\Delta Q^{\uparrow} \sim -\Delta\theta \tag{10.17}$$

10.1.1.3 Sea level change. The sea level changes that accompany ice-sheet variations can affect the carbon cycle in several significant ways. Three particular mechanisms involving air–ocean transfer have been widely discussed, and a fourth mechanism involving the terrestrial biosphere is discussed in the next section.

1. The coral reef hypothesis (Berger, 1982): When the sea level drops, the coral reefs ($CaCO_3$) formed or deposited on the previously submerged continental shelves and platforms (particularly in lower latitudes) are exposed, and by the reverse (weathering) reaction [Eq. (10.15)] they draw down CO_2 from the atmosphere. This component of the total weathering process, $W_c^{\downarrow}$, is discussed more fully in Section 10.4. Conversely, when the sea level rises in conjunction with melting ice sheets, coral reefs grow upward toward sea-level, releasing CO_2 in the surface waters and ultimately to the atmosphere. Thus the variations of the upward flux of CO_2 due to variations in sea

Figure 10-4 Estimate of the global distribution of the net upward flux of carbon from the ocean to the atmosphere. After Takahashi (1989).

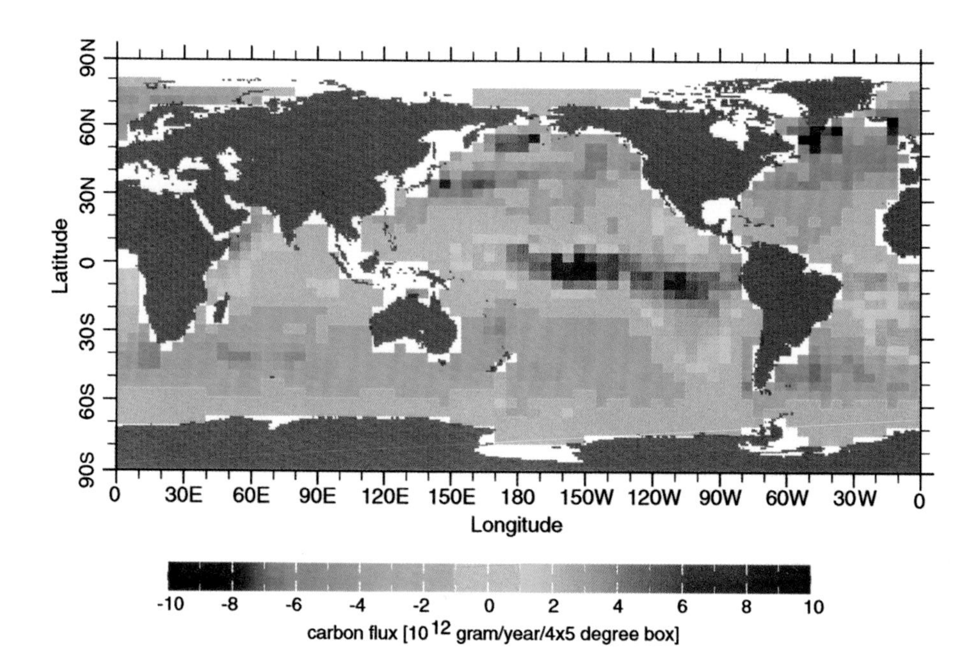

90N
60N
30N
0
30S
60S
90S
Latitude
0
30E
60E
90E
120E
150E
180
150W
120W
90W
60W
30W
0
Longitude
-10
-8
-6
-4
-2
0
2
4
6
8
10
carbon flux [10 12 gram/year/4x5 degree box]

level, are related to the CO_2 changes as follows:

$$\Delta(Q^{\uparrow} + W_c^{\uparrow}) \sim -\left(\frac{dI}{dt}\right) \sim \Delta\mu \tag{10.18}$$

representing another positive feedback. According to Opdyke and Walker (1992) and Munhoven and Francois (1996), this mechanism alone can account for a substantial fraction (60 ppmv) of the total CO_2 change that occurred between glacials and interglacials (see Figs. 3-6 and 8-2).

2. The nutrient-shelf hypothesis (Broecker, 1982): Not only are coral reefs exposed to weathering when the sea level drops, but also the nutrients that were deposited on the continental shelves by river flow from the adjacent continents. Due to erosion and runoff of the exposed beds a large supply of nutrients becomes available to the surface ocean waters, promoting productivity and hence a downdraw of CO_2 from the atmosphere. Once again Eq. (10.16) applies, i.e., $\Delta Q^{\uparrow} \sim \Delta\mu$, representing an enhancement of the coral reef positive feedback.

3. By changing the total volume of ocean water the concentrations of all its trace components will vary, with potentially significant effects on variables such as salinity.

10.1.1.4 Dust-borne iron and carbonate deposition. In addition to the sea level consequences of global ice mass changes discussed above, as noted in Section 3.5 there are also accompanying increases in aridity and atmospheric dust load with increased glaciation. This may result in an increased sprinkling of iron-rich dust particles on the surface of the ocean that can enhance productivity by iron fertilization and hence draw down CO_2 from the atmosphere (Martin, 1990). In addition, deposition of carbonate-rich dust increases surface alkalinity and hence decreases CO_2 further by the action of the "alkalininty pump" (see Section 10.1.1). Thus, we have two other possible positive feedbacks for CO_2 of the same form as Eq. (10.18).

10.1.2 Mathematical Formulation of the Ocean Carbon Balance

From a more formal viewpoint, the many biogeochemical processes that affect the amount of carbon (C_w) and alkalinity (A) in a volume of the uppermost layer of the ocean, both of which are major determinants of $Q^{\uparrow}$ in accordance with Eqs. (10.8) and (10.13), can be represented in the form of a dynamical system in which the nutrient content (measured by the amount of phosphate, denoted by P) is considered as an additional variable (Munhoven, 1997). Denoting these properties in a shallow upper layer by the subscript d, and in the deep ocean by the subscript D, the governing equations can be expressed in the following general form, which can be specialized for any number of connected ocean basin volumes (e.g., the Atlantic, Pacific, Indian and Southern Ocean, or in the integral over the whole world ocean):

Upper layer, d

$$\frac{dC_{wd}}{dt} = Z_d^{(c)} - Q^{\uparrow} + R^{(c)} - B_d^{(c)} - B_d^{(G)} \tag{10.19}$$

$$\frac{dA_{\mathrm{d}}}{dt} = Z_{\mathrm{d}}^{(A)} + R^{(A)} - 2B_{\mathrm{d}}^{(c)} + r_{(\mathrm{N:C})}B_{\mathrm{d}}^{(G)} \tag{10.20}$$

$$\frac{dP_{\mathrm{d}}}{dt} = Z_{\mathrm{d}}^{(P)} - U_{\mathrm{d}}^{(P)} \tag{10.21}$$

Deep ocean, D

$$\frac{dC_{\mathrm{wD}}}{dt} = Z_{\mathrm{D}}^{(c)} - B_{\mathrm{D}}^{(c)} - B_{\mathrm{D}}^{(G)} + V_{\mathrm{w}}^{\uparrow} \tag{10.22}$$

$$\frac{dA_{\mathrm{D}}}{dt} = Z_{\mathrm{D}}^{(A)} + r_{(\mathrm{N:P})}B_{\mathrm{D}}^{(G)} - 2B_{\mathrm{D}}^{(c)} \tag{10.23}$$

$$\frac{dP_{\mathrm{D}}}{dt} = Z_{\mathrm{D}}^{(P)} + r_{(\mathrm{P:C})}B_{\mathrm{D}}^{(G)} \tag{10.24}$$

where $Q^{\uparrow}$ is given by Eqs. (10.8) and (10.13), the net influx of any quantity $x = (C, A, P)$ by the circulation is given by

$$\begin{aligned} Z^{(x)} &= q_{\psi} \cdot \left[x(\mathrm{in}) - x(\mathrm{out})\right] \\ &= Z^{(x)}(\mathrm{in}) - Z^{(x)}(\mathrm{out}) \end{aligned} \tag{10.25}$$

Here q_{ψ} is the volume exchange rate between ocean volumes in $m^3\,s^{-1}$, $R^{(x)}$ is the rate of delivery of x by river flow, $B^{(c)}$ and $B^{(G)}$ are the rates of consumption of C_{w} by inorganic ($CaCO_3$) and organic biological productivity, respectively (negative values denoting carbonate, dissolution, and organic remineralization, respectively), and $U^{(P)}$ is the rate of uptake of phosphate by biological productivity. Munhoven (1997) assumes that the organic biological productivity is linearly related to the phosphorous uptake, i.e., $B^{(G)} = r(\mathrm{C:P})U^{(P)}$, where $r(x : y)$ denotes the mean ratio of x to y in seawater, and $U^{(P)}$ is linearly related to the phosphate inflow rate $Z^{(P)}(\mathrm{in})$. The inorganic $CaCO_3$ productivity, $B^{(c)}$, contains a component that is also proportional to $U^{(P)}$, supplemented by coral reef productivity, which is determined by sea level and its rate of change as described in Section 10.1.1. The system can be closed either by formulating additional physical equations for the ocean volume flux q_{ψ} and the river fluxes $R^{(x)}$, or by prescribing these quantities as forcing functions (Munhoven, 1997). Note that when summed over all domains, $\sum Z_i^{(x)} = 0$. A separate set of physical considerations must be invoked to calculate the carbonate dissolution rate of carbonate in the deep ocean and sediments ($-B_{\mathrm{D}}^{(c)}$); these considerations involve calculation of mean lysocline and carbonate compensation depths as well as diagenetic processes in the sediments.

A feedback diagram illustrating the connections between the variables, as part of a larger system involving ice sheet and deep ocean temperature variations, is shown in Chapter 12 (Fig. 12-2).

Another variable of interest is the oxygen (O_2) level of the ocean. Whereas O_2 may be considered a constant in the upper layer, in equilibrium with the atmosphere ($d[O_2]_{\delta}/dt = 0$), the variations in the deep ocean are dependent on ventilation by the overturning circulation $Z_{\mathrm{D}}^{(\mathrm{O}_2)}$, and consumption by organic remineralization $B_{\mathrm{D}}^{(G)}$, i.e.,

$$\frac{d[\mathrm{O}_2]_{\mathrm{D}}}{dt} = Z_{\mathrm{D}}^{(\mathrm{O}_2)} + r(\mathrm{O}_2 : \mathrm{C}) B_{\mathrm{D}}^{(G)} \tag{10.26}$$

By extending this type of formal structure to a fuller grid of ocean volumes in the context of an OGCM, adding further considerations of the isotopic ($\delta\,^{13}$C) properties of oceanic carbon, even more detailed models of the air–sea carbon flux can be examined (e.g., Maier-Reimer *et al.*, 1993).

10.1.3 A Parameterization for $Q^{\uparrow}$

To complement the above more detailed (though still highly approximate) treatments of the factors that quantitatively influence $Q^{\uparrow}$ we may take a much simpler global approach by assuming that the processes that can affect $Q^{\uparrow}$, represented by Eqs. (10.16), (10.17), and (10.18), can be formalized by the following general expression,

$$Q^{\uparrow} = r_0 + r_1 \widetilde{T}_{\mathrm{S}} - r_2 \dot{I} - r_3 \theta - \nu_{\mu} \mu \tag{10.27}$$

where r_1, r_2, and r_3 are positive constants, ν is a damping rate constant, and $\dot{I} = d(\sum \Psi)/dt$ is given by Eq. (9.69), i.e., $\dot{I} \sim -\Delta T_{\Psi}$. As already noted [e.g., Eqs. (10.16) and (10.18)], because $\widetilde{T}_{\mathrm{S}} \sim \mu$ and $\dot{I} \sim -\Delta\mu$, it is clear that opposing the normal negative feedbacks represented by $\nu_{\mu}\mu$ (e.g., the *piston pressure effect*) are substantial possibilities for positive feedback, particularly when there are significant planetary ice variations and associated sea level changes). These possibilities, together with Eq. (10.17), suggest that the total effect of the processes represented by Eq. (10.27) can be expressed in an alternate form in which the fast-response variables T_{S} and T_{Ψ} are eliminated,

$$Q^{\uparrow} = \kappa_{\mu}^{-1}\left[\beta_0(\widehat{\mu}) - K_{\mu}\mu - \beta_{\theta}\theta\right] \tag{10.28}$$

where $\kappa_{\mu} = \mu/C_{\mathrm{a}} = 0.47$ ppmv/GtC is a scaling factor, and we set $K_{\mu} = (\beta_1 - \beta_2\mu + \beta_3\mu^2)$ as the simplest canonical form of a variable dissipative rate constant that allows for the possible dominance of positive feedback for some lower range of μ when ice is likely to prevail. The broader implications of Eq. (10.28) in the context of dynamical systems analysis was discussed in Chapter 6. The coefficients β_1, β_2, β_3, and β_{θ} are free parameters, and $\beta_0(\widehat{\mu})$ is a slowly varying function of the tectonic mean value of CO_2 ($\widehat{\mu}$) (see Section 5.3) that can be determined from the long-term tectonic constraints, correcting for the possible inapplicability of $K_{\mu}(\widehat{\mu})$ when $\widehat{\mu}$ is in a high range of values indicative of ice-free conditions. As noted above, $Q^{\uparrow}$ includes the effects of the weathering of coral reefs exposed during ice-related sea level changes.

As a more general argument for proposing so simple a form for $Q^{\uparrow}$ as Eq. (10.28), we repeat a previous discussion by Saltzman and Maasch (1991). The difficulties of evaluating the fluxes for both continent and ocean are made apparent by the lack of sufficient understanding of these fluxes to be able to account for the deficit between the known anthropogenic input and the known increase in atmospheric CO_2, representing an unexplained loss of roughly 1 ppm per year to the continental and oceanic reservoirs (Takahashi, 1989). It is therefore clearly beyond our capabilities to account for the major CO_2 variations that occurred naturally in the Pleistocene, measured by gas samples in the Vostok core to be the order of 10^{-2} ppm per year (Barnola *et al.*, 1987). The plethora of plausible biological/chemical/physical scenarios and box models that have been advanced to account for these Pleistocene CO_2 variations (e.g., Broecker, 1982; Berger, 1982; Keir and Berger, 1983; Knox and McElroy, 1984; Toggweiller and Sarmiento, 1985; Wenk and Siegenthaler, 1985; Volk and Hoffert, 1985; Boyle, 1988b; Broecker and Peng, 1989) are testimony that the required fluxes are of such a low order of magnitude that almost all factors affecting the carbon cycle can be of relevance. We are, therefore, in much the same position with regard to the observed Pleistocene variations of CO_2 as we find ourselves in with regard to global ice mass and deep ocean temperature (Saltzman and Sutera, 1984)—that is, just as it is impossible at present to calculate the fundamental fluxes leading to net accumulation of ice and net changes in deep ocean temperature to the accuracy required by the observations over the Cenozoic (see Chapter 5), it is also impossible to calculate the fundamental CO_2 fluxes leading to net changes of CO_2 on these same time scales. Thus, we are faced with the need to construct, inductively, a meaningful law for modeling the slow variations of CO_2 based on matching the consequences of qualitatively plausible arguments (such as depicted in the above scenarios) with the observed CO_2 record. It is in this inductive spirit that we have proposed the parameterization given by Eq. (10.28). Nonetheless, we must recognize that it is of great value to pursue the development of more detailed models in which the ocean is spatially resolved and in which closure is achieved by independent dynamical equations for ΣCO_2, A, and nutrients (e.g., P) involving explicit consideration of the ocean circulation, biological productivity, and weathering products (see, e.g., Maier-Raimer *et al.*, 1993), a simplified version of which is represented by the Munhoven and Francois (1996) system, Eqs. (10.19)–(10.26). Such a development will be of particular value to the extent that new deductions are made that can account for, and predict, the past evolution of other measureable variations, such as of calcium carbonate deposition and phosphorous variations as indicated by cadmium/calcium variations (see Section 2.5.1).

10.2 TERRESTRIAL ORGANIC CARBON EXCHANGE, $W_G^{\uparrow}$

The exchange of carbon between the atmosphere and the organic material in the combined biospheric regolith, and rock pools (e.g., vegetation, soil, peat, methane hydrate, coal, and sedimentary kerogen in rocks) comprising the terrestrial biosphere was subdivided according to Eq. (10.5) into two components, $W_G^{\uparrow} = \widehat{W}_G^{\uparrow} + \Delta W_G^{\uparrow}$. The component $\Delta W_G^{\uparrow}$ represents shorter term processes on time scales ranging from living biomass successions ($\sim$100 y) to scales on which peat and other soil carbon masses can vary ($\sim$1000 y) under the influence of transient ice-age oscillations. $\widehat{W}_G^{\uparrow}$ represents

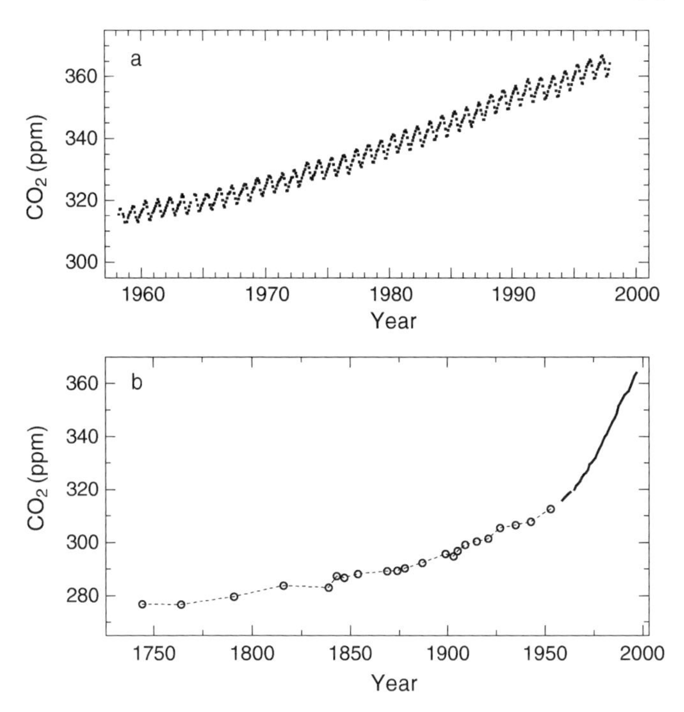

Figure 10-5 (a) Variation of the monthly average atmospheric CO_2 concentration (ppmv), observed continuously at Mauna Loa, Hawaii, over the past 30 y (Keeling *et al.*, 1989). (b) Variation of CO_2 over the past 250 y as indicated by air trapped in ice at Siple Station, Antarctica (circles) (Neftel *et al.*, 1985; Friedli *et al.*, 1986) and by the Mauna Loa measurements (solid line).

very long-term net fluxes that prevail over millions of years, representing the major burials (e.g., coal and sedimentary kerogen) and weathering that occur in conjunction with tectonic forcing. Although the reservoir of organic carbon in the biospheric pools that exclude rocks is much less than the total carbon in the ocean, it is still large in comparison with the atmosphere (see Fig. 10-2). A small percentage change due to perturbations in the net air–biosphere fluxes can significantly alter the atmospheric CO_2 content, potentially accounting for a large fraction of the observed variations on Pleistocene ice age time scales. In this section we emphasize the shorter time scale processes involving the terrestrial (continental) biosphere $\Delta W_G^{\uparrow}$, reserving Section 10.2.4 for a brief discussion of the longer term carbon burial processes $\widehat{W}_G^{\uparrow}$. The shorter term oceanic biosphere processes were included in $Q^{\uparrow}$ as the "organic biological pump."

The potential role of the terrestrial biosphere on atmospheric CO_2 variations is clearly illustrated by the presently observed seasonal variation of CO_2 (see Fig. 10-5a). It is seen that, along with the monotonic increase in the annual mean CO_2 level due to anthropogenic forcing, there is a roughly 10 ppm seasonal variation associated with the vegetative uptake and release of CO_2 over the extensive Northern Hemisphere land masses. This climatic effect is to be distinguished from the "CO_2 – fertilization" effect whereby higher values of CO_2 enhance the "greening" of the terrestrial biosphere, drawing down CO_2 (e.g., Friedlingstein *et al.*, 1995). Both of these negative feedback effects operate on relatively short time scales, but are opposed by other essentially positive feedbacks that occur on somewhat longer time scales (to be discussed in the next three subsections).

10.2.1 Sea Level Change Effects

In addition to the effects of sea level change on the air–sea carbon exchange described at the end of Section 10.1, the changes in shelf area exposed for vegetative growth can also have a significant effect on CO_2 variations. During the last glacial maximum, about 20 ka, for example, there were vast new shelf areas for vegetative growth and consumption of CO_2, particularly in the present-day regions of Southeast Asia and Australia (see Fig. 3-8). The reverse process occurs at times of ice-sheet decay, the net effect being yet another positive feedback over the ice-age cycle for which Eq. (10.18) is appropriate. It is recognized, however, that at the same time that continental shelf area is being made available for vegetative growth, other more poleward vegetated areas are being covered with glaciers and colder temperatures, which may be reducing vegetation worldwide (Adams *et al.*, 1990).

10.2.2 Thermal Effects

As a general rule chemical reactions (e.g., oxidation of carbon) occur at a faster rate during warm conditions than during cold conditions. Thus warming trends favor the upward flux of CO_2 from the regolith carbon pool (e.g., soil peat, and methane hydrates) and colder trends favor storage in this pool. Oscillations in the carbon flux due to this factor are most pronounced in higher latitudes, where the climatic swings in temperature and water availability are greatest. In particular, during moist cooling stages optimum conditions arise in these latitudes for storing carbon during transitions to peatland, and formation of methane hydrate (clathrate) in frozen soil (tundra) under anaerobic conditions. During warming stages the balance appears to shift toward a release to the atmosphere of the stored CO_2 and methane. The precise timing of the storage and efflux of carbon is still in question, however (e.g., Raich and Potter, 1995; Goulden *et al.*, 1998; Smith and Shugart, 1993; Waelbroeck *et al.*, 1997). These thermal effects probably influence even larger carbon storage and efflux processes in tropical land masses (e.g., Klinger, 1991).

We note also, that in addition to terrestrial processes, the storage and release of carbon in methane hydrate in shelf and deep ocean sediments under the influence of temperature variations may be of significance in the carbon balance of the ocean and hence the atmosphere.

10.2.3 Ice Cover Effects

Closely related to the above sea level and thermal effects are the ice cover effects of the large ice sheets that occurred in extensive high-latitude areas. When ice sheets grow they alter the climates in their advance, producing the cool moist conditions that favor the transition from forest to peatlands. According to the ideas of Franzén (1994), these peatlands may subsequently be covered by the advancing ice sheets, trapping large stores of carbon. In the process, the subglacial regolith containing the stored carbon is reworked, so that when the ice sheets retreat under warming conditions the newly exposed peat material is more vulnerable to oxidation. The resulting release of CO_2 to the atmosphere could last for several thousand years before a new period of peatland growth again occurs under cooling conditions. This proposed cycle of carbon storage and release forms the basis of yet another possible positive feedback that could also increase CO_2 during a time of high CO_2 (when ice sheets are retreating) and decrease CO_2 further during a time of low CO_2 (when ice sheets are growing) (see Fig. 8-2). This kind of scenario has been discussed by Franzén (1994) and co-workers (1996), Klinger (1991), and Klinger *et al.* (1996), as a potentially important internal contributor to the maintenance of the ice-age cycles of the Late Pleistocene. A similar positive feedback scenario with regard to the methane hydrate (clathrate) cycle has been discussed by MacAyeal and Lindstrom (1990).

In summary, many of the terrestrial biospheric effects (Sections 10.2.1–10.2.3) operating on the same time scales as the Late Pleistocene glaciation cycles have the potential for providing a positive feedback that can supplement the positive feedbacks involving air–sea interactions noted in Section 10.1 and represented in Eq. (10.28). Thus it is possible that to first order, $\Delta W_G^{\uparrow} \sim \Delta\mu$. Much work remains to be done on this subject, however, before definitive conclusions can be made regarding the timing and global nature of the complex processes involved.

In the next subsection we consider the longer term, tectonic time scale processes $\widehat{W}_G^{\uparrow}$ that involve deep burial of carbon (e.g., as coal, or as kerogen in deep ocean sediments) and in Section 10.4 we discuss the influence of vegetation on long-term weathering processes.

10.2.4 Long-Term Terrestrial Organic Burial, $\widehat{W}_G^{\downarrow}$

As on the shorter time scales discussed above, the longer term CO_2 variations can be influenced only by a net burial of carbon in sedimentary rock. In general, if organic matter produced photosynthetically, either on land or in the ocean, is not oxidized it can accumulate in several forms: on land as litter, peat, and ultimately under tectonic metamorphism as coal, and in the oceans as organic-rich sediments that become kerogen in black shales, methane clathrates, and more rarely oil, a process enhanced by anoxic conditions.

When these buried carbon deposits are exhumed by tectonic uplift, the exposure to atmospheric oxygen may result in a net flux of CO_2 to the atmosphere that may exceed the long-term global deposition ($\widehat{W}_G^{\uparrow} > 0$). In addition, the deep organic carbon deposits can be released to the atmosphere by volcanic degassing. An excellent discussion of the biogeochemical processes involved in the long-term organic burial cycles is given by Berner (1989).

At various stages in Earth history it appears that the burial of organic carbon has been a dominant process, e.g., the Permo-Carboniferous (360–240 Ma), when huge amounts of coal were formed under what were probably moist, highly vegetated conditions. An estimate of the burial rate over the Phanerozoic is given by Berner and Canfield (1989) based on the abundance of organic carbon in sedimentary rocks. At present, there is a rapid anthropogenic burning (oxidation) of the coal and oil reserves accumulated over millions of years, $A^{\uparrow}$, leading to the dramatic increase in atmospheric CO_2 shown in Fig. 10-5, which is a highly accelerated form of oxidative weathering. In the next section we discuss the somewhat slower process of planetary outgassing $V^{\uparrow}$, one component of which ($V_{\mathrm{G}}^{\uparrow}$) results from metamorphic/volcanic/diagenetic processes acting on buried organic material.

10.2.5 The Global Mass Balance of Organic Carbon

In light of the above discussion, and taking into account organic burial in the ocean sediments, the general continuity equation for the rate of change of Earth's total mass of organic carbon in all forms, C_{G}, can be written as,

$$\frac{dC_{\mathrm{G}}}{dt} = B_{\mathrm{G}}^{\downarrow} + B_{\mathrm{Gt}}^{\downarrow} - V_{\mathrm{G}}^{\uparrow} \tag{10.29}$$

where $B_{\mathrm{G}}^{\downarrow}$ is the difference between deposition of organic matter in the ocean sediments and its oxidation via weathering after uplift onto the continents (net oceanic burial in the sediments), $B_{\mathrm{Gt}}^{\downarrow}$ is the net terrestrial organic burial (i.e., excess of production over oxidative weathering near the surface), and $V_{\mathrm{G}}^{\uparrow}$ ($= V_{\mathrm{G}\mu}^{\uparrow} + V_{\mathrm{Gw}}^{\uparrow}$) is the net destruction of organic material in rocks due to metamorphic and diagenetic processes and consequent outgassing to the atmosphere ($V_{\mathrm{G}\mu}^{\uparrow}$) and oceans ($V_{\mathrm{Gw}}^{\uparrow}$).

10.3 OUTGASSING PROCESSES, $V^{\uparrow}$

Under the influence of tectonic movements and the heat and pressure at deeper levels, the carbonates (represented by $CaCO_3$) and silica (SiO_2) deposited in the ocean sediments can be transformed into metamorphic and igneous rock containing calcium silicates (generalized as $CaSiO_3$), producing CO_2 by the reverse *Urey reaction*,

$$CaCO_3 + SiO_2 \rightarrow CaSiO_3 + CO_2 \tag{10.30}$$

Similarly, organic material can be reworked to produce CO_2 by the production of reduced gases followed by their oxidation by atmospheric O_2. The overall reaction is the reverse of Eq. (10.14):

$$\mathrm{C(org)} + O_2 \rightarrow CO_2$$

This metamorphically produced CO_2 can then be vented to the atmosphere from volcanoes, springs, and simple seepage on land, and also to the ocean from the sea floor, particularly at the midocean ridges. The CO_2 released in the ocean finds its way to the atmosphere over relatively short time scales by increasing the total CO_2 (ΣCO_2) and hence the partial pressure of CO_2 near the surface, thereby increasing $Q^{\uparrow}$.

The contributions to this total outgassing to the atmosphere $V_{\mu}^{\uparrow}$ can therefore be resolved into a carbonate component $V_{\mu\text{C}}^{\uparrow}$ resulting from Eq. (10.30), and an organic component $V_{\mu\text{G}}^{\uparrow}$ resulting from Eq. (10.14), i.e.,

$$V_{\mu}^{\uparrow} = V_{\mu\text{C}}^{\uparrow} + V_{\mu\text{G}}^{\uparrow} \tag{10.31}$$

The variations of outgassing take place on all time scales, but as a further useful resolution in accord with Eq. (10.5) we can distinguish between the long-term, tectonic-mean flux $\widehat{V}^{\uparrow}$ and the higher frequency flux variations $\Delta V^{\uparrow}$, i.e.,

$$V^{\uparrow} = \widehat{V}^{\uparrow} + \Delta V^{\uparrow} \tag{10.32}$$

The most general method for estimating $V_{\mu}^{\uparrow}(t)$ is by assuming that the amount of CO_2 released to the atmosphere is proportional to the amount of volcanic ash and rock accumulated as a function of time as revealed by sedimentary and ice cores and stratigraphic analyses (e.g., Budyko *et al.*, 1987). To estimate the longer term fluctuations, $\widehat{V}^{\uparrow}$, it is usually assumed that the degassing rate is proportional to the rate of seafloor spreading, which is a measure of global metamorphic and volcanic decarbonation (Berner *et al.*, 1983).

To attempt a quantitative calibration of the variations implied by the above considerations it is necessary to establish the present level of degassing $V_{\mu}^{\uparrow}(0)$ as the most viable control point. Berner (1990, 1999) has adopted the value 6.7×10^{12} mol CO_2 y^{-1}, based on the rate of global degassing necessary to balance Ca–Mg silicate weathering. A good discussion of $V^{\uparrow}$ with emphasis on the Early Cenozoic is given by Kerrick and Caldeira (1993). In Section 10.6 we discuss Berner's (1994) representation of $\widehat{V}^{\uparrow}$ based on sea floor spreading variations.

By tectonic movements, igneous metamorphic rocks are steadily being lifted and exposed over long time periods at Earth's surface. It is by the weathering of these silicate rocks that the long-term (tectonic) carbon cycle is completed (see Fig. 10-6). In the next section we discuss this weathering process in more detail.

10.4 ROCK WEATHERING DOWNDRAW, $W^{\downarrow}$

As noted at the beginning of this chapter, the global weathering downdraw of atmospheric CO_2 can be resolved into two parts,

$$W^{\downarrow} = W_{\text{c}}^{\downarrow} + W_{\text{s}}^{\downarrow} \tag{10.33}$$

where $W_{\text{c}}^{\downarrow}$ represents the weathering of exposed rocks governed by the reverse reaction [Eq. (10.15)]. On the other hand $W_{\text{s}}^{\downarrow}$ represents the weathering of the metamorphic and igneous rocks (exposed by tectonic processes, e.g., orogenic plate collisions) that

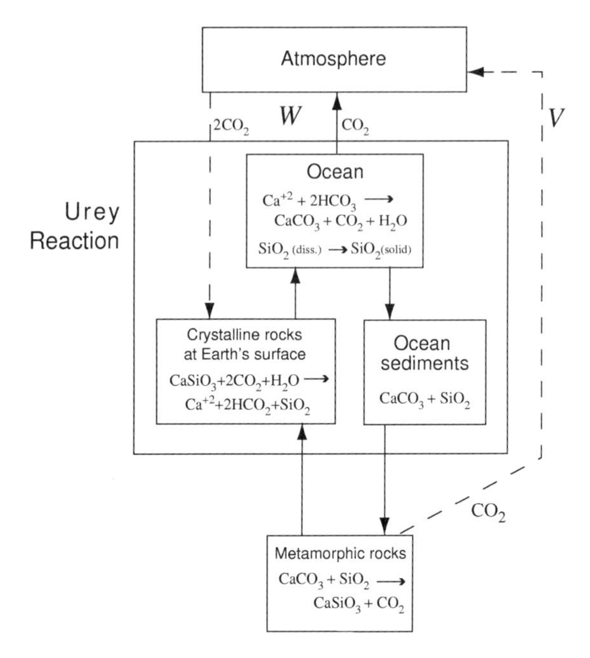

Figure 10-6 Schematic representation of the main geochemical links and reactions that determine the atmospheric CO_2 concentration on a time scale greater than 1 My. After Sarmiento and Bender (1994).

are relatively rich in calcium (and magnesium) silicates (generalized as $CaSiO_3$ and $MgSiO_3$). These exposed rocks are then vulnerable to weathering by the weathering reaction

$$CaSiO_3 + 2CO_2 + H_2O \rightarrow SiO_2 + Ca^{2+} + 2HCO_3^- \qquad (10.34)$$

whereby two units of carbon dioxide are removed from the atmosphere, producing dissolved SiO_3, Ca^{2+}, and HCO_3^-, which flow to the ocean in rivers. In the ocean the SiO_3 molecules are incorporated in silicate shells (diatoms and radiolaria) and deposited on the seafloor in sediments. At the same time, the calcium and biocarbonate ions are recombined in calcareous organisms, mainly in shallow layers, by the reaction given by Eq. (10.15), increasing the partial pressure of CO_2 in these waters. Assuming a total return of the CO_2 thus produced to the atmosphere, the net effect of all these processes is the CO_2 drawdown represented by the Urey weathering reaction, $CaSiO_3 + CO_2 \rightarrow CaCO_3 + SiO_2$. The $CaCO_3$ and SiO_2 deposits in the sediments are ultimately available for the metamorphic, volcanic, and tectonic processes described in Section 10.3 that can complete the carbon cycle over the tectonic time scale. This sequence is portrayed in Fig. 10-6.

The magnitude of the flux of carbon due to these carbonate and silicate weathering processes depends on several factors:

1. The tectonically forced continental exposure of carbonate and silicate rocks, which determines their area, elevation, and geographic location; for example, conti-

nental plate movements caused the exhumation of the Himalayan and New Guinea arcs, both containing silicate-rich rocks.

2. The lithology of the exposed silicates, including their grain size and minerology, both of which affect their reactivity.

3. Climatic factors, notably temperature and precipitation, both of which accelerate the weathering process (e.g., Kump *et al.*, 2000).

4. Changes in sea level (due, for example, to ice-sheet variations), which can expose or cover carbonate-rich continental shelves and platforms (cf. the "coral reef hypothesis," Section 10.1).

5. Vegetation, which is determined by both lithology and climate and which promotes weathering by accelerating CO_2 removal by producing carbon-containing acids that react with carbonate and silicate rock material. We may distinguish between the shorter term processes associated with soil and climate conditions, and very long-term effects associated with the evolution of plants. In particular it is believed that the emergence of large vascular plants (380–355 Ma) strongly enhanced weathering of silicate rocks (Algeo *et al.*, 1995). At a later stage the emergence of angiosperm (e.g., deciduous)-dominated ecosystems in the mid-Cretaceous ($\sim$100 Ma) from the previously gymnosperm (e.g., coniferous)-dominated systems further increased the weathering process (Volk, 1989a,b).

6. The effect of CO_2 on the rate of growth of vegetation, which provides a negative feedback, because vegetation increases weathering.

In general we can separate those factors that are largely externally imposed by tectonic and long-term evolutionary processes (i.e., item 1, the mineralogical component of item 2, and the rise of large vascular plants in item 4). These factors define the "weatherability," to be distinguished from the remaining factors that affect the weathering processes as internal variables dependent on climate and CO_2. In Section 10.6 we discuss Berner's (1994) representation of the carbonate and silicate weathering rates.

As noted in Chapter 2 (Section 2.4.4), independent methods for estimating the silicate weathering rates based on the strontium isotope ratio ($^{87}Sr/^{86}Sr$) in seawater as recorded in sedimentary cores have been proposed. In particular, Raymo *et al.* (1988) suggest that because of an assumed constant relationship between silicate rock and its Sr isotope content and the existence of some level of correlation with $\delta\,^{18}O$, the magnitude of this isotopic ratio can be used as a proxy for the chemical weathering rate and even as a more general proxy for climate. Because of many uncertainties (e.g., Berner and Rye, 1992; Caldeira, 1992) this method is still considered speculative, especially because there are large variations in the Sr isotopic composition of different silicate rocks. A further, also speculative, suggestion has been made that the osmium isotope record may be able to compliment the Sr isotope record to provide a better estimate of chemical weathering of silicate rocks (Reusch *et al.*, 1998; Oxburgh, 1998).

10.5 A GLOBAL DYNAMICAL EQUATION FOR ATMOSPHERIC CO_2

We have now discussed qualitatively all of the components contributing to the rate of change of atmospheric CO_2 as represented by Eq. (10.3). Introducing the decompositions of the total degassing flux [Eq. (10.31)] and weathering flux [Eq. (10.33)], and neglecting anthropogenic forcing ($A^{\uparrow}$), this equation takes the form

$$\kappa_{\mu}^{-1}\frac{d\mu}{dt} = Q^{\uparrow} + W_{\mathrm{G}}^{\uparrow} - W_{\mathrm{C}}^{\downarrow} - W_{\mathrm{S}}^{\downarrow} + V_{\mu\,\mathrm{C}}^{\uparrow} + V_{\mu\,\mathrm{G}}^{\uparrow} \tag{10.35}$$

where $\mu = \kappa_{\mu} C_{\mathrm{a}}$, and in accordance with the suggested representation, Eq. (10.28), the sea-to-air flux, $Q^{\uparrow}$, can be parameterized in the form

$$\kappa_{\mu} Q^{\uparrow} = \beta_0(\widehat{\mu}) - \left(\beta_1 - \beta_2\mu + \beta_3\mu^2\right)\mu - \beta_{\theta}\theta$$

Applying the resolution [Eq. (10.5)] of any variable x into a long-term tectonic-mean (e.g., 10-My average) component denoted by $\widehat{x}$, and a shorter term (e.g., the Plio-Pleistocene ice variations scale) departure denoted by Δx, Eq. (10.3) can be decomposed into the following pair of equations governing a quasi-statically equilibrated, very slowly varying tectonic-mean value ($\widehat{\mu}$) and a nonequilibrium dynamical departure from this slowly varying state ($\Delta\mu$). That is,

$$\left[\widehat{Q}^{\uparrow} + \widehat{W}_{\mathrm{G}}^{\uparrow} - \widehat{W}_{\mathrm{C}}^{\downarrow} - \widehat{W}_{\mathrm{S}}^{\downarrow}\right](\widehat{\mu}) + V_{\mu}^{\uparrow} = 0 \tag{10.36}$$

and, defining $\Delta\mu \equiv \xi$, and $\Delta\theta \equiv \vartheta$, for notational simplicity,

$$\begin{aligned}\frac{d\xi}{dt} &= b_1(\widehat{\mu})\xi - b_2(\widehat{\mu})\xi^2 - b_3\xi^3 - b_{\theta}\vartheta \\ &\quad + \kappa_{\mu}\Delta\left\{\left[W_{\mathrm{G}}^{\uparrow} - W_{\mathrm{C}}^{\downarrow} - W_{\mathrm{S}}^{\downarrow}\right](\xi) + V_{\mu}^{\uparrow}\right\}\end{aligned} \tag{10.37}$$

where $b_1(\widehat{\mu}) = (2\beta_2\widehat{\mu} - 3\beta_3\widehat{\mu}^2 - \beta_1)$, $b_2 = (3\beta_3\widehat{\mu} - \beta_2)$, $b_3 \equiv \beta_3$, $b_{\theta} \equiv \beta_{\theta}$ $(= \partial\mu/\partial\theta)$, and, in accordance with Eq. (10.31), $V_{\mu}^{\uparrow} = (V_{\mu\,\mathrm{C}}^{\uparrow} + V_{\mu\,\mathrm{G}}^{\uparrow})$. Here $\widehat{\mu}$ is to be determined from the long-term carbon balance driven by tectonics, the atmospheric component of which is given by Eq. (10.36). In the next section we discuss the additional carbon-cycle constraints, and the representations of $\widehat{W}^{\downarrow}$ and $\widehat{V}^{\uparrow}$, necessary for closure of this tectonic-mean CO_2 problem, based partly on Berner's (1991, 1994) GEOCARB model.

10.6 MODELING THE TECTONICALLY FORCED CO_2 VARIATIONS, $\widehat{\mu}$: LONG-TERM ROCK PROCESSES

To calculate the long-term evolution of atmospheric CO_2 due to the slow "rock" processes embodied in Eq. (10.36) requires that we express the weathering ($\widehat{W}^{\downarrow}$) and outgassing ($\widehat{V}^{\uparrow}$) of CO_2 as functions of the external tectonic influences and the prevailing mean values of CO_2. In addition, because the long-term mean flux between the ocean and atmosphere ($\widehat{Q}^{\uparrow}$) is difficult to determine independently, it is desirable to evaluate this term by considering the carbon balance of the ocean.

10.6.1 The Long-Term Oceanic Carbon Balance

The total amount of carbon in the ocean, C_{w}, is primarily in the form of dissolved inorganic carbon (DIC, i.e., ΣCO_2) given by Eq. (10.10), plus a much smaller amount in the form of dissolved organic carbon (DOC) and an even smaller amount of solid biotic and detrital carbon (see Fig. 10-2). The rate of change of C_{w} can be written in the form

$$\frac{dC_{\mathrm{w}}}{dt} = -Q^{\uparrow} + V_{\mathrm{w}}^{\uparrow} - B_{\mathrm{Gw}}^{\downarrow} - B_{\mathrm{C}}^{\downarrow} + R_{\mathrm{C}}^{\downarrow} + R_{\mathrm{S}}^{\downarrow} \tag{10.38}$$

where $V_{\mathrm{w}}^{\uparrow} = (V_{\mathrm{cw}}^{\uparrow} + V_{\mathrm{Gw}}^{\uparrow})$ is the outgassing mass flux of carbon from the ocean floor due to metamorphic/diagenetic reworking of carbonate rocks $V_{\mathrm{cw}}^{\uparrow}$ and organic rocks $V_{\mathrm{Gw}}^{\uparrow}$; $B_{\mathrm{Gw}}^{\downarrow}$ is the rate of removal of carbon from the ocean due to burial in the sediments of soft-tissue organic material representing the difference between organic production and oxidation in the ocean; $B_{\mathrm{C}}^{\downarrow}$ is the rate of removal of carbon from the ocean due to burial of hard-tissue carbonate material representing the difference between production and dissolution of $CaCO_3$; $R_{\mathrm{C}}^{\downarrow}$ and $R_{\mathrm{S}}^{\downarrow}$ are the river fluxes of bicarbonate ions (HCO_3^-) into the ocean resulting from the carbonate and silicate weathering reactions [Eqs. (10.15) and (10.34), respectively]. Note that $R_{\mathrm{S}}^{\downarrow} = W_{\mathrm{S}}^{\downarrow}$ but $R_{\mathrm{C}}^{\downarrow} = 2W_{\mathrm{C}}^{\downarrow}$ because in addition to the carbon atoms removed from the atmosphere represented by $W_{\mathrm{C}}^{\downarrow}$ in Eq. (10.35) another atom of carbon must also be removed from the exposed carbonate rock material in accordance with Eq. (10.15). Equation (10.38) fully conforms with Eqs. (10.19) and (10.22) if we note that for the whole ocean $C_{\mathrm{w}} = (C_{\mathrm{wd}} + C_{\mathrm{wD}})$, $(Z_{\mathrm{D}}^{(c)} + Z_{\mathrm{d}}^{(c)}) = 0$, $R^{(c)} = (R_{\mathrm{c}}^{\downarrow} + R_{\mathrm{s}}^{\downarrow})$, $B_{\mathrm{c}} = (B_{\mathrm{d}}^{(c)} + B_{\mathrm{D}}^{(c)})$, and $B_{\mathrm{Gw}} = (B_{\mathrm{d}}^{(G)} + B_{\mathrm{D}}^{(G)})$.

Because the equilibration time scale for the ocean is relatively short, we can express the tectonic-mean balance of oceanic carbon in a form similar to Eq. (10.36). Thus $d\widehat{C}_{\mathrm{w}}/dt \approx 0$ and invoking the above relations between river flow of carbon and weathering,

$$-\widehat{Q}^{\uparrow} + \widehat{V}_{\mathrm{w}}^{\uparrow} - \widehat{B}_{\mathrm{G}}^{\downarrow} - \widehat{B}_{\mathrm{C}}^{\downarrow} + \widehat{W}_{\mathrm{S}}^{\downarrow} + 2\widehat{W}_{\mathrm{C}}^{\downarrow} = 0 \tag{10.39}$$

10.6.2 The GEOCARB Model

If we now add Eqs. (10.36) and (10.39), thereby eliminating explicit consideration of $\widehat{Q}^{\uparrow}$, we obtain Berner's (1991, 1994) GEOCARB relation for this combined atmosphere–ocean domain,

$$\widehat{W}_{\mathrm{C}}^{\downarrow} + \widehat{W}_{\mathrm{G}}^{\uparrow} + \widehat{V}_{\mathrm{G}}^{\uparrow} + \widehat{V}_{\mathrm{C}}^{\uparrow} - \widehat{B}_{\mathrm{G}}^{\downarrow} - \widehat{B}_{\mathrm{C}}^{\downarrow} = 0 \tag{10.40}$$

where $\widehat{V}_{\mathrm{G}}^{\uparrow} = (\widehat{V}_{\mathrm{G}\mu}^{\uparrow} + \widehat{V}_{\mathrm{Gw}}^{\uparrow})$ and $\widehat{V}_{\mathrm{C}}^{\uparrow} = (\widehat{V}_{\mathrm{C}\mu}^{\uparrow} + \widehat{V}_{\mathrm{Cw}}^{\uparrow})$.

Noting that the burial rate of carbonates ($\widehat{B}_C^{\downarrow}$) includes a component due to the riverine flux of HCO_3^- produced by silicate weathering ($\widehat{W}_S^{\downarrow}$) as well as a component due to carbonate weathering ($\widehat{W}_C^{\downarrow}$), and assuming the existence of a long-term balance between these fluxes, we have an independent relation,

$$\widehat{W}_S^{\downarrow} = \left(\widehat{B}_C^{\downarrow} - \widehat{W}_C^{\downarrow}\right) \tag{10.41}$$

Taking into account all the factors discussed qualitatively in Sections 10.3 and 10.4, Berner (1991, 1994) proposes the following time-dependent representations for the weathering and outgassing fluxes:

$$\widehat{W}_C^{\downarrow}(t) = f_c(\widehat{T}) \cdot \underbrace{\left[f_{LA}(t) \cdot f_D(t) \cdot f_E(t) \cdot k_{wC} \cdot C_C(t)\right]}_{\mathcal{W}_c} \tag{10.42}$$

$$\widehat{W}_S^{\downarrow}(t) = f_s(\widehat{T}) \cdot \underbrace{\left[f_R(t) \cdot f_E(t) \cdot f_D^{0.65}(t)\right]}_{\mathcal{W}_s} \widehat{W}_S^{\downarrow}(0) \tag{10.43}$$

$$\widehat{W}_G^{\uparrow}(t) = f_D(t) \cdot f_R(t) \cdot k_{wG} \cdot C_G(t) \tag{10.44}$$

$$\widehat{V}_C^{\downarrow} = f_{SR}(t) \cdot f^*(t) \cdot \widehat{V}_C^{\uparrow}(0) \tag{10.45}$$

$$\widehat{V}_G^{\uparrow} = f_{SR}(t) \cdot \widehat{V}_G(0) \tag{10.46}$$

where $f_c(\widehat{T})$ and $f_s(\widehat{T})$ are dimensionless functions expressing all of the dependence of $\widehat{W}_C^{\downarrow}$ and $\widehat{W}_S^{\downarrow}$, respectively, on the global temperature $\widehat{T}$ and hence through Eq. (7.34) on the atmospheric CO_2, $\widehat{\mu}$, and solar luminosity (but excluding the feedbacks of ice and the deep ocean state on surface temperature). Also included is the feedback of CO_2 on plant growth and weathering. The terms $\mathcal{W}_C$ and $\mathcal{W}_S$ constitute the *weatherability* of carbonates and silicate rocks, respectively, being a function only of a variety of externally imposed tectonic factors, of which f_{LA} is the ratio of the carbonate land area to the present value, f_D is ratio of the river runoff per unit area to the present value, due only to paleogeographic (not CO_2) changes, f_E is a dimensionless parameter expressing the dependence of $W^{\downarrow}$ on soil biological activity due to land plants, e.g., angiosperm vs. gymnosperm dominance (= unity at present) with inclusion of some of the feedback effects of CO_2 on plant growth, f_R is the ratio of the weathering effect of mean continental relief to its present value, and k_{wC} and k_{wG} are the rate constants for weathering of carbonates and organic material, respectively. $C_C(t)$ and $C_G(t)$ are the prevailing global masses of carbon in the forms of carbonate rock and as organic matter, respectively.

The balance equation for variations of $C_G(t)$ is given by Eq. (10.29), and a similar balance equation for the variation of $C_C(t)$ can be written in the form

$$\frac{dC_C}{dt} = \widehat{B}_C^{\downarrow} - \widehat{W}_C^{\downarrow} - \widehat{V}_C^{\uparrow} \tag{10.47}$$

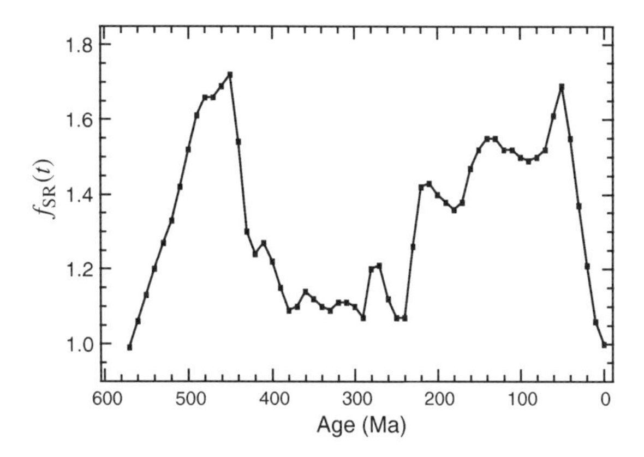

Figure 10-7 Estimate of the variation of the rate of degassing of CO_2 as determined by the ratio $f_{SR}(t)$ of the tectonic spreading rate to its present value in accordance with Eqs. (10.45) and (10.46). After Berner (1994).

expressing the fact that over long time periods the carbonate rock can be produced by a net $CaCO_3$ burial, $\widehat{B}_C^{\downarrow}$, and depleted by weathering, $\widehat{W}_C$, and metamorphic conversion to CO_2, $\widehat{V}_C^{\uparrow}$, represented by the Urey reaction, Eq. (10.30). Because of the long time scales involved in both organic and carbonate rock processes, the equations for C_G and C_C must take the nonequilibrium forms given by Eqs. (10.29) and (10.47), respectively.

The outgassing flux representations for $\widehat{V}_C^{\uparrow}$ and $\widehat{V}_G^{\uparrow}$ given in Eqs. (10.45) and (10.46) introduce two additional parameters: $f_{SR}(t)$ is the ratio of the spreading rate [e.g., as determined by Engebretson *et al.* (1992) and Gaffin (1987)] to the present spreading rate (see Fig. 10-7), and $f^*(t)$ is a parameter that weights the relative amount of deposition of carbonate in the deep sea (where it is more likely to be subducted and thermally decomposed) to the amount deposited on shallow continental shelves or platforms (where only much weaker, e.g., diagenetic, processes can break down the carbonates via Eq. (10.30) (see Volk, 1989b). These weaker processes appear to have dominated before about 150 Ma when calcareous foraminifera that could drop to the deep sediments first appeared, and more generally during times of high sea level, when shallow submerged platforms widely prevail. More extensive discussions and formulations of each of the above tectonic factors, f_x, are given by Berner (1991, 1994).

At this point we have a system of ten equations [Eqs. (10.40)–(10.47), (10.29), and (7.34b)] governing eleven variables ($\widehat{W}_C$, $\widehat{W}_G$, $\widehat{W}_S$, $\widehat{B}_G$, $\widehat{B}_C$, $\widehat{V}_C$, $\widehat{V}_G$, $\widehat{C}_G$, $\widehat{C}_C$, $\widehat{T}$, and $\widehat{\mu}$). Thus, we need one additional constraint to close the system. This can be obtained from the balance of isotopic carbon as measured by $\delta\,^{13}C$, which is a measure of the partition of carbon between inorganic ($CaCO_3$) and organic forms (see Section 2.4.3). Berner (1991, 1994) expresses this constraint in the form of three additional balance equations containing two additional variables, δ_C and δ_G, which are the mean $\delta\,^{13}C$ values of the total carbon mass of carbonate rock and organic material, respectively. These variables are governed by the following continuity equations for ^{13}C

mass in each of these reservoirs [$(\delta_C \cdot C_C)$ and $(\delta_G \cdot C_G)$], invoking $(^{13}C/^{12}C) \ll 1$:

$$\frac{d(\delta_C \cdot C_C)}{dt} = \delta_{BC} B_C^{\downarrow} - \delta_C \left(W_C^{\downarrow} + V_C^{\uparrow} \right) \tag{10.48}$$

$$\frac{d(\delta_G \cdot C_G)}{dt} = \delta_{BG} B_G^{\downarrow} - \delta_G \left(W_G^{\uparrow} + V_G^{\uparrow} \right) \tag{10.49}$$

where δ_{BC} and $\delta_{BG} = (\delta_{BC} - \alpha_C)$ are the time-dependent values of $\delta\,^{13}C$ for carbonates and organics being buried at any time t given by the observed $\delta\,^{13}C$ variations in calcareous fossils and limestones in shallow seawater (Berner, 1987, 1989). The factor α_C represents the fractionation between carbonates and organic matter.

To complete the $\delta\,^{13}C$ formulation, and finally close the full GEOCARB model for tectonically forced CO_2 variations due to long-term rock processes, we apply the constraint that $\delta\,^{13}C$ is conserved globally for the combined carbonate and organic pools, i.e., $d[(\delta_C \cdot C_C) + (\delta_G \cdot C_G)]/dt \approx 0$, implying from Eqs. (10.48) and (10.49) that

$$\delta_C \left(W_C^{\downarrow} + V_C^{\uparrow} \right) + \delta_G \left(W_G^{\uparrow} + V_G^{\uparrow} \right) - \delta_{BC} B_C^{\downarrow} - (\delta_{BC} - \alpha_C) B_G^{\downarrow} = 0 \tag{10.50}$$

Other isotopic balances of species such as strontium $(^{87}Sr/^{86}Sr)$ and osmium $(^{187}Os/^{186}Os)$ may provide additional constraints that may be of value in modeling long-term CO_2 variations (e.g., Berner and Rye, 1992; Reusch and Maasch, 1998) (see Section 2.4.4).

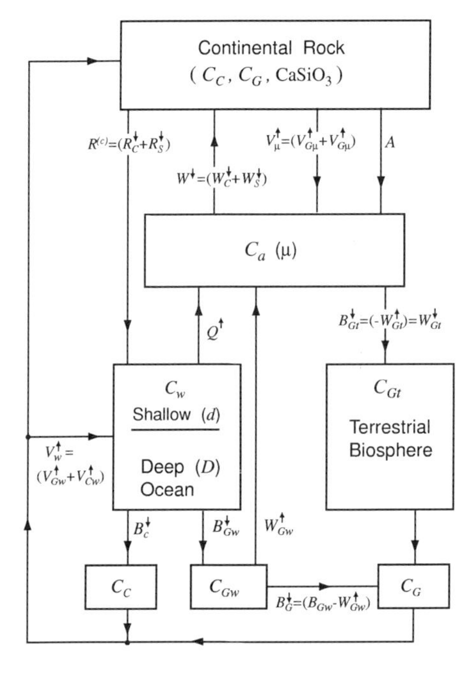

Figure 10-8 Box diagram showing the flow of carbon in all forms between the main reservoirs, as represented by the closed system, Eqs. (10.3), (10.29), (10.38), and (10.47), wherein all symbols are defined.

In Chapter 13 we discuss the results of calculations of $\widehat{\mu}$ for the Phanerozoic (past 600 My) made by Berner (1994) based on specifications of all the parameters of the GEOCARB system, and outline a procedure for including the additional influences on $\widehat{\mu}$ such as ice-sheet coverage and deep ocean state. In Chapter 15 we show how these estimates can be further corrected for the effects of higher frequency (Pleistocene-scale) variations, all as part of a fuller theory of paleoclimatic variations based on the resolution given by Eq. (10.5). Having already described the physics of ice-sheet coverage in Chapter 9, in the next chapter we complete our discussion of the main slow-response variables by developing physical relationships governing the deep ocean state.

10.7 OVERVIEW OF THE FULL GLOBAL CARBON CYCLE

Equations (10.3), (10.29), (10.38), and (10.47) form a closed system expressing conservation of carbon in all forms [C_a, $C_G = (C_{Gw} + C_{Gt})$, C_w, and C_C], operating on all time scales. In Fig. 10-8 we present a box diagram representing the flow of carbon in all the above forms between these main reservoirs (see Fig. 10-2). All of the symbols in this diagram conform with those identified in the above equations.

11

SIMPLIFIED DYNAMICS OF THE THERMOHALINE OCEAN STATE

The fundamental equations governing the behavior of the ocean were given in Chapter 4. Based on simplifications of these equations many models have now been formulated with the aim of predicting and accounting for the main circulations and other properties of the ocean [see reviews by Niiler (1992) and Haidvogel and Bryan (1992) as examples]. These models have ranged from full three-dimensional ocean general circulation models such as that of Semtner and Chervin (1992), to highly simplified basin-wide, zonal average (meridional plane) models, multibox models, and even one-box, globally averaged models. Separate models have also been constructed to account independently for the wind-driven and thermohaline components of the circulation as well as for the upper mixed layer (which we have here considered to be a part of the fast-response system; see Chapter 7).

As an ultimate goal, a theory of paleoclimatic evolution should rest on a three-dimensional ocean model coupled to three-dimensional models of the atmosphere, cryosphere, chemosphere, and biosphere in a time-dependent framework (see Fig. 5-1). Enlarging on the ideas already expressed in Section 5.1, we note, however, that because of the many uncertainties in the representation of the physics in more detailed models, and unavoidable model "drifts" that can easily obscure the true net fluxes across the ocean–atmosphere interface that give rise to the slow deep ocean changes in which we are interested (e.g., Fig. 3-3), it will be of value to try to formulate simpler low-resolution models that can encapsulate the most elemental deep ocean dynamical processes leading to the long-term variations. At the least, the results of such simpler models, when coupled with appropriately simplified models of the other slow-response variables, should provide a useful integral constraint on the long-term evolutionary behavior of full three-dimensional models that are required to describe the spatial distributions of the ocean state and fluxes.

The simplest models in this regard are the box-thermodynamic models with severely parameterized flux dynamics. In some sense, these are oceanic analogs of the atmospheric energy balance models described in Chapter 7, and, in fact, these simple atmospheric and oceanographic models have often been compatibly coupled to form a combined model (e.g., Birchfield, 1989). As in the atmospheric case, the two-box ocean model to be described in Section 11.4 reveals, in its most primitive form, the possibility for an instability that can drive paleoclimatic variability arising from the positive feedback role of salinity. Moreover, in the extreme limit, a one-box global model embodying the main results of the higher order models may be of relevance, as suggested in Section 8.3, and we shall expand on this suggestion. In particular, we shall suppose that the main effects of the ocean on the long-term variations of the climate system can be projected onto the behavior of the deep ocean temperature (θ), which is related to the mean depth and horizontal extent of the main thermoclines in the world ocean and the slow thermohaline circulations.

Before embarking on a discussion of these simple box models it is important, in assessing the merits of the approximations involved, to recognize the fuller nature of the circulations that influence the thermohaline state of the ocean (e.g., θ). These circulations are in fact extremely complex, owing their origin to a highly interactive mix of both mechanical wind-driven effects and density-driven thermohaline effects. An example of this complexity is the role of the rotation of Earth in tending to channel the circulation into relatively narrow "rivers" of flow within the ocean as a whole (e.g, the Gulf Stream), which account for a large portion of the combined fluxes due to both wind-driven and density-driven forcing. In spite of this complicated interactive nature of the flow, as a first approximation it is not unreasonable to break down the total flow into four main components:

1. The frictionally induced *wind-driven circulations* that tend to create upper level oceanic horizontal flows in the direction of the mean surface winds. In the presence of continental boundaries these take the form of the main gyres presently observed in the Atlantic and Pacific basins, for example, but if unimpeded by continents may also take the form of a strong circumpolar currents as in the present Southern Ocean. Associated with these main horizontal flows are divergent Ekman drift components that create large regions of upwelling and downwelling in the upper ocean, hence contributing to the full vertical overturning of the ocean. A fuller discussion of the dynamics of this mode is given in Section 11.3.1. Unlike this wind-driven component, the remaining three modes of circulation all owe their existence to the presence of density variations (i.e., to variations of temperature and salinity). These modes are as follows.

2. The *thermohaline circulation*, defined here as that component of the flow driven by horizontal pressure forces resulting from hydrostatically balanced horizontal density differences. These temperature and salinity differences have large meridional components between the tropics and polar regions, engendering a large meridional TH circulation within individual ocean basins; however, on the scale of the "world ocean" there can exist a significant TH circulation component between the ocean basins (e.g., Pacific and Atlantic) due to density differences between these ocean basins.

3. In addition to the large-scale hydrostatic TH circulation, relatively intense *localized bouyancy-driven (i.e., nonhydrostatic) convective circulation cells* induced by unstable vertical gradients of density, particularly in high latitudes where strong surface cooling and evaporation can occur.

4. Also in higher latitudes, where a strong horizontal density gradient and a large Coriolis parameter prevails, the possibility exists for *baroclinically unstable waves* in

which meandering horizontal motions occur in tandem with a pattern of upwelling and downwelling motions (Saltzman and Tang, 1975).

As noted, both the local bouyancy-driven convective overturnings and the baroclinically driven overturnings tend to occur in higher latitudes, and, moreover, occur on much shorter space and time scales than the grand TH circulations and wind-driven circulations. Thus, whereas these latter circulations are major features of the climatic-mean (100-y) ocean state, the former two modes tend to vanish in the climatic mean. However, these shorter scale overturnings result in a net increase in the density of deeper columns of water in high latitudes, which contribute importantly to the poleward gradient of density that drives the TH circulation. Thus, their steady vertical flux properties must be included (i.e., parameterized) in considering the climatic-mean gradients that drive the TH circulation. With this understanding, for paleoclimatic purposes we can consider the TH and wind-driven circulations as the two major components of the full climatic-mean ocean circulation. We shall denote the stream functions for the vertical-plane climatic-mean TH circulation and horizontal-plane wind-driven circulation by ψ and ϕ, respectively. In addition, we define the stream function for the vertical plane overturnings due to the divergent part of the wind-driven circulation in the Ekman layer by ψ_{E}. The stream function for the combined effects of the TH circulation and wind-driven overturning will be denoted by $\psi_{\mathrm{tot}} = \psi + \psi_{\mathrm{E}}$. A more detailed discussion of these modes is given in the following sections.

11.1 GENERAL EQUATIONS

The fundamental equations [Eqs. (4.1)–(4.7)] applicable to the "synoptic several-day-average" ocean variables can be restated in the Boussinesq form, wherein density is a variable only insofar as it modifies the vertical force balance (i.e., buoyancy). These equations are

$$\frac{d\mathbf{v}}{dt} = \left(\frac{\partial \mathbf{v}}{\partial t} + \mathbf{v}\cdot\nabla v + w\frac{\partial \mathbf{v}}{\partial z}\right) = \frac{1}{\rho^*}\nabla \mathcal{P} - f\mathbf{k}\times\mathbf{v} + \frac{1}{\rho^*}\frac{\partial \underline{\tau}}{\partial z} + \nu_{\mathrm{h}}\nabla^2\mathbf{v} \quad (11.1\mathrm{a})$$

$$\frac{dw}{dt} = \left(\frac{\partial w}{\partial t} + \mathbf{v}\cdot\nabla w + w\frac{\partial \mathbf{v}}{\partial z}\right) = -\frac{1}{\rho^*}\frac{\partial \mathcal{P}}{\partial z} - \frac{g}{\rho^*}(\rho-\rho^*) + \nu_{\mathrm{v}}\nabla^2 w \quad (11.1\mathrm{b})$$

$$\mathcal{P} \equiv (p - p_{\mathrm{H}}^*), \qquad \frac{\partial p_{\mathrm{H}}^\star}{\partial z} = -\rho^* g \quad (11.2)$$

$$\nabla\cdot\mathbf{v} + \frac{\partial w}{\partial z} = 0 \quad (11.3)$$

$$\frac{\partial T}{\partial t} + \mathbf{v}\cdot\nabla T + w\frac{\partial T}{\partial z} = K_{\mathrm{h}}^{(T)}\nabla^2 T + K_{\mathrm{v}}^{(T)}\frac{\partial^2 T}{\partial z^2} \quad (11.4)$$

$$\frac{\partial S}{\partial t} + \mathbf{v}\cdot\nabla S + w\frac{\partial S}{\partial z} = K_{\mathrm{h}}^{(S)}\nabla^2 S + K_{\mathrm{v}}^{(S)}\frac{\partial^2 S}{\partial z^2} \quad (11.5)$$

$$(\rho-\rho^*) = [-\mu_{\mathrm{T}}(T-T^*) + \mu_{\mathrm{s}}(S-S^*)] \quad (11.6)$$

where ρ and T denote values for seawater, $f = 2\Omega \sin\varphi$, $\underline{\tau} = (\tau_x \mathbf{i} + \tau_y \mathbf{j}) = \rho^* \nu_\mathrm{v} / \partial \mathbf{v} / \partial z$ is the horizontal eddy stress, ν_v and ν_h are the vertical and horizontal eddy diffusivities for momentum, $K_\mathrm{v}^{(T)}$ and $K_\mathrm{h}^{(T)}$ are the vertical and horizontal thermal eddy diffusivities [assuming the main internal heat source in the ocean is due to eddy conduction, i.e., $\rho q = \lambda \partial^2 T / \partial z^2$ in Eq. (4.4), where λ is the eddy conduction coefficient], and $K_\mathrm{v}^{(S)}$ and $K_\mathrm{h}^{(S)}$ are corresponding eddy diffusivities for salinity. In the equation of state, Eq. (11.6), we have neglected the small effect of isothermal compressibility μ_P. For convenience in this ocean-dynamics chapter we may alternatively take the vertical coordinate to be the depth Z, below sea level ($Z = 0$). Thus, we could replace ∂z by $-\partial Z$ in Eqs. (11.1)–(11.5). The bottom of the ocean is denoted by $Z = D$.

11.1.1 Boundary Conditions

Although the lateral walls of the ocean basins are complex sloping surfaces, it is common to simply assume the normal velocities vanish ($\mathbf{v}_n = 0$) and the tangential velocities $\mathbf{v}_\mathrm{T}$ may satisfy either "no-slip" ($\mathbf{v}_\mathrm{T} = 0$) or "free-slip" ($\partial \mathbf{v}_\mathrm{T} / \partial n = 0$) conditions. It is also assumed that the fluxes of heat and salinity vanish at these lateral walls, though in principle some allowance must be made for river fluxes.

If the bottom ($Z = D$) is assumed to be flat, $w_\mathrm{D} = 0$, and the horizontal stress $\underline{\tau}_\mathrm{D}$ can be represented by some drag law (e.g., $\underline{\tau}_\mathrm{D} = \rho C |\mathbf{V}_\mathrm{D}| \mathbf{V}$). The heat and salinity flux conditions take the forms $-\rho c K_\mathrm{v}^{(T)} (\partial T / \partial z)_\mathrm{D} = G^\uparrow$, and $K_\mathrm{v}^{(S)} (\partial S / \partial z)_\mathrm{D} = 0$, where $G^\uparrow$ is the upward geothermal flux ($\mathrm{W\,m^{-2}}$). This geothermal flux has been estimated to be only of the order of $10^2\ \mathrm{W\,m^{-2}}$, but can be higher at local benthic hot spots (e.g., the Red Sea).

At the surface ($Z = 0$), assumed to be flat and rigid (thereby excluding gravity waves), we have $w = 0$ and the horizontal wind stress is

$$\underline{\tau}_\mathrm{s} = \rho^* \nu_\mathrm{v} \left(\frac{\partial \mathbf{v}}{\partial z} \right)_\mathrm{s} \tag{11.7}$$

The downward heat flux on the ocean side of the air–sea interface, $H_\mathrm{ss}^\downarrow$, is given by Eq. (4.46), i.e.,

$$H_\mathrm{ss}^\downarrow = \left(H_\mathrm{s}^{(1)\downarrow} - H_\mathrm{s}^{(2)\uparrow} - H_\mathrm{s}^{(3)\uparrow} - L_\mathrm{v} E - L_\mathrm{f} M \right) = -\rho^* c K_\mathrm{v}^{(T)} \left(\frac{\partial T}{\partial z} \right)_\mathrm{s} \tag{11.8}$$

The upper boundary condition for the salinity flux is

$$\rho K_\mathrm{v}^{(S)} \left(\frac{\partial S}{\partial z} \right)_\mathrm{s} = S F^\uparrow \tag{11.9}$$

where $F^\uparrow = (E - P - \mathcal{R} - \rho_\mathrm{i}\, \partial h_\mathrm{i} / \partial t)$ is the net rate of removal of fresh water from the ocean surface in $\mathrm{kg/m^2\,s}$, $\mathcal{R}$ is the river flux into the ocean including the runoff from glaciers and $\rho_\mathrm{i}\, \partial h_\mathrm{i} / \partial t$ is rate of removal of freshwater by sea-ice production.

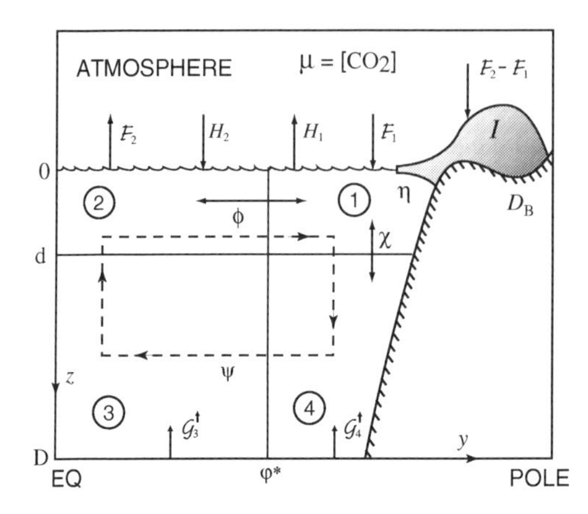

Figure 11-1 Schematic representation of a four-box ocean, showing the main modes of circulation between the boxes (ψ, ϕ, χ) and the main boundary fluxes of heat ($H, \mathcal{G}^{\uparrow}$) and freshwater ($\mathcal{F}$).

11.2 A PROTOTYPE FOUR-BOX OCEAN MODEL

We start our consideration of highly simplified models by separating an idealized world ocean into four elemental boxes representing near-surface and deep ocean states in low and high latitudes, neglecting the topographic and bathymetric asymmetries that my exist between the Northern and Southern Hemispheres (see Fig. 11-1). These neglected asymmetries can lead to very different wind stress and density gradients, resulting in different circulations in each hemisphere and the possibility for a cross-equatorial circulation. Thus, the results derived from models based on this idealized system can only be taken as illustrations of the possibilities arising solely from the fundamental radiative imbalance between the tropical and polar regions.

As shown in Fig. 11-1, the variables in the four elemental box regions will be designated by the subscripts 1, 2, 3, and 4. The depth of a relatively shallow top layer is d (typically assumed to be on the order of a few hundred meters embracing the upper mixed layer), and the depth of the deep layer is $D - d$, typically taken to be about 4 km (cf. Section 10.1.2). The latitude of the plane separating high and low latitudes, φ^*, can be taken to be roughly 45°. The volume of each box is denoted by V_n $(n = 1, 2, 3, 4)$, the total volume being denoted by $V = V_1 + V_2 + V_3 + V_4$. The main internal ocean variables to be identified in each box are the temperature T and salinity S.

Shown schematically in Fig. 11-1 are the stream functions for the net mean meridional overturning due to the combined TH circulation and wind-stress-induced overturning (ψ_{tot}), the horizontal wind-driven gyre circulations (ϕ), and vertical convective–baroclinic eddy circulations (χ), the sea-ice mass coverage (η), the net upward freshwater flux due to excess evaporation over precipitation ($\mathcal{F}^{\uparrow} = AF^{\uparrow}/\rho$, in $\text{m}^3\,\text{s}^{-1}$, $A =$ surface area of box) and the net downward heat flux at the ocean surface in $\text{W}\,\text{m}^{-2}$ ($H^{\downarrow}$). Although in Fig. 11-1 we represent the complete TH plus wind-driven circulation $\psi_{\text{tot}} = \psi + \psi_{\text{E}}$, in the simple models to be discussed we shall emphasize only the density-driven TH circulation component ψ. Also shown in the figure are schematic representations of the total ice-sheet mass (I), a typical subglacial bedrock depression (D_B), and atmospheric carbon dioxide (μ), all of which, in addition to the deep ocean properties, mainly in box 3, comprise the major slow-response variables of the climate system discussed in Chapter 8. Note that in Fig. 11-1 we include the possibility that there can be a net storage of freshwater in the cryosphere.

Given these latter slow-response variables (I, D_B, μ) and prescribed or parameterized forms of $\mathcal{F}^{\uparrow}$ and $\mathcal{H}^{\downarrow}$, the remaining variables ($T_{1,2,3,4}$, $S_{1,2,3,4}$, η, ϕ, χ, ψ) form a set that has been invoked in a large class of models of interdecadal–millenial scale fluctuations. Thus, for example, the four-box model is the basis for the study of Griffies and Tziperman (1995), and three-box versions, in which boxes 3 and 4 are combined into a single "deep ocean" box, have been studied by Birchfield (1989), and, in which boxes 1 and 4 have been combined into a single high-latitude polar box, have been studied by Nakamura *et al.* (1994) and Paillard (1995). Models emphasizing the upper two boxes, including in some cases sea-ice insulating effects and brine formation effects, with a more passive deep ocean, have been studied by Lohmann *et al.* (1996) and by Yang and Neelin (1997), including sea-ice effects (see also Saltzman, 1978).

Historically, the most significant reduction of the four-box system is the original two-box model of Stommel (1961) comprising of a single low-latitude box (2 and 3) and single high-latitude box (1 and 4). This minimal system, which as we have noted illustrates in its simplest form the possibility that the ocean contains a potential instability that can give rise to multiple equilibria, will be discussed in more detail in Section 11.4. It is not unless one considers three or more boxes, however, that the instability can give rise to free oscillatory behavior that might be identified with interdecadal through millennial scale periodicities (Ruddick and Zhang, 1996).

11.3 THE WIND-DRIVEN, LOCAL-CONVECTIVE, AND BAROCLINIC EDDY CIRCULATIONS

Before discussing the Stommel two-box model and the slow-response deep ocean state, both of which indirectly involve the thermohaline circulation (represented by the stream function, ψ), we first describe briefly the faster response circulations driven by wind stresses on the upper layer (i.e., the gyres) and by local convective and baroclinic instabilities confined mainly to the higher latitude ocean where destabilizing vertical and horizontal density gradients are a maximum. The climatic-average effects of these motions, which play a significant role in regulating the ultimate climatic distribution of the ocean properties, are represented schematically by the stream functions ϕ and χ shown in Fig. 11-1. Theoretical discussions of these modes of circulation are given by Pedlosky (1979), Gill (1982), and Stern (1975). Here we outline qualitatively the nature of these faster circulations and their potential effects on the slow-response deeper ocean behavior.

11.3.1 The Wind-Driven Circulation: Gyres and Upwelling

The surface winds that blow over the world ocean and drive the gyres form a part of the atmospheric general circulation that is itself forced to a significant extent by the thermal gradients prevailing at the ocean surface. These thermal gradients are, in turn, modified by the heat transport due to the gyre circulation. Thus, the behavior of this gyre component provides one example of the connectedness of the atmosphere and ocean that must ultimately be treated within the framework of a single coupled system. Nonetheless it is useful as a first approximation to view the wind-driven circulation as a forced response to a given wind stress field, and much of the gyre and upwelling theory is based on this approach.

Let us consider in more detail the upper box region of Fig. 11-1, $Z = 0$ to $Z = d$, which we further subdivide into a shallow "mixed layer" of depth d_E near the surface. This surface layer is dominated by turbulent friction such that a near-balance exists between the Coriolis force and the eddy viscous stress acting in horizontal planes. Thus Eq. (11.1a) is reduced to the simple Ekman layer balance,

$$\rho^* f \mathbf{k} \times \mathbf{v}_E = \frac{\partial \underline{\tau}}{\partial z} \tag{11.10}$$

where $\mathbf{v}_E$ is the so-called Ekman drift velocity. Averaging Eq. (11.10) through the depth of this mixed layer, from $Z = d_E$ (where $\tau \to 0$) to the top surface $Z = 0$ (where $\tau = \tau_s$, the surface wind stress), we obtain

$$\check{\mathbf{v}}_E = -\frac{1}{\rho^* f d_E} \mathbf{k} \times \tau_s \tag{11.11}$$

where the wiggly overbar denotes the vertical average. Assuming constant values of ρ^* and diffusivity ν_v in Eq. (11.7), the Ekman depth d_E can be shown to be proportional to $\sqrt{\nu_v/f}$, typically attaining a value of about 50 m. It is implied by Eq. (11.11) that the flow in this layer is to the right of the surface wind stress in the Northern Hemisphere and to the left in the Southern Hemisphere. Taking the horizontal divergence of Eq. (11.7), and applying the mass continuity Eq. (11.3) with the upper boundary condition $w = 0$, it follows that the vertical velocity at the base of the Ekman layer (w_E) forced by this "Ekman pumping" is given by

$$w_E = \frac{\mathbf{k} \cdot \nabla \times \underline{\tau}_s}{\rho^* f} \tag{11.12}$$

These upwellings and downwellings forced by the large-scale atmospheric wind systems represent an important mechanism for exchange of properties to and from the ocean surface. From a dynamical viewpoint the values of w_E generated at d_E can be taken as an upper boundary condition for the behavior of the deeper ocean below, at least down to the level d (several hundred meters), where $w_d \ll w_E$. In this deeper region, the motions are nearly geostropic, with a stable ocean stratification tending to inhibit turbulent friction except near coastal boundary layers. As a caveat, however, we note that the wind-driven circumpolar current around Antarctica, where no continental barriers are present, are associated with a forced vertical circulation extending to great depths. This forms a component of the so-called *Deacon cell* with possibly important implications for the more general thermohaline circulation (Toggweiler and Samuels, 1995).

Thus, following Stommel (1948), if we assume as the simplest case that in this layer $(d - d_E)$ a new balance exists in which the inertial terms are negligible $(d\mathbf{v}/dt = 0)$ and frictional forces that might be competitive near boundaries are represented in the simple Rayleigh form $(-\kappa\mathbf{v})$, Eq. (11.1a) becomes

$$\frac{1}{\rho^*}\nabla P + f\mathbf{k}\times\mathbf{v} + \kappa\mathbf{v} = 0 \tag{11.13}$$

By taking the curl $(\mathbf{k}\cdot\nabla\times)$ of this equation, noting that $\mathbf{k}\cdot\nabla\times[f\mathbf{k}\times\mathbf{v}] = (f\nabla\cdot\mathbf{v} + \mathbf{v}\cdot\nabla f) = (-f\,\partial w/\partial z + v\beta)$, where $\beta = \partial f/\partial y$, we obtain the vorticity balance in which $\zeta = \mathbf{k}\cdot\nabla\times\mathbf{v} = \partial v/\partial x - \partial w/\partial y$

$$v\beta + \kappa\zeta = f\,\frac{\partial w}{\partial z}. \tag{11.14}$$

Averaging Eq. (11.14) from d to d_E, denoted by a wiggly overbar, and applying the boundary conditions $w_d = 0$ and Eq. (11.12) we obtain

$$\breve{v}\beta + \kappa\zeta = f\,\frac{\mathbf{k}}{(d - d_E)\rho^*}\cdot\nabla\times\underline{\tau}_s \tag{11.15}$$

which, owing to the closeness of the sub-Ekman layer interior flow to geostropic horizontally nondivergent balance, can be expressed in terms of a stream function defined by $\mathbf{v} = \mathbf{k}\times\nabla\phi$, in the form

$$\beta\,\frac{\partial\phi}{\partial x} + \kappa\nabla^2\phi = \left[(d - d_E)\rho^*\right]^{-1}\mathbf{k}\cdot\nabla\times\underline{\tau}_s \tag{11.16}$$

This equation forms the basis of Stommel's (1948) model, the solution of which for a realistic meridional profile of the stress τ_s (i.e., tropical easterlies, midlatitude westerlies, and weak polar easterlies) can give a good first-order representation of the upper ocean gyre structure, as schematized for an idealized ocean basin in Fig. 11-2. Note the westward intensification of the currents (e.g., the Gulf Stream). This result follows because a zonal wind stress field of the kind invoked by Stommel for the main subtropical gyre (see Fig. 10-2) will induce southward water flow in the northern hemisphere over a long zonal spatial scale wherein $\nabla^2\phi$ is small and the friction term is negligible. Thus the main balance over this more easterly expanse of ocean is between the transport of Earth's vorticity (the β-effect) and the imposed wind stress curl, i.e., the Sverdrup (1947) balance,

$$\beta\breve{v} = -\frac{\mathbf{k}\cdot\nabla\times\underline{\tau}_s}{(d - d_E)\rho^*} \tag{11.17}$$

The northward mass flow required to balance this broad equatorward flow must take the form of a more narrow intense poleward current along the western boundary,

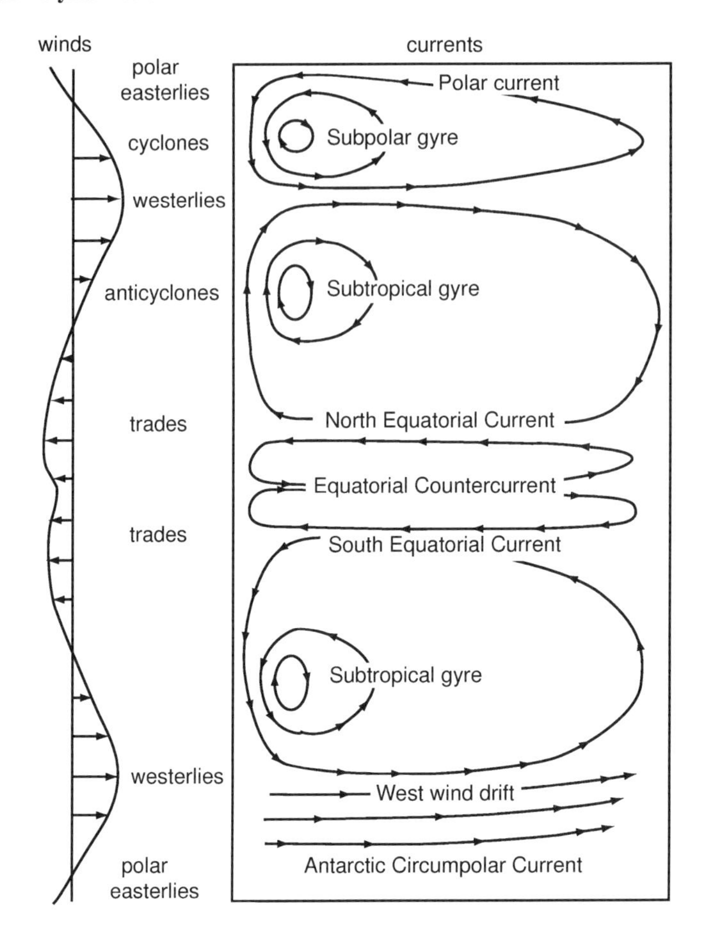

Figure 11-2 Main features of the surface layer wind-driven oceanic circulation.

having a short zonal length scale L_{B}. Over this short scale the friction term $\kappa\nabla^2\phi$ and β-term can achieve large values that nearly balance each other, while the wind stress forcing remains at its constant smaller value. This required balance implies that $L_{\mathrm{B}} \approx \kappa/\beta \approx 100$ km for realistic values. Many further elaborations of this theory have been made, for example, by the inclusion of the neglected inertial terms ($\mathbf{v}\cdot\nabla\mathbf{v} \neq 0$) by Munk (1950).

In Fig. 11-2, note also the convergence of the gyre flow in the subtropics requiring downwelling that deepens the thermocline, and divergence of the flow in the tropics and at high latitudes requiring upwelling that tends to shallow the thermocline and bring deeper water properties closer to the surface (e.g., colder water rich in nutrients and carbon). This effect is quite pronounced in coastal regions, where northerly and southerly winds associated with the subtropical high-pressure systems can drive Ekman drifts away from the coast, causing significant upwelling.

These coastal effects create an asymmetry in the east–west profile of the thermocline in the subtropics. Another major asymmetry occurs in the tropics where the easterly tradewinds drive water westward, piling up a deep warm water pool, notably in the western Pacific. Thus, except as altered by an El Niño weakening of the trades, the thermocline decreases almost linearly in depth from the west to the east coast, where cold deep water actually outcrops to the surface. [The probable existence of uniform warm pool temperatures in the western equatorial Pacific over the Late Cenozoic is of $\delta^{18}O$ records from that area, leading to the most credible record of planetary ice volume with a minimal temperature effect (see Section 2.4).] As a qualification regarding the complex equatorial flow we further note that in applying Eq. (11.17) it follows that the observed doldrum minimum in the trades gives rise to an equatorial countercurrent as pictured in highly idealized form in Fig. 11-2.

Given the prevailing poleward decrease of both temperature and salinity forced by the meridional profiles of surface heating and evaporation, the main subtropical gyre will transport both heat and salinity poleward. This poleward gyre transport is of the same order as that due to the thermohaline circulation to be discussed in the next section, being somewhat weaker than the TH transport in the present-day Atlantic but significantly larger in the Pacific (Semtner and Chervin, 1988). The possible effects on this transport of tectonically altered seaways in the past has been discussed, for example, by Barron and Peterson (1989) for the Cretaceous/Tethys circulation and by Haug and Tiedemann (1998) for the closing of the Isthmus of Panama in the late Cenozoic.

As a first approximation it seems reasonable that the magnitude of the gyre circulations will depend on the meridional gradient of surface temperature, which is a primary driver of the general circulation of the atmosphere, including the surface wind component.

11.3.2 Local Convective Overturnings and Baroclinic Eddy Circulations

The above wind-driven flows are essentially barotropic, not involving consideration of density variations or variations of the currents with depth. Such density gradients, which are steadily being generated by differential heating, give rise to two main forms of instabilities, both of which occur primarily in high latitudes under the influence of surface cooling:

1. Local deep overturnings wherein cold, brine-rich waters lose their buoyancy and sink in cellular patterns to deeper levels to be replaced by relatively lighter waters from below (e.g., Stern, 1975).
2. Baroclinic eddy circulations. Even for relatively stable vertical stratification, the strong horizontal density gradients between polar waters and the subtropical waters represent a baroclinic zone in which meander-like eddies form; these are characterized by both horizontal and deep vertical exchanges of mass and properties (e.g., Saltzman and Tang, 1975).

For both of these reasons, the higher latitude waters tend to be strongly mixed in the vertical, tending to make the combined boxes 1 and 4 in Fig. 11-1 more homogeneous.

This provides some justification for the many models in which these two boxes are combined into a single high-latitude box (e.g., Nakamura *et al.*, 1994).

11.4 THE TWO-BOX THERMOHALINE CIRCULATION MODEL: POSSIBLE BIMODALITY OF THE OCEAN STATE

The gradient of density between the polar region and the equatorial region creates pressure differences in horizontal planes that can drive a grand poloidal thermohaline circulation. These pressure differences follow as a consequence of the near-hydrostatic balance of the ocean, in which case Eq. (11.2) reduces to the form, $\partial P/\partial z = -g\delta\rho$. From this equation we can see that, given an initially uniform pressure at the bottom of the ocean, the pressure will decrease more strongly with height in the dense polar region than in the lighter tropical region, creating an upper level gradient that tends to drive water poleward in upper layers in accordance with Eq. (11.1). The equatorward return flow required by Eq. (11.3) would tend to produce a steady circulation shown in Fig. 11-3 similar to a *thermal circulation* in the atmosphere. As noted in Section 11.3.1, however, this TH circulation may be significantly modified by deep wind-driven effects, particularly in the Southern Ocean, which, among other effects, tends to create a more asymmetric, cross-equatorial circulation than would be implied by external radiative forcing (Toggweiler and Samuels, 1995).

11.4.1 The Two-Box System

In a significant paper Stommel (1961) showed how the difference in the contributions of temperature and salinity to this pole to equator density difference can give rise to an instability in the ocean that can result in multiple equilibrium states. Stommel's result was based on a two-box simplification of the four-box model described in Section 11.2 in which boxes 1 and 4, and boxes 2 and 3, are each combined to form single low- and high-latitude boxes extending from the top to the bottom of the ocean, separated from each other at a distance $y = L$ from the equator $y = 0$.

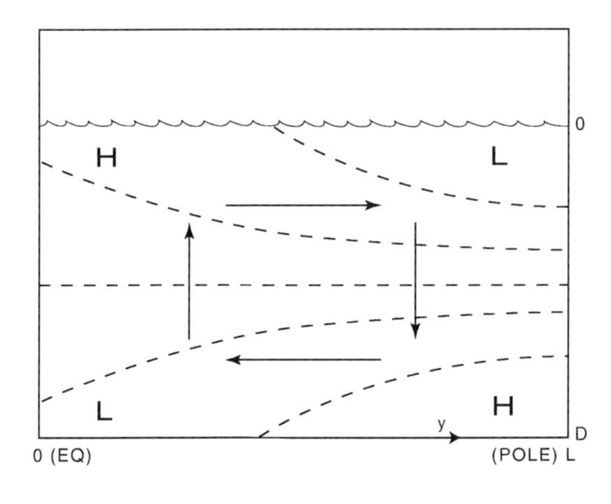

Figure 11-3 Schematic cross-section of the thermohaline circulation driven by pressure differences engendered hydrostatically by equator-to-pole density variations.

Here we shall start by considering a more detailed three-box version of Fig. 11-1 in which we retain a representation of the relatively faster response upper layer ($\Delta Z = d$), wherein the gradients of T and S forced by surface fluxes of heat and freshwater are a maximum; these fluxes can be approximated by distinct well-mixed values in each of boxes 1 and 2. The region below this upper level ($\Delta Z = D - d$) is represented by a single well-mixed deep ocean layer formed by combining boxes 3 and 4 of Fig. 11-1 (cf. Birchfield *et al.*, 1990). The properties of this large volume of deep ocean vary on a longer time scale in response to small net fluxes of heat and salt across $Z = d$. We designate the value of temperature and salinity in this deep ocean by θ and S_D, respectively. To reduce this three-box ocean to a Stommel-like two-box ocean we can define a high-latitude (polar) state

$$(T_p, S_p) \equiv \frac{(T_1, S_1)V_1 + (\theta, S_D)V_4}{V_p} \tag{11.18}$$

and a low-latitude (equatorward) state

$$(T_e, S_e) \equiv \frac{(T_2, S_2)V_2 + (\theta, S_D)V_3}{V_e}$$

where $V_p \equiv (V_1 + V_4)$ and $V_e \equiv (V_2 + V_3)$. Because heat and freshwater fluxes occur at the ocean surface it is reasonable to assume that most of the poleward change in T and S occurs mainly between the upper boxes (1 and 2), with relatively uniform values in the combined deep water box (3 and 4).

Averaging Eqs. (11.4) and (11.5) over the climatic-average period $\delta_c = 100$ y (thereby excluding explicit consideration of all subcentennial scale variability, including interdecadal variability), averaging spatially over each of the two volumes V_p and V_e, and applying the boundary conditions $w = 0$ at the surface ($Z = 0$) and bottom ($Z = D$) and $v = 0$ at the equator ($y = 0$) pole ($y = L$), we obtain the following set of equations governing the mean values of T and S in each volume (see Fig. 11-4).

$$V_p \frac{dT_p}{dt} = Q_T + \mathcal{H}_p^{\uparrow} + \mathcal{G}_p^{\uparrow} \tag{11.19a}$$

$$V_e \frac{dT_e}{dt} = -Q_T - \mathcal{H}_e^{\downarrow} + \mathcal{G}_e^{\uparrow} \tag{11.19b}$$

$$V_p \frac{dS_p}{dt} = Q_s - \mathcal{F}^{\downarrow} S_p \tag{11.20a}$$

$$V_e \frac{dS_e}{dt} = -Q_s + \mathcal{F}^{\uparrow} S_e \tag{11.20b}$$

where the well-mixed climatic-mean values of T and S in each box, n, are

$$(T_n, S_n) = \frac{1}{V_n} \iiint_n (T, S)\, dx\, dy\, dz. \tag{11.21}$$

Recalling our definitions [Eqs. (7.2) and (7.3)] of the zonal average $\langle x \rangle \equiv W^{-1} \int x\, dx$ (where in this case W is the width of an ocean basin) and of the departure representing mean gyres is $x_* \equiv x - \langle x \rangle$, the net thermal and salinity fluxes across a latitude wall at $y = L^*$ are

$$\begin{aligned} Q_{(T,S)} &= \int_0^D \int_0^W \left\langle \overline{v \cdot (T, S)} \right\rangle_{L^*} dx\, dz \\ &= Q_{(T,S)\psi} + Q_{(T,S)\phi} + Q^{\star}_{(T,S)} \end{aligned} \tag{11.22}$$

where $Q_{(T,S)\psi} = \iint \langle v \rangle \langle (T, S) \rangle\, dx\, dz$, $Q_{(T,S)\phi} = \iint \langle v_* (T, S)_* \rangle\, dx\, dz$, and $Q^{\star}_{(T,S)} = \iint \langle \overline{v^{\star} \cdot (T, S)^{\star}} \rangle\, dx\, dz$ are the respective contributions due to the mean thermohaline circulation $\langle v \rangle$, mean gyre flow $v_* = (\partial \phi / \partial x)$, and subclimatic-mean circulations $v^{\star}$, including phenomena such as baroclinic eddies and interdecadal fluctuations [see Eq. (5.3)].

The upward fluxes of sensible heat at the ocean surface $\mathcal{H}_n^{\uparrow}$, the geothermal flux at the bottom $\mathcal{G}^{\uparrow}$ (both in units of $m^3\, K\, s^{-1}$) and the freshwater flux $\mathcal{F}_n^{\uparrow}$ (in units of $m^3\, s^{-1}$), across the horizontal area A_n of each region, are

$$\left(\mathcal{H}_n^{\uparrow}, \mathcal{G}_n^{\uparrow}, \mathcal{F}_n^{\uparrow}\right) = \iint_{A_n} \left[\left(H_{SS}^{\uparrow}/\rho c\right), \left(H_D^{\uparrow}/\rho c\right), \left(F^{\uparrow}/\rho\right)\right] dx\, dy \tag{11.23}$$

Note that Eqs. (11.20a) and (11.20b) imply conservation of salt for the whole ocean (the validity of which over geologic time can only be assumed), but not of global mean salinity,

$$\breve{S} = \frac{V_e S_e + V_p S_p}{V} \tag{11.24}$$

which can be expected to vary as a consequence of changing storage in the cryosphere (i.e., $\mathcal{F}_e^{\uparrow} \neq \mathcal{F}_p^{\downarrow}$). That is,

$$\frac{d\Pi_{salt}}{dt} = 0 \tag{11.25}$$

where $\Pi_{salt} = \rho^* (V_p S_p + V_e S_e)$ and ρ^* is the standard seawater density. To demonstrate consistency with Eqs. (11.20a) and (11.20b), note that $\mathcal{F}_e^{\uparrow} = -dV_e/dt$ and $\mathcal{F}_p^{\downarrow} = dV_p/dt$.

If one assumed no net mean temperature change of the whole ocean, then $\mathcal{H}_p^{\uparrow} = \mathcal{H}_e^{\downarrow}$. As in the case of the freshwater balance on paleoclimatic time scales, however, we can expect small imbalances in the net surface heat flux that ultimately lead to slow variations in the mean temperature of the deep ocean (see Section 8.3). In general, $\mathcal{H}^{\uparrow}$ is a fast-response surficial variable, calculable from a GCM as a function of prescribed slow-response variables, i.e., $\mathcal{H}^{\uparrow} = \mathcal{H}^{\uparrow}(I, \mu, \theta)$.

In many studies $\mathcal{H}^{\uparrow}$ has been represented by a Newtonian approximation, the so-called "restoring condition" (Haney, 1971), i.e.,

$$\mathcal{H}_n^{\uparrow} = \Gamma\left(T_n - T_n^*\right) \tag{11.26}$$

where $(V_n\Gamma^{-1})$ is the time scale over which temperature departures are restored to fixed values T_n^* by heat fluxes. These fixed values would most appropriately represent climatological average temperatures in the surface mixed layer, but in this simple box model are taken to represent the full upper layer $\Delta Z = d$. For paleoclimatic considerations T_n^* should be considered to be very slowly varying quantities that are functions of the slow-response variables, i.e., $T_n^* = T_n^*(I, \mu, \theta)$.

On the other hand, there is no Newtonian-type restoring mechanism for salinity because the freshwater flux $\mathcal{F}_n^\uparrow$ is not a function of the prevailing value of S_n. Instead, $\mathcal{F}^\uparrow$ is a measure of the intensity of the global hydrologic cycle determined by the atmospheric general circulation and surface heat balance. In this regard, it is not unreasonable from GCM results (see Section 7.10) to expect that $\mathcal{F}^\uparrow$ is diminished during cold glacial states and enhanced during warm nonglacial states.

11.4.2 A Simple Model of the TH Circulation

In Section 11.3.1 we discussed the dynamics of the horizontal motions that can give rise to the poleward transport of heat and salinity represented by $Q_{(T,S)\phi}$ and $Q^\star_{(T,S)}$ in Eq. (11.22). We now discuss the dynamics of the two-dimensional TH circulation shown in Fig. 11-2 that gives rise to the flux $Q_{(T,S)\phi}$. Given the temperature and salinity distribution, this circulation is governed by the **j**-components of Eqs. (11.1) and (11.3), (11.2), and (11.6), i.e., in Cartesian coordinates $(dy = a d\phi)$,

$$\frac{dv}{dt} = -\frac{1}{\rho^*}\frac{\partial \mathcal{P}}{\partial y} - fu - \kappa v \tag{11.27}$$

$$\frac{dw}{dt} = -\frac{1}{\rho^*}\frac{\partial \mathcal{P}}{\partial z} - \frac{g}{\rho^*}(\rho - \rho^*) - \kappa w \tag{11.28}$$

$$0 = \frac{\partial v}{\partial y} + \frac{\partial w}{\partial z} \tag{11.29}$$

$$(\rho - \rho^*) = \left[-\mu_{\mathrm{T}}(T - T^*) + \mu_{\mathrm{S}}(S - S^*)\right] \tag{11.30}$$

where we have expressed the viscous terms in Eqs. (11.27) and (11.28) by a Rayleigh friction (κ = const.).

From the continuity Eq. (11.29) we can define a stream function, ψ, for the TH circulation,

$$v = -\frac{\partial \psi}{dz}, \qquad w = \frac{\partial \psi}{\partial y} \tag{11.31}$$

Cross-differentiating Eqs. (11.27) and (11.28) with respect to z and y, respectively, to eliminate $\mathcal{P}$, then subtracting Eq. (11.28) from Eq. (11.27), and applying Eq. (11.31), we obtain the vorticity equation for the TH circulation,

$$\frac{d\nabla^2\psi}{dt} = -f\frac{\partial u}{\partial z} + \frac{g}{\rho^*}\left(-\mu_{\mathrm{T}}\frac{\partial T}{\partial y} + \mu_{\mathrm{S}}\frac{\partial S}{\partial y}\right) - \kappa\nabla^2\psi \tag{11.32}$$

where the vorticity is

$$\nabla^2 \psi = \left(\frac{\partial v}{\partial z} - \frac{\partial w}{\partial y} \right) = -\left(\frac{\partial^2 \psi}{\partial y^2} + \frac{\partial^2 \psi}{\partial z^2} \right)$$

The Coriolis term $(-f\, \partial u/\partial z)$ is probably small compared to the density gradient and viscous terms, and is generally neglected (e.g., Cessi and Young, 1992) or is considered to be absorbed into the viscous terms (Sakai and Peltier, 1995); we shall neglect it here.

A solution satisfying the condition that the normal velocity on the boundaries vanishes (i.e., $\psi = 0$ on the boundaries) is of the form

$$\psi = \psi_{\max} \sin \frac{\pi}{L} y \cdot \sin \frac{\pi}{D} z \tag{11.33}$$

where $\psi_{\max}$ is the maximum value of ψ at $(L/2, D/2)$, measuring the strength of the TH circulation. This form represents a more symmetric overturning than pictured in Figs. 11-1 and 11-4, being similar to the idealized picture shown in Fig. 11-3. The governing equation for the variations of $\psi_{\max}$ is obtained by substituting Eq. (11.33) in Eq. (11.32), and integrating over the spatial domain $y = 0$ to L and $z = 0$ to D, yielding

$$\frac{d\psi_{\max}}{dt} = \frac{a}{\rho^*} (\rho_{\mathrm{p}} - \rho_{\mathrm{e}}) - \kappa \psi_{\max} \tag{11.34}$$

where $\rho_{\mathrm{p}} \equiv \rho(L)$, $\rho_{\mathrm{e}} \equiv \rho(O)$, and $a = [gLD^2/4(L^2 + D^2)]$, a constant.

Thus, the variations of $\psi_{\max}$ are driven by the density difference between the two volumes, under the retarding influence of linear dissipation. It is typically assumed that the viscous resistance represented by κ is large enough that a quasi-static equilibrium prevails $[d(\psi_{\max})/dt \approx 0]$ of the form

$$(\psi_{\max})_0 \approx -\frac{k_\psi}{\rho^*} \delta\rho \tag{11.35}$$

where $\delta\rho \equiv (\rho_{\mathrm{e}} - \rho_{\mathrm{p}})$, $k_\psi = a/\kappa$, and $\rho = \rho(T, S)$ in accordance with Eq. (11.6). Note that this assumption tends to place the circulation ψ in the category of a relatively "fast-response" variable (in comparison, e.g., with the deep ocean temperature θ; see Section 8.3), adjusting on a short time scale to an imposed density difference. On the other hand, because salinity has no intrinsic self-regulating dissipative property, it can fall in the category of a slow-response variable, especially if positive feedbacks occur, as will be discussed in more detail in Section 11.4.4. The other slower response variables that ultimately control the variations of the density difference include the deep ocean state represented by θ, as well as ice sheet coverage, the greenhouse gas level, and slow external radiative forcing (see Chapter 8).

11.4.3 Meridional Fluxes

Concerning the horizontal fluxes Q_T and Q_S represented by (11.22), Stommel (1961) makes the approximation

$$Q_T \approx Q_{\psi T} = |q_\psi|\delta T \tag{11.36a}$$

$$Q_S \approx Q_{\psi S} = |q_\psi|\delta S \tag{11.36b}$$

where $\delta T \equiv (T_e - T_p)$, $\delta S = (S_e - S_p)$, and q_ψ is the "volume exchange flux" at $y = L/2$ due to the TH circulation [i.e., the rate of water volume exchange, in Sverdrups (10^6 m^3 s^{-1}), across the latitude wall at $y = L/2$]. With the boundary conditions $\psi(D) = 0$, $\psi(D/2) = \psi_{max}$, the poleward part of the volume exchange (assumed to be positive in the upper branch of the circulation, as pictured in Fig. 10-3) is given by

$$\begin{aligned} q_\psi &= \int_0^W \int_{D/2}^D v \, dz \, dx \\ &= \int_0^W \int_{D/2}^D \frac{\partial \psi}{\partial z} \, dx \, dz \\ &= W \psi_{max} \end{aligned}$$

where ψ_{max} [$= (\psi_{max})_0$] is given by Eq. (11.35) and W is the width of the ocean basin. The absolute sign $|q|$ in Eq. (11.36) indicates that the flux of heat or salt is independent of the direction of the circulation between the boxes. Stommel (1961) identifies q_ψ as a pipelike loop connecting V_e and V_p (see Fig. 11-4).

More generally, however, in accordance with Eq. (10.22) the gyre flux contribution, Q_ϕ, and to a lesser extent the horizontal eddy flux, $Q^\star$, are also of importance, the

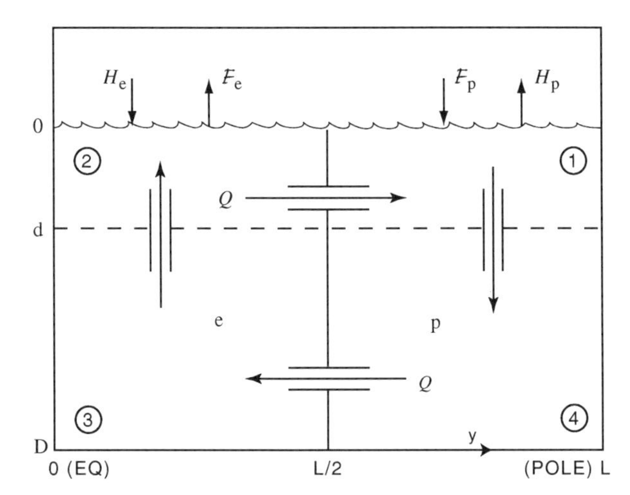

Figure 11-4 The two-box ocean system comprising a low-latitude box (e) and a polar box (p) connected by a volume exchange Q.

combination of which, denoted by $Q^{\star}_{\phi}$, we can also approximate in the form

$$\mathcal{Q}^{\star}_{\phi(T,S)} \equiv \left[\mathcal{Q}_{\phi(T,S)} + \mathcal{Q}^{\star}_{(T,S)}\right] = \left|q^{\star}_{\phi}\right| \delta(T, S) \tag{11.37}$$

where $q^{\star}_{\phi}$ is the combined gyre/eddy volume exchange between $V_{\rm e}$ and $V_{\rm p}$, which we shall assume is mainly due to the gyres with a smaller contribution due to the baroclinic eddies and smaller scale horizontal eddy mixing.

This volume exchange $q^{\star}_{\phi}$ is not forced by the salinity differences that contribute strongly to $\delta\rho$, but is forced in an indirect way by surface temperature differences and vertical heat fluxes that drive the atmospheric general circulation, including the surface wind systems. Thus, it is plausible to express $q^{\star}_{\phi}$ as a function only of the temperature difference, $\delta T = (T_{\rm e} - T_{\rm p})$, i.e.,

$$q^{\star}_{\phi} = \frac{k_{\phi}}{\rho^{*}} \mu_{\rm T} \delta T \tag{11.38}$$

where k_{ϕ} is a constant having the same units as k_{ψ}.

11.4.4 Dynamical Analysis of the Two-Box Model

The basic result concerning the potential instability and bimodality of the ocean can be demonstrated with the application of the following further simplifying approximations:

1. The geothermal flux $\mathcal{G}^{\uparrow}$ is neglected relative to the other terms in Eqs. (11.19a) and (11.19b).

2. The thermal restoring time scale $(V_n \Gamma^{-1})$ is assumed to be relatively short ($\sim$10 y) so that the temperatures $T_{\rm e}$ and $T_{\rm p}$ are maintained at very nearly the fixed values $T_{\rm e,p}(I, \mu, \theta)$; from Eqs. (11.18a) and (11.18b) this implies that

$$\delta T \equiv (T_{\rm e} - T_{\rm p}) \approx \delta T^{*} \qquad \text{(a constant)} \tag{11.39}$$

Thus the thermal state of the ocean is fixed if the slow-response variables I, μ, and θ are held constant, and there is no need to consider the thermal equations Eqs. (11.19a) and (11.19b); i.e., all of the dynamics of the ocean is now embodied in the way salinity varies relative to the fixed δT^{*} as governed by Eqs. (11.20a) and (11.20b).

3. The freshwater flux terms in Eqs. (11.20a) and (11.20b) are linearized by setting $\mathcal{F}_{\rm (p,e)} \cdot \mathcal{S}_{\rm (p,e)} = \mathcal{F}_{\rm (p,e)} \cdot \breve{S}$, where $\breve{S}$ is the global mean value of salinity [Eq. (11.24)].

4. As a final simplification, if we assume equal low- and high-latitude volumes, $V_{\rm e} = V_{\rm p} \equiv V$, the equation governing the salinity difference $\delta S \equiv (S_{\rm e} - S_{\rm p}) \sim (-S_{\varphi})$ can be written in the form Eq. (11.20b) – Eq. (11.20a):

$$\frac{V}{2} \frac{d(\delta S)}{dt} = \breve{S}\mathcal{F}^{\uparrow} - |q_{\psi}|\,\delta S - \left|q^{\star}_{\phi}\right|\delta S \tag{11.40}$$

where, using Eqs. (11.30), (11.38), and (11.39),

$$q_\psi = k_\psi\left(\mu_T\,\delta T^* - \mu_S\,\delta S\right) \tag{11.41a}$$

$$q_\phi^\star = k_\phi \mu_T\,\delta T^* \tag{11.41b}$$

Thus, introducing Eq. (11.41) into Eq. (11.40) we obtain

$$\frac{V}{2}\frac{d(\delta S)}{dt} = \breve{S}\mathcal{F}^\uparrow - K_S(\delta S)\cdot\delta S \tag{11.42a}$$

where the damping coefficient $K_S(\delta S)$ is given by,

$$\begin{aligned} K_S(\delta S) &= \left(k_\phi\mu_T\left|\delta T^*\right| + k_\psi\left|\mu_T\,\delta T^* - \mu_S\,\delta S\right|\right) \\ &= \mu_T\left|\delta T^*\right|\left[k_\phi + k_\psi\left(1 - \mu_S\,\delta S/\mu_T\,\delta T^*\right)\right] \end{aligned}$$

Note that for fixed values of $\mathcal{F}^\uparrow$ and δT^* a range of these values may exist wherein an increase of δS will weaken K_S to a point that allows a further increase of δS, i.e., a positive feedback that may give rise to an instability. As noted in Section 11.4.2 this would imply that δS would have to be considered a slow-response variable (see Section 6.1).

More formally, if we define the following nondimensional variables

$$\begin{aligned} s &= \left(\frac{\mu_S}{\mu_T\,\delta T^*}\right)\delta S \\ (\Psi, \Phi) &= \left(\frac{1}{k_\psi\mu_T\,\delta T^*}\right)\left(q_\psi, q_\phi^\star\right) \\ \Pi &= \left[\frac{\mu_S\breve{S}}{k_\psi\mu_T^2(\delta T^*)^2}\right]\mathcal{F}^\uparrow \\ t^* &= \frac{2k_\psi\mu_T\,\delta T^*}{V}\,t \end{aligned}$$

the governing equations [Eqs. (11.40) and (11.41)] become

$$\frac{ds}{dt^*} = \Pi - |\Psi|s - |\Phi|s \tag{11.42b}$$

$$\Psi = (1 - s) \tag{11.43a}$$

$$\Phi = (k_\phi/k_\psi) \equiv K_\phi \tag{11.43b}$$

The quantity Π is here considered to be an external parameter representing the ratio of the freshwater fluxes to the temperature gradient, which is a function of $y = (I, \mu, \theta; R, h, V, \mathcal{W})$ determined by the atmospheric general circulation via a GCM (see Chapter 7), but is not directly determined by the salinity gradient as measured by the free variable s. For any value of Π the steady-state (equilibria) of the system,

Eqs. (11.42) and (11.43), governing s and hence the TH circulation Ψ via Eq. (11.43) are as follows:

1. When $s < 1$ (i.e., a "direct" thermally dominated TH circulation, $\Psi > 0$, with sinking cold dense water in high latitudes), two equilibria can exist if $\Pi < (1 + K_\phi)^2/4$,

$$s^{(1)} = \left(1 - \Psi^{(1)}\right) = \frac{1}{2}\left[(1 + K_\phi) - \sqrt{(1 + K_\phi)^2 - 4\Pi}\,\right]$$

$$s^{(2)} = \left(1 - \Psi^{(2)}\right) = \frac{1}{2}\left[(1 + K_\phi) + \sqrt{(1 + K_\phi)^2 - 4\Pi}\,\right]$$

2. When $s > 1$ (i.e., a reverse, salinity-dominated, TH circulation, $\Psi < 0$, with sinking salty water in low latitudes), an additional equilibrium exists for all $\Pi > K_\phi$,

$$s^{(3)} = \left(1 - \Psi^{(3)}\right) = \frac{1}{2}\left[(1 - K_\phi) + \sqrt{(1 - K_\phi)^2 + 4\Pi}\,\right]$$

The dependence of the equilibrium values of the TH circulation, $\Psi^{(n)}$, on the forcing Π is shown in Fig. 11-5, which is the bifurcation diagram for this ocean system (cf. Fig. 7-3 for the atmospheric EBM). By an eigenvalue analysis it is readily shown that $\Psi^{(1)}$ and $\Psi^{(3)}$ are stable (solid curves), separated by an unstable $\Psi^{(2)}$ (dashed line), with $\Psi^{(1)}$ and $\Psi^{(2)}$ merging at a "saddle node" when $\Pi = (1 + K_\phi)^2/4$ [see Section 6.3, Eq. (6.16), and Fig. 6-5a)]. The possibility for an instability and multiple

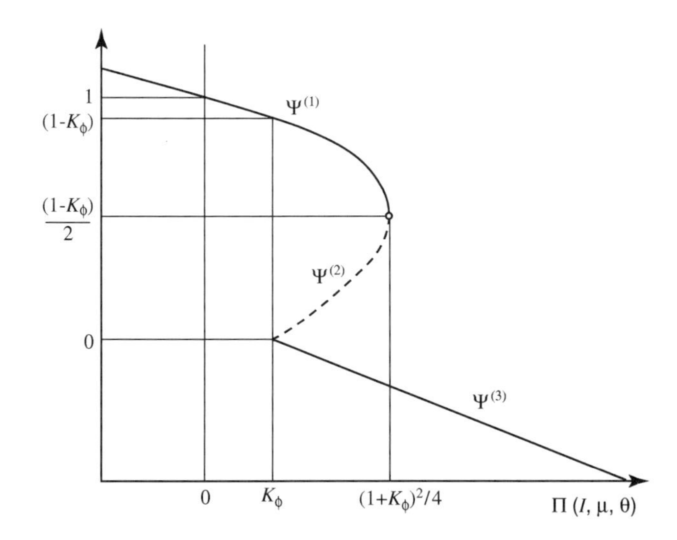

Figure 11-5 Bifurcation diagram showing the equilibrium values of the nondimensionalized volume flux by the TH circulation Ψ for the two-box ocean system, as a function of the "external" forcing function $\Pi(I, \mu, \theta)$.

equilibria of the ocean due to salinity, as illustrated by this simple model, is the central result of this analysis.

Thus, for the same value of forcing within the limits $K_\phi < \Pi < (1 + K_\phi)^2/4$ the system can admit stable equilibria either with a strong thermally direct circulation $\Psi > 0$ corresponding to a small salinity difference between equator and pole $s = (1 - \Psi)$, or another stable equilibrium with a weak direct circulation, $\Psi < 0$, associated with a large salinity difference. This latter salinity-dominated mode $\Psi^{(3)}$, which can be realized only if the salinity forcing can overcome the steady gyre flux, i.e., $\Pi > K_\phi$, might be identified with warm climatic periods of the past, e.g., the Cretaceous and Early Cenozoic when the Tethys Sea provided a large area of excess evaporation in low latitudes (see Figs. 3-1 and 3-3). This will be discussed further in Chapter 13 (Section 13.5).

The possible existence of multiple equilibria raises the question of possible rapid transitions between these modes as a factor in paleoclimatic change. Such transitions can arise as part of the more complete multivariable system involving ice and carbon dioxide, in the form of sustained oscillations, or can arise by slow forcing of Π to the bifurcation points $\Pi = K_\phi, (1 + K_\phi)^2/4$, or by random forcing that causes a tunneling between $\Psi^{(1)}$ and $\Psi^{(3)}$. This latter possibility can be augmented by the "stochastic resonance" mechanism described in Chapter 7.

We note also that many two-dimensional zonal mean models and three-dimensional ocean models (OGCMs) have now been studied that can be tuned to replicate in greater detail these potential bimodal properties and reveal their sensitivity to atmospheric parameters. The first was made by Bryan (1986) and was followed, as examples, by Manabe and Stouffer (1988), Marotzke *et al.* (1988), Wright and Stocker (1991), and Fichefet and Hovine (1993). An example of a specific paleoceanographic application to glacial-maximum states was provided by Bigg *et al.* (1998). In Chapters 12 and 13 we discuss this possible role of ocean instability in a more time-dependent context as part of the fuller theory of paleoclimatic change.

At this point, it is worth emphasizing again that it is the ultimate role of the full-fledged OGCMs to include properly all of the coupled interactive circulation components that we have artificially separated for illustrative purposes. Such OGCMs form an essential component of a full time-dependent climate system model (CSM) as outlined at the beginning of Chapter 5 (see Fig. 5-1). However, we have also noted in that chapter that these more detailed models are still not capable of calculating the nonequilibrium net flux of heat across the ocean surface on the order of 10^{-1} W m^{-2}, as required by paleoclimatic observations of the variations of θ. Nor are these models able to account definitively for what is presently known about the grand circulation, and the subgrid-scale variability of the ocean, without the assignment of several free parameters. Thus, as we have said, it is not unreasonable to begin to approach the problem of paleoclimatic variations with the simplest models, which may require fewer free parameters than the more complex models. Moreover, these simpler models can at the least serve to illustrate the possibilities for long-term behavior, including possible instabilities that may ultimately be captured more rigorously by the full three-dimensional models.

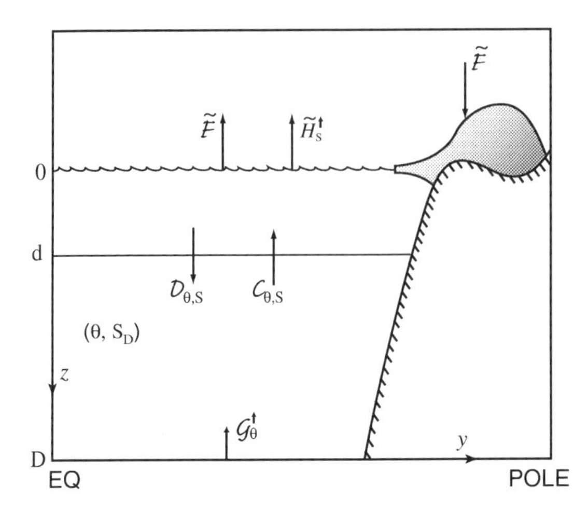

Figure 11-6 Schematic representation showing the main fluxes ($\mathcal{D}_{\theta,S}$, $\mathcal{C}_{\theta,S}$, $\mathcal{G}_\theta^{\uparrow}$) determining the deep ocean thermal state (θ).

11.5 INTEGRAL EQUATIONS FOR THE DEEP OCEAN STATE

In the previous section we described the first-order results of a two-box model driven mainly by the gradients of temperature and salinity in a relatively shallow layer ($\Delta Z = d$) of the full deep high- and low-latitude volumes V_e and V_p. The deep ocean portions of these boxes were treated as relatively passive domains in accordance with Eqs. (11.18a) and (11.18b). We now consider more explicitly the much slower response properties of this deeper layer by forming another two-box model in which boxes 3 and 4 of Fig. 11-1 are combined to represent a single deep ocean box of volume $V_D = (V_3 + V_4)$ beneath a single shallower upper layer box of volume V_d comprising of boxes 1 and 2 of Fig. 11-1 (see Fig. 11-6). The deep layer is characterized by a mean temperature $\theta = (T_3 + T_4)/2$, and a mean salinity $S_D = (S_3 + S_4)/2$. We assume this upper box to be part of the fast-response surficial system governed by a mixed-layer GCM (see Chapter 6).

11.5.1 The Deep Ocean Temperature

In a manner similar to the derivation of Eq. (11.19), an equation governing the rate of change of θ is obtained by averaging Eq. (11.4) over the deep ocean volume V_D and applying the boundary conditions. The following equation is obtained:

$$\frac{d\theta}{dt} = -\mathcal{C}_\theta^{\uparrow} + \mathcal{D}_\theta^{\downarrow} + \mathcal{G}_\theta^{\uparrow} \tag{11.44}$$

where $\mathcal{C}_\theta^{\uparrow}$, $\mathcal{D}_\theta^{\downarrow}$, and $\mathcal{G}_\theta^{\uparrow}$ are defined and interpreted as follows [as in Eq. (4.43), a tilde will denote an area average over the world ocean]:

The rate of decrease of θ due to the global upward advective flux of heat over the world ocean across the base of the upper layer is given by

$$\mathcal{C}_\theta^{\uparrow} \equiv \frac{\widetilde{H}_{\mathrm{d1}}^{\uparrow}}{\rho c(D-d)} = \frac{1}{V_D} \iint \overline{(wT)}_{\mathrm{d}}\, dx\, dy \tag{11.45}$$

This flux is a consequence of all large-scale vertical motions ranging from the grand thermohaline circulation (ψ) to the more localized and more rapid deep convective overturnings (χ), including those associated with baroclinic eddies that occur mainly in higher latitudes. In principle, a mainly salinity-driven circulation, such as has been postulated to give rise to "warm saline bottom water" (Brass *et al.*, 1982), or a wind-driven circulation, might lead to a much reduced or possibly negative value of $\mathcal{C}_\theta^\uparrow$.

The rate of increase of θ due to the downward diffusion of heat by small-scale vertical eddy motions across the base of the upper layer at $Z = d$ is given by

$$\mathcal{D}_\theta^\downarrow \equiv \frac{\widetilde{H}_{\mathrm{d2}}^\downarrow}{\rho c(D-d)} \equiv \frac{1}{V_\mathrm{D}} \iint K_\mathrm{v}^{(T)} \left(\frac{\partial T}{\partial z}\right)_\mathrm{d} dx\, dy \tag{11.46}$$

where $\widetilde{H}_{\mathrm{d2}}$ is the net downward diffusive heat flux at $z = d$. These processes typically transfer heat downward from the warmer upper layer, particularly in low and middle latitudes.

The rate of increase of θ due to the upward geothermal flux of heat across the ocean–sediment interface is given by

$$\mathcal{G}_\theta^\uparrow \equiv \frac{\widetilde{H}_\mathrm{D}^\uparrow}{\rho c(D-d)} = \frac{1}{V_\mathrm{D}} \iint K_\mathrm{v}^{(T)} \left(\frac{\partial T}{\partial z}\right)_\mathrm{D} dx\, dy \tag{11.47}$$

where $\widetilde{H}_\mathrm{D}^\uparrow$ is the globally averaged upward geothermal flux of heat in $\mathrm{W\,m^{-2}}$ [see Eq. (4.50)].

Thus, Eq. (11.44) can be written, alternately, in terms of the net heat fluxes across $Z = d$

$$\frac{d\theta}{dt} = \frac{1}{\rho c(D-d)}\left(\widetilde{H}_\mathrm{d}^\downarrow + \widetilde{H}_{\mathrm{d2}}^\downarrow + \widetilde{H}_\mathrm{D}^\uparrow\right) \tag{11.48}$$

Due to the limited heat capacity of the relatively shallow upper layer, which is an order of magnitude smaller than the deep layer, $H_\mathrm{d}^\downarrow$ [$= H_{\mathrm{d1}}^\downarrow + H_{\mathrm{d2}}^\downarrow$] should be of the same order as the net heat flux at the ocean surface, $\widetilde{H}_\mathrm{S}^\downarrow$, that results from shortwave and longwave radiation and sensible and latent heat exchanges with the atmosphere. Thus, from Eq. (11.48)

$$\left|\widetilde{H}_\mathrm{S}^\downarrow\right| \sim \left|\widetilde{H}_\mathrm{d}^\downarrow\right| \sim \rho c D\left|\frac{d\theta}{dt}\right| \tag{11.49}$$

It would appear from present paleoceanographic evidence (Chappell and Shackleton, 1986) that θ increased by about 2°C in the most rapid interval of change between glacial maximum to minimum conditions (i.e., $|d\theta/dt| \sim 2°\mathrm{C}/10^4$ y). Using the values of ρ, c, and D from Table 4-3, this implies that $\widetilde{H}_\mathrm{d}^\downarrow \approx \widetilde{H}_\mathrm{S}^\downarrow$ must be known to the order of at least $10^{-1}\ \mathrm{W\,m^{-2}}$, a requirement that is impossible to satisfy given our level of ability to measure and parameterize the surface fluxes (cf. Section 5.1). It will therefore be necessary to introduce at least one free (or "adjustable") parameter to calculate the variations of θ.

11.5.2 The Deep Ocean Salinity

The same procedure applied to Eq. (11.5) yields the following equation for the rate of change of the mean salinity in the deep ocean volume V_D:

$$V_D \frac{dS_D}{dt} = -\mathcal{C}_S^{\uparrow} + \mathcal{D}_S^{\downarrow} \tag{11.50}$$

where

$$\mathcal{C}_S^{\uparrow} = \iint (\overline{wS})_d \, dx\, dy \tag{11.51}$$

is the rate of increase of S_D due to net downward advective fluxes of saline water across $z = d$ due to the thermohaline circulation (ψ) and deep local convective overturnings (χ), and

$$\mathcal{D}_S^{\downarrow} = \iint K_v^{(S)} \left(\frac{\partial S}{\partial z}\right)_d dx\, dy \tag{11.52}$$

is the rate of increase of S_D due to eddy diffusion across $Z = d$. Because salinity is generally a maximum in the upper layer, particularly in low and middle latitudes, the fluxes represented by this term would tend to increase S_D. On the other hand, a thermally direct TH circulation with sinking in high latitudes (where salinities are low due to excess precipitation) would tend to decrease S_D ($\mathcal{C}_S^{\uparrow} > 0$), while a reverse TH circulation with sinking of warm saline water in low latitudes produced by excess evaporation would tend to increase S_D ($\mathcal{C}_S^{\uparrow} < 0$). The potential multiple equilibrium properties of the deep ocean salinity state, in concert with the thermal state and TH circulation (ψ) described in Section 11.4, are nicely illustrated in the more complete three-box model of Birchfield (1989).

In the upper layer ($\Delta z = d$) the variations of mean salinity S_d are given by

$$V_d \frac{dS_d}{dt} = \mathcal{C}_S^{\uparrow} - \mathcal{D}_S^{\downarrow} + \mathcal{F} S_d \tag{11.53}$$

where $\widetilde{\mathcal{F}} \equiv \mathcal{F}_e^{\uparrow} - \mathcal{F}_p^{\downarrow} = dV_d/dt$ is the net global upward freshwater flux associated with accumulation of freshwater in the cryosphere. Thus, the net salinity change of the whole ocean, due to sea level changes associated with ice sheet variations, is

$$\frac{d\breve{S}}{dt} = -\frac{\widetilde{\mathcal{F}}\widetilde{S}_d}{V} \tag{11.54}$$

where, complementary to Eq. (11.24), $\breve{S} = (V_D S_D + V_d S_d)/V$, and use is made of the fact that whereas V_d can change due to $\mathcal{F}$-induced sea level changes, the deep ocean volume is fixed ($dV_D/dt = 0$). Note that although the global salinity may change in accordance with Eq. (11.54), the total mass of salt in the ocean, $\Pi_{salt} = \rho_w(V_D S_D + V_d S_d)$, is conserved, i.e., $\Pi_{salt}/dt = 0$ [cf. Eq. (11.25)].

11.6 GLOBAL DYNAMICAL EQUATIONS FOR THE THERMOHALINE STATE: θ AND S_φ

As we have suggested in Section 8.3 (see Fig. 8-3), it seems plausible to project the full slow behavior of the deep ocean on the single variable θ. This is because the variations of θ are the consequence of the continued action of the faster response TH circulation, which is simultaneously influencing the deep salinity in tandem with θ. Thus, a strong TH circulation, though steadily opposed by diffusion from above, engenders relatively low values of both θ and S_D, while the continued action of a thermally indirect TH circulation tends to produce relatively high values of θ and S_D. In each of these end-member states density increases with depth, as required by convective stability considerations. The possibility for a new form of instability associated with lateral gradients of density primarily due to salinity gradients that can drive the system toward these end-member states was discussed in Section 11.4.

We now specialize Eq. (11.44) governing the slow variations of θ by suggesting some simple parameterizations of $C_\theta^\uparrow$ and $D_\theta^\downarrow$. From our previous discussion in this chapter it can be assumed that the rate of upward convective heat flux due to both the thermohaline (TH) circulation and the more local baroclinic overturnings is strongly related to the surface density gradient between low and high latitudes, ρ_φ. As noted, this assumption is the basis for most zonal-average and box models of the TH circulation as represented by Eq. (11.35). The meridional density gradient is driven, on a most fundamental level, by the gradient in solar heating, which creates warmer tropical waters and colder polar waters and by the salinity variations induced by evaporation and condensation fields forced by the primary temperature and atmospheric circulation patterns. In general, the variations of ρ are more strongly determined by temperature in the warmer low latitudes and by salinity in the colder high latitudes, with the pressure dependence being negligible.

Following the discussion in Section 11.4, as a first approximation we may set

$$C_\theta^\uparrow = \gamma_1 \rho_\varphi \tag{11.55}$$

where $\rho_\varphi = (\partial \langle \rho_s \rangle / \partial \varphi) = (\mu_S \langle S_\varphi \rangle - \mu_T \langle T_\varphi \rangle)$, γ_1 $(= \partial\theta/\partial\rho_\varphi)$ is a free parameter, and the angular brackets define a zonal average [Eq. (7.3)]; for simplicity we shall henceforth drop the brackets. In turn, T_φ can be related to the global slow-response variables I, μ, θ and slow external forcing $R(\varphi)$ by a first-order expansion of the forms given by Eqs. (5.12) and (7.34).

On the other hand, as noted above, the poleward salinity gradient S_φ might have to be regarded as another slow-response variable governed by a dynamical equation. In particular, from the analysis given in Section 11.4.4 we saw that the equilibrium structure for $\delta S \sim -S_\varphi$ (Fig. 11-5) represents a combination of a saddle node bifurcation, admitting an unstable branch in the regime of a thermally direct circulation ($s < 1$), and a thermally indirect, stable, branch in the regime of a thermally indirect salinity-driven circulation ($s > 1$). This combination would seem to be representable by the single "imperfect" form, Eq. (6.18), of the generic cubic Eq. (6.13) (see Section 6.2).

Thus, we may suppose that the variations of S_φ can be repsented in the form

$$\frac{dS_\varphi}{dt} = j_0 + \Pi_{S\varphi} - K_S \cdot S_\varphi \tag{11.56}$$

where $\Pi_{S\varphi} = [\breve{S} F_\varphi^\uparrow - j_1 T_\varphi]$, representing the combined salinity-gradient generation due to the freshwater flux gradient $[F_\varphi^\uparrow \sim (E-P)_\varphi]$ and due to the meridional surface salinity flux forced by the thermally driven part of the thermohaline circulation ($\sim T_\varphi$). The salinity-gradient feedbacks are embodied in the nonlinear damping coefficient, $K_S = (j_2 - j_3 S_\varphi + j_4 S_\varphi^2)$; $j_0 \cdots j_4$ are constants. $F_\varphi^\uparrow$ and T_φ, and hence $\Pi_{S\varphi}$, are fast-response variables that can be related to $y\ [= (I, \mu, \theta; R)]$ by equations of the form given by Eq. (7.34), i.e., $(F_\varphi, T_\varphi) = \sum k_y^{(F_\varphi, T_\varphi)} \cdot y$, where $k^{(F^\uparrow, T_\varphi)}$ are sensitivity coefficients that can be determined from SDM and GCM experiments. From the experiments performed thus far, it would appear that $F_\varphi^\uparrow$ is relatively insensitive to the ice-sheet mass (see e.g., Fig. 7-1) and increases a small amount with increasing CO_2. Apparently the increase in evaporation due to higher surface temperatures tends to be offset by an opposing decrease due to higher atmospheric vapor pressure and weaker wind speeds.

In postulating Eq. (11.56) we are putting in a simple mathematical form the idea that the pole-to-equator salinity gradient ($-S_\varphi$) is steadily being driven to an increased value by the freshwater flux gradient ($-F_\varphi^\uparrow$), to reduced values by the thermally direct thermohaline circulation engendered by high values of T_φ, and that for a range of values of S_φ there may be enough positive feedback to engender the possibility for an instability.

In a manner similar to our representation of $\mathcal{C}_\theta^\uparrow$ appearing in Eq. (11.44), it is reasonable to suppose that the simplest approximation for the net downward diffusive flux, $\mathcal{D}_\theta^\downarrow$, is of a Newtonian form, i.e., proportional to the difference between the mean surface temperature $\widetilde{T}_s$ and deeper ocean temperature θ. Thus,

$$\mathcal{D}_\theta^\downarrow = \nu_\theta \left(\widetilde{T}_s - \theta\right) \tag{11.57}$$

where ν_θ is a constant and $\widetilde{T}_s$ is given by an equation of the form, Eq. (7.34), i.e.,

$$\widetilde{T}_s = \widetilde{T}_s^* + \mathcal{B}^{(\widetilde{T}_s)} \ln(\mu/\widehat{\mu}) + \sum_y k_y^{(\widetilde{T}_s)} \cdot \Delta y \tag{11.58}$$

In essence, ν_θ^{-1} roughly represents the time it would take for the ocean to warm up to its mean surface value if the TH circulation were shut off. Thus from Eqs. (11.55) and (11.57), Eq. (11.44) can be written in the form

$$\frac{d\theta}{dt} = -\gamma_1(\mu_S S_\varphi - \mu_T T_\varphi) + \nu_\theta\left(\widetilde{T}_s - \theta\right) + \mathcal{G}_\theta^\uparrow + \omega_\theta \tag{11.59}$$

where T_φ and $\widetilde{T}_s$ are related to all of the slow-response variables by the GCM-determined sensitivity functions $k_y^{(T\varphi)} = (\partial T_\varphi/\partial y)$, and $k_y^{(\widetilde{T}_s)} = (\partial \widetilde{T}_s/\partial y)$, respectively, and ω_θ represents stochastic forcing due to all random higher frequency phenomena not adequately represented by the parameterizations.

Note that it is likely that $(\rho_\varphi)' \sim -\theta'$, and $\widetilde{T}_s' \sim \theta'$, both of which introduce positive feedbacks. If these positive feedbacks can dominate the negative feedbacks (e.g., as represented by $-\nu_\theta\theta$) the possibility exists that the deep ocean behavior might introduce an instability that can drive internal oscillations on paleoclimatic time scales. This would be an expression of the instability illustrated by the Stommel model (Section 11.4), owing its origin to the same salinity vs. temperature effect, as shown by Birchfield's (1989) three-box model. In this latter model the deep ocean temperature and salinity variations imply simultaneous changes in the horizontal upper-layer density gradients; that is, a cold deep ocean ($\theta' < 0$) implies a strong direct TH circulation that delivers excess salinity to high latitudes accompanied by reduced rainfall, while a warm deep ocean ($\theta' > 0$) implies a weak TH circulation delivering less salinity from low latitudes at the same time rainfall is increased in high latitudes.

Thus, the salinity gradient equation, Eq. (11.56), and deep ocean temperature equation, Eq. (11.59), can be coupled with the similar dynamical equations for the ice-sheet variables (e.g., I, D), and $CO_2(\mu)$ discussed in Sections 9.11 and 10.5 to form a dynamical system governing the main slow-response variables. This coupling to form a closed set representing the "center manifold" of the climate system (see Section 5.3) is the subject of the next part of this book, starting with Chapter 12.

PART III

Unified Dynamical Theory

12

THE COUPLED FAST- AND SLOW-RESPONSE VARIABLES AS A GLOBAL DYNAMICAL SYSTEM

Outline of a Theory of Paleoclimatic Variation

As noted in Section 5.1, the quantitative calculation of paleoclimatic evolution requires an unattainable level of accuracy in determining net fluxes across the boundaries of the ice sheets and ocean. For this reason the problem must be approached in a more inverse or "phenomenological" manner, by tuning as few free parameters as possible, each of which represents an unknown net rate constant that can decide the sign and magnitude of the small differences between large opposing fluxes. Because the paleoobservations in time and space that must be accounted for by this inverse procedure are so complex (as illustrated, e.g., by the proxy time series of global ice mass shown in Fig. 1-4), it seems appropriate to begin with a "global" approach to the problem in the form of a low-order dynamical system governing the largest scale features of the slow-response fields. The global integral results obtained could then serve to constrain a more detailed evolutive solution in which the three-dimensional geographic features of the ice sheets and deep ocean state are coupled to an atmospheric GCM, which is our ultimate goal. The connection between such a low-order dynamical system "control" model and the fuller three-dimensional model was discussed in Chapter 5 and portrayed in Figs. 5-1 and 5-4, and will be discussed further in Chapter 17. In light of all the difficulties, to this point there have been no attempts to obtain such a three-dimensional evolutive solution beyond the past 100-ky ice-age cycle, in which all of the slow-response variables (e.g., CO_2) are computed rather than prescribed; most discussions have been concerned with the global aspects of the problem posed by records such as shown in Fig. 1-4. Henceforth, in this chapter we shall focus on the development of a general dynamical model governing the global features of the slow-response variables, which can be specialized to yield most of the global models proposed to date. Based on this general system we shall describe the framework for a unified theory of climatic change over the Phanerozoic (with emphasis on past 5 My), embracing many of the ideas presented in the more specialized models.

We will begin by first summarizing the set of equations developed in the previous chapters that form a closed system governing both the fast-response variables (e.g., T_S) and the slow-response climatic variables with which they are equilibrated. Using a flow diagram we then pictorialize the physical interactions represented by these equations showing more clearly the main feedbacks represented. In order to reduce the general climate system to its "center manifold," we follow with an elimination of the fast-response variables (X) to form a dynamical system governing only the slow-response (control) variables (Y). In accord with the structured approach discussed in Section 5.4 we next decompose the set of the equations into an externally forced tectonic-mean part ($\widehat{Y}$) and a subset governing the departures (ΔY). The questions raised by potentially strong positive feedbacks that may affect each of the variables in this system, any of which can lead to instabilities, bring to the fore the material discussed in Chapter 6 regarding the elements of bifurcation theory and dynamical systems analysis that will be of importance when applied to any model of paleoclimatic variability.

12.1 THE UNIFIED MODEL: A PALEOCLIMATE DYNAMICS MODEL

We now list the set of equations derived in Chapters 7 to 11, forming a closed system in which the long-term changes in Earth's global climate are projected onto the dynamical behavior of only a few "prognostic" variables [the ice-sheet group $I \;(= \sum_j \Psi_j)$, D_B, W_B, $\mathcal{S}_I$ (Section 9.11), atmospheric CO_2 denoted by μ (Section 10.5), and the thermohaline ocean state, θ and S_φ (Section 11.6)] to which the fast-response variables governed by a GCM are equilibrated. This system, which constitutes what we might call a "paleoclimate dynamics model" (or PDM) is composed of the following equations (omitting the stochastic forcing terms):

Ice Sheets (Section 9.11):

$$\frac{d\Psi_j}{dt} = a_0 - a_1 T_\Psi - \nu_I \Psi_j - \mathcal{C}_I(\Psi_j, D_B) - \mathcal{S}_I \qquad \left(I = \sum_j \Psi_j\right) \tag{12.1}$$

$$\frac{dD_B}{dt} = eH - \nu_D D_B \qquad \left(H = c_H \Psi^{1/5}\right) \tag{12.2}$$

$$\frac{dW_B}{dt} = \mathcal{M}_B(\Psi) - \nu_W W_B \tag{12.3}$$

$$\frac{d\mathcal{S}_I}{dt} = k_w W_B - \nu_B \mathcal{S}_I \tag{12.4}$$

where c_H is a constant and $e = (\rho_i/\rho_b)\nu_D \approx 0.25\nu_D$ (see Section 9.8).

The total ice-sheet mass for the globe is the sum of the individual ice sheets, $I = \sum_j \Psi$, and the diagnostic relations for $\mathcal{C}_I$ and $\mathcal{M}_B$ are given by Eqs. (9.54) and (9.66), respectively.

Carbon Dioxide (Section 10.5):

$$\frac{d\mu}{dt} = \kappa_\mu Q^\uparrow(\mu, \theta) + \kappa_\mu\left(V_\mu^\uparrow - W_\mu^\downarrow\right) \tag{12.5}$$

where $\kappa_\mu Q^\uparrow(\mu, \theta) = [\beta_0(t) - (\beta_1 - \beta_2\mu + \beta_3\mu^2)\mu - \beta_\theta\theta]$, $V_\mu^\uparrow = (V_{\mu C}^\uparrow + V_{\mu G}^\uparrow)$, and $W_\mu^\downarrow = (W_C^\downarrow + W_S^\downarrow - W_G^\uparrow)$ (see Chapter 10).

The Ocean State (Section 11.6):

As given by Eq. (11.59),

$$\frac{d\theta}{dt} = -\gamma_1(\mu_S S_\varphi - \mu_T T_\varphi) + \nu_\theta\big(\widetilde{T}_{sw} - \theta\big) + \mathcal{G}_\theta^\uparrow \tag{12.6}$$

and, as given by Eq. (11.56),

$$\frac{dS_\varphi}{dt} = j_0 + \Pi_{S\varphi} - K_S \cdot S_\varphi \tag{12.7}$$

where $\Pi_{S\varphi} = [\breve{S} F_\varphi^\uparrow - j_1 T_\varphi]$ and $K_S = (j_2 - j_3 S_\varphi + j_4 S_\varphi^2)$.

The set of prognostic equations, Eqs. (12.1)–(12.7), involve some coefficients that are at least partially constrained by glaciological/asthenospheric or oceanic physics (e.g., κ_I, ν_I, ν_θ, ρ_I, ρ_b, ν_D, c_H), plus other relatively unconstrained, i.e., "free" or "adjustable," parameters (e.g., a_1, γ_1, β_i, j_i) that can be determined inversely as the values necessary to account for the many degrees of freedom involved in the complex observed paleoclimatic variability such as portrayed in Fig. 1-4a.

To close the PDM system we must invoke the fast-response, diagnostic, component of the system, the GCM, governing the surface temperatures T_Ψ, T_φ, and T_{sw}, and the surface freshwater flux $F^\uparrow \approx \delta(E - P)$ as a function of the slow-response variables that can now include S_φ, i.e., μ, I, θ, S_φ, and slow-forcing variables $\{F_i\} = \{R(\mathcal{S}), \Omega, U^\downarrow, h, G^\uparrow, V^\uparrow, \mathcal{W}\}$ (see Section 5.4.1). To first order we adopt the following representations,

The GCM (Sections 7.9 and 11.6):

$$T_\alpha = \widehat{T}_\alpha(0) + \Delta_t \widehat{T}_\alpha + \Delta T_\alpha \tag{12.8a}$$

$$F^\uparrow = F^\uparrow(0) + \Delta_t F^\uparrow + \Delta F^\uparrow \tag{12.8b}$$

where

$$\Delta_t \widehat{X} = \mathcal{B} \ln\big[\widehat{\mu}/\widehat{\mu}(0)\big] + \sum_y k_y^{(X)}\big(\widehat{y} - \widehat{y}(0)\big) \qquad \big(X = T_\alpha, F^\uparrow\big)$$

$$\Delta X = \mathcal{B} \ln(\mu/\widehat{\mu}) + \sum_y k_y^{(X)}(y - \widehat{y})$$

$y = (I, \theta, S_\varphi; F_i)$, and $\mathcal{B}$ and $k_y = [(\partial \widehat{T}/\partial y)]$ are constants. Thus, the GCM is a major component of this system, providing not only the three-dimensional atmospheric and surficial climate (denoted by X in Chapter 5) in equilibrium with the slow-response state (Y) of any geologic period, but also by systematic experiments, providing the sensitivity functions (e.g., $\mathcal{B}$, k_y) by which closure of the slow-response dynamical system can be achieved (see Fig. 5-4). This determination of the sensitivity functions constitutes Problem 1 posed in Section 5.2.

12.2 FEEDBACK-LOOP REPRESENTATION

A schematic diagram of the complete feedback system implied by these equations is shown in Fig. 12-1, emphasizing the potential destabilizing role of the carbon cycle discussed in Chapter 10, the interrelations of the ice-sheet variables (I, D, $\mathcal{C}_I$, W_B, and $\mathcal{S}_I$) described in Chapter 9, and the connection of the main slow-response variables ($\mu, I, \theta, S_\varphi$) with the fast-response variables representing the surface climatic state and the oceanic circulations governed by a coupled AGCM and OGCM. The composite surface climatic state is denoted by the symbol "Σ," which includes all atmospheric and surface state variables (e.g., T, $\mathbf{V}$, P, E, $\varphi_i, \ldots$). As a rough categorization, large values of Σ represent a warm climate characterized by high temperature and snowline

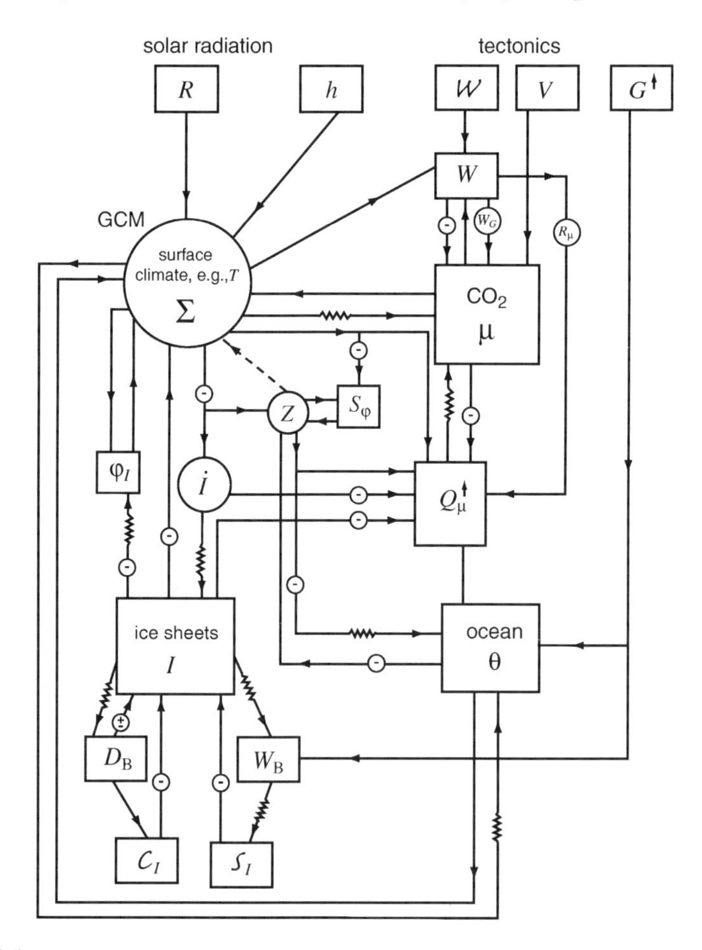

Figure 12-1 Representation of the dynamical system described in Section 12.1 in the form of a feedback-loop diagram showing the main couplings and interactions between the ice, ocean, and atmospheric variables under the influence of external forcing. A wiggly arrow denotes a time-delayed (inertial) influence of changes in one variable on another, in the direction of the arrow. See text (Section 12.2) for further discussion.

latitude φ_i, weak temperature gradients and winds, and a strong hydrologic cycle [e.g., large spatial variance of $(E - P)$], as distinct from a cold climate in which opposite conditions prevail. In addition, the strength of the ocean circulation is denoted by Z, which includes the thermohaline circulation (ψ), the gyre circulation (ϕ), and more localized convective and baroclinic circulations (χ), i.e., $Z = (\psi, \phi, \chi)$. In general, a "warm climate" (high Σ) tends to equilibrate with a weak oceanic circulation (low Z).

In these feedback diagrams a barbed link connecting two variables signifies that a change in one variable leads causally, in the direction of the arrow, to a change in the other variable. This change may be either of the same sign or opposite sign (denoted by a minus). A wiggle in the link signifies an inertial phase lag (or delay) in the response, representing a "prognostic" relationship; otherwise the response is essentially "instantaneous," representing a "diagnostic" relationship. Any closed set of links, which can be followed around in the direction of the arrows, constitutes a feedback loop. If there is an odd number of negative links in a loop it represents a negative feedback; this means that a change of a given sign in any variable in the loop will be opposed, tending to produce oscillatory behavior that can, however, be unstable as well as stable. On the other hand, if there is an even number of negative links (or none) the loop represents positive feedback that tends to reinforce a change in any variable in the loop, in the direction of the change. This can lead either to some larger finite value of the variable or to an unstable growth of the variable to an infinite value.

In general there are competing positive and negative feedback loops involving any prognostic variable. To form a mathematical dynamic model in the absence of quantitative measures of the strength of these loops a qualitative judgment must be made concerning the dominance of one over the other. A test of the validity of the judgment is the agreement of the output with the observational evidence, though even in this case the right answer might be achieved for the wrong reason. In the case of dominant positive feedbacks leading to a first-order instability, it is plausible from conservation requirements to demand that the negative feedbacks become dominant in the higher orders as the system departs markedly from the unstable equilibrium. This was discussed in more detail in Chapter 6.

The presence of damping due to fast-response negative feedback dissipative processes is assumed for each variable, though not shown in the figure; in our equations these possibilities are represented formally by ν_y $(y = I, D, W_B, \mu, \theta)$. The two main sources of external forcing of the system are radiative forcing (R), which includes both solar constant changes $\widetilde{R}$ and Earth-orbital (Milankovitch) forcing δR; and tectonic forcing, which alters the continent–ocean distribution h, CO_2 weatherability $\mathcal{W}$, and volcanic outgassing $V^{\uparrow}$.

Note that the basic energetical drive of the system is provided by the influence of external solar radiative forcing on the fast-response climatic system (the physics of which is represented, for example, by a GCM). As a further step, we have assumed that most of the relevant fast-response behavior, including the hydrologic cycle, can be associated with the induced surface temperature field, T, and $(E - P)$ field, through the full GCM solutions.

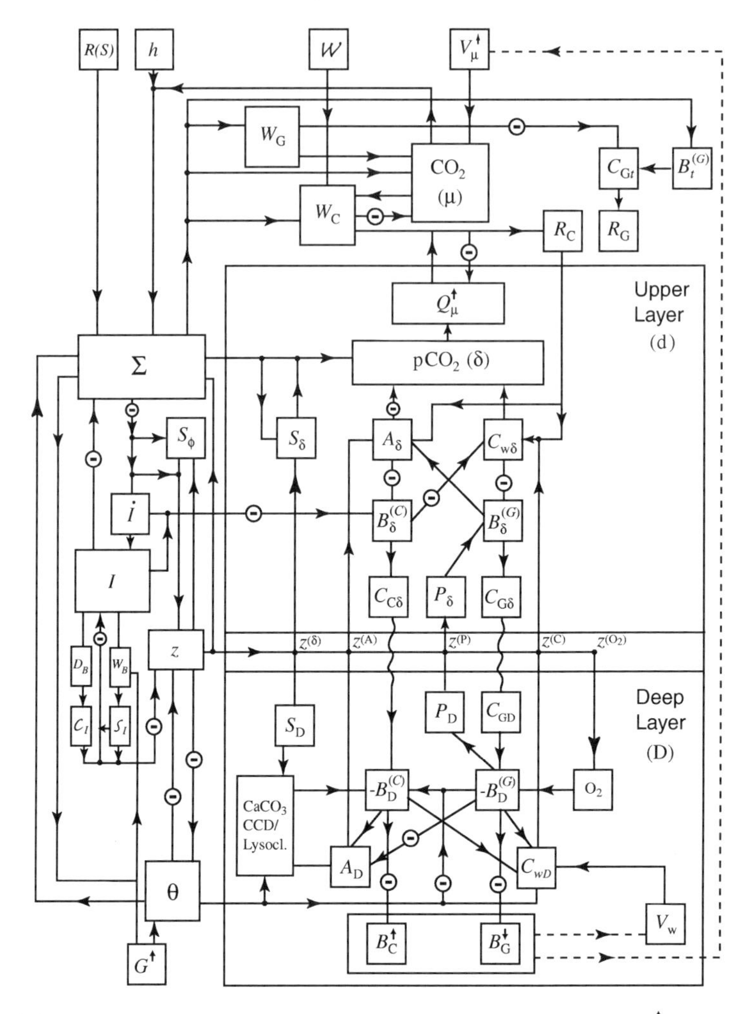

Figure 12-2 Box diagram showing the air–sea carbon flux process that determines $Q_\mu^\uparrow$ in Fig. 12-1 based on the system [Eqs. (10.19)–(10.26)] described in Chapter 10.

A more detailed representation of the air–sea carbon flux processes contained within the box, $Q_\mu^\uparrow$, based on the system given by Eqs. (10.19)–(10.26), is shown in Fig. 12-2. Whereas Fig. 12-1 is a feedback-loop diagram possessing the properties described above, Fig. 12-2 is more in the nature of a conventional "box-diagram" showing the fluxes of carbon between various reservoirs in the ocean (related by the processes $B^{(C)}$ and $B^{(G)}$) as they affect the surface pCO_2 and hence $Q_\mu^\uparrow$, under the influence of all the other physical factors that were represented in Fig. 12-1. Another discussion of the distinction between these two types of diagrams in the context of the carbon cycle is given by Berner (1999).

12.3 ELIMINATION OF THE FAST-RESPONSE VARIABLES: THE CENTER MANIFOLD

By substituting the diagnostic equations [Eqs. (12.8a) and (12.8b)] into the dynamical equations for ice-sheet mass [Eq. (12.1)], deep ocean temperature [Eq. (12.6)], and the salinity equation [Eq. (12.7)] in the manner of Eq. (5.14), we obtain the following alternate forms of these two equations, in which all explicit reference to the fast-response variables is removed [i.e., the so-called "adiabatic elimination," Haken (1983)]:

$$\frac{d\Psi_j}{dt} = \widehat{a}_0 - a_1\left\{\mathcal{B}^{(\Psi)}\ln\left(\frac{\mu}{\widehat{\mu}(0)}\right) + k_\theta^{(\Psi)}\theta + \sum_i k_i^{(\Psi)}\left[F_i - \widehat{F}_i(0)\right]\right\} - K_\Psi\Psi_j - \mathcal{C}(\Psi, D) - \mathcal{S} \tag{12.9}$$

where $\widehat{a}_0 = a_0 - a_1[\widehat{T}_\psi(0) - k_{\mathrm{I}}^{(\Psi)}\widehat{\Psi}(0) - k_\theta^{(\Psi)}\widehat{\theta}(0)]$, and $K_\Psi = (\nu_{\mathrm{I}} + a_1 k_{\mathrm{I}}^{(\Psi)})$;

$$\frac{d\theta}{dt} = \widehat{c}_0 - c_{\mathrm{I}}I - c_\mu\ln\left(\frac{\mu}{\widehat{\mu}(0)}\right) - \sum_i c_i\left[F_i - \widehat{F}_i(0)\right] - \gamma_1\mu_{\mathrm{S}}S_\varphi - K_\theta\theta + \mathcal{G}_\theta^{\uparrow} \tag{12.10}$$

where $I = \sum\Psi_j$, $\widehat{c}_0 = \{\nu_\theta\widehat{\widetilde{T}}_{\mathrm{sw}}(0) - \gamma_1\mu_{\mathrm{T}}\widehat{T}_\varphi(0) - \sum c_y\cdot[\widehat{I}(0), \widehat{\theta}(0)]\}$ $(y = I, \theta)$, $c_y = (\gamma_1\mu_{\mathrm{T}}k_y^{(T_\varphi)} - \nu_\theta k_y^{(\widetilde{T}_{\mathrm{s}})})$, $c_\mu = (\gamma_1\mu_{\mathrm{T}}\mathcal{B}^{(T_\varphi)} - \nu_\theta\mathcal{B}^{(\widetilde{T}_{\mathrm{s}})})$, and $K_\theta = [\nu_\theta(1 - k_\theta^{(\widetilde{T}_{\mathrm{s}})}) + \gamma_1\mu_{\mathrm{T}}k_\theta^{(T_\varphi)}]$ and

$$\frac{dS_\varphi}{dt} = \widehat{j}_0 + \Pi_{S_\varphi}(y) - j_1S_\varphi + j_2S_\varphi^2 - j_3S_\varphi^3 \tag{12.11}$$

where $y = (I, \mu, \theta, S_\varphi; F_{\mathrm{i}})$, $\widehat{j}_0 = \{j_0 + \Pi_{S_\varphi}[\widehat{y}(0)]\}$.

Note that due to the ice-albedo feedback we have $k_{\mathrm{I}}^{(\Psi)} < 0$, and we can also expect that $k_\theta^{(\widetilde{T}_{\mathrm{s}})} > 0$ and $k_\theta^{(T_\varphi)} > 0$, affording the possibility that, as in the case of CO_2 (see Section 10.1), positive feedbacks may dominate the behavior of Ψ and/or θ as well as of S_φ for some ranges of their values. The ramifications of these possibilities will be discussed in the next section. When these three prognostic equations are coupled with the other prognostic equations governing ice sheet basal processes and carbon dioxide, Eqs. (12.2)–(12.5), we arrive at an alternate set that constitutes our proposed "slow" or "center" manifold of the global climate system, to which all the faster response (e.g., atmospheric) climatic fields are attracted (as governed by a GCM). Note that in developing the carbon dioxide equation [Eq. (12.5)] in accordance with Eq. (10.28), we implicitly incorporated the elimination of the fast-response variables [e.g., of T_{s} in Eq. (10.16)].

In essence, these dynamical statements are our proposed "equations of motion" of the slow-response climatic trajectory, placing in a formal dynamical structure some of the leading hypotheses regarding the cause and behavior of the ice ages: (1) the orbital hypothesis (Croll, 1864; Milankovitch, 1930), (2) the carbon dioxide hypothesis (Tyndall, 1861; Arrhenius, 1896; Chamberlin, 1899; Plass, 1956), and (3) the bedrock depression/calving catastrophe hypothesis (Ramsay, 1925; Pollard, 1982), (4) the basal sliding hypothesis (Wilson, 1964; Weertman, 1969; Budd, 1975; Oerlemans, 1982a,b), and (5) the thermohaline circulation hypothesis (Chamberlin, 1906; Stommel, 1961; Weyl, 1968). In addition to these individual hypotheses, the full set of equations, taken as a whole, essentially represents a new hypothesis: namely, that ice variations may be the consequence of a dynamical interaction between all of these physical influences, especially if positive feedbacks lead to instability of the system. This possibility will be discussed next.

12.4 SOURCES OF INSTABILITY: THE DISSIPATIVE RATE CONSTANTS

We have just shown that after substituting for the fast-response variables (e.g., T) the intrinsic dissipative processes, measured by ν_y appearing in the dynamical equations, are modified by the sensitivity functions $\mathcal{B}$ and k_y. As noted above, for both variables (Ψ, θ) the new damping rate constants K_y are smaller than ν_y as a result of positive temperature feedbacks represented by $k_\mathrm{I} < 0$ and $(\mathcal{B}, k_\theta^{(\widetilde{T}_\mathrm{s})}, k_\theta^{(T_\varphi)}) > 0$, tending therefore to destabilize the system. Such a possible destabilization was already incorporated in the CO_2 equation, Eq. (12.5), and the salinity equation, Eq. (12.7), by expressing the effective rate constants as $K_\mu = (\beta_1 - \beta_2\mu + \beta_3\mu^2)$ and $K_\mathrm{S} = (j_2 - j_3 S_\varphi + j_4 S_\varphi^2)$. Similarly, the modified rate constants K_Ψ for ice might also require a nonlinear form that admits a range of values of I, within which positive feedbacks can dominate $(K_\psi < 0)$, but which are constrained by conservation requirements to remain bounded. That is, at the cost of additional free parameters (ϕ_1, ϕ_2, ϕ_3) one could postulate the same generic form as was suggested for μ and S_φ, i.e.,

$$K_y = \phi_1 + \phi_2 y + \phi_3 y^2 \tag{12.12}$$

The cubic nature of damping, $K_y y$, implied by this coefficient, arises in many areas of physics, usually called the Landau (1944) form when only one variable is involved, but generalizes to the so-called Landau–Hopf form when more than one variable is involved, thereby permitting a bifurcation to oscillatory behavior. A discussion of the mathematical–physical foundations and consequences of this general cubic form of damping was given in Chapter 6 as part of a broad overview of relevant aspects of dynamical systems analysis. We now remark briefly on the physical nature of the positive feedback processes involved for ice sheets and the deep ocean, recalling that a more detailed discussion of such processes for the carbon cycle was already given in Chapter 10.

In the case of the ice sheets $(y = \Psi)$, for example, in addition to the need to allow for an increasing glacial response time as ice sheets grow (that should ultimately be limited by the "ice desert effect" resulting from reduced snowfall and lower temperatures at higher ice elevations and by an increased discharge creep velocity), there is an added ice-albedo effect represented in K_I due to the added expanse of snow and sea-ice cover associated with increased ice-sheet mass. Many ice-age models have placed a main potential source of instability in the ice equation operating through this positive ice-albedo feedback; that is, greater ice coverage causes higher surface shortwave solar reflectivity, which causes colder surface conditions and more extensive ice coverage. This effect is probably amplified by the high dust loads during glacial priods. Thus, the positive ice-albedo feedback, embodied in K_I, lengthens the response time of the ice sheets, prolonging the existence of ice sheets in spite of the dissipative process measured by ν_I. This illustrates the general role of positive feedback: if dominant, it can be a source of instability; if it is large, but not dominant over other dissipative processes (negative feedbacks), it can still be very important by significantly increasing the response-time to a larger value than would be implied by the explicit dissipative processes alone.

Similar arguments have also been made for the behavior of the ocean state. In this regard we have noted that the dissipative rate constant for salinity K_S may give rise to a positive feedback that can destabilize the system. For example, $S_\varphi \ll 0$ will probably imply lower than average ocean surface salinity in higher latitudes, which in turn would imply a weaker thermohaline circulation. This would then tend to reduce the transport of salt poleward, further decreasing S_φ and weakening the TH circulation. The net result would be an increase of the thermally direct part of the thermohaline circulation, which would tend to cool the deep ocean. These feedbacks are illustrated in Fig. 12-1 (see box labeled S_φ). The simplified physics underlying this scenario was discussed in Section 11.4 and is the basis for several models of the deep ocean, containing the possibility for instability and bimodality (possible end-member states being similar to the ones shown in Fig. 8-3), e.g., Manabe and Stouffer (1988).

Thus, we must recognize, that some positive feedback exists for ice variability alone (e.g., the ice-albedo feedback), and possibly for the salinity gradient and thermohaline circulation leading in some models to instability and multiple equilibrium states; both of these sources of instability might be important in driving a natural oscillation of a form exhibited by the proxy data. However, as was discussed more fully in Chapter 10, because of the many possible sources of positive feedback in the carbon cycle, especially when μ is small and ice is more prevalent in the climate system [see Section 10.1 and Saltzman (1987b) and Saltzman and Maasch (1988, 1991)], we have guessed that the behavior of CO_2 is the most likely source of major instability in the slow-response climate system, as was suggested by Plass (1956), and we will expand on this possibility in the illustrative model to be described in Chapter 14. Thus, in this illustrative case, we shall explicitly employ the form, Eq. (12.12), only for CO_2, as already expressed in Eq. (12.5), treating K_Ψ and K_S as constants; we remain aware, however, of the possibility that these qualities might also be better represented in the form given by Eq. (12.12), implying new potential sources of instability.

As noted in Section 12.2 regarding the feedback-loop structure shown in Fig. 12-1, in addition to the possible instability that can arise due to the physics of each individual slow-response variable [e.g., ($I \to T \to I$), ($\mu \to T \to \mu$)], there are other sources for instability due to positive feedback loops between the individual variables that can dominate over dissipative processes within some ranges. Thus, for example, in Fig. 12-1 we find the positive carbon cycle loop ($\mu \to T \to I \to Q_\mu^\uparrow \to \mu$). Other possible sources of instability are introduced by the potential for ice sheet calving ($I \rightarrow D_B \rightarrow \mathcal{C}_I$) and basal melting-surge processes ($I \rightarrow W_B \rightarrow \mathcal{S}_I$) that can result in a catastrophic collapse of an ice sheet when critical thresholds are reached.

12.5 FORMAL SEPARATION INTO TECTONIC EQUILIBRIUM AND DEPARTURE EQUATIONS

Following the approach suggested in Section 5.4 we now resolve the PDM system [Eqs. (12.2)–(12.5), (12.9), (12.10), and (12.11)] in accordance with Eq. (5.5), i.e.,

$$y = \widehat{y} + \Delta y$$

where $y = (I, \mu, \theta; F_i)$, $\widehat{y}$ is the ultra-long-term, roughly 10-My average, state wherein the climatic variables $(\widehat{I}, \widehat{\mu}, \widehat{\theta})$ are in equilibrium with geologic tectonic rock processes and slow variations in external forcing $\widehat{F}$, i.e., $d\widehat{y}/dt = 0$. Thus the set of equations governing the tectonic-mean ice mass $\widehat{I}$, carbon dioxide $\widehat{\mu}$, and deep ocean temperature $\widehat{\theta}$, respectively, is as follows:

1. Ice mass ($I = \sum_j \Psi_j$),

$$\widehat{a}_0 - a_1 \mathcal{B}^{(\Psi)} \ln\left(\frac{\widehat{\mu}}{\widehat{\mu}(0)}\right) + \sum_i k_i^{(\Psi)} \left[\widehat{F}_i - \widehat{F}_i(0)\right] - K_\Psi^* \widehat{\Psi}_j = 0 \tag{12.13}$$

where K_Ψ^* is an augmented damping constant that includes the dissipative effects of $\mathcal{C}_I$ and $\mathcal{S}_I$, and over the long 10-My time scale we neglect the details of basal ice-sheet behavior represented by D_B and W_B.

2. Carbon dioxide, $\widehat{\mu}$,

$$\widehat{Q}^{\uparrow}\left(\widehat{\mu}, \widehat{\theta}\right) + \widehat{V}_\mu^{\uparrow} - W_\mu^{\downarrow}(\widehat{\mu}) = 0 \tag{12.14a}$$

As discussed in Section 10.6, because $\widehat{Q}^{\uparrow}$ is difficult to estimate it is desirable to combine this atmospheric equation with the corresponding ocean carbon balance, Eq. (10.39), thereby eliminating this term. This leads to the fundamental Eq. (10.40) for the GEOCARB model (Berner, 1994),

$$\widehat{W}_C^{\downarrow}(\widehat{T}) + \widehat{W}_G^{\uparrow} + \widehat{V}_G^{\uparrow} + \widehat{V}_C^{\uparrow} - \widehat{B}_G^{\downarrow} - \widehat{B}_C^{\downarrow} = 0 \tag{12.14b}$$

which, when coupled with Eqs. (10.41)–(10.50) and (7.34), forms a closed system for $\widehat{\mu}$ from which a solution, $\widehat{\mu}(t)$, can be obtained. Note that it follows from Eqs. (12.14) and (10.28) that $\beta_0(\widehat{\mu}) = \{(\beta_1 - \beta_2\widehat{\mu} + \beta_3\widehat{\mu}^2)\widehat{\mu} + \beta_\theta\widehat{\theta} - \kappa_\mu[V_\mu^{\uparrow} - W_\mu^{\downarrow}(\widehat{\mu})]\}$.

3. Themohaline ocean state, $\widehat{\theta}, \widehat{S}_\varphi$,

$$\widehat{c}_0 - c_I\widehat{I} - c_\mu \ln\left(\frac{\widehat{\mu}}{\widehat{\mu}(0)}\right) + \sum_i c_i\left[\widehat{F}_i - \widehat{F}_i(0)\right] + \gamma_1\mu_S\widehat{S}_\varphi$$
$$- K_\theta\widehat{\theta} + \widehat{\mathcal{G}}_\theta^{\uparrow} = 0 \tag{12.15}$$

$$\widehat{j}_0 + \Pi_{S_\varphi}(\widehat{y}) - K_S(\widehat{S}_\varphi) \cdot \widehat{S}_\varphi = 0 \tag{12.16}$$

where $K_S(\widehat{S}_\varphi) = (j_2 - j_3\widehat{S}_\varphi + j_4\widehat{S}_\varphi^2)$. If, in accordance, with the suggestion made in the last subsection, we chose to assume a diminished role for a salinity-driven instability, we could set K_S = a constant. In this case $\widehat{S}_\varphi = K_S^{-1}[\widehat{j}_0 + \Pi_{S_\varphi}(\widehat{y})]$, which upon expansion via GCM experiments in the manner of Eq. (12.8) and substitution in Eq. (12.5) would lead to modified values of $\widehat{c}_0$, c_I, c_i and K_θ.

To recover the fast-response fields that are in equilibrium with $\widehat{y}$, we must invoke the GCM sensitivity relationships embodied in the relations given by Eqs. (12.8a) and (12.8b), where $\widehat{T}$ and T_φ are determined with reference to an arbitrary fixed state that we may take to be the present-day values denoted by $\widehat{T}(0)$ and $T_\varphi(0)$, i.e.,

$$\left(\widehat{T}, T_\varphi\right) = \left[\widehat{T}(0), T_\varphi(0)\right] + \mathcal{B}^{(T,T_\varphi)} \ln\left(\frac{\widehat{\mu}}{\widehat{\mu}(0)}\right) + \sum_y k_y^{(T,T_\varphi)}\left[\widehat{y} - \widehat{y}(0)\right] \tag{12.17}$$

where $y = (I, \theta; F_i)$. In accordance with Eq. (12.8) and the discussion in Section 7.9, the ultra-long-term variations of $\widehat{T}$ and T_φ that accompany the changes in $\widehat{I}$, $\widehat{\mu}$, $\widehat{\theta}$, and S_φ are also functions of all the processes included as external forcing, F_i. That is, e.g., $\widehat{T} = \widehat{T}(\widehat{I}, \widehat{\mu}, \widehat{\theta}; \widehat{F}_i)$, where $\widehat{F}_i = \{R(\widehat{\mathcal{S}}), \Omega, \widehat{U}^\downarrow, \widehat{h}, G^\uparrow, \widehat{V}^\uparrow, \widehat{\mathcal{W}}\}$. This tectonic-mean component comprising Problem 3 posed in Section 5.2 will be discussed more fully in Chapter 13.

If we subtract Eqs. (12.11)–(12.13) from the full set [Eqs. (12.2)–(12.5), (12.8), and (12.9)], we obtain a dynamical system governing the departures ΔY, which we can view as "corrections" due to all the internally driven phenomena (e.g., fluxes within and between the ocean, atmosphere, and ice masses). To simplify the notation we have already set $\Delta\mu \equiv \xi$ and $\Delta\theta \equiv \vartheta$ (see Section 10.5), and we shall henceforth further set $\Delta\Psi_j \equiv \zeta_j$ $(\Delta I = \sum_j \zeta_j)$, and $\Delta S_\varphi \equiv \sigma$.

Concerning the dependence of ΔY on external forcing F_i, we shall assume at this stage that the variations of Ω are neglectable ($\Delta\Omega = 0$). However, due to the special importance of Earth-orbital insolation changes on temperature we shall explicitly represent only this insolation forcing in the ice-sheet and deep ocean temperature equations to follow. Similarly, because of the possible importance of CO_2 outgassing ($V^\uparrow$) and weathering ($W^\downarrow$), even on glacial–interglacial time scales, and because of our previous inclusion of thermal effects in the parameterization [Eqs. (10.27) and (10.28), we shall explicitly include only $\Delta V^\uparrow$ and $\Delta W^\downarrow$ in the CO_2 equation governing ξ [see Eq. (10.37)]. Thus, the set governing ΔY can be written in the following form, which includes the equation for carbon dioxide already derived in Section 10.5 as well as the approximation $\ln(\mu/\widehat{\mu}) = \xi/\widehat{\mu}$, valid near $\widehat{\mu}$:

$$\begin{aligned}\frac{d\zeta}{dt} &= -a_1\left[\frac{\mathcal{B}^{(\Psi)}}{\widehat{\mu}}\xi + k_\theta^{(\Psi)}\vartheta + k_R^{(\Psi)}\Delta R + \sum_i k_i^{(\Psi)} \cdot \{\Delta F_i\}\right] \\ &\quad - K_I\zeta - \mathcal{C}(\zeta_j, D_B) - \mathcal{S}_I \qquad \left(\Delta I = \sum_j \zeta_j\right)\end{aligned} \tag{12.18}$$

$$\frac{dD_B}{dt} = eH - \nu_D D_B \qquad \left(H = c_H \zeta_j^{1/5}\right) \tag{12.19}$$

$$\frac{dW_B}{dt} = \mathcal{M}_B(\Psi) - \nu_w W_B \tag{12.20}$$

$$\frac{d\xi}{dt} = K_\xi(\widehat{\mu}) \cdot \xi - b_\theta \vartheta + \kappa_\mu \Delta\left[V_\mu^\uparrow - W_\mu^\downarrow(\widehat{\mu}, \xi)\right] \tag{12.21}$$

$$\frac{d\vartheta}{dt} = -c_I \sum_j \zeta_j - \frac{c_\mu}{\widehat{\mu}} \xi + \gamma_1 \mu_S \sigma - c_R \Delta R + \sum_i c_i\{\Delta F_i\} - K_\theta \vartheta \tag{12.22}$$

$$\frac{d\sigma}{dt} = \Pi_{S_\varphi}(\vartheta, \xi, \Delta I, \Delta F_i) + K_\sigma\left(\widehat{S}_\varphi\right) \cdot \sigma \tag{12.23}$$

where $K_\xi = [b_1(\widehat{\mu}) - b_2(\widehat{\mu})\xi - b_3\xi^2]$, $b_1(\widehat{\mu}) = (2\beta_2\widehat{\mu} - 3\beta_3\widehat{\mu}^2 - \beta_1)$, $b_2 = (3\beta_3\widehat{\mu} - \beta_2)$, and $b_3 = \beta_3$, and $K_\sigma = [l_1(\widehat{S}_\varphi) - l_2(\widehat{S}_\varphi) \cdot \sigma - l_3\sigma^2]$, $l_1(\widehat{S}_\varphi) = (2j_3\widehat{S}_\varphi - 3j_4\widehat{S}_\varphi^2 - j_2)$, $l_2 = (3j_4\widehat{S}_\varphi - j_3)$, and $l_3 \equiv l_4$.

Equations (12.18)–(12.23) can be specialized to represent the underlying physics involved in most models of the glacial–interglacial cycles of the Plio-Pleistocene (see Chapter 14), a special example of which is more fully developed in Chapter 15.

In the next chapter we discuss in more detail theoretical aspects of the long-term evolution of the tectonic-mean state, the global nature of which is assumed to be governed by Eqs. (12.13)–(12.17).

13

FORCED EVOLUTION OF THE TECTONIC-MEAN CLIMATIC STATE

As noted in Section 12.5, the determination of the ultra-long-term variations of the tectonic-mean climatic state depends on a knowledge of all the astronomical and solid Earth changes with which they are equilibrated. These include changes in the solar constant ($\mathcal{S}$) as it affects external radiative forcing $R(\mathcal{S})$, the rate of rotation of the Earth (Ω), the long-term exposure to bolides and cosmic dust (U), tectonic changes in continental topography and oceanic bathymetry (h), changes in the geothermal flux ($G^{\uparrow}$), and changes in volcanism ($V^{\uparrow}$) and carbonate/silicate weatherability ($\mathcal{W}$).

The set of statements given in Chapter 12 governing the response of the tectonic-mean state variables to this long-term forcing can be summarized in the following compact form using the notation introduced in Section 5.4:

$$\widehat{X} \equiv \{\widehat{x}_i^{(F)}\} = f(\widehat{Y}; F) \tag{13.1}$$

where $\widehat{X}$ is the set of fast-response variables that we identify most closely with a description of "climate" as governed, e.g., by a GCM; the values $\widehat{Y} \equiv \{\widehat{I}, \widehat{\mu}, \widehat{\theta}, \ldots\}$ are the mean slow-response variables, and $F = \{F_i\} = \{\widehat{R}(\mathcal{S}), \Omega, \widehat{U}, \widehat{h}, \widehat{G}^{\uparrow}, \widehat{V}, \widehat{\mathcal{W}}, \ldots\}$ is the set of contributions to long-term external forcing.

Some GCM solutions for past geologic periods of special interest, for which particular components of $\widehat{Y}$ and F could be prescribed from geologic or astronomical observations, were discussed in Section 7.11 and further relevant references for the pre-Quaternary period can be found, e.g., in Parrish (1998, pp. 249–250). We shall not dwell on these special snapshot GCM simulation studies, but rather will be more concerned with the general physical expectations from variations in the external forcing $\widehat{F}$, and with the structuring of an evolutive theory for these different paleoclimatic states. In essence, we are now dealing with Problem 3 posed in Section 5.2 (see Saltzman, 1990). In this regard we must recognize that a complete theory of long-term climate change requires a deduction of all the components of $\widehat{Y}$ (as well as of $\widehat{X}$) as free variables, with only the purely external forcings $\widehat{F}$ prescribed (see Section 5.2). These changing values of $\widehat{Y}$, as governed, e.g., by the coupled system , Eqs. (12.10)–(12.13), can have at least as large an effect on $\widehat{X}$ as $\widehat{F}$. Before discussing this response of $\widehat{Y} = (\widehat{\mu}, \widehat{I}, \widehat{\theta})$ to $\widehat{F}$ (Sections 13.4 and 13.5), in the following three sections we first discuss in a general way the direct effects of the monotonic changes in solar luminosity, $\mathcal{S}$, and the rate of rotation of Earth, Ω, believed to have occurred over the history of Earth (Section 13.1), the slow changes in continent–ocean topography, h, driven by mantle convection and associated plate tectonics (Section 13.2), and the intermittent dust-loading effects of volcanism, $V^{\uparrow}$, and cosmic processes, U (Section 13.3). In Section 13.4 we discuss, in somewhat more detail, the best developed aspect of tectonic forcing in terms of a time-dependent theoretical model, namely, the effects of $\mathcal{S}(t)$, $h(t)$, $V_{\mu}^{\uparrow}(t)$, $G^{\uparrow}(t)$, and silicate weatherability $\mathcal{W}(t)$ on multimillion-year average CO_2 states $(\widehat{\mu})$ as represented by the GEOCARB model of Berner (1994).

It is clear that external forcing changes of the kind discussed in this chapter must be paramount in causing the marked climatic regime changes that have occurred, even over the past 5 million years when the set of Earth-orbital variations were relatively stationary. In fact, it is fair to say that the ultimate "cause" of the onset of the present glacial epoch at about 2.5 Ma, and the marked change in character of the ice-age variations within this epoch at about 900 ka, must be due to such longer term changes in forcing that are independent of the climatic state itself. In this chapter we present a brief summary of ideas regarding these slow external changes, which have such important implications for the onset and maintenance of the late Cenozoic ice ages to be considered in the next chapter.

13.1 EFFECTS OF CHANGING SOLAR LUMINOSITY AND ROTATION RATE

13.1.1 Solar Luminosity ($\mathcal{S}$)

In Fig. 1-6a we showed the change in the solar constant believed to have occurred throughout the age of the Earth. A rough estimate of the thermal effects of this monotonic change can be obtained by assuming a linear sensitivity, $k_{\mathcal{S}}^{(T)} \equiv \partial \widetilde{T}_{S}/\partial \mathcal{S} = 0.13°C/W\,m^{-2}$, as suggested by numerical solutions of a GCM, the NCAR CCM3 (Kothavala *et al.*, 2000). Setting $\mathcal{S}(t) = \mathcal{S}(0) + ct$, where $\mathcal{S}(0) = 1368\ W\,m^{-2}$ (present value) and $c = 68 \times 10^{-9}\ W\,m^{-2}\,y^{-1}$ (see Section 1.3), we obtain the estimate of the increase in mean surface temperature shown in Fig. 13-1. It is implied that if this were the only factor influencing the mean surface temperature, Earth would have been in a cold, subzero, glacial state for a long stretch of its early history. This poses what has been termed the "faint-early-sun parodox," because there is little evidence for such a prolonged glacial state. Thus, it seems evident that other heat sources must have been dominant, the most often cited guesses being an enhanced geothermal or volcanic flux; an enhanced greenhouse atmosphere (CO_2, CH_4, and NH3), discussed in Section 13.4; the effects of a more oceanic planet in which, as shown in Fig. 1-8, the land fraction is much reduced relative to its present value (Section 13.2); and/or the effects of a greater rate of rotation of the Earth, to be discussed next.

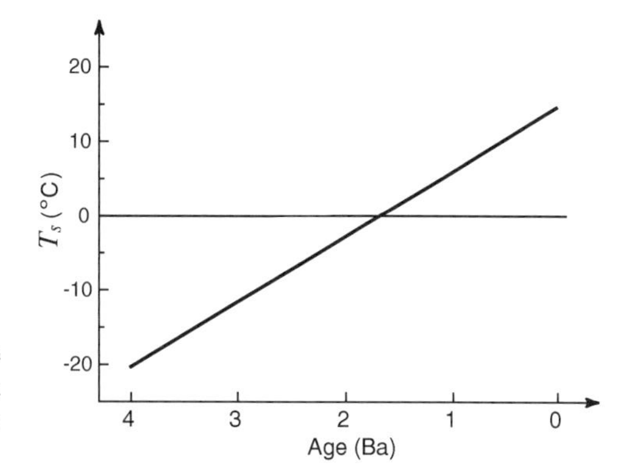

Figure 13-1 Estimated variation of the mean surface temperature over the age of Earth due only to the increase in solar luminosity shown in Fig. 1-5.

13.1.2 Rotation Rate (Ω)

As discussed in Section 1.3 and shown in Fig. 1-6b, it has been inferred that the rate of the Earth's rotation has been decreasing steadily over time under the influence of tidal-frictional torques exerted by other celestial bodies. This implies a steadily increasing length of the day (LOD), estimated to have varied from 14 hours at 4 Ga to our present value of 24 hours.

Using GCMs, Hunt (1979) and Jenkins *et al.* (1993) have explored the ramifications of such a change in Ω. In accordance with dynamical theory [see summary by Saltzman (1962)] and laboratory experiments (Fultz *et al.*, 1959), the results show that for the earlier fast rotation rates the horizontal scales of the motions are reduced (e.g., a narrower Hadley cell and higher wavenumber baroclinic eddies), the meridional temperature gradient is larger, and according to Jenkins *et al.* (1993) cloud cover is reduced, leading to a somewhat warmer surface state that may partially offset the effects of the "fainter sun." It would follow also that for higher rotation the amplitude of the diurnal cycle of surface temperature would weaken at the same time the frequency increases (Saltzman and Ashe, 1976).

13.2 GENERAL EFFECTS OF CHANGING LAND–OCEAN DISTRIBUTION AND TOPOGRAPHY (h)

In Fig. 3-1 we showed a sequence of maps of the evolving continent–ocean distribution believed to have occurred over a period of 500 My (Scotese and Golonka, 1992). Some discussion of the response to this changing continent–ocean distribution, including references to earlier GCM and SDM studies, was already given in Section 7.11.

The studies cited there and in Crowley and Burke (1998), corresponding to selected time periods during which significantly different arrangements of continent and ocean prevailed, demonstrate many of the expected features of the surface temperature distribution. For example, land areas have warmer summers and colder winters than ocean areas, especially in interior regions, making snow formation in these interior regions highly vulnerable to summer melting. Also, land masses near the pole (e.g., Gondwana) are favored sites for snow accumulation.

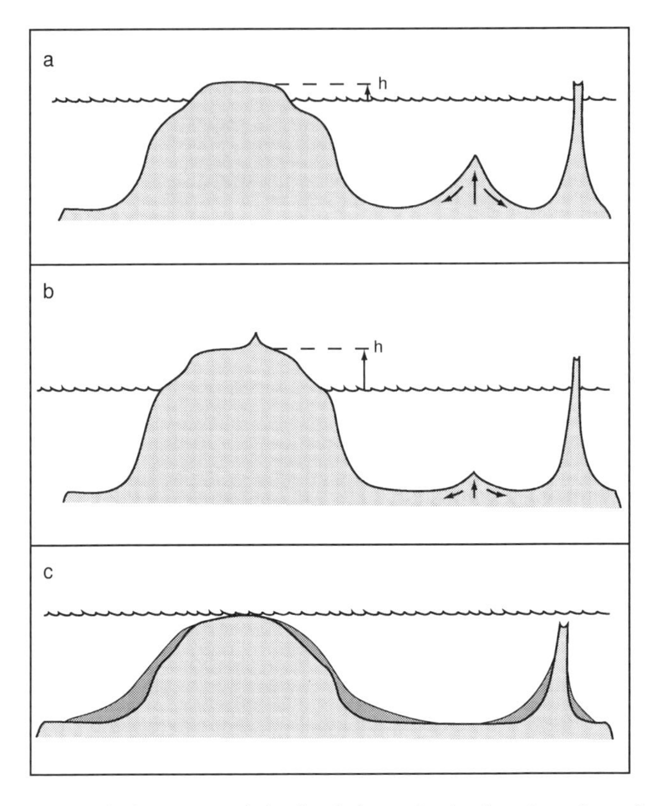

Figure 13-2 Scenarios for long-term variation in relative sea level and continental area due to (a) intense sea-floor spreading, (b) weak sea-floor spreading, and (c) no tectonic activity or sea-floor spreading.

The purely land vs. ocean results must be amended when one considers continental topography, a feature that is difficult to estimate from geologic evidence (Scotese and Golonka, 1992). Extensive systematic studies have yet to be made taking this topography into account, but all of the generalities regarding phenomena such as rain-shadow and blocking effects that influence present climate are undoubtedly valid for ancient topographic patterns as well. Excellent, more detailed, reviews of paleoorographic and paleobathymetric effects studied to date are given in the collected work edited by Crowley and Burke (1998). In the following discussion we shall consider in general terms some of the broadest climatic implications of the tectonically induced surface boundary conditions.

The relative amounts and location of continent and ocean at any time can be viewed as the outcome of a competition between tectonic plate movement and associated mountain-building and sea-floor spreading processes versus climate-influenced erosional processes that tend to degrade and remove the mountains and fill the continental shelves and ocean basins with their sedimentary residue. In Fig. 13-2 (a and b), we show in a highly schematic form two plausible extreme outcomes of these tectonic processes, assuming (1) that erosion is steadily being offset by mountain building, (2) that the total volume of ocean water is constant, and (3) that on the long

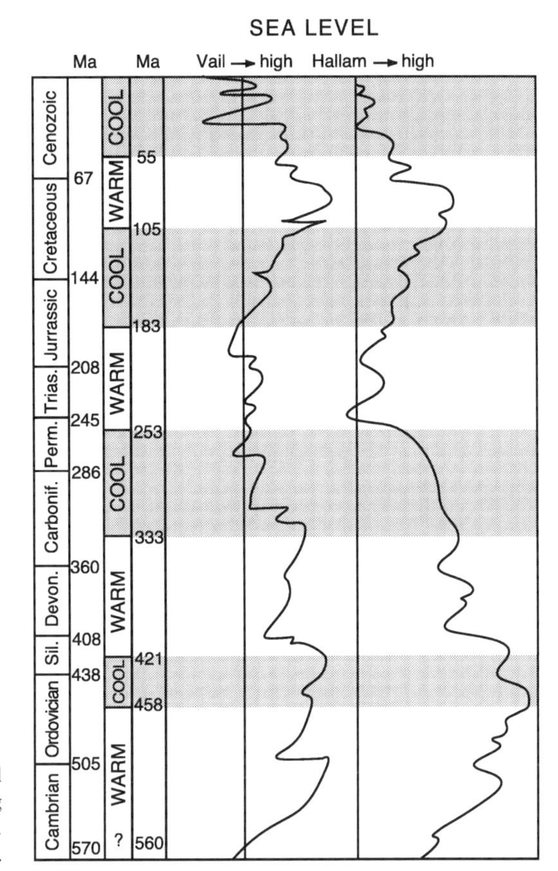

Figure 13-3 Estimates of sea level variation over the Phanerozoic, showing times of cool and warm global conditions. After Vail *et al.* (1977) and Hallam (1984).

tectonic time scale the rate of sea-floor spreading determines the ocean basin capacity and hence the sea level [see e.g., Williams *et al.* (1998), Chapter 6]. Estimates of sea level change over the Phanerozoic based largely on these considerations are shown in Fig. 13-3. Figure 13-2a represents the case in which sea-floor spreading is very large, raising the floor of the ocean and causing not only high sea level relative to the continents (smaller land/sea ratio), but also increases the geothermal heat flux and volcanic outgassing fluxes to the atmosphere and ocean. This scenario may be representative of conditions during the mid-Cretaceous breakup of Pangea at about 100 Ma (see Fig. 3-1). On the other hand, Fig. 13-2b represents the case in which sea-floor spreading is much reduced, resulting in (1) a lowered ocean floor and sea level relative to the continents (larger land/sea ratio), (2) decreased geothermal and volcanic flux activity, and (3) perhaps a higher mountain topography as an end result of earlier sea-floor-driven continental convergence and collision. This scenario may be representative of conditions at about 300 Ma (supercontinent Pangea) or even our present state of relatively low sea level and a tendency for the conglomeration of Eurasia, Africa, and the Australia-Indonesia land masses (see Fig. 3-1).

As was noted, in both cases (a and b, Fig. 13-2), it was tacitly assumed that erosional loss of continental topography is being roughly offset by tectonic mountain-building processes. In Fig. 13-2c we illustrate the hypothetical situation in which all tectonic forcing driven, e.g., by mantle convection ceases, leaving only the erosion and weathering processes to prevail. In this case we would remain with an all-ocean Earth in which the eroded continental material would appear as sedimentary layers filling the ocean basins, particularly on the submerged shelves and continental flanks. As was shown in Fig. 1-7 it has been estimated (Hargraves, 1976) that the fraction of Earth's surface occupied by land has steadily increased over the age of Earth, probably most related to the scenario shown in Fig. 13-2a, because tectonic activity is believed to have been large in the earliest stages of Earth.

In GCM experiments conducted by Jenkins *et al.* (1993) it was shown that the lower surface albedo of an all-water planet would lead to a 4K higher mean surface temperature $\widetilde{T}_s$ than at present, even taking clouds into account. They suggested that such a more oceanic Earth, coupled with the higher-Ω effect described above, could help compensate for the faint-early-sun effect. Although the model treated considered a "swamp ocean with no circulation or heat capacity," as noted earlier by Ramsay (1925), it still seems plausible for the following reasons that a more oceanic state, such as represented by case (a) of Fig. 13-2, would indeed be warmer than a more continental Earth:

1. High continents are sites of lower surface temperatures that are conducive to glaciation, especially if the continents are in higher latitudes.
2. Although continents may deflect ocean currents poleward, thereby ameliorating global temperature (e.g., the Gulf Stream), the blocking effects of continental masses also act strongly to reduce heat transport by the ocean and especially the atmosphere (e.g., the Asiatic and North American mainlands).
3. As noted, the more intense sea-floor-spreading rate that is a primary cause of higher sea level can be expected to be accompanied by a larger oceanic geothermal flux and more active volcanic outgassing, particularly of CO_2, at the same time weathering downdraw of CO_2 is reduced due to lower topography (see Chapter 10).
4. The albedo of ocean surface is less than that of land. A summary of some of these global properties for cases (a) and (b) is given in Table 13-1.

Of special importance for paleoclimatic variations are the effects of changing ocean bathymetry, particularly as it relates to a changing distribution of seaways and sills (e.g., Mikolajewicz *et al.*, 1993; Whitehead, 1998). The examples mentioned at the end of Section 11.3.1 include (1) the opening of the Drake Passage at about 38 Ma, believed to be a critical factor influencing the world ocean circulation and Antarctic glaciation (Toggweiller and Samuels, 1995; Kennett, 1981), and (2) the closing of the Isthmus of Panama between 8 and 2.5 Ma, leading to the onset of a dominant Gulf Stream (e.g., Haug and Tiedemann, 1998), before which (e.g., in the Cretaceous) an unrestricted westward Tethys current could encircle the globe in the tropics (e.g., Barron and Peterson, 1989). The final closing of this Isthmus near 2.5 Ma is often cited as a possible cause of the onset of the Pleistocene glacial mode. In studies by Collins

Table 13-1 Qualitative Summary of Global Properties for Two Tectonic States[a]

Global properties	Case a (100 Ma?)	Case b (0, 300 Ma?)
Sea-floor spreading rate		
Volcanism outgassing ($V^{\uparrow}$)		
Geogthermal flux ($G^{\uparrow}$)		
Sea level		
Temperature (T)	High	Low
Water vapor (ε)		
Atmospheric CO_2 (μ)		
Thermocline depth and extent (θ)		
Continental height above sea level (h)		
Glaciation (I)		
Rock weatherability ($\mathcal{W}$)		
Meridional temperature gradient (T_{φ})	Low	High
Atmospheric and oceanic circulation ($\mathbf{V}$)		
Land to sea ratio		
Closeness of continents to pole		

[a] For two cases, as shown in Fig. 13-2 (a and b).

et al. (1996) and Frank *et al.* (1999) it is argued that the closing actually occurred much earlier (before 5 Ma) and hence could not be a direct cause of Northern Hemisphere glaciation. As an additional example, it has been suggested that the formation of the Greenland–Iceland–Faroes sill structure was a key factor in establishing the North Atlantic as the site of intense cold deep water formation. Conversely, the Bering–Aleutian sill has in all likelihood inhibited the exchange of colder Arctic waters and North Pacific waters, with implications for the rate of throughflow of freshwater to the North Atlantic (Goosse *et al.*, 1997).

13.3 EFFECTS OF LONG-TERM VARIATIONS OF VOLCANIC AND COSMIC DUST AND BOLIDES

Earth's atmosphere is steadily being loaded with aerosol ejecta from volcanoes ($V^{\uparrow}$), as well as cosmic material from comets and asteroids ranging in size from fine dust to very sporadic massive bolides (U). This dust loading is steadily being washed out to Earth's surface, particularly in the troposphere, leaving a variable amount in the atmosphere, depending on the rate of loading.

There has undoubtedly been great variability in the magnitude of all the above processes over Earth's history, which may play some role in determining its cli-

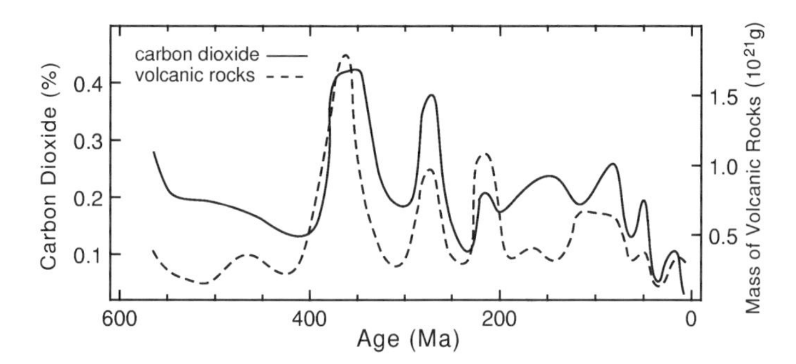

Figure 13-4 Estimate of the variations in CO_2 and the mass of volcanic rock. After Budyko (1977).

matic history. For example, Fig. 13-4 shows one estimate of volcanic activity made by Budyko (1977) based on a global survey of volcanic rock and other related geologic features. Similarly, the occurrence of cosmic dusting of the atmosphere appears to have varied significantly over time, possibly even exhibiting a 100,000-y periodicity related to the tilting of the plane of Earth's orbit into and out of the dust band that envelops the sun (Kortenkamp and Dermott, 1998).

The possible connections between this ubiquitous loading of volcanic and cosmic dust and the climate is still a matter of conjecture. In some scenarios the potential role of this dust as cloud nucleii is emphasized, whereas in others the direct effect on the radiation balance is suggested. But, even in this latter case questions persist concerning the dominance of the cooling effect of shortwave reflection (albedo) versus the warming effect of absorption. It seems likely that for mammoth loadings connected with major volcanic eruptions the albedo effects will dominate, as seems to have been the case in the "year-without-a-summer" following the 1815 Tambora explosion.

In this latter connection, far less ubiquitous than the cosmic dusting of Earth's atmosphere, are the irregularly spaced impacts of Earth by large cosmic bodies (bolides) that create huge explosions, filling the atmosphere with huge amounts of Earth material. Significant evidence exists that such impacts have occurred on the ultralong time scales we are considering in this chapter, which may have had major effects on climatic and biospheric evolution. Strong evidence exists, for example, that a major impact of this kind occurred at about 65 Ma [the Cretaceous (K)–Tertiary (T) boundary], causing, among other effects, the demise of the dinosaurs. An increased interest in such phenomena has emerged in connection with potential manmade explosions that could result in a "nuclear winter."

In addition to the aerosol loading of the atmosphere, volcanic activity leads also to the emission of chemical species (e.g., greenhouse gasses; see Fig. 10-7) and heat (which we can include as a part of the geothermal flux $G^{\uparrow}$). The effects of volcanic outgassing of CO_2 were discussed in Chapter 10, and will be further considered in the next section.

13.4 MULTIMILLION-YEAR EVOLUTION OF CO_2

As noted above, the best developed part of the theory of ultra-long-term climatic evolution relates to the tectonically driven variations of atmospheric CO_2 as influenced by external factors [$R(\mathcal{S})$, h, and $V^{\uparrow}$] as well as by silicate weatherability factor $\mathcal{W}$. A first estimate of these CO_2 variations, which we denote by $\widehat{\mu}_G$, can be obtained from the GEOCARB system described in Section 10.6.2, representing the equilibrated balance of processes formulated in Section 10.6. Assuming that $\widehat{\mu}_G$ is a dominant influence on temperature over the Phanerozoic, we can, in principle, iteratively couple this solution with the remaining tectonic-mean equilibrium equations [Eqs. (12.13) and (12.15)] to obtain first estimates of the concomitant variations of global ice amount $\widehat{I}$ and deep ocean temperature $\widehat{\theta}$, and hence iteratively improve the estimates of $\widehat{\widehat{T}}_s$, as given by Eq. (12.17). In principle, we can further improve this solution for $\widehat{\widehat{T}}_s$ by taking into account the effects of changes in Earth's topography and bathymetry (e.g., seaways) measured by h. Thus, in Fig. 13-5 we outline the portion of the full feedback-flow diagram shown in Fig. 12-1 that includes the links to $\widehat{I}$ and $\widehat{\theta}$ by which Berner's GEOCARB solution for CO_2 ($\widehat{\mu}_G$) can be extended iteratively to obtain a more complete solution, $\widehat{\mu}$. In the Sections 13.4.1 and 13.4.2 we discuss, respectively, the GEOCARB solution for $\widehat{\mu}_G$ and $\widehat{\widehat{T}}_s(t)$, and the first-order consequences for $\widehat{I}$ and $\widehat{\theta}$. In Section 13.5 we describe the possible effects of a salinity-induced instability of the thermohaline circulation (see Section 11.4), represented in Fig. 13-5 by Z and S_φ, and the general effects of continent/ocean/seaway changes (h). Still excluded are the effects of air/sea carbon flux $Q_\mu^{\uparrow}$ and ice-sheet surges that are relevant on shorter than tectonic time scales (see Chapters 15 and 16). An ultimate objective would be to account, at least roughly, for the curves shown in Figs. 1-3 and 3-3. This could possibly be supplemented by a more detailed solution for the tectonic-mean state over the Late Cenozoic, thereby providing the values of $\widehat{\mu}$ appearing in Eqs. (12.18), (12.21), and (12.22) that are required to close the theory of the ice-age departures (to be discussed in Chapter 15).

13.4.1 The GEOCARB Solution

In Fig. 13-6 we show Berner's (1994) solution for the evolution of the tectonic-mean (i.e., 10-My average) amount of CO_2 over the Phanerozoic, given all of the geologically inferred variations in the external forcing functions [e.g., weatherability $\mathcal{W}$, and the tectonic "spreading rate" $f_{SR}(t)$, shown in Fig. 10-7, that determines the volcanic output of CO_2 ($\widehat{V}_C^{\uparrow} + \widehat{V}_G^{\uparrow}$]. Note, in particular, the extremely high values in the earlier Phanerozoic, and the minimum in the Permo-Carboniferous and in the Late Cenozoic, which is the culmination of a generally declining trend over the past 200 My. This solution should be more reliable than Budyko's (1977) earlier estimate shown in Fig. 13-3, which was based mainly on volcanic activity.

The change in mean global temperature $\widehat{\widehat{T}}_s$ associated with this change in $\widehat{\mu}$, as well as with variations in $\widehat{I}$, $\widehat{\theta}$, and solar constant changes $\mathcal{S}$, is given by Eq. (12.17)

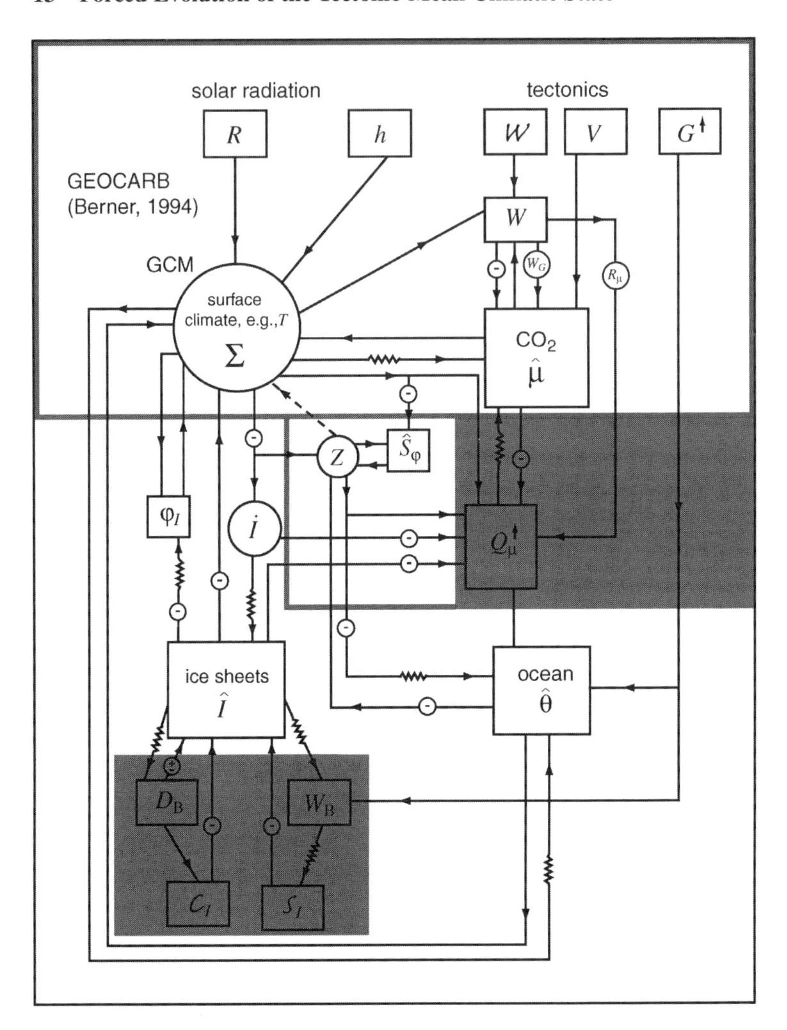

Figure 13-5 Portion of the full feedback-flow diagram shown in Fig. 12-1, including Berner's (1994) GEOCARB model (top part), plus links to $\widehat{I}$ and $\widehat{\theta}$ by which this model can be iteratively extended to obtain a more complete solution.

in the explicit form,

$$\widehat{\widehat{T}}_{\rm s} = \widehat{\widehat{T}}_{\rm s}(0) + \mathcal{B}\ln\left(\frac{\widehat{\mu}}{\widehat{\mu}(0)}\right) + k_{\rm I}^{(T)}\left[\widehat{I} - \widehat{I}(0)\right] + k_{\theta}^{(T)}\left[\widehat{\theta} - \widehat{\theta}(0)\right] + k_{\rm S}^{(T)}\left[\widehat{\mathcal{S}} - \widehat{\mathcal{S}}(0)\right] \tag{13.2}$$

In Fig. 13-7 we show the effects only of the CO_2 variations given in Fig. 13-6 plus the effect of solar constant changes given in Fig. 13-1, where $\mathcal{B} = 6°\mathrm{C}$ (Oglesby and Saltzman, 1990a) and $k_{\rm S}^{(T)} = 0.28°\mathrm{C}\ (\mathrm{W/m^2})^{-1}$ [the CCM1 value obtained by Marshall *et al.* (1994)] and $[\mathcal{S}(t) - \mathcal{S}(0)] = (68 \times 10^{-9}\ \mathrm{W/m^2\ y})t$ (see Section 13.1).

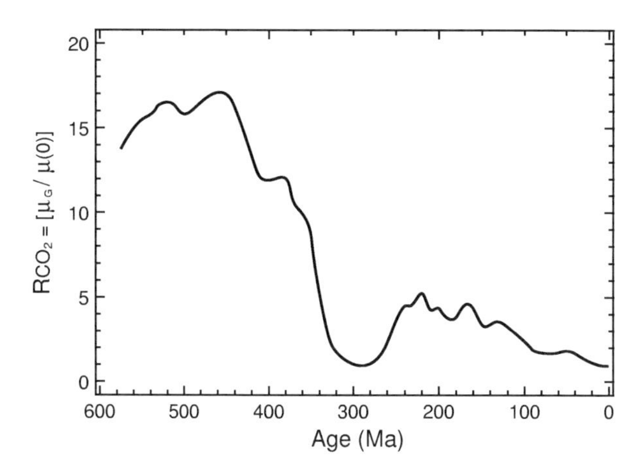

Figure 13-6 Berner's (1994) solution for the variation of carbon dioxide over the Phanerozoic due only to tectonic forcing on a million-year time scale ($\widehat{\mu}$) expressed as a ratio to the present value [$\widehat{\mu}(0)$]. Note the coincidence of low values of CO_2 and the cold episodes at about 300 Ma and at present, shown in Fig. 1-3.

A comparison of the minima in this temperature curve and the magnitude of glaciation revealed by geologic evidence (see Chapter 3) represented by the shaded areas at the bottom of Fig. 13-6 (see Fig. 3-2) reveals good agreement, especially with regard to the two main ice episodes (the Permo-Carboniferous centered at about 300 Ma and the present Late Cenozoic era). This lends support to the idea that CO_2 variations are a significant factor in climatic change, and, in particular, can account for a good deal of the variability shown in Fig. 1-3.

Also shown in Fig. 13-7 is the variation of temperature that would occur due only to the solar constant variations represented by the last term of Eq. (13.2). An important role for CO_2 in resolving the faint-early-sun paradox is suggested.

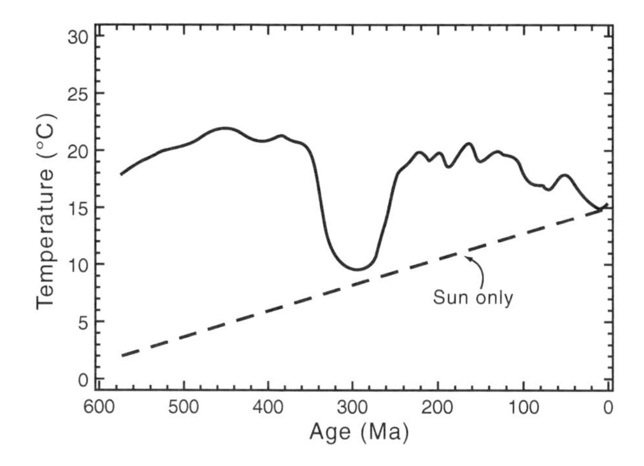

Figure 13-7 Temperature variation over the Phanerozoic due to the combined effects of the CO_2 changes shown in Fig. 13-6 and the solar luminosity changes shown in Fig. 13-1. After Berner (1994).

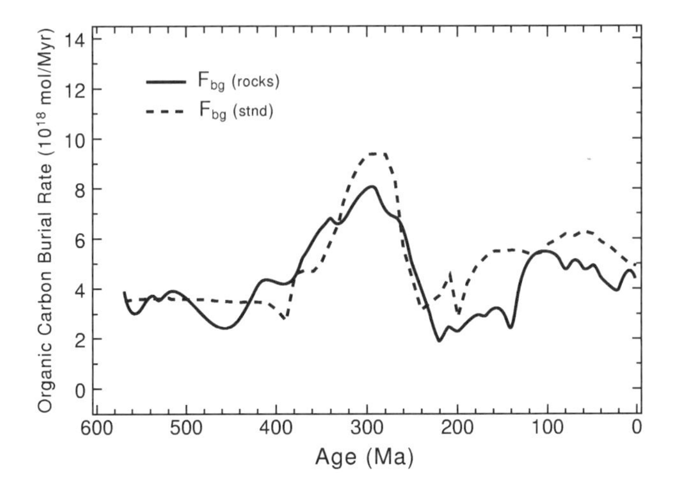

Figure 13-8 Variation of the organic burial rate (C_G) over the Phanerozoic obtained from the GEOCARB model (Berner, 1994).

In addition to the CO_2 variations, the GEOCARB model also yields the concomitant time variations of all the other dependent variables in the system. For example, the solution obtained for organic burial $C_G(t)$ is shown in Fig. 13-8, which compares favorably with geologic evidence for this quantity.

The GEOCARB results we have just described omit any explicit reference to the effects of ice-sheet or deep ocean variations. These are not only of intrinsic interest (see, e.g., the deep ocean paleotemperature curve, Fig. 3-3), but may also have a feedback influence on the CO_2 variations themselves (see Fig. 13-5). In particular, when $\widehat{\mu}$ (and hence $\widehat{\widehat{T}}_s$) fall below some critical value it is plausible that ice mass and ocean cooling begin to be significant, requiring that to determine $\widehat{\mu}$ some account be taken of the set of positive feedbacks associated with changing marginal ice extent, sea level, thermocline structure, and ocean solubility and productivity discussed in Chapter 10. This brings to the fore the necessity to consider the neglected terms depending on I, θ, and the effects of $Q^{\uparrow}$ (which can also introduce atmosphere/ocean exchanges of CO_2 on a smaller time scale than 10 My, to be discussed in Chapter 14.) Thus, a correction to the GEOCARB-type solution is suggested, which is based on the more general, nonequilibrium representation of CO_2 variations governed by the full PDM given in Section 12.1. The nature of this correction can be surmised from the observed behavior of the system during the relatively well-documented Late Cenozoic cold-mode period. By specializing the full model for this particular period we can obtain a first-order theory of this last glacial epoch, as well as determine consistent values of the free parameters (coefficients) that can improve the representation of the longer term tectonically forced Phanerozoic changes. In the next subsection we discuss the first-order implications of the CO_2 changes on the evolution of global ice mass ($\widehat{I}$) and deep ocean temperature ($\widehat{\theta}$).

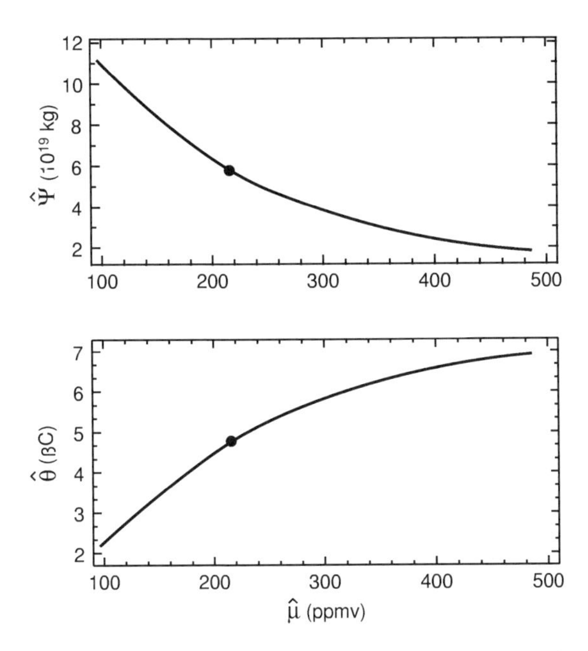

Figure 13-9 Solution obtained by Saltzman and Maasch (1990) for the dependence of ice mass ($\widehat{\Psi}$) and mean deep ocean temperature ($\widehat{\theta}$) on changes of carbon dioxide ($\widehat{\mu}$). After Saltzman and Maasch (1990).

13.4.2 First-Order Response of Global Ice Mass and Deep Ocean Temperature to Tectonic CO_2 Variations

Given the tectonic-mean CO_2 variations ($\widehat{\mu}$), such as obtained from the GEOCARB model, one can calculate the accompanying variations of $\widehat{I}$ and $\widehat{\theta}$ implied by the relations given by Eqs. (12.13) and (12.15), respectively. Thus, for simplicity, letting n equal the number of ice sheets prevailing at any time, assumed to be of roughly the same mass Ψ, i.e., $\widehat{I} = n\widehat{\Psi}$, and neglecting all direct external forcing of $\widehat{I}$ and $\widehat{\theta}$, i.e., $\widehat{F_i}$ and $\widehat{G}_\theta^\uparrow$ (but retaining all the external forcing that determined $\widehat{\mu}$), Eqs. (12.13) and (12.15) can be solved for $\widehat{I}(\widehat{\mu})$ and $\widehat{\theta}(\widehat{\mu})$ in the forms

$$\widehat{\Psi} = A_1 - A_2 \ln \frac{\widehat{\mu}}{\widehat{\mu}(0)} \tag{13.3a}$$

$$\widehat{\theta} = B_1 - B_2 \ln \frac{\widehat{\mu}}{\widehat{\mu}(0)} \tag{13.3b}$$

where A_1, A_2, B_1, and B_2 are functions of the fixed parameters ($K_\Psi^*, \widehat{a}_0, a_1, k_i^{(\Psi,\rho,w)}$, $K_\theta, \widehat{c}_0, c_\mu, c_I$).

In Fig. 13-9 we show, as an example, the diagnostic relations $\widehat{\Psi}(\widehat{\mu})$ and $\widehat{\theta}(\widehat{\mu})$ obtained from Eqs. (13.3a) and (13.3b) by Saltzman and Maasch (1990). To obtain these relations the free parameters (e.g., a_1, c_μ, c_I) are taken as the values required to account for the Pleistocene variations (e.g., the major 100-ky cycle) following the procedures to be discussed in Chapter 15. In accord with Fig. 13-5, these values of $\widehat{\Psi}$ and $\widehat{\theta}$ can be used iteratively to converge upon an improved estimate of $\widehat{\mu}$, and of $\widehat{\widehat{T}}_s$, using the complete form of Eq. (13.2). The aim would be to account more fully for the curve $\widehat{\widehat{T}}$ shown in Fig. 1-3, and for the $\widehat{\theta}$ curve shown in Fig. 3-3 for the Cenozoic. A significant further improvement in such estimates must result from properly adding the effects of tectonic changes in $h(\lambda, \varphi)$ to Eq. (13.2), explicitly measuring, e.g., the role of the many seaway changes that occurred over Earth history on the global ocean circulation (see Section 13.2).

As noted above, another related improvement might be realizable by including the possibility for a salinity-driven instability of the thermohaline circulation (Section 11.4). In the next section, however, we reconsider this possibility in the context of the evolution of the tectonic-mean state.

13.5 POSSIBLE ROLE OF SALINITY-DRIVEN INSTABILITY OF THE TECTONIC-MEAN STATE

In Section 11.4 we described a simple two-box model of the salinity-driven thermohaline circulation, admitting the possibility for an instability that could give rise to bimodal states. Although the reality of this bimodality property is highly speculative (and, in fact, we will chose to neglect it in favor of a carbon-cycle-driven instability in our discussion of the glacial–interglacial cycles to be given in Chapter 15), we shall here illustrate a possible scenario wherein a steplike change of $\widehat{\theta}$, perhaps similar to that for the mid-Miocene shown in Fig. 3-3, might be explained as a consequence of such an ocean instability (see Fig. 11-5). Thus, we now include in our thinking the loops involving the TH circulation (Z) and the salinity gradient (S_φ) shown in Fig. 13-5. In particular, in Section 11.4 a nondimensional quantity Π was defined as a function of the ratio of the freshwater flux $\mathcal{F}^{\uparrow}$ and the meridional temperature difference δT^*. This ratio would appear from GCM experiments to increase with the CO_2 concentration and, in particular, with the tectonically forced CO_2 concentration, $\widehat{\mu}_G(R, h, V^{\uparrow}, \mathcal{W})$, which can be viewed as a proxy for purely external influences, along with the direct effects of slow tectonic changes (h) in the form of marginal sea formation leading to evaporative basins in lower latitudes (e.g., the Tethys Sea in the Cretaceous). Thus, from a long-term paleoclimatological perspective, we can take $\Pi = \Pi_{ext}(\widehat{\mu}_G; h)$ as an external forcing function. Then, from the response curve (Fig. 11-5), we can determine the values of $\widehat{S}_\varphi$ and $\widehat{\psi}_{max}$ that are in equilibrium with Π_{ext}. These values can, in turn, be used in Eq. (12.15) to determine a more complete tectonic-mean state $(\widehat{I}, \widehat{\mu}, \widehat{\theta})$ when coupled with Eqs. (12.13), (12.14b), and (12.15) (see Section 13.4.2). For example, from Eq. (12.15), written in the form

$$\widehat{\theta} = \frac{1}{K_\theta}\left\{\widehat{c}_0 - c_I\widehat{I} + c_\mu \ln\left(\frac{\widehat{\mu}}{\widehat{\mu}(0)}\right) - \gamma_1\mu_s\widehat{S}_\varphi + c\left[\widehat{R} - \widehat{R}(0)\right] + \sum c_{h_i}\left[h_i - h_i(0)\right]\right\} \tag{13.4}$$

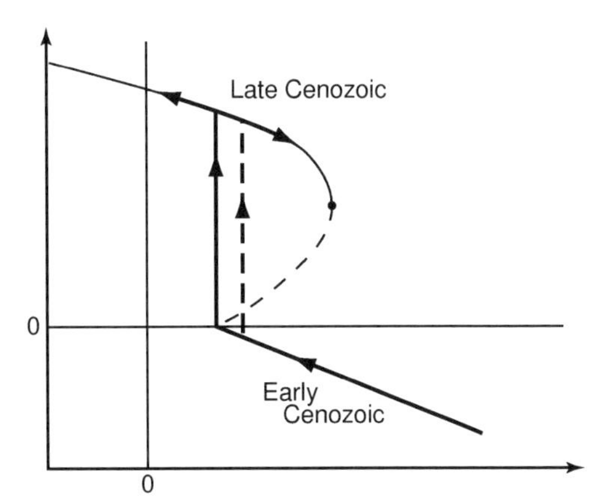

Figure 13-10 Possible path of the equilibrium state of the thermohaline state of the ocean ($\widehat{\psi}_{\text{max}}$, $\widehat{S}_{\text{P}}$) from the Early to Late Cenozoic as a function of the forcing function $\Pi(\widehat{\mu}_{\text{G}}, h)$ based on the two-box model bifurcation diagram shown in Fig. 11-5.

we see that $\widehat{\theta}$ is a function not only of $\widehat{I}$, $\widehat{\mu}$, $\widehat{R}$, and $h_i(\lambda, \varphi)$ but also as a decreasing function of the salinity gradient $\widehat{S}_{\varphi}$.

Starting in the Late Cretaceous (70 Ma) we might conjecture that the equilibrium path followed by the tectonic-mean climate system during the Cenozoic was as qualitatively portrayed by the arrows in the version of Fig. 11-5 shown in Fig. 13-10. That is, a high value of $\widehat{\mu}_{\text{G}}$ and well-developed tropical evaporative basins near the Cretaceous/Tertiary (K/T) boundary initially place the system in the warm-saline-bottom-water (WSBW) mode characterized by a weak $\widehat{S}_{\varphi}$, a high value of $\widehat{\theta}$, and little ice ($\widehat{I} \approx 0$). The ensuing net tectonic reduction of CO_2 shown in Fig. 13-6, and plate movements that reduce the Tethys Sea, steadily weaken this warm mode, as shown by the arrow, quasi-statically moving the equilibrium to reduced WSBW and colder temperatures. Coupled with the opening up of the Atlantic Ocean (see Fig. 3-1) and circumpolar Southern Ocean and tectonic uplift in Antarctica, the stage is set for more massive ice formation on Antarctica (Oglesby, 1989) that could drive the system to the other side of the unstable portion of the equilibrium curve, i.e., to the upper stable branch, perhaps even before the bifurcation point, as shown by the dashed path in Fig. 13-10. As indicated, it would appear that the system has resided near this colder, stronger, direct TH circulation and salinity gradient branch throughout the Late Cenozoic to the present time. The glacial–interglacial oscillations about this equilibrium will be the subject of the next two chapters.

13.6 SNAPSHOT ATMOSPHERIC AND SURFICIAL EQUILIBRIUM RESPONSES TO PRESCRIBED $\widehat{y}$-FIELDS USING GCMS

As mentioned in Section 7.11, several GCM simulation experiments have been performed to determine the atmospheric and surficial climatic states $\widehat{X}$ that would be in equilibrium with various combinations of prescribed external forcing (e.g., $\widehat{\mu}_B$, $h(\lambda, \varphi)$, R) appropriate for different times in the past. Repeating our comments in the introduction to this chapter, we shall not go beyond the references already given in Section 7.11 except to note again that a more recent listing of relevant references is given by Parrish (1998, pp. 249–250).

14

THE LATE CENOZOIC ICE-AGE DEPARTURES

An Overview of Previous Ideas and Models

In the previous chapter we discussed the ultra-long-term variations of the tectonic-mean climatic state. It was suggested that the tectonically driven cooling of the system over the Late Cenozoic could bring Earth to a point at which ice formation and oscillation become a dominant part of paleoclimatic variability (see Chapters 1 and 3). In this chapter we shall review the ideas and models proposed to account for the departures from the tectonic-mean state that represent this Late Cenozoic "glacial epoch." For this purpose we shall use the feedback diagram (Fig. 12-1), repeated here to outline as special cases the main physical ingredients of the models that have been proposed. Of particular interest will be the dominant near-100-ky-period oscillation in the late Pleistocene, the onset of this main oscillation at about 900 ka, and the earlier onset of the glacial epoch at about 2.5 Ma (see Section 1.2). Thus we are returning to the basic challenge posed in the Chapter 1 (Section 1.4).

14.1 GENERAL REVIEW: FORCED VS. FREE MODELS

Before examining the models as specializations of the PDM system, Eqs. (12.1)–(12.8), and the feedback diagram shown in Fig. 12-1, we first make some more general remarks. On the most fundamental level, we note that all climatic variability is due to either (1) variable external forcing (e.g., Milankovitch Earth-orbital changes) or (2) instability of the internal system that may arise from steady external forcing. From all indications an explanation of the observed paleoclimatic variability involves a complex combination of both of these possibilities.

From a historical viewpoint there has been an ebb and flow of popularity of several major hypothesis, all following the first field-based demonstrations that Earth had indeed experienced at least one great ice age (Agassiz, 1840). As noted in Section 12.3, these include major hypotheses involving the role of Earths' orbit, and carbon dioxide, as well other hypotheses regarding bedrock depression, calving and basal sliding of ice sheets, and the thermohaline ocean state. More recently, the potential role of internal instability of the whole system due to dominant positive feedback loops has been advanced as a major hypothesis. Good discussions of these historical aspects, particularly with regard to the two leading hypotheses (Earth-orbital and carbon dioxide) are given, e.g., by Imbrie (1982) and Maasch (1992).

As discussed in Chapter 1 (Section 1.4), the impossibility of accounting for the Pleistocene ice record as a simple quasi-static linear response to the known Earth-orbital forcing is apparent from Fig. 1-4, showing the fluctuations in insolation at high Northern Hemisphere latitudes compared with the $\delta\,^{18}O$ record of ice variations. We saw that the dominant changes of ice in the late Pleistocene (past 1 My) are of a period near 100 ky, although the forcing is mainly of the precessional and obliquity periods, $\sim$20 and $\sim$40 ky, respectively. Moreover, in addition to the fairly abrupt onset of the Pleistocene glacial epoch at about 2.5 Ma, another fairly abrupt transition in the period, amplitude, and mean value of the ice fluctuations is seen in Fig. 1-4 to have occurred at about 900 ka, despite the fact that the character of the forcing was unchanged over the full Pleistocene and before. We conclude from all these facts that the ice variations are governed by a nonlinear dynamical (i.e., time-dependent) system, and the orbital forcing is not a sufficient condition for the existence of the major near-100-ky oscillations.

A basic unresolved question that remains, however, is whether the orbital forcing is a necessary condition for the 100-ky oscillations, and one way to classify previous models of the Pleistocene ice variations is according to the assumed answer to this question. In Tables 14-1 and 14-2 we list some early contributions in these two categories, respectively.

14.1.1 Models in Which Earth-Orbital Forcing Is Necessary

As seen in part C, Table 14-1, many, if not most, modeling groups have assumed that Earth-orbital forcing is indeed necessary for the 100-ky oscillations, i.e., if forcing were removed the major 100-ky-period glacial–interglacial oscillations would not occur. According to these models the main fluctuations of the ice ages represent the output of a nonlinear forced oscillator driven by orbital variations, which, however, would have to be supplemented either by some nonorbital (e.g., tectonic) forcing or some free internal or stochastic behavior in order to be able to yield the above noted transitions near 2.5 and 0.9 Ma. Included in this group are models in which (1) slow periodic Milankovitch variations under the influence of stochastic perturbations cause the system to transition between multiple equilibrium states corresponding to glacials and interglacials (e.g., Sutera, 1981; Benzi *et al.*, 1982), (2) the near-20-ky precessional variations nonlinearly produce 100-ky relaxation oscillations of ice sheets (e.g., Oerlemans, 1980a), or 100-ky transitions between bimodal glacial/interglacial states when certain thresholds are reached (Lindzen, 1986), and (3) combination tones of precessional forcing act in concert with a near-10-ky-period free oscillation to produce a 100-ky-period oscillation by “entrainment” (Ghil and Le Treut, 1981). It has also been proposed (Muller and MacDonald, 1995) that near-100-ky-period fluctuations in the inclination of the orbital plane in conjunction with the location of cosmic dust bands might be able to generate the observed major ice-age variations (see Section 1.3). On the basis of results from the models listed in Table 14-1, it is generally

Table 14-1 Examples of Time-Dependent Models of Pleistocene Variations: Orbital Forcing Necessary

---

A. Basic linear model
- Hasselmann (1976), Lemke (1977) [$R \neq 0$]
- Imbrie and Imbrie (1980) [ε small, $I' = \varepsilon F$]
- Fong (1982) [ε large, $I' = F$]

B. More complex models; no dominant 100-ky response, but capable of simulating 20- and 40-ky responses to orbital forcing
1. Energy balance models
   - Suarez and Held (1976)
   - Schneider and Thompson (1979)
2. Ice-sheet models
   - Weertman (1976)
   - Andrews and Mahaffy (1976)
   - Pollard *et al.* (1980)
   - Budd and Smith (1981)
   - Huybrechts (1990)
3. Coupled EBM/SDM ice-sheet models
   - Pollard (1983b)
   - Berger *et al.* (1988b)
   - Budd and Rayner (1990)
   - Lindstrom (1990)

C. Forced 100-ky response (variables treated are listed in square brackets)
1. Nonlinear rectification of 20/40-ky response
   - Birchfield and Weertman (1978) [I]
   - Imbrie and Imbrie (1980), Snieder (1985) [I]
   - Lindzen (1986) [I, T]
2. Ice-sheet models including bedrock depression
   - Oerlemans (1980b) [I, D_B]
   - Birchfield *et al.* (1981), Birchfield and Grumbine (1985) [I, D_B]
   - Ghil and Le Treut (1981), Le Treut and Ghil (1983) [I, D_B]
   - Pollard (1982, 1983a,b) [I, D_B]
   - Peltier (1982), Hyde and Peltier (1985) [I, D_B]
   - Watts and Hayder (1983) [I, D_B]
3. Stochastic resonance energy balance model [$T(+F, R)$]
   - Sutera (1981)
   - Benzi *et al.* (1982)
   - Nicolis (1982)
   - Matteucci (1989)

Table 14-2 Examples of Time-Dependent Models of Pleistocene Variations: Orbital Forcing Not Necessary (Free Oscillations)[a]

A. No dominant ~100-ky response
- Eriksson (1968) [I, T, μ linear]
- Sergin (1979), Sergin and Sergin (1976) [I, D_B, θ]
- Saltzman (1978) [sea-ice extent (η), θ]
- Källén *et al.* (1979), Ghil (1984) [I, T]
- Saltzman (1982), Saltzman and Moritz (1980), Saltzman *et al.* (1981) [η, $\theta \sim \mu$]
- Ghil *et al.* (1987) [I, θ]

B. ~100-ky response
- Saltzman *et al.* (1982) [I, θ]
- Oerlemans (1982a,b) [I, D_B, $W(T_B)$, F]
- Saltzman and Sutera (1984) [I, χ, θ]
- Saltzman *et al.* (1984) [I, χ, θ]
- Saltzman (1987) (Sutera, 1987; Maasch, 1988) [I, μ, θ]
- Maasch and Saltzman (1990) [I, μ, θ, $+F$]
- Saltzman and Maasch (1990) [I, μ, θ, $+F$] (Cenozoic)
- Chalikov and Verbitsky (1984, 1990) [I, D_B, θ)
- Saltzman and Verbitsky (1992, 1993, 1994a) [I, μ, θ, D_B, $+F$]
- Saltzman and Verbitsky (1993) [I, $\mathcal{S}_I$, W_B]

[a] In this table χ represents a marine ice-shelf component.

accepted that the near-20- and 40-ky observed fluctuations of planetary ice mass can be explained by precessional and obliquity forcing. On the other hand, questions remain in trying to explain the near-100-ky oscillations of the Late Pleistocene as a forced response using the models listed in Table 14-1; none of these is able to account for the mid-Pleistocene transition near 0.9 Ma, or the onset of the ice epoch at about 2.5 Ma, without *ad hoc* assumptions external to the model itself.

14.1.2 Instability-Driven (Auto-oscillatory) Models

As an alternative, the possibility exists that the 100-ky oscillation does not owe its existence to variable external forcing, but instead is the consequence of a bifurcation to a free oscillation that is driven by an instability of the climatic system that is being forced by steady or ultra-low-frequency (e.g., tectonic) variations. The existence of abrupt changes in the climate record is indeed an indication that some internal instability must be present, perhaps activated by slow forced changes in some critical parameter.

In order for such a free oscillator to exist the ice mass must interact dynamically with at least one additional slow-response variable. This variable might simply be the evolving continental location of the ice-sheet mass as it moves toward its water vapor source, here represented by its equatorward ice-edge latitude φ_I (Tanner, 1965). Another candidate is the bedrock depression beneath the ice sheet (Oerlemans, 1980b), but this appears to require a much larger bedrock response time than is appropriate to achieve a purely free oscillation independent of forcing (Peltier, 1982). As described in Section 12.4, a perhaps more plausible candidate than these for this additional variable is atmospheric CO_2, which, in turn, is likely to be controlled by the thermohaline ocean state. This possibility was suggested qualitatively by Plass (1956), illustrated quantitively by Saltzman and Moritz (1980) and Saltzman *et al.* (1981), and given credence by the ice-core observations of Delmas *et al.* (1980), Neftel *et al.* (1985), Barnola *et al.* (1987), and Sowers *et al.* (1991) [as well as by the more speculative $\delta\,^{13}C$-derived inferences of Shackleton and Pisias (1985)]. Another possible candidate is the thermohaline state of the ocean as embodied in the variations of the salinity gradient S_φ (e.g., Stommel, 1961). In part B, Table 14-2, we list a sequence of models developed to illustrate these free-oscillation possibilities; in part A, Table 14-2, we list earlier models that demonstrated the possibility for free oscillatory behavior of the climate system of periods much shorter than 100 ky (perhaps applicable to millennial-scale oscillations).

14.1.3 Hierarchical Classification in Terms of Increasing Physical Complexity

In the remainder of this chapter we will illustrate the basic differences in the main classes of models described in Section 14.1.1 (listed in Table 14-1) from another viewpoint, i.e., as specializations of the full dynamical system reviewed in Chapter 12. We do this by showing the portion of the complete feedback-flow diagram (Fig. 12-1) that is appropriate for each class of models as special cases (see Fig. 14-1). Note that the Berner (1994) model for long-term tectonic CO_2 changes was depicted in this manner in Chapter 13 (Fig. 13-4).

The succession of different models of the Late Cenozoic ice epoch will be represented by boxed domains of Fig. 14-1, labeled 1–10. Each of these domains sequentially includes additional feedback loops involving new variables or processes, with both forced and free variability possible.

14.2 FORCED ICE-LINE MODELS (BOX 1, FIG. 14-1)

In the earliest models no attempt was made to deal with the high-inertia ice sheets; instead, a simple identification was made between regions of summer snow cover (as measured by the "snow-line" latitude, which we shall denote by φ_I) and the existence of the ice sheets. Thus, only the albedo of a "pancake ice sheet" was taken into account. The most primitive versions of such "ice-line" models, described in Section 7.7, are those in which the ice line is tied to a fixed value of the mean annual or summer temperature in equilibrium with changes in Earth-orbital forcing, as deduced from an EBM (see Box 1 of Fig. 14-1). We may further resolve these models into two classes: (1) those admitting only stable equilibria and (2) those allowing for unstable equilibria and stochastic resonance.

1. **Stable equilibria.** The first examples of papers based on this model are those of Suarez and Held (1976, 1979), followed by the work of Schneider and Thompson (1979). These models established that the radiative forcing due to Earth-orbital variations have the potential to account for major swings of the ice line that could conceivably be identified with the observed ice variations of near-20- and 40-ky-periods. However, even taking into account the expected lack of phase agreement due to the small inertia of surface temperature and snowcover compared to ice mass, there is still little correspondence of the results with the observed $\delta\,^{18}$O-derived measurements, which, as shown in Fig. 1-4, are dominated by the near-100-ky-period variability.

One of the characteristics of the observed $\delta\,^{18}$O record of ice variations is a redlike spectrum forming the background for the maxima of periods near 100, 40, and 20 ky. As pointed out by Hasselmann (1976) and Lemke (1977), when stochastic forcing is added to an ice-line model (e.g., an EBM) containing one stable equilibrium, such a realistic "red spectrum" is obtained, but few of the observed details in the observed time evolution shown in Fig. 1-4 are accounted for.

2. **Unstable equilibria/the stochastic resonance model.** A notable addition to the ice-line theory is possible by applying a combination of Milankovitch and stochastic forcing, providing that a bimodality in the thermal response to an EBM is postulated; such a bimodality is based on an assumed internal instability, usually attributed to a strong positive snowcover–albedo feedback. As described in Section 7.6, this leads to the possibility for "stochastic resonance" (Benzi *et al.*, 1981, 1982; Nicolis, 1982) that can amplify the weak near-100-ky-period eccentricity forcing component of Earth-orbital forcing, yielding a solution in closer agreement with observations. However, the existence of such multiple stable equilibria for surface temperature or ice cover is not supported by results from GCMs (Saltzman *et al.*, 1997) (see Section 7.8). The feedback diagram for this type of model is the same as is shown in Fig. 14-1 (Box 1), with the proviso that positive ice-albedo feedback now makes T_S a slow-response variable subject to an internal instability, which is now dominant; as noted in Section 12.2, stochastic forcing is assumed to apply to T_S as well as to all other variables, but is not shown in the figure.

14.3 ICE-SHEET INERTIA MODELS

14.3.1 The Simplest Forms (Box 2)

More realistic models are possible if the inertia of ice sheets is taken into account. In the simplest form it is assumed that the presence of real ice sheets can be taken into account by a linear equation for the global ice mass I, in which ice-sheet physics and explicit connection to atmospheric or surficial physics are neglected, i.e.,

$$\frac{dI}{dt} = -\frac{1}{\varepsilon_I} I + F_I \tag{14.1}$$

where ε_{I} is a dissipative time constant for ice sheets, and following Milankovitch (1930), F_{I} is the high-latitude summer radiative forcing (including a combination of eccentricity, obliquity, and precessional frequencies) that is assumed to control the survival and accumulation of winter snowfall. These models are represented in the feedback diagram shown in Fig. 14-1 (Box 1 plus Box 2), examples of which are described by Calder (1974), Imbrie and Imbrie (1980), and Snieder (1985).

With the Imbrie and Imbrie (1980) model it was easy to show that if ε_{I} is taken as a constant having a reasonable value (say 10 ky), and two free parameters are assigned in the representation of F_{I}, the ~20/40-ky-period oscillations could be replicated as in the ice-line models, but the phase difference between the forcing and response could now be more realistic. In addition, it was shown that by assigning an additional free parameter in the form of a variable time constant ε_{I} that depends on the sign of dI/dt, a nonlinear system is obtained by which some additional power could be gained in the near-100-ky-period range (but also an excessive amount in a period of 400 ky), and there was also a suggestion that the observed "sawtooth" variation characteristic of some glacial cycles could also be obtained. The Snieder (1985) study illustrated the manner in which nonlinear interactions in the Imbrie and Imbrie (1980) model could give rise to the added amplitude of the 100-ky cycle.

In the case when ε_{I} is taken to the large limit, say greater than 100 ky, Eq. (14.1) reduces to

$$\frac{dI}{dt} \approx F_{\mathrm{I}} \tag{14.2}$$

which seems to represent the case treated by Fong (1982) emphasizing the role of the near-100-ky-period eccentricity variations. The consequences of this relation are clearly at variance with the observations portrayed in Fig. 1-9, as discussed by Saltzman (1986).

14.3.2 More Physically Based Ice-Sheet Models: First Applications

Forced ice oscillation models based on the physics of ice-sheet behavior discussed in Sections 9.3 and 9.4 (including simple bedrock depression effects, but not yet including calving or basal sliding processes) were developed by Weertman (1964, 1976), who introduced a simple relation between the glacial firn line (i.e., the line separating interior regions of accumulation from peripheral regions of ablation) and Earth-orbital radiative variations measured by a fluctuating "equilibrium" or "snow" line (see Section 9.4). These studies were extended by Birchfield (1977) and Birchfield and Weertman (1978). It was suggested in this latter study that rectification of the Milankovitch forcing due to nonlinearity of the ice-sheet model could bolster the amplitude of the 100-ky-period response to some extent.

Figure 14-1 The feedback-flow diagram shown in Fig. 12-1, partioned by the boxed domains labeled 1–10, which represent sequential additions to a fuller dynamical model of paleoclimatic variability. The evolution of models of increasing complexity involving these added domains and feedbacks is discussed in Sections 14.2–14.6.

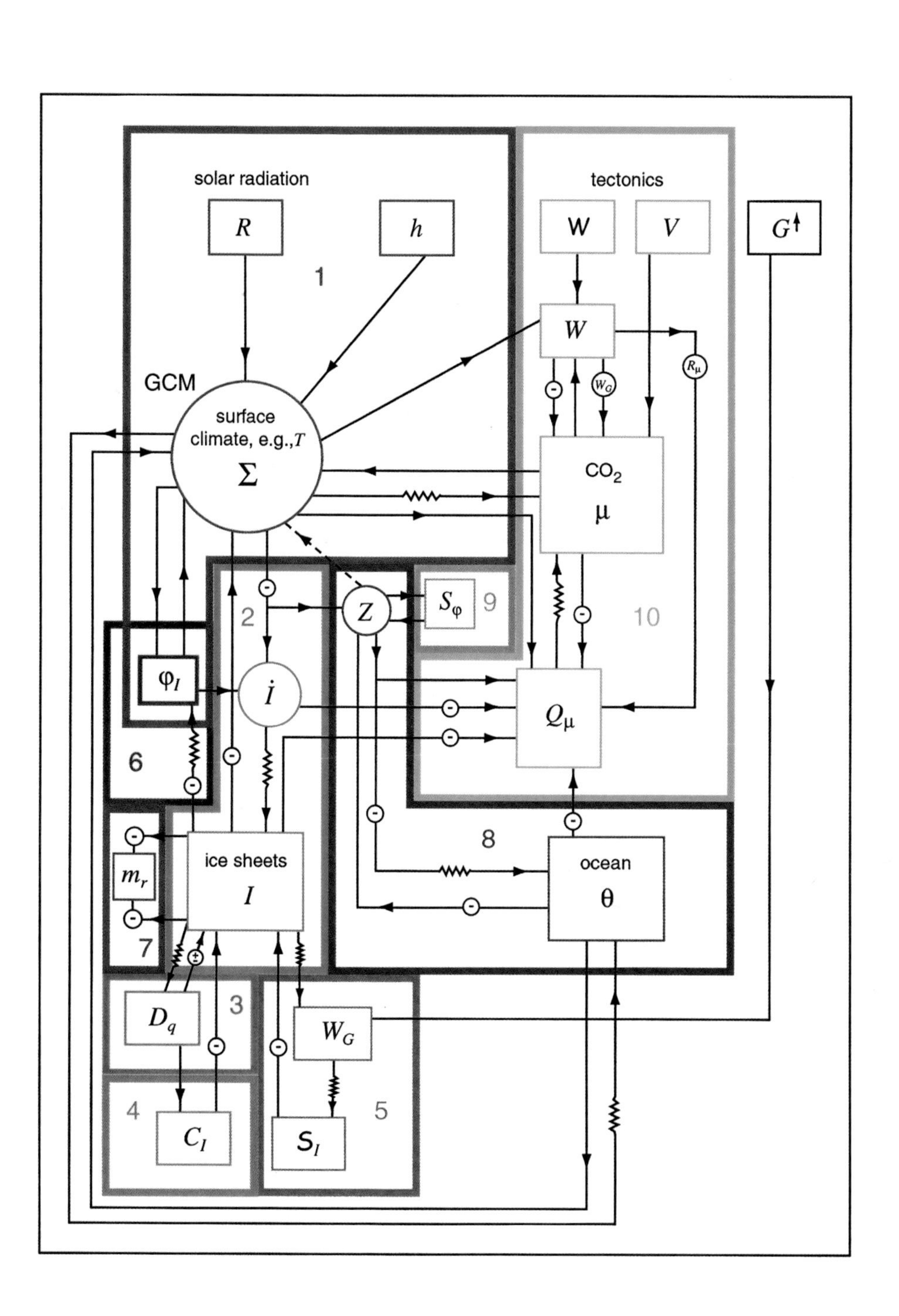

solar radiation
R
h
1
GCM
surface
climate, e.g., T
Σ
tectonics
W
V
G
W
CO2
μ
2
Z
S
9
10
φI
İ
Qμ
6
8
ice sheets
I
ocean
θ
mr
7
3
Dq
WG
4
CI
5
SI

An important further step was made by Pollard (1978, 1982, 1983a,b), Pollard *et al.* (1980), and Oerlemans (1980a, 1981a,b), who coupled an ice-sheet model with an EBM to determine the accumulation–ablation profile as a function of the surface temperature of the ice sheet. At roughly the same time, other forced models incorporating this feature, allowing also for some free internal variability, were developed that could generate some 100-ky-period variance. One such study was made by Oerlemans (1980b), based on a feedback between the ice sheet and its bedrock depression [a general idea first suggested by Ramsay (1925)]; however, as noted above, this required an overly large response time for bedrock, about 30 ky, to achieve a 100-ky period. Other studies initiated by Källén *et al.* (1979) and extended by Ghil and Le Treut (1981) and Le Treut and Ghil (1983) were driven by the ice-albedo feedback, leading to free oscillatory behavior and "combination tones" of the precessional (19/23-ky period) forcing that could be "entrained" by (or "frequency-locked" with) these free oscillations leading to an amplification of the 100-ky-period cycle. The free oscillations in this model were obtained by postulating that warmer ocean conditions lead to increased ice accumulation [requiring a plus rather than minus sign in the ($T \rightarrow \dot{I}$) link in Fig. 14-1], while, in turn, increased ice leads to colder ocean temperatures; this scenario would be more likely to prevail when the ice sheets are continental, not yet being marine bounded, where warmer ocean temperatures would enhance the ablation/calving process. An early suggestion of this kind of cycle was described by Stokes (1955), and is further discussed on the basis of North Atlantic paleoobservations by Ruddiman *et al.* (1980).

Yet another series of studies was made at about this time by Peltier (1982), Peltier and Hyde (1984), and Hyde and Peltier (1985) in which Milankovitch forcing coupled with bedrock depression effects on the ice sheet profile could also produce an amplified 100-ky-period response. In the Peltier (1982) study it is also noted that relaxation–oscillations involving the ice sheet and its bedrock depression might be possible, but not of a 100-ky period.

14.3.3 Direct Bedrock Effects (Box 3)

The main role of bedrock depression in the above models (represented by Box 3 in Fig. 14-1) is the lowering of the height of the ice-sheet by roughly a quarter to a third of its thickness (see Section 9.8). The consequence of this lowered ice-sheet height on net snow accumulation is ambiguous, however, most models postulating a net decrease due to a dominating effect of warmer ice surface temperatures (e.g., Weertman, 1964; Oerlemans, 1980a; Peltier, 1982), but some postulating a net increase due to a dominating effect of increased accessibility to moisture (the reverse of the ice-elevation "desert" effect). Thus, it is not clear whether bedrock depression will feed back in a positive or negative way on ice mass, as indicated by the plus/minus sign in Fig. 14-1.

As noted above, with an unrealistically large response time ε_D of $\sim$30 ky, and assuming a negative feedback of D_B on I, Oerlemans (1980b) could obtain a periodic response to orbital forcing of the required 100 ky. As pointed out by Peltier (1982) and Peltier and Jiang (1996), a more appropriate value of ε_D is about 3 ky, in which case this mechanism by itself cannot produce the 100-ky-period response. However, in conjunction with the calving effect to be discussed next, Pollard (1982) showed that bedrock depression could account fairly well for the observed ice volume variations over the past few hundred thousand years.

14.3.4 Bedrock-Calving Effects (Box 4)

The variable $\mathcal{C}_{\mathrm{I}}$ (Box 4) in Fig. 14-1 represents the addition of a potentially significant consequence of bedrock depression, also suggested by Ramsay (1925) and developed more quantitatively by Pollard (1983a,b), i.e., the effect of the bedrock depression, D_{B}, on the calving rate (see Section 9.9). This offers the possibility for a longer period fluctuation and more rapid deglaciations than would otherwise be possible. Further applications of this calving mechanism, coupled as in Pollard's case with an EBM, were made by Watts and Hayder (1984) and Hyde and Peltier (1985). As shown by Saltzman and Verbitsky (1993), this mechanism can also lead to ice-sheet fluctuations having a near-40-ky period, thereby supplementing the effects of obliquity forcing.

14.3.5 Basal Meltwater and Sliding (Box 5)

Model group 5 in Fig. 14-1 represents the further addition of basal water (W_{B}) and sliding ($\mathcal{S}_{\mathrm{I}}$). The potentially important role of this mechanism in ice-sheet behavior was noted by Wilson (1964) and Weertman (1966) and applied by Oerlemans (1982a,b) to the ice-age oscillation problem. It may be, however, that this mechanism is most important in accounting for the 5- to 10-ky-period oscillation known as Heinrich events, to be discussed in Chapter 16.

14.3.6 Ice Streams and Ice Shelf Effects

For marine-bounded ice sheets the process of basal sliding is closely connected with the formation of ice streams that carry ice to the ocean, often terminating in floating ice shelves (see Section 9.7). An early attempt to incorporate these shelves in a simple model of Pleistocene glacial cycles was made by Pollard (1983a,b), following a key discussion of their potential role by Denton and Hughes (1981). A more highly simplified dynamical systems study in which ice shelf mass was considered to be an important variable that could interact with the ocean state and ice sheets to produce the 100-ky cycle as a free oscillation was discussed by Saltzman *et al.* (1982) and Saltzman and Sutera (1984). This ice shelf variable (represented by χ in Table 14-2) is not represented separately in Fig. 14-1, but should be considered to be an intrinsic part of the total ice mass I.

14.3.7 Continental Ice-Sheet Movement (Box 6)

On continental areas ice stream movements take the form of "lobes" that usually penetrate toward lower latitudes. More generally, whether by such ice stream movements or by normal creep processes, ice-sheet masses tend to redistribute by moving equatorward "toward their moisture source" from their initial growth area in higher latitudes. This introduces a new connection of the ice mass I to the variable labeled φ_I in Fig. 14-1, which is the latitude of the ice extent. One idea that has been discussed (e.g., Tanner, 1965) is that as I increases and hence φ_I decreases, the ice sheet will reach a point of collapse due to the intensified calving and ablation engendered by higher radiation and warmth. A cycle of regrowth of the ice cap at higher latitudes can then ensue. More detailed models coupling an ice sheet with the atmospheric climate should intrinsically include this possible mechanism, even in two-dimensional (φ, h_I) models (e.g., Berger *et al.*, 1990; DeBlonde and Peltier, 1991a) and certainly in three-dimensional (λ, φ, h_I) models, to be discussed next. From these studies it would appear that although this mass shift process is indeed operative, it cannot by itself account for the 100-ky cyclical evolution.

14.3.8 Three-Dimensional (λ, φ, h_I) Ice-Sheet Models

Almost all of the above models of the ice sheets were essentially two-dimensional (φ, h_I) idealizations of the vertically integrated equations discussed in Section 9.3. More geographically detailed (λ, φ, h_I) models governing the varying thickness of the North American and Eurasian ice sheets were subsequently applied to the ice-age oscillation problem, exemplified by Budd and Smith (1981) and Hughes (1996) using an ice-sheet model uncoupled to an atmospheric climate model, and by DeBlonde and Peltier (1991b), Marsiat (1994), and Tarasov and Peltier (1997a,b) using a coupled EBM/ice-sheet model. [This latter coupled model has been applied by Hyde *et al.* (1999) to Pangean boundary conditions in the Permo-Carboniferous (~300 Ma).] In none of these three-dimensional studies, however, is an attempt made to solve for more than the last cycle of glaciation.

14.4 THE NEED FOR ENHANCEMENT OF THE COUPLED ICE-SHEET/ATMOSPHERIC CLIMATE MODELS

The physical ingredients included in all the models discussed thus far should apply not only to the past 900 ky, during which time the 100-ky-period oscillation is a dominant, but also to the earlier Pleistocene and Pliocene, during which time there was little 100-ky-period variance. Clearly some new ingredient is necessary, the longer term variation of which can explain the difference in behavior across a relatively sharp time of transition. This highlights one of the main criticisms of the orbitally forced ice-sheet models, namely, their insufficiency as an explanation of the ice ages (see Chapter 1). Thus, for this reason alone, it is clear that the other slow-response variables forming part of the full dynamical system, Eqs. (12.1)–(12.8), portrayed in Fig. 14-1 (μ, θ, S_φ) should be considered. An equally important reason to consider the other variables is simply their intrinsic importance in a full description of the climatic state (e.g., the mean temperature of the deep ocean, θ). By considering these additional variables, a much richer set of possibilities for internal instability and free oscillation of the slow-response climate system is permitted. This seems especially relevant in the case of carbon dioxide (μ), to which surface temperature is as sensitive for the lower values that prevailed during the ice ages (see Oglesby and Saltzman, 1990a) as the Earth-orbital changes (the only factor considered in the models discussed up to now). Moreover, from the discussion in Chapter 13 it is known that a strong connection exists between CO_2 and glaciation over tectonic time scales (see Figs. 13-6 and 13-7), and it also appears that there exists a causative relation between variations of ice (I) and CO_2 as revealed by the Vostok measurements (see Fig. 8-2). For these reasons it is natural to explore the possibilities for modeling the Late Cenozoic changes based on the introduction of CO_2 as a key new variable, which by the nature of the dynamics of CO_2 embodied in Eq. (12.5) also requires the consideration of the thermal state of the ocean, represented by θ. Therefore, in the following sections we continue with our review of the hierarchy of models that will now embrace consideration of all the other dependent variables, portrayed in Fig. 14-1, including, in addition to μ, θ, and S_φ, the mass of regolith beneath the ice sheets, m_r, to be discussed next in Section 14.5.1, now being considered as a possibly important variable (Clark, 1994).

14.5 ICE-SHEET VARIABLES COUPLED WITH ADDITIONAL SLOW-RESPONSE VARIABLES

We now review models in which the additional variables, m_r, θ, and S_φ, are included, as portrayed in Boxes 7, 8, and 9, respectively, of Fig. 14-1. Models in which carbon dioxide (μ) are introduced as a key additional variable are discussed separately in Section 14.6.

14.5.1 Regolith Mass, m_r (Box 7)

In Section 9.6 we described a mechanism proposed by Clark and Pollard (1998) whereby the successive growth and decay of ice sheets may lead to the progressive removal of regolith mass over several ice-age cycles. According to the hypothesis this might ultimately lead to a regolith-denuded bedrock base on which larger ice sheets could grow without the supposed sliding effects of the soft sediment layer. It is proposed that this may be able to account for the mid-Pleistocene transition at about 900 ka. The possibility for such an effect, dependent on the new variable m_r, is represented in Fig. 14-1 by a new, separate, feedback loop delineated as Box 7; the model by Clark and Pollard (1998) is the first example of a dynamical model incorporating this feature. As noted at the close of Section 9.6, there are indications, however, that this mechanism may not be operative as envisioned; in addition, in Antarctica the regolith material is completely undisturbed after the growth and passage of large ice sheets, which may simply slide over the regolith without substantial disturbance.

14.5.2 The Deep Ocean Temperature θ (Box 8)

To a limited extent this deep ocean variable, described more fully in Section 8.3, was taken into account in a somewhat artifical way in some of the models already discussed. For example, Källén *et al.* (1979) allow the thermal relaxation time in their EBM to have values much larger than would be appropriate for the combined atmosphere and subsurface boundary (e.g., mixed ocean) layer. This leads to a separate prognostic equation for temperature that might, inconsistently, be identified with the deep ocean rather than the surface temperature from which the longwave radiation is calculated.

A more explicit introduction of the deep ocean temperature was made by Sergin (1979) based on his work in the early 1970s. This was the first major study showing how an ice-sheet, deep ocean thermal system coupled with an atmospheric model could exhibit long-period oscillations independent of any external forcing. A more elaborate model of this sort, improving upon the ice-sheet model used by Sergin and introducing a more detailed atmospheric and surficial represented based on the zonally averaged SDM of Saltzman and Vernekar (1971), was studied later by Chalikov and Verbitsky (1990).

At about the same time as Sergin's work was being performed, a systematic series of studies involving θ as a key variable was initiated by Saltzman (1978), Saltzman and Moritz (1980), and Saltzman *et al.* (1981). These followed upon the qualitative suggestions by Weyl (1968) and Newell (1974) that the deep ocean temperature should vary in a counterintuitive way with ice variations, this being due to the insulating properties of sea ice in high latitudes (large sea-ice coverage inducing warmer deep ocean temperatures). It was also assumed in these studies that atmospheric CO_2 would vary in tandem with θ, leading to a positive feedback that could introduce an instability and lead to auto-oscillations of a relatively short period (1.5 ky) (see Sections 6.5 and 6.6). This scenario was subsequently extended to apply to the higher inertia ice shelves attached to the ice sheets, rather than sea ice, in an attempt to account for the 100-ky-period oscillation as a free response (Saltzman *et al.*, 1982, 1984; Saltzman and Sutera, 1984, 1987). In subsequent papers (Saltzman and Maasch, 1988; Maasch and Saltzman, 1990) the variable θ was identified with a deficit in North Atlantic Deep Water (NADW).

Newly emerging paleooceanographic evidence (e.g., Chappell and Shackleton, 1986; Birchfield, 1987) began to refute this inverse relationship between ice cover and bulk deep ocean temperature and it became clear that the deficit in NADW during the ice ages was accompanied by a large increase in much colder and less saline water (Antarctic Bottom Water) (e.g., Boyle and Weaver, 1994). Therefore, in succeeding models, starting with Saltzman and Maasch (1990), a thermally direct effect of ice coverage (high I leading to colder θ) was postulated. In addition, just before this time it was realized (Saltzman, 1987a,b; Saltzman and Maasch, 1988) that, because of all the positive feedback in the carbon cycle, CO_2 could plausibly be assumed to be a free slow-response variable, not diagnostically coupled to θ as in the previous studies (see Chapter 10). Models emphasizing the role of CO_2 as such a significant new variable will be discussed in Section 14.6, but we first discuss some models in which the salinity gradient S_φ is introduced as a new thermohaline state variable.

14.5.3 The Salinity Gradient S_φ (Box 9)

The earliest studies involving the salinity gradient-driven thermohaline circulation, made in the context of the Late Cenozoic climate oscillations and based on the concepts described in Section 11.4, were attempts to account for millennial scale [e.g., Heinrich (H), Younger Dryas (YD), and Dansgaard–Oeschger (D–O)] oscillations. The longer period Heinrich oscillations clearly require a connection to the massive ice sheets, thereby involving the variable I as well as θ and S_φ, whereas the D–O oscillations usually involve only ocean–atmosphere models. Examples of these studies include Birchfield (1989), Birchfield *et al.* (1994), Paillard (1995), Paillard and Labeyrie (1994), and Sakai and Peltier (1995, 1996, 1997).

Paillard (1998) invokes the bimodal property of the simple TH circulation models described in Section 11.4 to illustrate the possibility that this property can provide the instability (see Section 12.4) to drive the main 100-ky oscillation through threshold effects. In another study by Brickman *et al.* (1999), the interaction of the salinity-driven TH circulation, governed by the Wright and Stocker (1991) model, with Milankovitch forcing is explored; the interaction with ice sheets is neglected, however, so that the feedbacks involve only the components $(R, T, Z, \theta, S_\varphi)$ in Fig. 14-1. It seems likely that the role of the salinity gradient-driven TH circulation will be found to be of primary importance in accounting for the ice-age oscillations and should be explored more fully in the near future. Up to this time greater attention has been given to the role of carbon dioxide, which we discuss next.

14.6 CARBON DIOXIDE, μ (BOX 10)

14.6.1 Earlier History

The key geochemical reactions whereby CO_2 is maintained and varies on tectonic time scales, described in Chapter 10 (Sections 10.1–10.4), were first identified by Ebelman (1845, 1847, 1855) and Högbom (1894) [see Berner (1995) and Berner and Maasch (1996) for excellent reviews]. As early as 1861 Tyndall was already speculating about the possible connection between the infrared greenhouse properties of this gas and also water vapor (both of which properties he measured) and the Late Pleistocene ice ages earlier revealed by Agassiz (1840). The first quantitative discussions of this possibility, using very simple climate models governing surface temperature, were made by Arrhenius (1896) following upon the carbon cycle work of his colleague, Högbom. Excellent reviews of Arrhenius' work are given in the special anniversary issue of *Ambio* (Vol. 26, 1997) and in Crawford's (1996) biography. Not long after, Chamberlin (1899) promoted specific scenarios regarding the roles of orogenic activity and chemical weathering in determining the CO_2 level and hence the climate over the Phanerozoic [a subject treated in more quantitative detail by Walker *et al.* (1981)

and Berner (1994); see Section 13.4]. In conjunction with Arrhenius' work, Chamberlin's writings made the CO_2 theory of the ice ages widely accepted, superceding, at least briefly, the Croll version of the Earth-orbital hypothesis. With Milankovitch's more quantitative studies in the 1920s and 1930s the Earth-orbital hypothesis became more prominent as an explanation of the glacial–interglacial cycles, especially because no evidence was available that CO_2 actually varied in a cyclical manner. In spite of this lack of geological evidence, in a significant theoretically based study Plass (1956, 1961), an expert in the physics of infrared radiation, presented the most cogent argument yet made for the relevance of CO_2 variations in paleoclimatic variations, including an interesting suggestion that ocean–atmosphere interactions could result in an instability that could drive free oscillations of the carbon cycle and hence drive the ice-age oscillations [see Weart (1997) for a review]. However, with the growing popularity of the Milankovitch Earth-orbital hypothesis, bolstered by significant geological evidence [e.g., Hays *et al.* (1976); see also Imbrie (1982) and Berger and Loutre (1997) for reviews], the CO_2 hypothesis remained relatively dormant until the 1980s. This lack of acceptability was enhanced by some notable theoretical studies concerning possible auto-oscillatory carbon cycle behavior by Eriksson and Welander (1956) and Eriksson (1963, 1968), which, though pioneering in their application of dynamical systems methods to a geophysical/chemical system, seemed to negate the potency of this mechanism as an explanation of the ice ages.

14.6.2 Quantitative Revival of the Carbon Dioxide Hypothesis

In a remarkable series of field and analytical studies of the air trapped in ice cores drilled in Greenland and Antarctica, it became clear that CO_2 did indeed vary on the same time scales as the Late Pleistocene ice ages [Berner *et al.* (1980), Delmas *et al.* (1980), Neftel *et al.* (1982); the most recent 420-ky record from the Vostok core is reported by Petit *et al.* (1999)]. Other biogeochemical evidence was provided by Shackleton and Pisias (1985). From a dynamical viewpoint, these findings provided the first signs of confirmation of the basic idea implied by Plass (1956) and the model proposed by Saltzman and Moritz (1980) in which it was postulated that CO_2 should be considered an important independent variable in accounting for climate system variations. After several studies in which CO_2 continued to be considered a fast-response variable diagnostically related to θ, as in the above study, CO_2 was formally included as a crititcal slow-response variable that could possses sufficient positive feedbacks to drive free oscillations of the system by Saltzman (1987a,b). This variable was then included in successively improved models by Saltzman and Maasch (1988, 1990, 1991), and Saltzman and Verbitsky (1993, 1994a), representing the set of models embracing Box 10 of Fig. 14-1. The last three of the above studies were based on the dynamical CO_2 equation, Eq. (10.35), described in Section 10.5.

The importance of considering CO_2 as a dependent variable in the full paleoclimate dynamics PDM system is supported by several factors already alluded to in previous sections:

1. The sensitivity of the high-latitude summer surface temperature, on which the survival and growth of ice sheets depend, is of the same order for CO_2 variations in the Pleistocene range as it is for the Milankovitch Earth-orbital changes (Section 7.9.1).
2. The covariability of CO_2 and ice mass over the first two 100-ky cycles strongly suggests a causative relation between the two (Section 8.2).
3. The covariability of tectonic CO_2 variations measured by Berner (1994) and the major Phanerozoic episodes of ice presence further suggests the relevance of CO_2 (Section 13.4).
4. The slow tectonic variations of CO_2 just mentioned, supported by the work of Raymo *et al.* (1988) and Cerling *et al.* (1997), represent the leading possibility to account for the apparent gradual cooling and ice formation during the Late Cenozoic that may have brought the climate system to the thresholds of the main transitions to the ice epoch at about 2.5 My and the major 100-ky-period oscillation at about 0.9 My.

The natural accommodation of this last possibility in a self-consistent way was first demonstrated by Saltzman and Maasch (1990, 1991) in an attempt to explain the mid-Pleistocene transition, and was extended by Saltzman and Verbitsky (1994a) to account also for the ice epoch transition of $\sim$2.5 My. These models will be discussed in more detail in Chapter 15. Several other investigators have proposed a slow CO_2 decline as the mechanism that might accomplish the mid-Pleistocene transition (e.g., Birchfield and Ghil, 1993), but without including any dynamical representation of CO_2 in the model. Recognizing the importance of CO_2 in accounting for the glacial–interglacial cycles, other models have included the Vostok CO_2 curve, as a prescribed external forcing function rather than as an internal free variable (e.g., DeBlonde and Peltier, 1993; Tarasov and Peltier, 1997a,b; Gallée *et al.*, 1992; Berger *et al.*, 1998b).

At this point it seems to be increasingly recognized that CO_2 should be considered on at least equal footing with Milankovitch variations in accounting for the glacial–interglacial cycles. If, in fact, it turns out that the tectonic decrease in CO_2 over the Late Cenozoic provided the threshold state for the initiation of the major ice buildups, it would then deserve recognition as the "cause" of the ice epoch and its oscillations. Thus, it may be appropriate to parallel Imbrie's (1982) table of key publications in the "astronomical theory of the Pleistocene ice ages" with a similar table for contributions to the "carbon dioxide theory" based on our discussion in this section (see Table 14-3).

14.7 SUMMARY

In this chapter we have reviewed the sequence of models in which new physical ingredients have been successively added to form an ever more complete representation of the full slow-response climate system (i.e., the "center manifold"). We have seen that the main components involve the ice sheets and their bedrock and basal properties, coupled with forced and free variations of carbon dioxide, operating on an Earth characterized by a high-inertia deep thermohaline ocean that can store carbon and heat, all under the influence of steady and variable (e.g., Milankovitch) external radiative forcing. From a more formal viewpoint, we have indicated how, by systematically deleting

Table 14-3 Key Publications in the Evolution of the Carbon Dioxide Theory of the Pleistocene Ice Ages

Developmental phase	Contributions
Era of 19th century pioneers (1845–1899)	Ebelmen (1845, 1847, 1855), Tyndall (1861), Högbom (1894), Arrhenius (1896), Chamberlin (1899)
Revival and debate (1900–1969)	Callendar (1938), Plass (1956, 1961), Eriksson (1963, 1968)
Modern observational revival (1980–present)	Berner *et al.* (1980), Delmas *et al.* (1980), Neftel *et al.* (1982), Shackleton and Pisias (1985), Petit *et al.* (1999)
Modern quantitative theoretical revival (1980–present)	Saltzman and Moritz (1980), Walker *et al.* (1981), Berner *et al.* (1983), Saltzman (1987), Saltzman and Maasch (1988, 1990, 1991), Berner (1991)

variables from the full PDM system, Eqs. (12.1)–(12.8), we can in principle recover the essential physics included in most of the models already treated, going all the way back to the most primitive ice-line models (Sections 7.7 and 14.2) and the simplest ice inertia models (Section 14.3.1).

In the next chapter we describe, and discuss the process of solution of, an illustrative case in which almost all of these components are present and harmonized in a self-consistent, "unified" dynamical model from which a good deal of the global-mean variability discussed in Chapters 1–3 can be deduced, with some additional side predictions. We have used the term "unified" here, and in the heading for Part III of this book starting with Chapter 12, in three main contexts: (1) as a unification of most of the main hypotheses individually proposed to explain the ice ages [e.g., Earth-orbital variations influencing the solar radiation distribution vs. greenhouse gas (CO_2) variations influencing terrestrial radiation], (2) as a unification of the two possible modes of exciting variability in the climate system (i.e., variable external forcing vs. instability-driven free internal behavior due to steady forcing), and (3) as a unification of the longer term tectonic variations described in Chapter 13, with the departures from these tectonically driven states discussed in this chapter, which together form the observational picture shown in Fig. 1-3.

In Section 14.1 and Tables 14-1 and 14-2 we summarized some of the earlier contributions discussed in this chapter to the two main lines of development of a dynamical theory of the Late Cenozoic ice-age departures; note that the physical components successively included in each of the models, as they appear in Fig. 14-1, are entered in brackets next to each reference.

15

A GLOBAL THEORY OF THE LATE CENOZOIC ICE AGES

Glacial Onset and Oscillation

We now return to our general PDM system governing the departures from the tectonic-mean state, summarized in Section 12.5 [Eqs. (12.19–(12.23)], specializing this set to obtain a model of the form studied by Saltzman and Maasch (1991) and extended by Saltzman and Verbitsky (1993). This model is more general than any of the special cases described in Chapter 14, because it will simultaneously include ice, CO_2, the deep ocean thermal state, and bedrock effects, as internally interactive components under the influence of both known Earth-orbital (Milankovitch) variations and a probable slow tectonically forced drawdown of CO_2 over this period. As such, the model can form the basis for a unified explanation of the ice ages embracing most of the major theories offered to date. This combined theory will be seen to be capable of accounting, from the same set of statements, not only for the observed ice-age oscillations, but also for their onset and regime transitions; in addition, potentially observable predictions of past variations of the CO_2 levels and ocean states are made.

Before proceeding to this discussion of the departure system governing the global ice-age oscillations, we first recall that the tectonic-mean states about which these departures occur are governed by the system given in Section 12.5. This system forms the basis for a first-order theory of the slow long-term evolution of the global system represented by $(\widehat{I}, \widehat{\mu}, \widehat{\theta}, \ldots)$ that was discussed in Chapter 13. In particular, we noted in Section 13.4 that the best developed part of this tectonic-mean theory concerned the evolution of the CO_2 level $(\widehat{\mu})$ and its implied surface temperature change $\widehat{T}_s$ [e.g., the GEOCARB model (Berner, 1994)]. Because this model depends only on a specification of purely external factors $(h, R, V^{\uparrow}, \mathcal{W})$ we can consider this solution for $\widehat{\mu}$ to be a proxy for these external tectonic forcings, which we shall denote by $\widehat{\mu}_B(h, R, V^{\uparrow}, \mathcal{W})$. In principle, as noted in Section 13.4, we can iteratively improve this solution by taking into account the effects of the implied changes in ice mass $(\widehat{I})$ and deep ocean temperature $(\widehat{\theta})$ on $\widehat{T}_s$ along with those due to $\widehat{\mu}$ and R, to yield a better estimate of $\widehat{\mu}$, and $\widehat{T}_s$ in accordance with Eq. (13.2). It was seen that $\widehat{\mu}$ and the associated temperature record $\widehat{T}_s$ exhibit particularly low values during the known periods of cold temperature and glaciation (e.g., the Permo-Carboniferous, and Late Cenozoic, which is our central concern in this chapter. When $\widehat{\mu}$ (and hence $\widehat{T}_s$) falls below some critical value it is plausible that ice mass and ocean cooling begin to be significant, requiring that some account be taken of a new set of positive feedbacks associated with changing marginal ice extent, sea level, thermocline structure, and ocean solubility and productivity (Saltzman and Maasch, 1988). This brings into consideration the neglected terms depending on I, θ, and the effects of $Q_{\mu}^{\uparrow}$, which can introduce atmosphere/ocean exchanges of CO_2 on a time scale smaller than 10 My. Thus, a further correction to the GEOCARB-type solution is suggested, which is based on the nonequilibrium behavior of the ocean/atmosphere carbon system, a suggested form of which is Eq. (10.37). The nature of this correction can be estimated from the behavior of the system during the relatively well-documented Late Cenozoic cold-mode period.

15.1 SPECIALIZATION OF THE MODEL

We begin by listing the set of approximations made to simplify the more general PDM represented by Eqs. (12.18)–(12.23) and portrayed as a feedback-loop diagram in Fig. 12-1:

1. That the growth and decay of ice over the past 5 My occur primarily in the form of two main ice sheets over North America (ζ_1) and Eurasia (ζ_2) (see Fig. 15-1), each of which has the same mass, $\zeta/2$, i.e., $\zeta_1 + \zeta_2 = \zeta$. Thus, in spite of the uncertainties, we are assuming in this simplified model that Antarctica (Ψ_A) and Greenland (Ψ_G) remain relatively constant at their tectonic equilibrium values so that $\Delta I = \zeta$.
2. That possible surges of ice measured by $\mathcal{S}_I$, due to variations in W_B, can be neglected to first order ($\mathcal{S}_I = 0$), their effects being incorporated in the dissipative coefficient K_I. Thus we regard melting at the base of the ice sheets, emphasized by Oerlemans (1982a,b) as a key factor in the main 100-ky oscillation, as a more secondary process the effects of which can be embodied in the time constant for ice sheets as a negative feedback factor. On the other hand, basal melting processes are undoubtedly crucial for higher frequency ice variations such as the Heinrich events, to be discussed in Chapter 16.
3. That insolation changes due to Earth-orbital variations are important mainly insofar as they cause ice-sheet variations through their influence on high-latitude summer temperatures T_Ψ, and have relatively little influence on annual/global average temperatures $\widetilde{T}$ [e.g., $c_R = 0$ in Eq. (12.22)].
4. That the direct effects of CO_2 changes on the deep ocean temperature θ are small compared to the effects of ice mass changes, which have strong direct effect on deep water formation [$c_\mu = 0$ in Eq. (12.22)].
5. That, in view of our lack of detailed knowledge regarding the Late Cenozoic evolution of the tectonically driven CO_2 level, we assume as the simplest approximation consistent with the Berner (1994) model and the Raymo and Rau (1994) measurements, that $\widehat{\mu}$ is linearly decreasing over this period, i.e.,

$$\widehat{\mu}(t) = \widehat{\mu}(0) + \left(\frac{d\widehat{\mu}}{dt}\right)t \qquad (15.1)$$

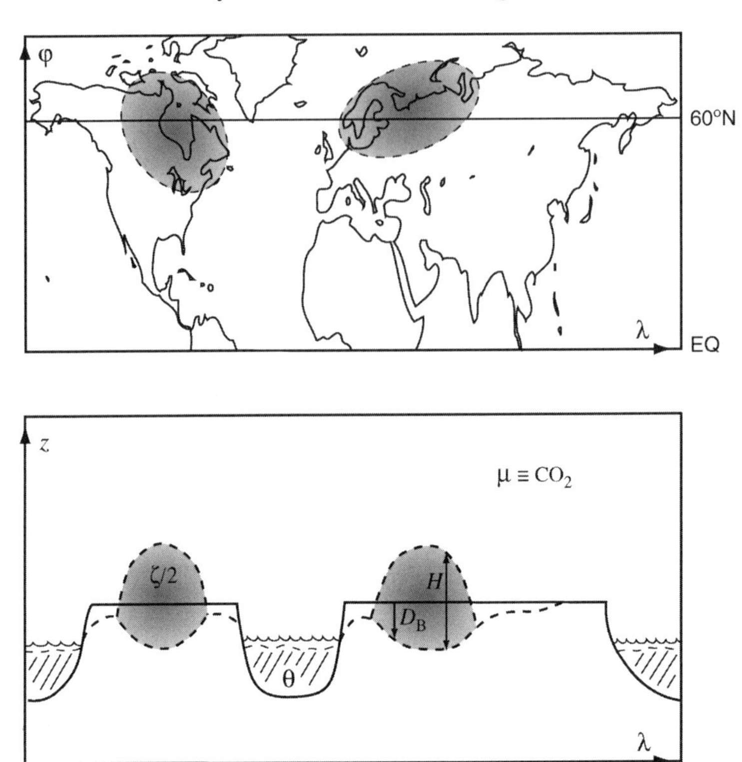

Figure 15-1 Schematic representation of an idealized two-ice-sheet system; geographic distribution (top), and vertical cross-section at about 60°N.

where $\widehat{\mu}(0)$ is a mean value representative of the last few glacial cycles and $(d\widehat{\mu}/dt)$ is the mean rate of change of the tectonic-equilibrium CO_2. Over this same period the associated equilibrium ice mass and ocean state would then vary in accordance with Eq. (13.3).

6. That, although there is good reason to consider the long-term effects of the ocean salinity gradient on the thermohaline circulation and all the other paleoclimatic variables, we shall here neglect this interaction ($\dot{\sigma} = 0$) and focus only on the thermal aspect of the ocean state as represented by θ.

7. That, in view of our lack of knowledge of volcanic outgassing of CO_2 and geochemical weathering rates on the glacial–interglacial time scale, we set $\Delta(V_\mu^{\uparrow} - W_\mu) = 0$, though there has been a suggestion that the CO_2 weathering drawdown may intensify during glacial periods (Munhoven and Francois, 1996), which would augment the positive feedback processes included in the representation of $Q_\mu^{\uparrow}$.

Thus, the system given by Eqs. (12.18)–(12.23) is reduced to the following system of dynamical equations governing the departures from the tectonic-mean state, ζ, D_B, ξ, ϑ, a portrayal of which in terms of the feedbacks retained from those shown in

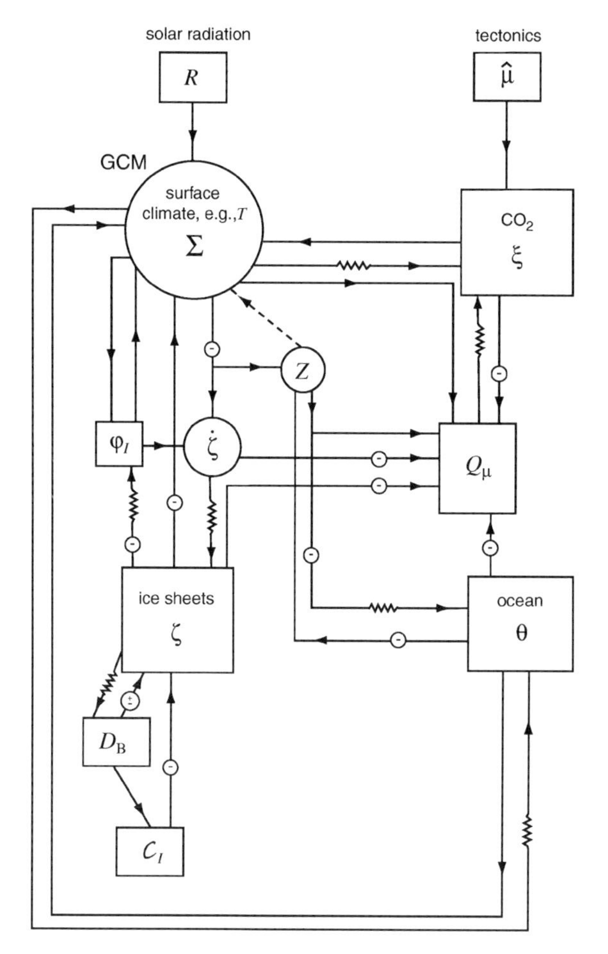

Figure 15-2 Feedback-flow diagram for the model [Eqs. (15.2)–(15.5)], as a simplification of the full system portrayed in Fig. 12-1.

Fig. 12-1 is given in Fig. 15-2.

$$\frac{d\zeta}{dt} = -a_1\left[\frac{\mathcal{B}^{(\Psi)}}{\widehat{\mu}}\xi + k_\theta^{(\Psi)}\vartheta + k_{\mathrm{R}}^{(\Psi)}\Delta R\right] - K_{\mathrm{I}}\zeta - \mathcal{C}_{\mathrm{I}}(\zeta, D_{\mathrm{B}}) \qquad (15.2)$$

$$\frac{dD_{\mathrm{B}}}{dt} = eH - \nu_{\mathrm{D}} D_{\mathrm{B}} \qquad \left(H = c_{\mathrm{H}}\zeta^{1/5}; \quad e = 0.25\nu_{\mathrm{D}}\right) \qquad (15.3)$$

$$\frac{d\xi}{dt} = K_\xi(\xi; \widehat{\mu})\xi - b_\theta\vartheta \qquad (15.4)$$

$$\frac{d\vartheta}{dt} = -c_{\mathrm{I}}\zeta - K_\theta\vartheta \qquad (15.5)$$

where $K_\xi = [b_1(\widehat{\mu}) - b_2(\widehat{\mu})\xi - b_3\xi^2]$, $b_1(\widehat{\mu}) = (2\beta_2\widehat{\mu} - 3\beta_3\widehat{\mu}^2 - \beta_1)$, $b_2 = (3\beta_3\widehat{\mu} - \beta_2)$, $b_3 = \beta_3$, and $\mathcal{C}_{\mathrm{I}}(\zeta, D_{\mathrm{B}}, \alpha_{\mathrm{B}}, Z^*)$ is given by Eq. (9.54), where $\Psi = \zeta/2$.

In these equations $\widehat{\mu}$ is considered to be a very slowly varying external "control" parameter that responds quasi-statically to tectonic forcing, F_{μ}, e.g., in accordance with the GEOCARB model (Section 13.4), and hence is a proxy for F_{μ}. In particular, we assume the form given by Eq. (15.1).

The system contains seventeen parameters, eight of which are constrained either by atmospheric physics governed by a general circulation model ($\mathcal{B}, k_{\theta}, k_{\mathrm{R}}$), glaciological and asthenospheric physics ($K_{\mathrm{I}}, \rho_{\mathrm{i}}, \rho_{\mathrm{b}}, \nu_{\mathrm{D}}, c_{\mathrm{H}}$), or ocean physics ($K_{\theta}$). This leaves nine unconstrained, or "free" parameters ($a_1, \alpha_{\mathrm{B}}, Z^*, \beta_1, \beta_2, \beta_3, b_{\theta}, c_{\mathrm{I}}, d\widehat{\mu}/dt$), which are at our disposal for fitting the system to conform with observations such as shown in Fig. 1-4, given the known external radiative forcing $\Delta R(t)$ shown, e.g., in Fig. 1-8. Note that five of these (a_1, α_{B}, b_3, c_{I}, and Z^*) are required by physical considerations to be positive. It will be a continuing challenge to constrain these free parameters further by additional physics and thereby reduce their number. As it turns out, the parameters involved in the bedrock-calving function (α_{B}, Z^*) can explain only a very small amount of the variance of ice mass and can be neglected, leaving only seven free parameters.

It is clear from the development of the dynamical equations that we are dealing with a highly qualitative physics containing unknown rate constants that can only be treated as free parameters. It remains as our task to determine whether it is possible to assign values to these free parameters that can account for all of the rich variability that accompanied the ice ages, and simultaneously make potentially verifiable predictions of variability and covariability not yet discovered in the geologic record. This poses a type of "inverse problem" that is not easily posed and solved by classical methods, being especially difficult because of its strong nonlinearity that must embrace drastic regime changes within a single model.

We next describe a procedure for accomplishing this task based on the discussions in Saltzman and Maasch (1991) and Saltzman and Verbitsky (1993), showing that the system can indeed yield solutions in good agreement with the observed record over the past several million years, such as shown in Fig. 1-4. We shall go through this process in some detail because it represents a formal inverse methodology that should find application in other complex dynamical system modeling problems where free parameters (i.e., rate constants) must be predicted to account for time-series observations.

15.2 THE 100-KY OSCILLATION AS A FREE RESPONSE: DETERMINATION OF THE ADJUSTABLE PARAMETERS

The values adopted for the physically constrained parameters are given in Table 15-1. We now outline the procedure by which we can assign values to the free parameters of the system that can yield a solution replicating the proxy ice and CO_2 variations portrayed in Figs. 1-4 and 3-5. One established feature of these variations is the dominance of near-100-ky-period ice and CO_2 oscillations during at least the past half-million years. In view of the absence of sufficiently strong forcing at this period, and the absence of the 100-ky-period oscillations earlier in the record in spite of similar external forcing, it is likely that this oscillation is unforced, arising instead as a consequence of an internal instability. Such an instability can be activated when an external

Table 15-1 Physically Constrained Parameters

Parameter	Value
$\mathcal{B}$	11°C
k_θ	0.5
k_R	$0.1°C\ (W\ m^2)^{-1}$
K_I	$1.0 \times 10^{-4}\ y^{-1}$
ν_D	$(3\ ky)^{-1}$
c_H	$1\ m^{1/2}$
K_θ	$2.5 \times 10^{-4}\ y^{-1}$
ρ_i/ρ_b	0.25

(control) parameter (e.g., the tectonically forced value of CO_2, $\widehat{\mu}$) achieves a critically low range of values, and we shall make this assumption. Other possibilities that depend on the existence of Earth-orbital forcing as a necessary condition have been proposed (see Chapter 14), but these proposals all have some shortcomings as discussed, and in any event must also be supplemented by some other unspecified external control to activate the oscillations only during the Late Pleistocene.

Using Eqs. (15.1)–(15.5) we now determine the values of the adjustable parameters needed to generate a free (unforced) oscillatory solution, independent of any bedrock depression effects, having the near-100-ky-period characteristic of the Late Pleistocene. Thus, at this stage we neglect Earth-orbital forcing ($\Delta R = 0$) and calving instabilities ($\mathcal{C}_I = 0$) [and fix the tectonically induced CO_2 concentration at the mean value known from the Vostok core measurements to have prevailed over the past 200 ky ($\widehat{\mu} \equiv \widehat{\mu}^* = 253$ ppm)]. Note that with $\mathcal{C}_I = 0$, bedrock depression variations (D_B) are decoupled from the dynamical system, and Eq. (15.3) becomes an auxiliary equation to determine the lagged bedrock response to the variations in the mass of the overlying ice sheets.

The general procedure for achieving this solution is described in Maasch and Saltzman (1990), and, with more explicit reference to our present system, in Saltzman and Maasch (1991). Following closely this latter paper we start by taking the response times for the ice sheets and ocean state to be $K_I^{-1} = 10$ ky and $K_\theta^{-1} = 4$ ky, respectively, and then seek to determine the remaining six coefficients (i.e., a_1, b_1, b_2, b_3, b_θ, c_I) that can give a plausible free oscillatory solution for ζ, ξ, and ϑ of a period near 100-ky, as observed in the Late Pleistocene. These coefficients are assumed to be positive; in particular, by assuming $b_1 > 0$, we allow the possibility for a rapid departure of CO_2 from an unstable equilibrium, which can drive the free oscillation (Saltzman and Maasch, 1988).

15.2.1 Nondimensional Form

To facilitate the analysis the system can be nondimensionalized by introducing the transformations: $t = \lambda_0 t^*$, $\zeta = \lambda_1 X$, $\xi = \lambda_2 Y$, $\vartheta = \lambda_3 Z$, and $\Delta R = |R|\mathcal{R}$, where

t^*, X, Y, and Z are nondimensional variables, $|R|$ is the standard deviation of ΔR, $\mathcal{R}(t)$ is the nondimensionalized high-latitude insolation having unit variance, and λ_0, λ_1, λ_2, and λ_3 are scaling parameters that determine the period and amplitude of any periodic solution. If we choose $\lambda_0 = K_\mathrm{I}^{-1}$, $\lambda_1 = a_1\widehat{\mathcal{B}}/\widehat{\mu}(K_\mathrm{I}b_3)^{1/2}$, $\lambda_2 = (K_\mathrm{I}b_3)^{1/2}$, and $\lambda_3 = a_1\widehat{\mathcal{B}}c_\mathrm{I}/\widehat{\mu}K_\theta(K_\mathrm{I}b_3)^{1/2}$, the dynamical system becomes

$$\frac{dX}{dt^*} = -X - Y - vZ - u\mathcal{R}(t^*) \tag{15.6}$$

$$\frac{dY}{dt^*} = -pZ + rY - sY^2 - Y^3 \tag{15.7}$$

$$\frac{dZ}{dt^*} = -q(X + Z) \tag{15.8}$$

where $p = a_1\widehat{\mathcal{B}}b_\theta c_\mathrm{I}/\widehat{\mu}K_\mathrm{I}^2K_\theta$, $q = K_\theta/K_\mathrm{I}$, $r = b_1(\widehat{\mu})/K_\mathrm{I}$, $s = b_2(\widehat{\mu})/(K_\mathrm{I}b_3)^{1/2}$, $u = k_\mathrm{R}^{(\Psi)}|R|(b_3/K_\mathrm{I})^{1/2}$, and $v = k_\theta^{(\Psi)}a_1\widehat{\mathcal{B}}c_\mathrm{I}/\widehat{\mu}K_\mathrm{I}k_\theta$.

With this transformation the original set of eight coefficients in Eqs. (12.5)–(12.7) have been replaced by four scaling parameters (λ_0, λ_1, λ_2, and λ_3) and four "structure-determining" parameters (p, q, r, and s), with the two remaining parameters u and v being functions of the others. In practice it may be more convenient to assign a value to u or v *ab initio* and then show *a posteriori* that these are consistent with estimates of $k_\mathrm{R}^{(\Psi)}$ and $k_\theta^{(\Psi)}$.

Note that the parameter λ_0 is equal to our assumed time constant for ice-sheets K_I^{-1} and q is the ratio of this ice-sheet time constant to the deep ocean time constant K_θ^{-1}. We can view the scaling parameters, λ_1, λ_2, and λ_3, as proxies for initial conditions. Thus, we remain with three adjustable parameters (p, r, s), to obtain the desired periodic structure, and three parameters (λ_1, λ_2, and λ_3) to scale the amplitudes.

We also note that p, r, and s are functions of the tectonic equilibrium state (i.e., of $\widehat{\mu}$) and therefore must vary over the full Cenozoic. It is this variability that can provide the bifurcation to the Late Pleistocene ~100-ky oscillation. These bifurcation properties are discussed in Section 15.4 as part of a "structural stability" analysis of the free solution. Our goal at this point, however, is to find the equilibria and their "internal stability" properties (Ghil and Childress, 1987) for a fixed value of $\widehat{\mu}$ $(= \widehat{\mu}^*)$ with a view to locating a single unstable region of the (p, q, r, s)-parameter space that can admit a 100-ky-period free oscillation.

15.2.2 Internal Stability Analysis to Locate a Free 100-ky-Period Oscillation in Parameter Space

Following Maasch and Saltzman (1990), we determine the equilibria of the unforced X–Y–Z system, Eqs. (15.5)–(15.7), as a function of the model parameters (p, q, r, s). Clearly, at least one equilibrium exists for $(X, Y, Z) = (0, 0, 0)$ that we identify as the "tectonic" equilibrium state. Because our system is cubic, two other real equilibria can exist that we call $(X_0^{(1)}, Y_0^{(1)}, Z_0^{(1)})$ and $(X_0^{(2)}, Y_0^{(2)}, Z_0^{(2)})$. Given these equilibria we can calculate the eigenvalues in order to determine their linear stability properties (see

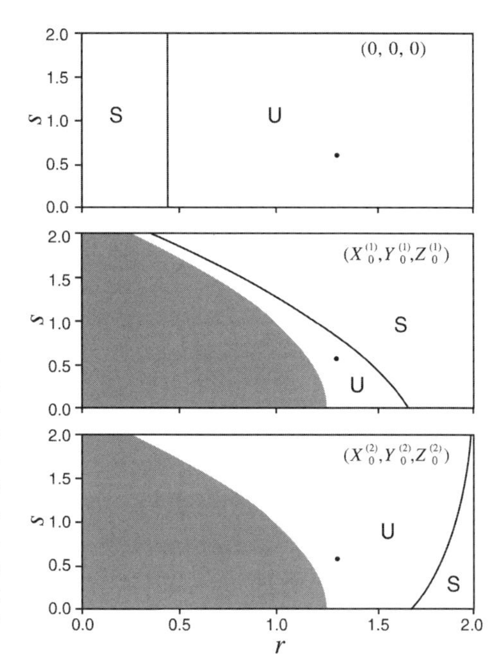

Figure 15-3 Stability of three possible equilibrium points of the unforced system [Eqs. (15.6)–(15.8)], as a function of the parameters r and s, for assigned values $p = 1.0$, $q = 2.5$, and $v = 0.2$. Top, middle, and lower panels show the stable (S) and unstable (U) regions corresponding to the equilibria at $(X, Y, Z) = (0, 0, 0)$, (X_0^1, Y_0^1, Z_0^1), (X_0^2, Y_0^2, Z_0^2), respectively; the shaded region denotes the presence of only a single equilibrium at (0, 0, 0). The dot represents the point in r–s parameter space for the special solution described in this section.

Chapter 6). With this information we are in a position to find the regions of parameter space that can yield time-dependent solutions compatible with the observations (e.g., ~100-ky-period oscillations).

As shown in Fig. 15-3 we find by numerical experiment that for the chosen values, $p = 1.0, q = 0.25$ (i.e., $K_\theta^{-1} = 4$ ky), and $v = 0.2$, there exists a range of values of r and s for which unstable equilibria (and hence the possibility of free oscillatory behavior) exist. Note that for relatively larger values of r and s multiple stable equilibria exist. By further numerical experimentation we find that free oscillations indeed tend to exist for all values of r and s not admitting a stable equilibrium that can serve as an attractor. The region of oscillatory behavior, and the period of the oscillation, are shown in Fig. 15-4; note in particular the existence of regions of r–s parameter space where near-100-ky-period free oscillations are possible. More generally, it is possible to map out all of these stability properties as a four-dimensional function of p and q as well as r and s.

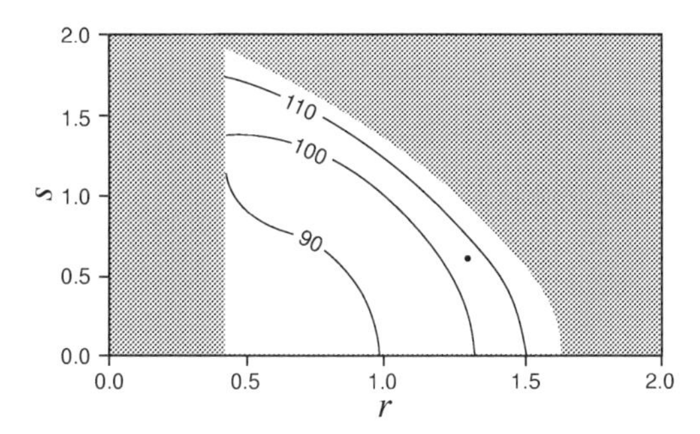

Figure 15-4 Region of r–s parameter space for which free oscillatory solutions exist. Numbered lines denote the period of the solution in units of 10^3 y (ky). The dot represents the parameter values chosen for our special solution, as described in the text.

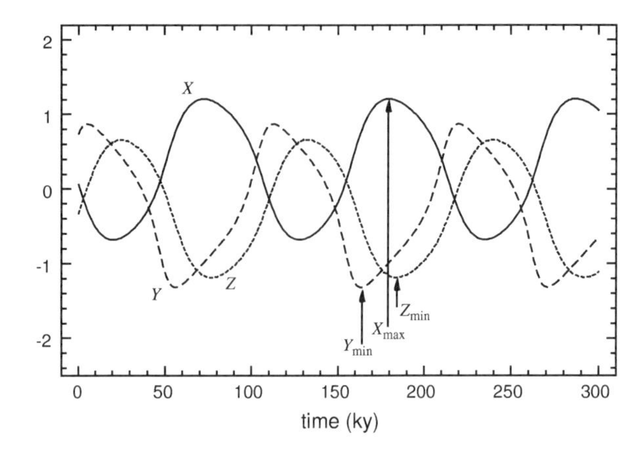

Figure 15-5 Free (unforced) solution of the dynamical system [Eqs. (15.6)–(15.8)] for the parameters given, showing the concomitant oscillations of ice (X), carbon dioxide (Y), and ocean temperature (Z), having a period of $\sim 10^5$ ky, and their relative phases.

In Fig. 15-5 we show the last 300 ky of the free solution for X, Y, and Z obtained by setting $u = 0$ starting from arbitrary initial conditions at 1 Ma. This solution corresponds to the parameter values $r = 1.3$ and $s = 0.6$ (shown by the dot in Figs. 15-3 and 15-4). As expected, the free solution is characterized by an oscillation of near-100-ky period of the ice mass (proportional to X), coincident with phase-lagged variability of both carbon dioxide (Y) and deep ocean temperature (Z).

15.3 MILANKOVITCH FORCING OF THE FREE OSCILLATION

We next force the above free periodic system by the known Earth-orbital insolation variations ($\Delta R \neq 0$) to determine the extent to which this forcing can bring the solution closer to the observations, particularly with regard to the phase (or chronology) of the solution (e.g., the transition to the present interglacial state from the glacial state at roughly 20 ka).

In Fig. 15-6a we show the nondimensionalized variations of incoming July solar radiation at 65°N due to orbital variations, which is assumed to be the main external radiative forcing acting on the system over the Late Pleistocene. The response of the system to this forcing, for $u = 0.5$ [with all other parameters held fixed at the previously prescribed values ($p = 1.0$, $q = 2.5$, $r = 2.5$, $s = 0.6$, $v = 0.2$) that led to the free 100-ky-period oscillations shown in Fig. 15-5], is shown for 500 ky in Figs. 15-6 (b, c, and d for X, Y, and Z, respectively). For comparison, the SPECMAP (Imbrie *et al.*, 1984) δ^{18}O proxy for ice mass is shown by the dashed curve in Fig. 15-6b, and the Vostok estimate of atmospheric carbon dioxide is shown by the dashed curve in Fig. 15-6c. It is seen (1) that under the "pacemaker" influence of the orbital forcing a fairly realistic phase-lock is imposed on the free solution for at least the past four glacial terminations, (2) that some of the near-20- and 41-ky fluctuations in the data now appear in the solutions, and (3) that the observed tendency toward relatively

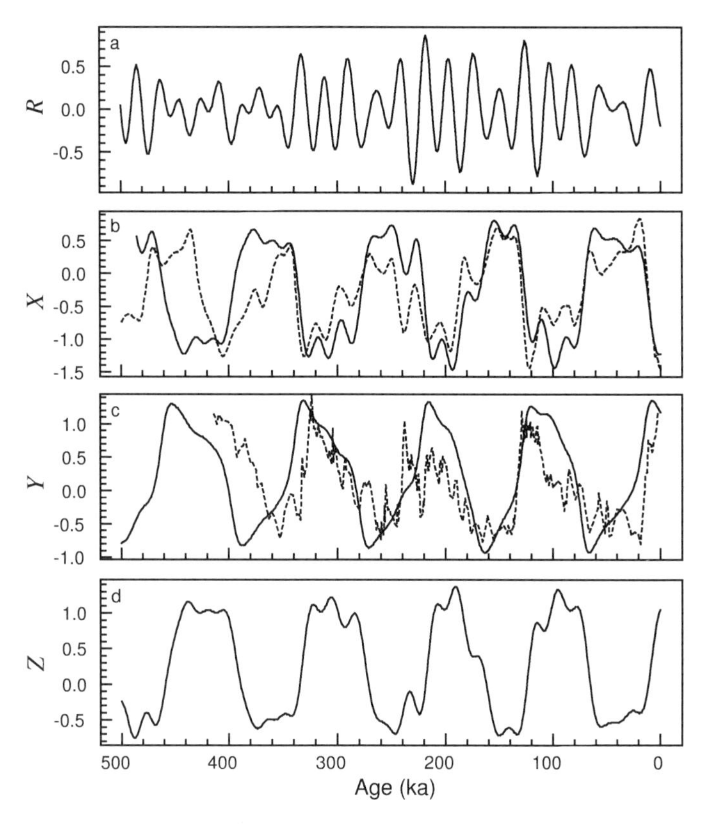

Figure 15-6 Solution of the dynamical system [Eqs. (15.6)–(15.8)] for the past 500 ky for the same parameter values used to obtain the free solution shown in Fig. 15-4, but now subject to the Earth-orbital radiative forcing shown in part a. The responses for ice (X), carbon dioxide (Y), and ocean temperature (Z) are shown in parts b, c, and d, respectively. For comparison, the dashed curve in b is the SPECMAP $\delta\,^{18}O$ estimate of ice variations and the dashed curve in c is the Vostok core estimate of CO_2 variation.

rapid transitions is now exhibited, which goes along with (4) a reduction of the phase difference between CO_2 and ice mass on the 100-ky time scale, in better agreement with the observations (e.g., Sowers *et al.*, 1991). In general, the fit to the SPECMAP and Vostok data is fairly good except in the fifth cycle, but with further considerations this fit can be improved (see below). Our goal here, however, is not to achieve a perfect fit but merely to achieve one that is robust and qualitatively credible, and that we hope will continue to be valid on this basis in the face of inevitable improvements in observations based, for example, on further ice and sediment cores.

From Fig. 15-6 we find that the ranges of X, Y, and Z are roughly $\delta X = 2.3$, $\delta Y = 2.3$, and $\delta Z = 2.1$, respectively. We assume that these ranges correspond to the geologically inferred ranges of I, μ, and θ (i.e., $\delta\Psi \approx 4.6 \times 10^{19}$ kg, $\delta\mu \approx 100$ ppm, and $\delta\theta \approx 2$°C), and that the tectonic equilibrium values ($\widehat{\Psi}^*$, $\widehat{\mu}^*$, and $\widehat{\theta}^*$) applying at about 250 ka are $\widehat{I}^* = 4.9 \times 10^{19}$ kg, $\widehat{\mu}^* = 253$ ppm, and $\widehat{\theta}^* = 5.2$°C. Taking a GCM-derived value, $\widehat{\mathcal{B}} = 11$°C, it then follows that $\lambda_0 = K_I^{-1} = 10$ ky, $\lambda_1 = 2.0 \times 10^{19}$ kg,

$\lambda_2 = 52.5$ ppm, and $\lambda_3 = 0.9°$C, from which we can then calculate all the coefficients of Eqs. (15.1)–(15.4): $a_1 = 8.7 \times 10^{14}$ kg (°C y)$^{-1}$, $K_I = 1.0 \times 10^{-4}$ y^{-1}, $b_1 = 1.3 \times 10^{-4}$ y^{-1}, $b_2 = 1.1 \times 10^{-6}$ (ppm y)$^{-1}$, $b_3 = 3.6 \times 10^{-8}$ (ppm^2 y)$^{-1}$, $b_\theta = 5.6\times10^{-3}$ ppm (°C y)$^{-1}$, $c_I = 1.2\times10^{-23}$ °C (kg y)$^{-1}$, and $K_\theta = 2.5\times10^{-4}$ y^{-1}. The values chosen for u (= 0.5) and v (= 0.2) imply reasonable values of the sensitivity coefficients $k_\theta = 0.5$ and $k_R = 0.2$ as estimated by GCM experiments.

15.4 STRUCTURAL STABILITY AS A FUNCTION OF THE TECTONIC CO_2 LEVEL

In this model the only external "control" variable that can regulate the onset of this near-100-ky-period cycle is the tectonic CO_2 forcing F_μ, which can change the value of $\widehat{\mu}$. We now determine the structural stability of the free solutions just obtained [i.e., the changes in the equilibria and their "internal stability properties;" see Ghil and Childress (1987) and Chapter 6] as a function of $\widehat{\mu}$. This would establish the capability of the model to admit bifurcations into the different regimes exhibited by the Late Cenozoic record (see Fig. 1-3). To obtain the equilibria we solve the steady-state form of Eqs. (15.2), (15.4), and (15.5), continuing to neglect orbital variations in radiative forcing and calving castrophes ($\Delta R = \mathcal{C}_I = 0$), which now represent perturbations of the equilibria (with bedrock depression D_B decoupled from the system).

Using the notation established in Section 5.3 (i.e., $x = x_0 + x'$, where x_0 is the steady-state value) we have the steady-state equilibrium set,

$$-a_1 \frac{\mathcal{B}^{(\Psi)}}{\widehat{\mu}} \xi_0 + k_\theta^{(\Psi)} \vartheta_0 - K_I \zeta_0 = 0 \tag{15.9}$$

$$b_1(\widehat{\mu})\xi_0 - b_2(\widehat{\mu})\xi_0^2 - b_3\xi_0^3 - b_\theta\vartheta_0 = 0 \tag{15.10}$$

$$-c_I\zeta_0 - K_\theta\vartheta_0 = 0 \tag{15.11}$$

This set can be combined to obtain the fundamental cubic equation governing ξ_0 as a function of $\widehat{\mu}$,

$$b_3\xi_0^3 + b_2(\widehat{\mu})\xi_0^2 - A_1(\widehat{\mu})\xi_0 = 0 \tag{15.12}$$

where

$$A_1(\widehat{\mu}) = \left[b_1(\widehat{\mu}) - \frac{b_\theta a_1 c_I \mathcal{B}}{\widehat{\mu}_1(K_I K_\theta - a_1 c_I k_\theta)} \right]$$

From the solution of Eq. (15.12) for $\xi_0(\widehat{\mu})$ we can obtain $\zeta_0(\xi_0)$ and $\vartheta_0(\xi_0)$, i.e.,

$$\zeta_0(\xi_0; \widehat{\mu}) = \frac{a_1 \widehat{\mathcal{B}} \xi_0(\widehat{\mu})}{\widehat{\mu}(K_I - a_1 k_\theta c_I K_\theta^{-1})} \tag{15.13a}$$

$$\vartheta_0(\xi_0; \widehat{\mu}) = c_I \zeta_0(\xi_0; \widehat{\mu}) K_\theta^{-1} \tag{15.13b}$$

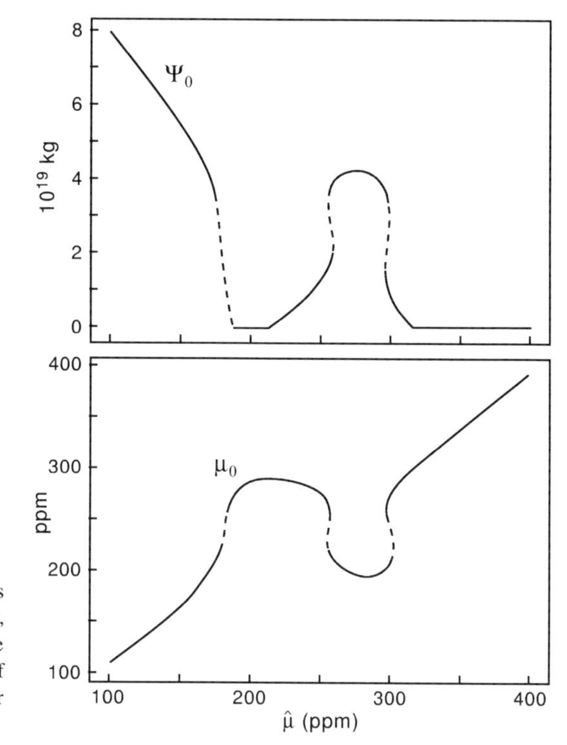

Figure 15-7 Equilibrium of ice mass Ψ_0 and carbon dioxide μ_0 ($= \widehat{\mu}_0 + \xi_0$), as a function of tectonically forced value of carbon dioxide μ_0. Dashed portions of curves denote unstable equilibria. After Saltzman and Verbitsky (1993).

A solution for a modified form of this system is given in Saltzman and Verbitsky (1993), given coefficient values similar to those established above as necessary to achieve the main Pleistocene 100-ky-period oscillatory regime. The solutions for $\mu_0 = \widehat{\mu} + \xi_0(\widehat{\mu})$ and $\zeta_0(\widehat{\mu})$ are shown in Fig. 15-7. In this figure, the solid portions of the curves indicate stable equilibria and the dashed portions indicate unstable equilibria obtained by a standard eigenvalue analysis (see Chapter 6). The transitions between these portions represent critical points where basic changes (bifurcations) in the behavior of the system occur.

Note that at the S-shaped dashed curve near $\widehat{\mu} = \widehat{\mu}^*$ ($= 253$ ppm) three unstable equilibria exist, as required to admit the free oscillation. However, the $\widehat{\mu}$ range in which these three unstable equilibria exist is quite small, suggesting that the 100-ky-period free oscillation might be a transient regime (as indeed it appears to be in the longer term record). In fact, it is shown in Saltzman and Verbitsky (1993) that if we apply the time-dependent model with $\widehat{\mu}$ fixed at $\widehat{\mu}^*$, including also the possibilities for calving instabilities ($C_{\mathrm{I}} \neq 0$) as well as Earth-orbital forcing (ΔR), we obtain a chaotic record of intermittent transitions between three modes: the nearby upper (cold-mode) and lower (warm-mode) stable regimes and the near-100-ky-period oscillatory mode (see Fig. 15-8). These represent temporary episodes of periodic order amid chaos.

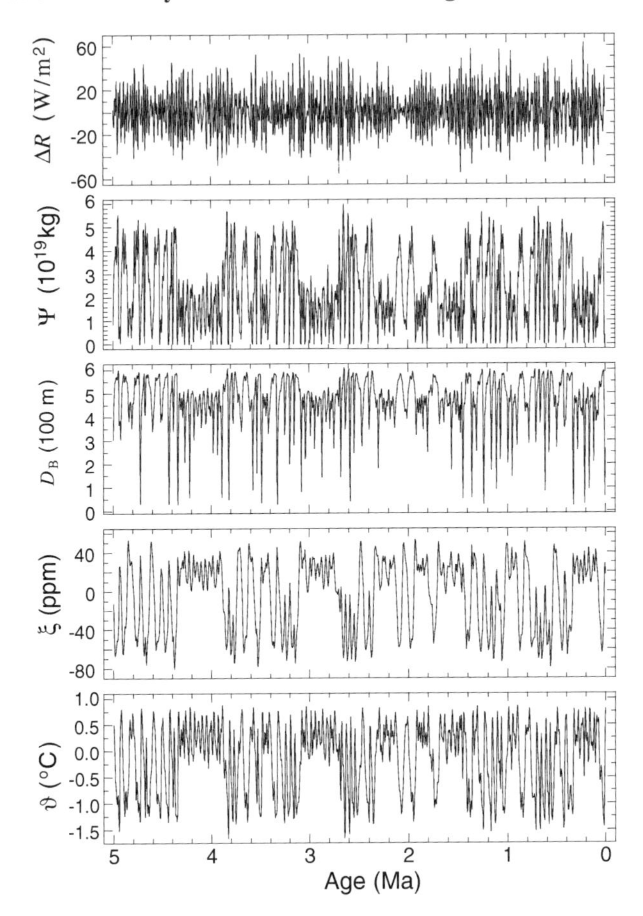

Figure 15-8 Top panel: variation of July insolation at 70°N due to Earth-orbital (Milankovitch) changes over the past 5 My (Berger, 1978a,b). Succeeding panels: response of the climatic dynamical system to the above radiation forcing for parameter values listed in Tables 15-1 and 15-2, with $\mu = \widehat{\mu}$, in the order of ice mass (Ψ), bedrock depression (D_B), carbon dioxide departure (ξ), and mean ocean temperature departure (ϑ), respectively.

15.5 A MORE COMPLETE SOLUTION

If, instead of fixing the tectonically forced value of carbon dioxide at the constant value $\widehat{\mu}^*$ (as in the three-mode intermittent solution shown in Fig. 15-8) we allow slow changes in $\widehat{\mu}$, the possibility arises that the structural changes exhibited in Fig. 15-7 can result in a temporal separation of the three separate modes in a manner that can replicate the observations shown in Fig. 1-4.

Because we can only make very speculative inferences regarding the effects of volcanic/metamorphic outgassing and weathering downdraw over the past 5 My (e.g., Walker *et al.*, 1981; Berner, 1991; Raymo and Ruddiman, 1992), we have asumed that $\widehat{\mu}$ has the simplest linear form within the bounds of such speculations that can lead to the sequence of bifurcations observed in the record, Eq. (15.1). In particular, we

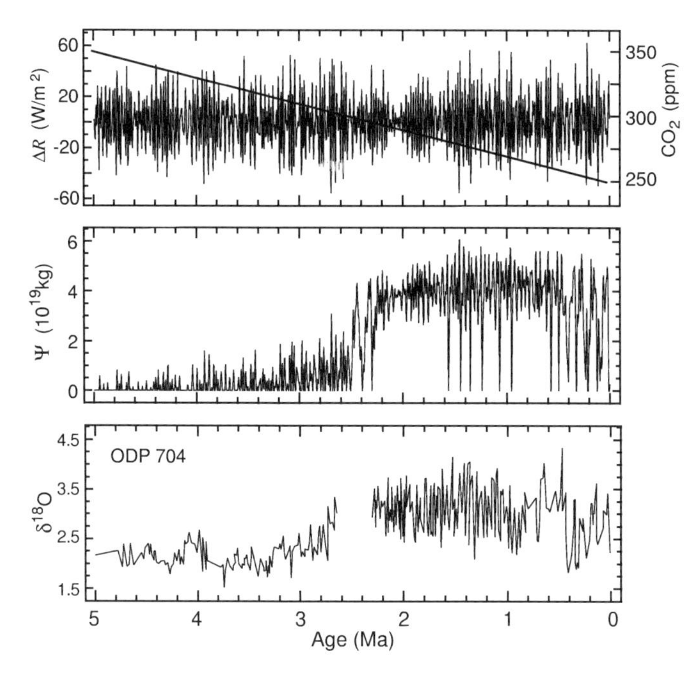

Figure 15-9 Time-dependent solution for ice mass over the past 5 My of the dynamical system (middle panel), when forced by both the orbital radiation variations and an assumed linear, tectonic, decrease in CO_2, shown in the top panel, compared with a $\delta\,^{18}O$ record at ODP 704, as discussed by Hodell (1993), shown in the bottom panel.

assume that $\widehat{\mu}$ decreases linearly by 100 ppm over the past 5 My, from 350 ppm at $t = 5$ Ma to a present value $\widehat{\mu}(0) = 250$ ppm, i.e., $\widehat{\mu}(t) = [350 \text{ ppm} - 20 \text{ (ppm/My)}]$ (see Fig. 15-9, top).

In Fig. 15-9 (top) we show the prescribed Milankovitch radiative changes at 70°N, ΔR (Berger, 1978a,b, Berger and Loutre, 1991), which together with $\widehat{\mu}(t)$ represent the two external forcing functions imposed in our model. Assigning the set of coefficients and parameters listed in Table 15-2, we obtain the 5-My solution of the system shown in Fig. 15-9 for the variations of the ice mass (Ψ), compared with the $\delta\,^{18}O$ record at the South Atlantic core 704 (Hodell, 1993). We note, however, that other $\delta\,^{18}O$ records, which might be more representative of ice variations than core 704, display a lower mean value of $\delta\,^{18}O$, between 2.5 and 0.5 My BP, and higher peaks of $\delta\,^{18}O$ during the 100-ky cycles of the past 0.5 My (e.g., core 806 shown in Fig. 1-4). Nonetheless, all $\delta\,^{18}O$ ice records show the existence of three distinct regimes over the past 5 My, each separated by relatively clear transitions (bifurcations), and each characterized by different average values, frequencies, and amplitudes; for all these records it is not unreasonable that at least seven free parameters would be necessary to replicate the complex variability of Ψ alone. In addition, however, the full solution predicts the evolution of CO_2 (μ) and the ocean state variable (ϑ) over the past 5 My (Fig. 15-10) (to be discussed in Section 15.6).

Table 15-2 Assigned Values of Free Parameters

Parameter	Value
a_1	8.0×10^{14} kg (°C y)$^{-1}$
β_1	6.12×10^{-3} y^{-1}
β_2	2.6×10^{-5} (ppm y)$^{-1}$
β_3	3.6×10^{-8} (ppm^2 y)$^{-1}$
b_θ	5.9×10^{-3} ppm (°C y)$^{-1}$
c_I	1.2×10^{-23} °C (kg y)$^{-1}$
α_B	20 m y^{-1} (Pollard, 1983a,b)
Z_0	475 m
$d\widehat{\mu}/dt$	20 ppm y^{-1}

The solution shown in Fig. 15-9 demonstrates that the imposition of the slow tectonic forcing of CO_2 transforms what would otherwise be the chaotic, intermittent, solution shown in Fig. 15-8 into an organized sequence of clearly defined regimes separated by well-defined transitions (bifurcations). The ice regimes and transitions have a close correspondence with the observed IRD and $\delta\,^{18}$O behavior, particularly the major regime changes near 2.5 and 0.5 Ma (later in this solution than the near-0.9-Ma

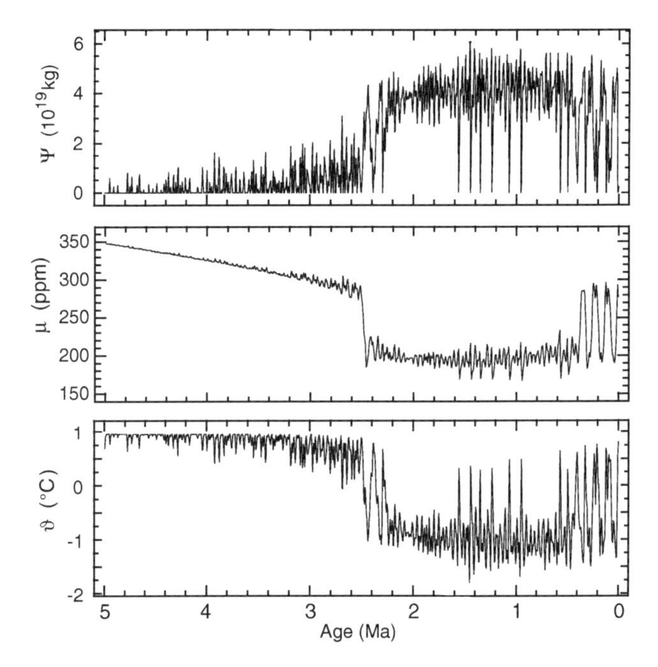

Figure 15-10 Predicted variations of CO_2 (μ) and ocean temperature (ϑ) accompanying the solution for ice mass shown in Fig. 15-9 and repeated in the top panel here.

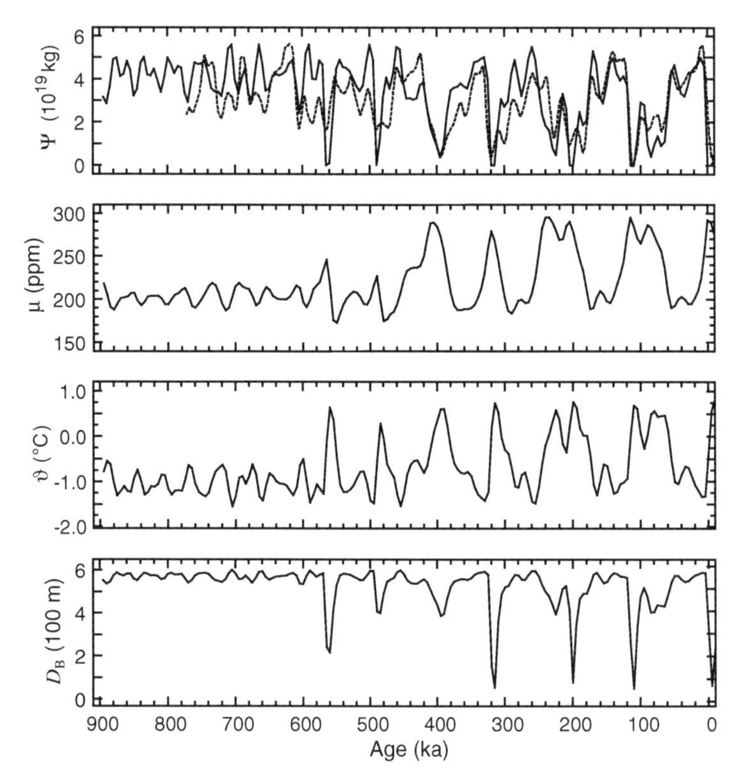

Figure 15-11 More detailed view of the solution shown in Fig. 15-10 over the past 900 ky for Ψ, μ, ϑ, and D_B subject to the radiative forcing as well as the slow tectonic decrease in CO_2 shown in the top panel of Fig. 15-9.

observed transition), which separate a very low-amplitude Milankovitch forced response in the Early Pliocene (5–2.5 Ma), a higher amplitude of ice fluctuations in the Late Pliocene/Middle Pliestocene (2.5–0.5 Ma) resulting from a combination of Milankovitch forcing and oscillations due to calving instabilities, and the Late Pleistocene (0.5–0 Ma) in which the major 100-ky-period oscillation driven by the carbon cycle instability emerges as the dominant variability (influenced also by calving instabilities). The spectra for these separate regimes agree fairly well with similar spectra obtained for the $\delta^{18}O$ variations of DSDP site 552 (Shackleton *et al.*, 1988). Note that with this model the glacial onset $\sim$2.5 Ma is driven by the slow tectonic decrease of CO_2, and hence does not invoke the closing of the Isthmus of Panama, the effect of which has been questioned (see Section 3.2).

As in Saltzman and Verbitsky (1993), in Fig. 15-11 we present in more detail the solution over the past 1 My, showing with higher resolution the variations of ice mass (Ψ) given in Fig. 15-11, compared with the SPECMAP reconstruction of Ψ (Imbrie *et al.*, 1984), and in Fig. 15-12 we present in even greater detail the past 400 ky of the solution, showing insolation forcing (ΔR), the variations of ice mass (Ψ) compared with the SPECMAP reconstruction, and the variations of carbon dioxide (μ) compared with

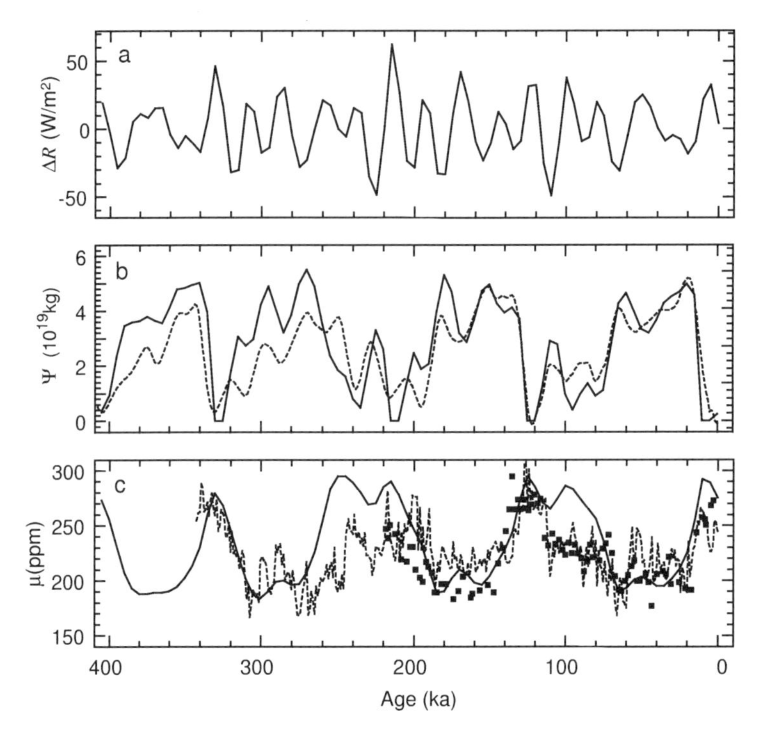

Figure 15-12 (a) Variations of July insolation at 70°N due to Earth-orbital forcing over the past 400 ky. (b) Expanded view over the past 400 ky of the time-dependent solution for ice mass, Ψ (solid curve), compared with the SPECMAP $\delta^{18}O$ reconstruction, scaled in terms of ice mass (dashed curve). (c) Time-dependent solution for total carbon dioxide μ (solid curve), compared with Vostok CO_2 data of Barnola *et al.* (1987) (squares) and the Shackleton and Pisias (1985) $\delta^{13}C$ estimates of CO_2 (dashed curve), which are scaled to the Vostok data.

both the Vostok (Barnola *et al.*, 1987) and Shackleton and Pisias (1985) estimates. A considerable amount of agreement is found between the solution for Ψ and μ and the observations, and, in particular, the phase difference between μ and Ψ is in good agreement with recent estimates (Sowers *et al.*, 1991; Shackleton *et al.*, 1992). Figure 15-12 is particularly instructive, showing that the longest period (~100 ky) fluctuations are associated with the internally generated CO_2 fluctuations while the phase and the shorter period (~20/40 ky) fluctuations are associated with the externally imposed Earth-orbital radiative variations and with calving instability (cf. Rooth *et al.*, 1978). This is even more clearly illustrated in Fig. 15-13, which shows the solution trajectory in the $(\Psi - \mu)$ phase-plane (solid line) compared with the observed trajectory (dashed), given previously in Fig. 8-2. In essence, this model, in which internal CO_2 variations play a vital role in providing the terminations of the main near-100-ky cycle, explains the paradox that this 100-ky cycle is most robust at times when the Milankovitch forcing, represented by the 400-ky envelope of amplitude, is a minimum (see Fig. 1-7), e.g., around 400 ka (MIS 11) and at present (MIS 1).

In the solution for Ψ we also find the signature of calving instabilities that contribute to the rapid deglaciation that terminate the Laurentide and Fennoscandinavian

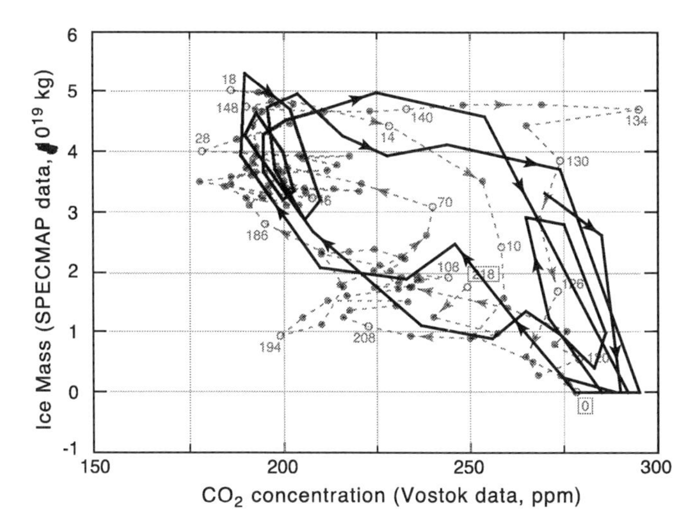

Figure 15-13 Trajectory of the solution in the phase plane of global ice mass and CO_2 (solid curve), compared with the observed trajectory shown in Fig. 8-2.

ice sheets. These instabilities cause an earlier collapse of ice mass than would be predicted by the insolation and CO_2 variations alone, leading to an improved solution for these deglaciations than is obtained in the absence of a bedrock-calving mechanism.

15.6 PREDICTIONS

As noted in Section 5.1, in addition to accounting for the known observations, a theory must contain some predictions that are, in principle, observable or verifiable, and that should guide the search for new knowledge. In our case, we predict the long record of carbon dioxide shown in Fig. 15-10 that extends backward in time beyond the 420-ky Vostok record, as well as the accompanying deep ocean state record over this longer period. We can expect that geological and geochemical evidence will become available to determine these longer term CO_2 and ocean state variations, placing the theory "at risk" in these regards. Note that the free parameters were tuned to account for the qualitative behavior only of the long-term δ^{18}O-derived global ice mass record shown in Fig. 1-4.

It may be of interest in connection with our remarks in Section 14.6 that in our earlier studies of this ice-age problem it was predicted that CO_2 should play an important role in generating oscillatory behavior (Saltzman and Moritz, 1980; Saltzman *et al.*, 1981). Upon verification that CO_2 indeed varied along with the ice oscillations (Delmas *et al.*, 1980; Berner *et al.*, 1980) this new CO_2 data instantaneously became a part of the "known" observations to be accounted for by any model purporting to govern the ice-age variations (see Fig. 1-5). The comparison of the CO_2 solution obtained by Saltzman and Maasch (1990) with the Vostok CO_2 measurements going back 420 y is shown in Fig. 15-14. Another major prediction, yet to be verified, is the

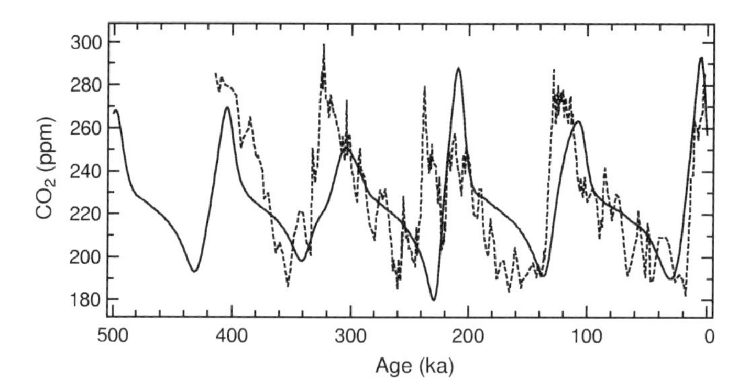

Figure 15-14 Comparison of the CO_2 solution obtained by Saltzman and Maasch (1990) (solid curve), with the Vostok CO_2 measurements going back 420 ky (Petit *et al.*, 1999) (dashed curve).

rapid decrease of CO_2 near the onset of Northern Hemisphere glaciation at ~2.5 Ma, followed by an extended period of low CO_2 until the onset of the major 100-ky oscillation in the Late Pleistocene (see Fig. 15-10). The comparison of the θ solution with the observations of T_S (K708-1), taken as a proxy of θ in accordance with Section 8.3 (Fig. 8-4), is shown in Fig. 15-15.

Another class of predictions from a model (that may, however, be unverifiable) is the projected response of the model to changes in external forcing (e.g., structural stability; see Chapter 6). It is possible to account, at the same level of accuracy, for the same observations with two models that have different sensitivities. As one example of this kind of prediction, in Fig. 15-16 we show the results of neglecting Milankovitch forcing altogether in the model discussed in this chapter, replacing it with stochastic forcing. We see that, for this model under these circumstances, the same general character of bifurcations and transition to the 100-ky-period oscillation would be obtained due to the CO_2 effect.

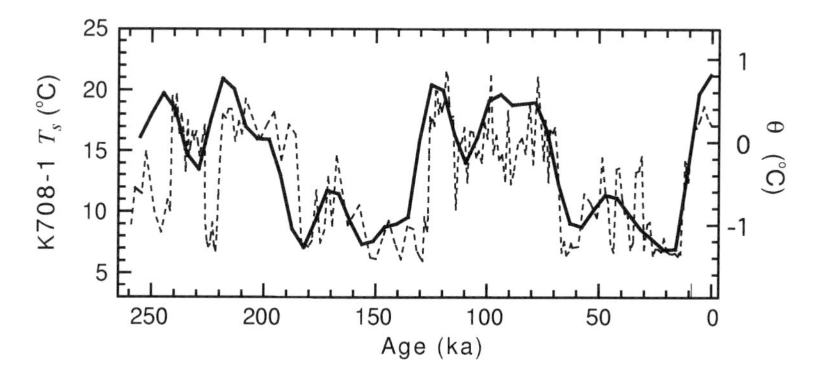

Figure 15-15 Time evolution over the past 250 ky of the proxy measure of the ocean state, T_S (K708-1) shown in Fig. 8-4 (dashed line), compared with the solution for θ shown in Fig. 15-11 (solid line).

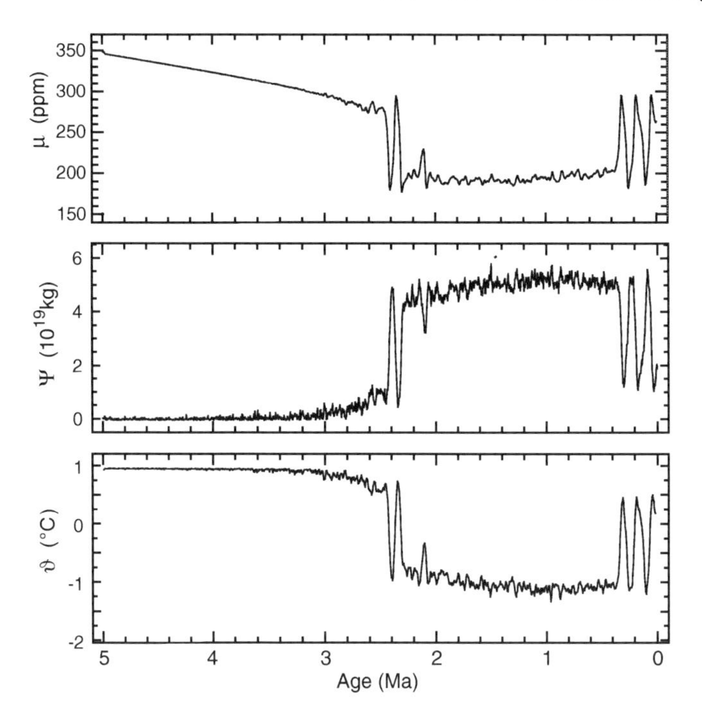

Figure 15-16 Solution over 5 My for CO_2 (μ), ice mass (Ψ), and ocean temperature departure (ϑ) when Milankovitch Earth-orbital forcing is replaced by stochastic forcing.

15.7 ROBUSTNESS AND SENSITIVITY

Another important goal of any theory is not only to reduce the number of free parameters needed to explain a given body of observations, but also to widen the range of values that such parameters can exhibit within which the theory is still valid. Moreover, because all systems are subject to the presence of noise, it is important that the solution be able to survive this presence; the possibility also exists that such noise is a necessary component of the theory, for example, by providing the mechanism for transition between alternate stable equilibria as in the stochastic resonance theory (e.g., Benzi *et al.*, 1981; Nicolis, 1982) (see Section 7.6).

Thus, a sensitivity analysis of the solution to variations in parameter values and levels of stochastic perturbation is an integral part of any complete theory. At this time only a limited amount of systematic objective study of this aspect of the theory has yet been done. This applies as well to the model described above, though some elementary aspects of the formalism required were discussed and applied by Saltzman and Moritz (1980), Saltzman and Sutera (1984), and Saltzman *et al.* (1984), and to some extent also in the later papers dealing with this model.

More recently Monte Carlo chain methods have been proposed to optimize the values of the free parameters with respect to the observations, also providing estimates of the confidence intervals for these values. In general a determination of those parameters (rate constants) to which the solution is most sensitive will expose those parts of the theory that most require further study and improvement from a physical viewpoint.

15.8 SUMMARY: A REVIVAL OF THE CO_2 THEORY OF THE ICE AGES

We have constructed a dynamical system model of the Northern Hemisphere ice-sheet and bedrock evolution and concomitant CO_2 and ocean state variations. With the assignment of nine free parameters, this system can simultaneously account qualitatively for the main regime transitions over the past 5 My and for much of the detailed behavior over the past few hundred thousand years from a single set of constraints. The model is forced externally only by known orbitally induced variations of summer insolation in high Northern Hemisphere latitudes and by an assumed slow variation of tectonic forcing of CO_2, forming a consistent first-order theory of the internal mechanisms governing the major climatic fluctuations over this extended period. In essence, this model represents a unification of the two major theories of the ice ages: the CO_2 theory (in which longwave radiation is altered by the greenhouse effect) and the Milankovitch theory (in which the distribution of shortwave radiation is altered by Earth-orbital changes), supplemented by a new third major theory resting on the possible role of internal instability.

When these slow-response "carriers" of the long-term climatic evolution are specified, the accompanying distributions of the fast-response atmospheric and surface-state variables can be deduced from an established general circulation model, which, as the best available model of these fast variables, constitutes the final component of a theory of the full climate (see Fig. 5-4).

In more descriptive terms, the sequence of processes implied by the model is as follows:

1. Tectonic forcing, including volcanism, metamorphism, and weathering processes, induces long-term variations in CO_2 that establish a low mean value of CO_2 over the Late Pleistocene (e.g., Chamberlin, 1899; Walker *et al.*, 1981; Berner, 1991; Raymo and Ruddiman, 1992). This state favors cold temperatures and ice formation.
2. When CO_2 reaches a low enough level, ice sheets begin to form over North America and Eurasia, which are linearly influenced by Earth-orbital (Milankovitch) variations on the near-20- and 40-ky-period time scales through their direct effect on high-latitude temperature, giving rise (with an inertial phase lag) to the oscillations of these periods observed in the record (Imbrie *et al.*, 1992).
3. The slow growth of ice sheets that accompanies the general cooling leads to a reduction of sea level, enhanced cold deep water production, and a shallowing and reduction in the latitudinal extent of the thermocline.
4. At a critical point, all of the above changes activate the positive feedbacks inherent in the oceanic carbon pumps, reducing the partial pressure of CO_2 in the ocean surface layer significantly and hence in the atmosphere, on a relatively short time scale. A new quasi-equilibrated state is achieved, characterized by low CO_2, large ice sheets, and a cold ocean.

5. The shallow thermocline of this new state and relatively strong thermohaline circulation promote a net leakage of CO_2 back to the atmosphere that overcomes the continuing slow tectonic CO_2 decrease, gradually raising the atmospheric CO_2 level, warming the surface, and melting ice.

6. A stage is reached at which the slow rise in CO_2 returns the system to the critical point at which the ice decrease, sea level rise, and general warming deactivate the carbon pumps, causing atmospheric CO_2 to rise sharply, terminating the main ice sheets. At the same time, however, the warming ocean (thermocline deepening) begins to restrict the release of CO_2 to the atmosphere—the whole system cannot reach a stable equilibrium and, given the time constants of the ice sheets and deep ocean, a near-100-ky oscillation is generated, featuring the following cyclical sequence of processes:

(a) Starting, arbitrarily, at a time of maximum continental ice mass (Ψ large), such a glacial state is associated with extensive sea ice formation and cooling of adjacent ocean water that promotes the growth of the cold-mode ocean, shallowing the thermocline and diminishing its poleward extent (decreasing θ).

(b) In this cold-mode state, a more symmetric thermohaline circulation, more energetic surface winds, and a shallow thermocline promote the upwelling of carbon-rich waters, particularly in low latitudes but also in higher latitudes where local vertical turbulent mixing is enhanced.

(c) The increasing atmospheric CO_2 values, in turn, initiate a decrease of ice (start of the deglaciation phase).

(d) The increase in CO_2 is further accentuated by positive feedbacks, an obvious one being due to the warming of the ocean surface and of continental land masses, which tend to release CO_2 to the atmosphere. Other examples of such positive feedbacks are due to the rise in sea level associated with the melting of ice, which is accompanied by worldwide coral reef growth growth and a general increase in the ratio of calcareous to siliceous plantonic productivity (Archer and Maier-Reimer, 1994), further increasing atmospheric CO_2 (Berger, 1982; Opdyke and Walker, 1992). In addition, rising sea level promotes the removal of nutrients by deposition on flooded continental shelves (Broecker, 1982), which goes along with a general decrease in oceanic productivity during interglacial conditions (Maslin *et al.*, 1995), both of which reduce the strength of the "biological CO_2 pump."

(e) In the face of sharply increasing CO_2 levels and bedrock/calving processes associated with rising sea level, a rapid collapse of the remaining ice caps occurs, leading ultimately (perhaps after a brief episode of meltwater-induced cooling in the North Atlantic, e.g., the Younger Dryas) to the warming of the oceans as a whole and reestablishment of a more salinity-driven circulation similar to the present one.

(f) This "warm-mode" ocean state, accompanying reduced glaciation, is characterized by a deepening and more extensive thermocline, and weak surface winds, inhibiting the efflux of CO_2 from the oceans and decreasing the atmosphere CO_2 level.

(g) The decreasing values of CO_2, in turn, promote the regrowth of glaciers, returning the climate system to phase (a) above.

(h) The chronology for this 100-ky-period cyclic behavior is set by the near-100-ky-period envelope of Milankovitch forcing, which imposes a "phase-lock" on the cycle by constructive interference of the maxima (and minima) in high-latitude incoming radiation of near-20- and 40-ky periods with maxima (and minima) of CO_2.

In summary, a set of values of the unknown parameters (rate constants) of the general system, Eqs. (15.2)–(15.5), has been determined and can account for the observations in a physically plausible way by assuming that positive feedbacks in the carbon cycle can provide the instability necessary to drive a free oscillation. As a side consequence of this theory, predictions of the longer term (5-My) behavior are made that include the mechanism for the onset of the ice ages at about 2.5 Ma and the initiation of the strong near-100-ky-period oscillation about 0.9 million years ago. In agreement with the conclusion expressed by Imbrie *et al.* (1992), it would appear that the near-20- and 41-ky-period variations are linearly forced by Earth-orbital changes. However, we suggest more strongly than Imbrie *et al.* (1993) that the main 100-ky-period oscillation is internally driven by an instability, probably residing in the behavior of the carbon cycle (Plass, 1956), with Milankovitch forcing playing the more secondary role of setting the phase of this oscillation. Thus, according to this theoretical model, the answer to the question, "Would we have ice ages if there were no Earth-orbital variations?" is a clear "Yes." It is implied by this model that the "cause" of the ice ages is the tectonic reduction of CO_2 to a low enough threshold (bifurcation) value to permit the growth of the Northern Hemisphere ice sheets and ultimately the internal excitation of the main free oscillator. We thereby return to the essential elements of the carbon dioxide theory of the ice ages, the history of which was reviewed in Section 14.6.

We must be aware, however, that the theory we have described constitutes neither a unique nor complete explanation of the ice ages, even for its global-mean aspects. It is certainly possible that other scenarios involving different mixes of internal ingredients and instabilities may provide alternate explanations of the same phenomena. For example, the possibility cannot be ruled out that a seat of instability might also lie in the thermohaline behavior of the ocean, and other plausible models might be constructed to illustrate this. Nonetheless, we think it is fair to say that it has yet to be demonstrated that any other model can explain as much of the observed variability over the past 5 My (including the main bifurcations) with fewer free parameters. Each of the succession of the reference papers, leading up to the most recent contributions, have in fact aimed at either reducing the number of free parameters or increasing the amount of variance explained.

In the next chapter we discuss the dynamics of paleoclimatic variability on time scales shorter than 20 ky, down to about 1 ky (i.e., millennial-scale oscillations).

16

MILLENNIAL-SCALE VARIATIONS

In Sections 3.4–3.6 we provided an overview of some of the variability that occurred on 1- to 10-ky time scales, the most notable of which have been identified as Heinrich (H) oscillations (6- to 10-ky period) and Dansgaard–Oeschger (D-O) oscillations (1- to 3-ky period) (see Figs. 3-7 and 3-9). The H-oscillations are clearly associated with the presence of ice sheets, their main signature being the ice-rafted debris in sediment cores derived from icebergs that are massively discharged from the ice-sheets peripheral to the North Atlantic (mainly the North American Laurentide). On the other hand, the D-O temperature oscillations recorded in the Greenland and Antarctic ice cores occur in the Holocene, i.e., even in the absence of the Laurentide and Scandinavian ice sheets. A striking example of a millennial-scale oscillation that might be identified with both H and D-O oscillations is the Alleröd/Bölling–Younger–Dryas sequence that occurred within the span of a few thousand years (see Fig. 3-7b). A schematic representation of the two scales of variability represented by H and D-O events are shown by the two minor peaks on the short-period flank of the main glacial–interglacial maximum in the idealized spectrum given in Fig. 1-2.

A longer term record of millenial-scale (1–12 ky) variations of North Atlantic sea surface temperature (SST), compared to that shown in Fig. 3-6, going back 500 ky, is shown in Fig. 16-1, along with the associated record of ice volume as measured by $\delta^{18}O$ (McManus *et al.*, 1999). These SST variations are replicated in amplitude by IRD fluctuations (not shown). It is seen that at times of highest ice volume (marine isotope stages 2, 4, 6, 8, and 10) there is always a high amplitude of IRD and SST variability in the Heinrich scale. A more detailed discussion of the H-events during both of the past two glacial maxima is given by van Kreveld *et al.* (1996).

That these H-events might have a more global manifestation than the North Atlantic alone is illustrated in Fig. 16-2, showing the coincidence with the H-events of relatively wet conditions across the southern United States, exemplified by the peaks in a *Pinus* pollen record in Florida (Grimm *et al.*, 1993), and with glacier advances and retreats in South America and New Zealand in phase with the Laurentide Heinrich surge

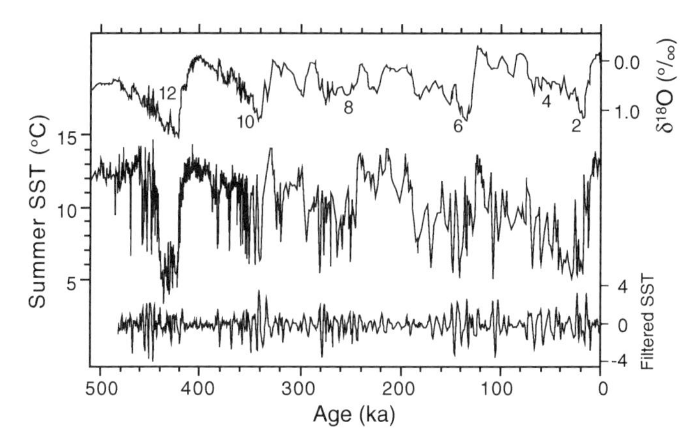

Figure 16-1 Records of summer sea surface temperature over 500 ka compared with $\delta^{18}O$ record of ice mass, showing the presence of high-amplitude oscillations during all periods of high ice mass (even-numbered marine isotopic stages). After McManus *et al.* (1999).

record. Good discussions of these global synchroneities is given by Broecker (1994) and Denton *et al.* (1999). One possible cause of some of these synchroneities is the destructive effect of rising sea level due to H-surges on Antarctic and other ice shelves and their feeder ice streams. There are also indications that rapid fluctuations of CO_2 on the order of 50 ppm occurred on the same time scale as the H-events (Anklin, 1997) which might account for some of the synchroneity shown in Fig. 16-2.

In addition, there has also been much speculation that the freshwater released during an H-event could stabilize the ocean enough to inhibit the thermohaline circulation (i.e., NADW production) (Maslin *et al.*, 1995; Seidov and Maslin, 1999). This could contribute to the worldwide effects and play some role in the cyclical behavior of the H-events.

A schematic pictorialization of the probable sequence of some of these events accompanying an H-event is shown in Fig. 16-3 [cf. Section 3.4 and Hostetler *et al.* (1999)]. It is believed that the cold SST during the H-event interval shown in the figure and discussed in Section 3.4 may be responsible for the survival of icebergs and their IRD loads, originally calved off North America, well into the eastern North Atlantic where they were first identified (Heinrich, 1988). Although unlikely (Andrews *et al.*, 1994), it has even been suggested that the periodic presence of this cold North Atlantic surface water (due to some independently forced behavior of the ocean circulation) might be the cause of the H-events, allowing a normal steady production of icebergs to achieve a survival peak in these cold SST intervals.

Progressing to shorter time scales, from the longer time scales discussed in Chapter 13 for the tectonic-scale (multimillion year) climate variations, and in Chapters 14

Figure 16-2 Geographically widespread geologic evidence for synchronous climate change over the past 70 ky. (a) GISP2 oxygen isotopes, (b) midlatitude snow line in Northern and Southern Hemisphere, (c) number of lithic fragments per gram of sediment from DSDP site 609, (d) bioturbation index from Santa Barbara Basin, (e) % *pinus* pollen record from Lake Tulane, Florida, and (f) Taylor Dome oxygen isotopes.

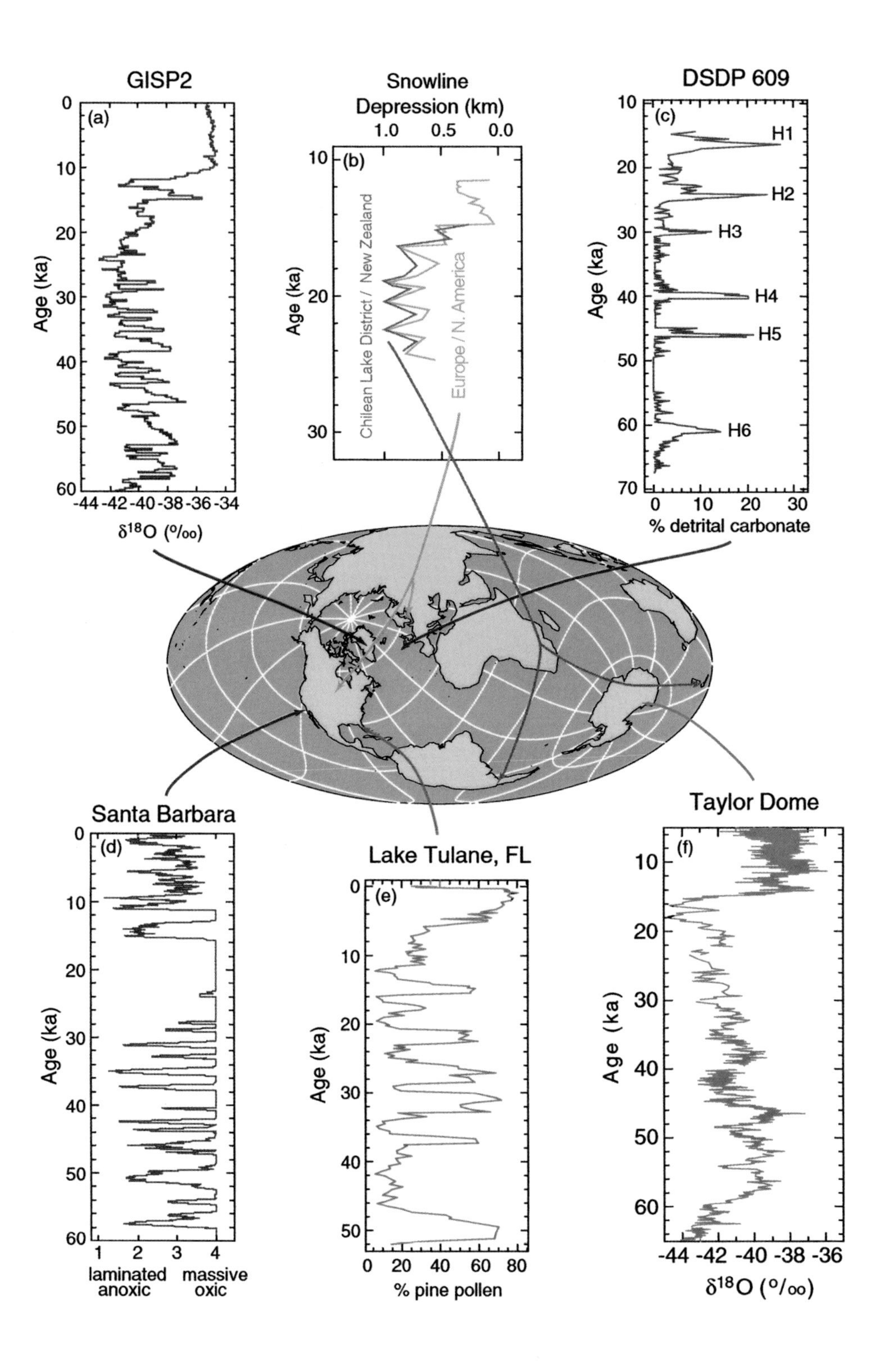

GISP2
(a)
Age (ka)
δ18O (o/oo)
Snowline Depression (km)
(b)
Chilean Lake District / New Zealand
Europe / N. America
DSDP 609
(c)
H1
H2
H3
H4
H5
H6
% detrital carbonate
Santa Barbara
(d)
laminated anoxic
massive oxic
Lake Tulane, FL
(e)
% pine pollen
Taylor Dome
(f)

and 15 for the ice age (20 ky–1 my) variations, we next discuss theoretical aspects of these millennial-scale (1–20 ky) variations, starting with the Heinrich oscillations.

16.1 THEORY OF HEINRICH OSCILLATIONS

It would appear likely that the basic driving mechanism for the H-oscillations resides in the internal physical behavior of ice-sheets that can lead to periodic basal melting and surging. Models to this effect, which we shall discuss here, have now been advanced by Verbitsky and Saltzman (1993, 1994, 1995c), MacAyeal (1993), and Saltzman and Verbitsky (1996) (see also Payne (1995)]. It is also likely that this internal behavior is supplemented to some degree by the feedback effects of these events on the sea surface conditions of the North Atlantic, which may, for example, influence the accumulation rate on the ice-sheet. To analyze the conditions for ice-sheet surging as the trigger for the H-events we now draw upon our discussion of ice-sheet dynamics in Chapter 9, to which the reader is referred for all definitions of symbols.

In Section 9.5 we noted that if the basal temperature (T_B) reaches the pressure melting point (T_M), given by Eq. (8.4), i.e., $T_M = 273.16 - (7.52 \times 10^{-8})\rho_i g h_I$ (in degrees K), a lubricating layer of liquid water forms that can lead to sliding or surging of a peripheral portion of an ice-sheet, usually in the form of ice streams. Based on the scale analysis in Section 9.2, it was found that this basal temperature is given by Eq. (9.45), a simplified version of which is given by Verbitsky and Saltzman (1994) in a form revealing the contributions of the geothermal flux T_B (geoth), basal boundary friction T_B(frict), and internal advection of cold upper surface ice to the basal boundary layer T(adv) [cf. Eq. (9.68)]:

$$T_B(t) = T_B(\text{geoth}) + T_B(\text{frict}) + T_B(\text{adv}) \tag{16.1}$$

where

$$\begin{aligned} T_B(\text{geoth}) &= \frac{1}{\rho_i c_i a} G^{\uparrow} \\ T_B(\text{frict}) &= \frac{g}{2c_i}\left(\frac{a}{k}\right)^{1/n} H^{1+1/n} \\ T_B(\text{adv}) &= \frac{g}{2c_i}\left(\frac{a}{k}\right)^{1/n} H^{1+1/n} \end{aligned}$$

and $t^* = H/a$ is a characteristic time it would take for ice to flow from the top of an ice sheet to the base, delivering there its upper surface temperature.

If $T_B = T_M$ the melting rate is given by Eq. (9.66), i.e.,

$$\mathcal{M}_B = \frac{1}{\rho_i L_f}\left\{G^{\uparrow} + \rho_i ca\big[T_H(t - H/a) - T_M\big] + 1.5\rho_i gaH\left(\frac{aH}{k}\right)^{1/n}\right\}$$

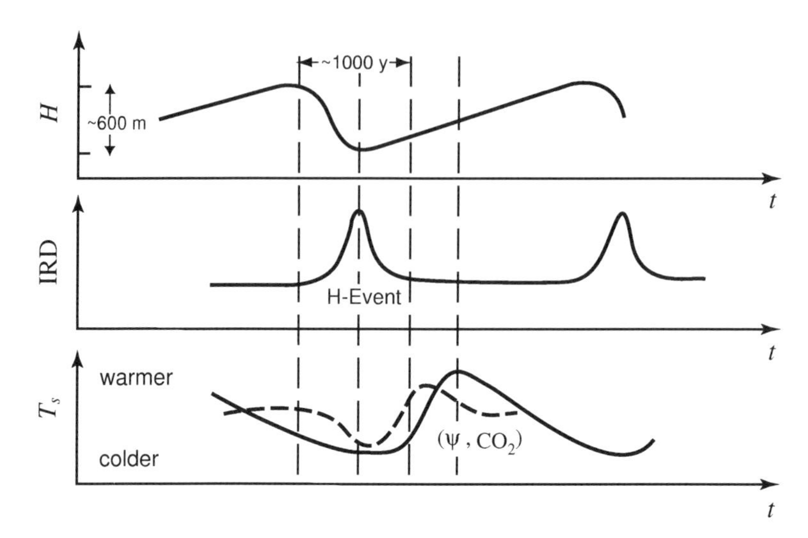

Figure 16-3 Schematic representation of a probable sequence of events accompanying a Heinrich oscillation, involving the height of an ice-sheet H, the deposition of ice-rafted debris (IRD), and sea surface temperature as measured by % *N. pachyderma*-left foraminifera (solid curve) and possible strength of the TH circulation (ψ) and CO_2 oscillations (dashed curve).

16.1.1 The "Binge–Purge" Model

Perhaps the most frequently cited model of periodic ice surge behavior on millennial time scales is the so-called binge–purge model introduced by MacAyeal (1993). The essence of this model is illustrated in Fig. 16-4, showing the envisioned sequence of stages involved in the H-event cycle of growth (binge) and ultimate surge (purge).

In this model the basal temperature is determined mainly by a balance between a steady upward geothermal flux $G^{\uparrow}$ toward the base, and an upward conductive flux $C^{\uparrow}$ away from the base that is determined by the vertical profile of temperature in the ice-sheet, as shown on the left of the figure. The magnitudes of these fluxes are represented by the lengths of the arrows. In these temperature profiles the black dot indicates the pressure melting point temperature T_{M}.

Thus starting from the top (binge) profile of the ice-sheet we have $T_{\mathrm{B}} < T_{\mathrm{M}}$ (ice frozen to ground), but $G^{\uparrow} > C^{\uparrow}$ so that $(dT_{\mathrm{B}}/dt) > 0$. A point is then reached when, as the ice-sheet continues to grow under a steady accumulation rate (a), $T_{\mathrm{B}} = T_{\mathrm{M}}$ and liquid basal water is formed. This, in turn, leads to sliding and frictional heating (denoted by F), which further increases the melting rate to a point where sufficient basal water is available to allow a full-fledged surge (i.e., "purge"). The ensuing reduction of the height of the ice-sheet steepens the temperature profile to a point where $G^{\uparrow} < C^{\uparrow}$ so that $dT_{\mathrm{B}}/dt < 0$. This ultimately cools the basal temperature so that $T_{\mathrm{B}} < T_{\mathrm{M}}$ and the "binge" condition with which we started is restored, permitting a repeat of the cycle.

As we have noted, one distinguishing element of this model is the absence of any influence of the internal advective process, i.e., $T_{\mathrm{B}}(\mathrm{adv})$. We now discuss the validity of this assumption.

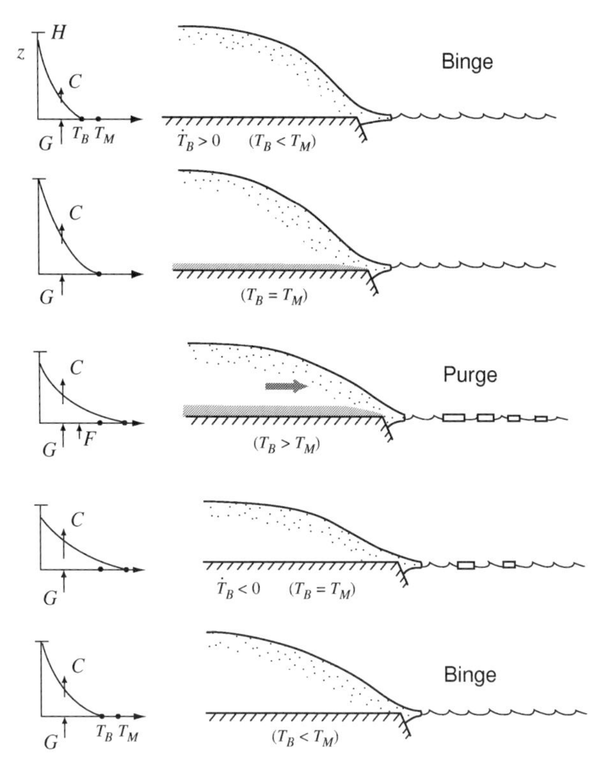

Figure 16-4 Pictorialization of the sequence of stages involved in the MacAyeal (1993) binge–purge model. Vertical profiles of temperature in the ice-sheet, and main vertical fluxes of heat due to conduction (C), geothermal flux ($G^{\uparrow}$), and friction (F), the magnitudes of which are indicated by the lengths of the arrows; T_{B} is the basal temperature and T_{M} is the pressure melting point.

16.1.2 Scale Analysis of the Factors Influencing T_{B}

Following the discussion by Verbitsky and Saltzman (1995a) (see Sections 9.2 and 9.5), two nondimensional numbers (K_{f} and K_{g}) can be defined that measure the relative magnitudes of the two factors involved in the above "binge–purge" model, the frictional component and the geothermal component, respectively, to the neglected advective component. Using the scaling definitions given in Table 9-1, these numbers are

$$K_{\mathrm{f}} = \frac{|T_{\mathrm{B}}(\mathrm{frict})|}{|T_{\mathrm{B}}(\mathrm{adv})|} = \frac{g}{2\gamma c}\left(\frac{a}{k}H\right)^{1/n} = \frac{\rho g^2}{2\gamma c}\left(\frac{1}{\mu_{\mathrm{i}}k}\right)^{1/n} \frac{V^{(2n+3)/n}}{A^{(5n+7)/2n}}$$

and

$$K_{\mathrm{g}} = \frac{|T_{\mathrm{B}}(\mathrm{geoth})|}{|T_{\mathrm{B}}(\mathrm{adv})|} = \frac{G^{\uparrow}k}{\lambda a \gamma H} = \frac{\mu_{\mathrm{i}} k G^{\uparrow} A^{(5n+7)/2}}{\lambda \gamma (\rho g)^n V^{(2n+3)}}$$

where $|T_{\mathrm{B}}(\mathrm{adv})| = \gamma H$ (γ = atmospheric lapse rate).

Using the values given in Tables 9-2 and 9-3, K_f and K_g can be estimated and are listed in the last three columns of Table 9-3 for two values of the rheological constant ($n = 1$ and 3) (see Verbitsky and Saltzman, 1995a). It is seen that in all cases the advective effect is dominant over the geothermal effect, being more nearly balanced by the frictional effect, and this is particularly true for the ice-age Laurentide, which probably spawned the Heinrich events. Thus, the binge–purge model inadequately represents the full physics governing basal temperature, and hence melting, in the peripheral regions of an ice-sheet from which sliding occurs and to which this scale analysis applies (see Section 9.2). This type of model might be more applicable, however, near the base of the central part of an ice-sheet where minimal advection from above occurs and horizontal friction-producing velocities are small so that the geothermal flux becomes more dominant, and, in fact, is a likely explanation for the basal lakes that are observed in central Antartica.

In considering the H-events, however, we are led to consider a more complete model in which geothermal fluxes and friction continue to play a role, but the advective conveyance of thermal properties from the surface to the base may play an even more important role. In the following discussion the full expression for T_B is considered in trying to account for the observed Heinrich events. It will be seen that because the advective effect depends on the temperature at the top of the ice-sheet (T_H), which is advected at a time lag $t^* = H/a$ to the base, the history of other climatic factors that influence T_H is now relevant, unlike the binge–purge model.

16.1.3 Diagnostic Analysis

In the above discussions it is implied that a necessary (but perhaps not sufficient) condition for a surge event is the existence of a lubricating layer of water resulting from basal ice melting when $T_B = T_M$. We now present the results of a first-order diagnostic calculation by Verbitsky and Saltzman (1995a) concerning the evolution of T_B relative to T_M for the Laurentide over the past 140 ky. In particular, we evaluate $T_B(t)$ from Eq. (16.1), and $T_M(t)$ from Eq. (8.4) applied to the scale ice-sheet height H, i.e.,

$$T_M(t) = 273.16 - (7.52 \times 10^{-8})\rho_i g H(t) \qquad (K) \tag{16.2}$$

In these formulas $H(t)$ is prescribed from the SPECMAP $\delta^{18}O$ ice volume record $V(t)$ similar to that shown in Fig. 1-4b, assuming half of the total ice volume is due to the Laurentide, and using Eq. (9.25) with $n = 3$, i.e.,

$$H(t) = \left[\frac{\mu_i a}{(\rho g)^3}\right]^{1/10} \left(\frac{V(t)}{2}\right)^{1/5}$$

In addition, the evolution of $T_H(t - t^*)$, on which $T_B(\text{adv})$ depends in accordance with Eq. (16.1), is assumed to be a function of (1) the mean North Atlantic SST, $T_S(t - t^*)$, as approximated by the record at site K708-1 shown in Fig. 8-4, and (2) the influence of the CO_2 variations as estimated from Eq. (7.33), using the Vostok $\mu(t)$

curve shown in Fig. 3-6. Thus,

$$\begin{aligned} T_B(\text{adv}) &= T_b(t - t^*) \\ &= T_s(t - t^*) - \gamma H(t - t^*) + \frac{\mathcal{B}}{\widehat{\mu}}\left[\mu(t - t^*) - \widehat{\mu}\right] \end{aligned} \quad (16.3)$$

Assigning the values given in Table 9-2, and assuming, as a first approximation, a constant accumulation rate $a = 13.8\ \text{cm y}^{-1}$, we obtain the results shown in Fig. 16-5.

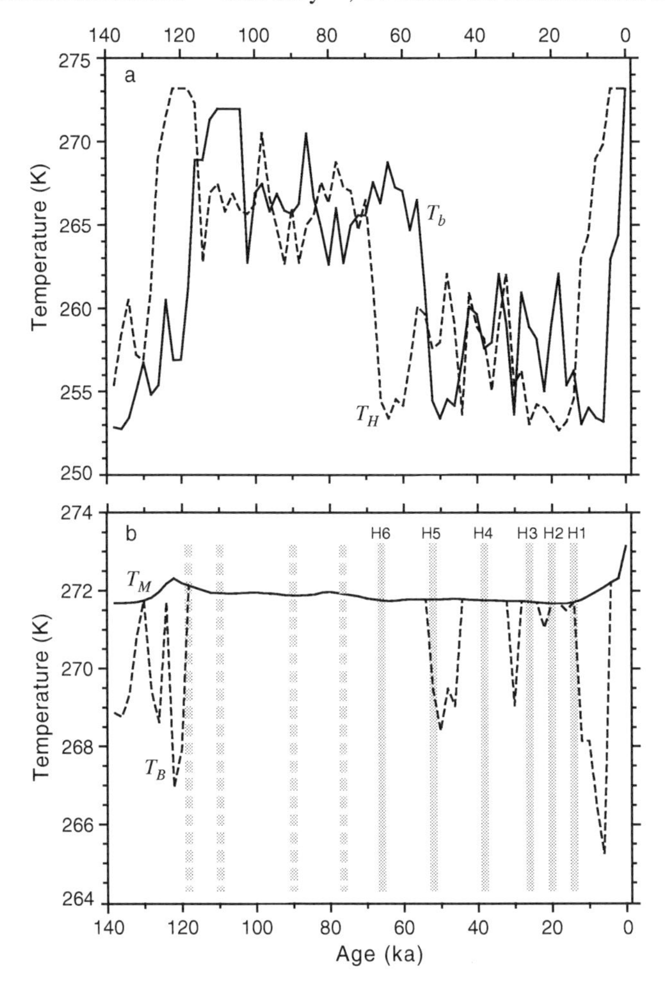

Figure 16-5 (a) Theoretical time variations of the mean temperature (in Kelvin) at the top of the Laurentide ice-sheet, $T_H(t)$ and at the top of the basal boundary layer $T_b(t) = T_H(t - t^*)$, shown by the dashed and solid curves, respectively. (b) Mean basal temperature $T_B(t)$ and pressure melting temperature $T_M(t)$ shown by the dashed and solid curves, respectively. Ice-sheets are vulnerable to surging or streaming activity when $T_B = T_M$. The six documented Heinrich IRD events are indicated by the vertical bars labeled H1–H6 (Bond *et al.*, 1993) and other major IRD events that occurred earlier are indicated by the dashed vertical bars (Grousset *et al.*, 1993). After Saltzman and Verbitsky (1996).

Note that the times when the system is vulnerable to surging occur when $T_B = T_M$, and the observed times of the six well-documented Heinrich events (Bond *et al.*, 1993) are denoted by H1, H2, ..., H6. We see that, according to our model, the ice-sheets were often vulnerable to a surge instability with the notable exceptions of cold (frozen to base) episodes centered near 10, 28, 50, and 130 ka. In particular, all of the observed Heinrich events are located at a time when $T_B = T_M$. Newer analyses based on calendar age rather than radiocarbon age dating (see Section 2.6) have even revealed that H5 occurred at 57 ka rather than 52 ka, and that H7 occurred at 130 ka, and at both times $T_B = T_M$. The analysis also predicts the possibility for some surge activity in the long period between 70 and 120 ka, which seems to be in agreement with the analysis of Grousset *et al.* (1993) and Chapman and Shackleton (1999) showing that major ice rafting episodes occurred in this period (dashed vertical lines) in spite of smaller ice-sheets.

16.1.4 Dynamical Analysis: A Simple Heinrich-Scale Oscillator

In the previous section it was shown that there is some degree of consistency between diagnostic calculations of T_B in relation to T_M and the observed Heinrich events that are presumed to occur only when $T_B = T_M$. In order to account for the quasi-cyclical evolution of these events, however, we must explicitly introduce the physics of the melting process that occurs when $T_B = T_M$, as represented by Eq. (9.66), in a closed dynamical model. The simplest such model can be obtained by specializing the PDM formulated in Section 12.1 by considering only the internal dynamics of an ice-sheet represented by Eqs. (12.1), (12.3), and (12.4) (see also Section 9.11). We thereby consider only the second-order variations of ice-sheet mass due to surge events, excluding all the first-order effects that drive the main evolution of the ice-sheets, which were discussed in Chapter 15. In terms of our feedback-loop diagram (Fig. 12-1) we are now considering only the loop $[I \rightarrow W_B \rightarrow \mathcal{S}_I \rightarrow I]$ excluding all the other links to I.

In particular, following Saltzman and Verbitsky (1996), we assume that the scale area A_I occupied by the Laurentide ice-sheet and the accumulation rate $a \approx a_0 - a_1 T_\psi$ were roughly constant over the past 70 ky, during which time the Heinrich purges occurred. Then neglecting any bedrock-calving instabilities ($\mathcal{C}_I = 0$) and using the scaling relation $\Psi = \rho_i A_I H$, we can write the system of Eqs. (12.1), (12.4), and (12.3), respectively, in the form

$$\frac{dH}{dt} = a - \frac{1}{\tau_H} H - \delta_s \tag{16.4}$$

$$\frac{d\delta_s}{dt} = \alpha W_B - \frac{1}{\tau_s} \delta_s \tag{16.5}$$

$$\frac{dW_B}{dt} = \mathcal{M}_B(H; G^\uparrow) - \frac{1}{\tau_w} W_B \tag{16.6}$$

where

$$\mathcal{M}_B = \frac{1}{\rho_i L_f} G^{\uparrow} - \rho_i c_i a\left[T_H(t - t^*) - T_M\right] + 1.5\rho_i g a H\left(\frac{aH}{k}\right)^{1/n} \tag{16.7}$$

when $T_B = T_M$, and zero if $T_B < T_M$; $\delta_S = (\rho_i A_I)^{-1}\mathcal{S}_I$ (representing the rate of ice thickness loss due to the surge flux); $\alpha = (\rho_i A)^{-1}k_w$; and $\tau_H\ (= \nu_I^{-1})$, $\tau_S\ (= \nu_B^{-1})$, and $\tau_W\ (= \nu_W^{-1})$ are characteristic time constants for ice creep, surging, and basal water discharge, respectively. These latter two time constants are only weakly constrained, but plausibility arguments have been made by MacAyeal (1993) that τ_S is of the order of 1 ky, and by Oerlemans and Van der Veen (1984) that (in accordance with a diffusive approximation for basal water flow) τ_W is also of this same order (1 ky). The rate constant α in Eq. (16.5), relating the surge changes to the basal liquid water amount, is even more weakly constrained. We consider τ_S, τ_W, and α, which must appear in some form in all models of ice sheet surging, to be free parameters to be determined as a (nonunique) set of predicted values required to account for the observed variations (in our case the Heinrich oscillations). To close the system we must add the formulas, Eqs. (16.1) and (16.3), for T_B and T_M. A schematic feedback-loop diagram showing the cyclical coupling of the main variables of the model is given in Fig. 16-6.

It may be recognized that the ice volume $V(H)$, the surge loss (δ_S), and basal water amount (W_B) are also the main variables of the more complex model described by Oerlemans (1982b), in which he demonstrated the possibility for oscillatory behavior, albeit for a much longer time scale (100 ky) than Heinrich events (10 ky). To illustrate the possibility that our present much simpler system can exhibit fluctuations of the

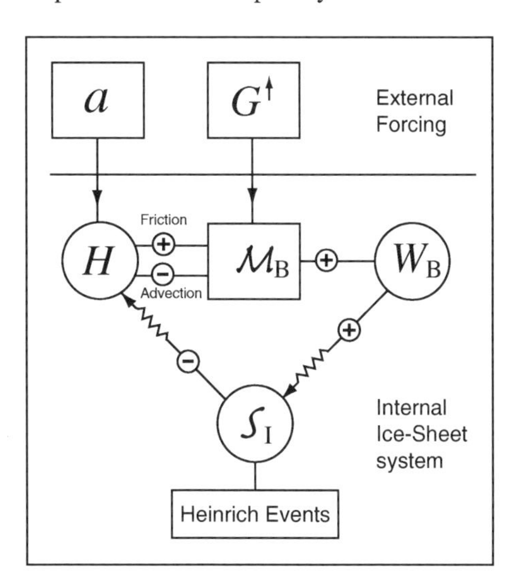

Figure 16-6 Schematic feedback-loop diagram for the model system [Eqs. (16.4)–(16.7)] showing the internal coupling between the height of an ice-sheet (H), basal water depth (W_B), and surge loss of ice thickness ($\mathcal{S}_I$) under the influence of steady forcing due to net snow accumulation (a) and geothermal flux ($G^{\uparrow}$).

Heinrich time scale, we take the following plausible values of physical constants and parameters: $\rho_{\rm i} = 917\ {\rm kg\,m^{-3}}$, $n = 3$, $g = 9.8\ {\rm m\,s^{-2}}$, $c = 2 \times 10^3\ {\rm J\,kg^{-1}/°C}$, $k = 10^{-6}\ {\rm m^2\,s^{-1}}$, $G^{\uparrow} = 0.04\ {\rm W\,m^{-2}}$, $\gamma = 10^{-2}\ {\rm °C\,m^{-1}}$, and $a = 14\ {\rm cm\,y^{-1}}$ (the low accumulation rate characteristic of the central parts of an ice-sheet, where we presume ice trajectories, delivering the surface temperature $T_{\rm H}$ to the bottom, originate), and $\tau_{\rm H} = 50$ ky. For the more weakly constrained parameters we assign the set of values, $\tau_{\rm S} = \tau_{\rm W} = 1.5$ ky, and $\alpha = 10^{-5}\ {\rm y^{-2}}$.

The solution is shown in Fig. 16-7 starting from an arbitrary initial condition. We note that an oscillation of roughly a 14-ky period occurs, characterized by lagged variations of $T_{\rm H}$, $T_{\rm B}$, $T_{\rm B}$, H, and $\mathcal{S}_{\rm I}$ that go along with episodes of melting ($\mathcal{M}_{\rm B} > 0$) when $T_{\rm B} = T_{\rm M}$. The results are not qualitatively changed for small changes in the parameters.

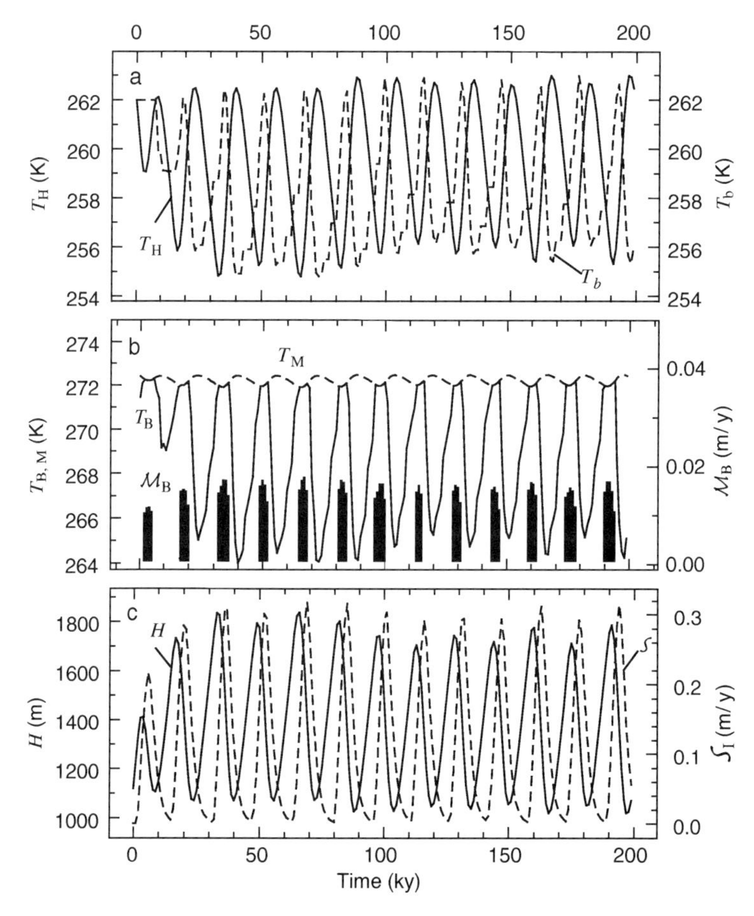

Figure 16-7 Time-dependent solution of the dynamical system, showing (a) the fluctuation of the upper surface temperature ($T_{\rm H}$) and the temperature at $z = z_{\rm b}(T_{\rm b})$, (b) the pressure melting point ($T_{\rm M}$), the basal temperature ($T_{\rm B}$), the basal melting rate ($\mathcal{M}_{\rm B}$) as black bars, and in the bottom panel the surge discharge in units of ice thickness loss ($\mathcal{S}_{\rm I}$) and ice-sheet thickness H.

Thus, using a dynamical model based on fundamental thermomechanical properties of an ice-sheet, we have shown that such a model can exhibit periodic fluctuations of roughly the same scale as observed Heinrich events. This model provides, in its simplest form, the essential physical processes governing the coupled variation of ice volume, mean basal water amount, and the surge of ice, and exposing clearly the free parameters that are likely to appear in any more detailed model. The system is influenced mainly by the internal thermodynamic processes in ice-sheets and the instability engendered by basal friction heating, as regulated by both cold advection from the upper ice surface and a much weaker influence of geothermal heating. In this respect, the mechanism involved is akin to that in Oerleman's (1982b) surge model, but only to the discharge phase of MacAyeal's (1993) binge–purge model.

It is clear that many improvements and extensions can be made to this simple model, some of which are obvious:

1. Making the accumulation rate (a) a variable dependent on climatic (e.g., SST) changes forced by the Heinrich events. This should involve consideration of changes in air flow due to changes in height of the ice-sheet and a treatment of the thermohaline circulation changes such as discussed by Birchfield (1989), Paillard and Labeyrie (1994), Paillard (1995), and Seidov and Maslin (1999). Thus, in terms of the feedback diagram, Fig. 12-1, new negative links should be added between S_I and both Σ and S_φ.
2. Considering, in more observational and theoretical detail, the implied basal water discharge process implied by W_B, the role of regolith deformation, and the more detailed processes involved in ice stream–ice shelf dynamics (see Chapter 9).
3. Coupling these H-events dynamically to the main longer term evolution of the ice-sheets on the one hand, and the higher frequency D-O type events (e.g., the Bond cycles shown in Fig. 3-7b) on the other hand (see, e.g., Paillard, 1995).

In the next section we discuss these higher frequency millennial oscillations.

16.2 DYNAMICS OF THE D-O SCALE OSCILLATIONS

Many ideas have been suggested regarding the cause of the 1- to 3-ky scale fluctuations that characterize the climate record, notably during the Holocene. Because, as in the Heinrich case, there is no known external forcing on this scale it seems clear that some internal instability, most likely involving the behavior of the ocean, is involved. In Fig. 11-1 we schematically represented in an equator-to-pole cross-sectional box model the main variables that have been invoked in models to account for this variability. All of the ocean variables have been defined and discussed in Chapter 11.

In particular (T_2, S_2) and (T_1, S_1) are the low- and high-latitude mixed-layer temperatures and salinities, respectively; η is the sea-ice extent; ψ and ϕ are the stream functions for the thermohaline and gyre circulations, respectively; θ is the deep ocean temperature; μ is the CO_2 concentration; and we have also included the longer term ice-sheet variables, I and D, that are involved in ice-age and H-event variations.

If we denote the set of these variables by

$$\Phi_i = \{T_1, T_2, S_1, S_2, \eta, \phi, \psi, \theta, \mu, I, D_B\}$$

each component of which is a climatic mean (i.e., δ = 100 y) value, we may resolve this set in the form (see Section 5.3), $\Phi = \Phi_0 + \Phi'$, where Φ_0 denotes an equilibrium value that can be identified with a 1-My average state and Φ' represents the departure of the climatic mean state from Φ.

The governing dynamical equations for this set can be expressed in the general form,

$$\frac{d\Phi' i}{dt} = \Sigma_{j\neq i} a_{ij}(\Phi_0) \cdot \Phi'_j - b_i \Phi'_i - N_i(\Phi_{0j}, \Phi' j) + F'_i + \omega_i \tag{16.8}$$

where the first term represents the sum of the linear terms, not including the dissipative terms represented in the second term ($b_i > 0$), N_i represents a set of nonlinear terms that may involve all the variables, F'_i represents all modes of external forcing, and ω_i represents additive stochastic forcing.

As a specialization of this set of equations we guess from physical considerations that the main linear terms involved are as portrayed in Table 16-1 where the a_{ij} are all assumed to be positive. In this table η'(ins) and η'(alb) represent the competing effects of sea-ice insulation and sea-ice albedo on surface temperature, respectively, and η'(brine) represents the effect of seasonal sea-ice production on surface salinity. We see that several main positive feedback loops are possible, all of which have been invoked as possible sources of instability that can drive millennial-scale oscillations. In the event that the positive feedbacks do indeed dominate over the dissipative terms,

Table 16-1 Specialization of the System [Eq. (16.8)][a]

$\dot{\Phi}'_i$	a_{ij}	$\sum_{j\neq i} \Phi_j$
T'_1	(> 0)	$+T'_2 + \eta'(\text{ins}) - \underline{\eta'(\text{alb})} + \psi' + \phi' + \underline{\underline{\underline{\mu'}}}$
T'_2		$+T'_1 + \eta'(\text{alb}) - \psi' - \phi' + \theta' + \underline{\underline{\underline{\mu'}}}$
S'_1		$+S'_2 + \eta'(\text{brine}) + \underline{\underline{\psi'}} + \phi' - \mu'$
S'_2		$+S'_1 - \underline{\underline{\psi'}} - \phi' + \mu'$
η'		$-\underline{T'_1} - \mu'$
ψ'		$+T'_2 - T'_1 + \theta' + \underline{\underline{S'_1}} - \underline{\underline{S'_2}}$
ϕ'		$+T'_2 - T'_1$
θ'		$+T'_1 + T'_2 - \psi'$
μ'		$\underline{\underline{\underline{+T'_1 + T'_2}}} - \eta' + \psi' + \phi' - \theta'$

[a] In terms of physically plausible linear terms, not including the dissipative terms. Potential positive feedbacks are indicated by the sets of underlined terms.

$b_i \Phi'_i$, the nonlinear terms N_i must become important at higher order to prevent run-away conditions and preserve the conservation principles as observed. As examples of such positive feedbacks in Table 16-1 the single-underlined terms represent the ice-albedo (or baroclinicity) feedback, the double-underlined terms represent the salinity feedback, and the triple-underlined terms represent the CO_2 feedback. Just a few of the many recent examples of models based on these instabilities are Birchfield *et al.* (1990), Yang and Neelin (1993), Griffies and Tziperman (1995), Zhang *et al.* (1995), Paillard (1995), Chen and Ghil (1995), and Sakai and Peltier (1997). In Sections 6.5 and 6.6 we discussed in detail properties of an earlier model for millennial-scale oscillations based on a two-component (η–θ) system driven by the CO_2 instability, with the sea-ice insulation effect providing the harmonic behavior as summarized by Eq. (6.25).

17

CLOSING THOUGHTS: EPILOGUE

In this last part of the book (see Chapters 13 and 15 in particular) we have illustrated the application of the program suggested in Chapter 5 regarding a structured approach to a dynamical theory of paleoclimatic evolution. As we have tried to make clear, these special applications can only be taken as early efforts, building upon what we regard as a basic foundation and framework of this structure. In the following discussion we elaborate on the fuller problem posed by the paleoclimatic record.

17.1 TOWARD A MORE COMPLETE THEORY

The ultimate goal of climatic theory is to account not only for the global ice mass and thermal evolution we have emphasized here, but to account for the full three-dimensional geographical distributions of all the climatic variables describing the atmosphere, oceans, and terrestrial biosphere, as well as the location and topography of the ice sheets. These aspects will require the explicit use of atmospheric, biospheric, and oceanic general circulation models coupled with three-dimensional ice-sheet models. In this regard, we can identify several levels in the hierarchy of complexity of such more complete models, all of which utilize the vertically integrated $[h_I(\lambda, \theta)]$ ice-sheet model described in Section 9.3 (see also Section 9.10).

In the simplest of such models, the two-dimensional ice-sheet behavior is coupled only to an atmospheric EBM, or to a more detailed atmosphere SDM in which land surface and ocean mixed-layer and sea-ice effects are included in highly parameterized forms. Good examples of this type of model that have been applied to the evolution of Late Pleistocene ice-age states include the zonal average LLN model of the Louvian–La-Neuve group (e.g., Gallée *et al.*, 1991, 1992; Berger *et al.*, 1998b), especially as coupled to the Marsiat (1994) ice-sheet model (Calov and Marsiat, 1998); the Toronto group (e.g., DeBlonde *et al.*, 1992; Tarasov and Peltier, 1997, 1999); and the Australian group (Budd *et al.*, 1998).

Now under development are even more complete models, still governing the climatic-mean variables using statistical-dynamical parameterizations of fluxes in the atmosphere and oceans, but including representations of the marine cryosphere, deep ocean, biosphere, and carbon cycle. Such models have been termed an "Earth-system model of intermediate complexity," or an EMIC. To our knowledge only two models of this type have as yet been applied to a limted extent in a time-dependent paleoclimatic mode (Chalikov and Verbitsky, 1990) [see also Sergin (1979) and Berger *et al.* (1998b)], but other models under development will probably find such an application [e.g., the Potsdam CLIMBER model, Ganopolski *et al.* (1998)]. It would appear that models at this level of complexity are most appropriate for the study of paleoclimatic evolution.

Proceeding to the most detailed complexity level, in the earliest stages of development are several models in which a full-fledged AGCM governing synoptic-mean variables is coupled asynchronously to models of the full ocean (OGCM), cryosphere, biosphere, and chemosphere (including the carbon cycle) in a three-dimensional geographic model to form a more complete climate system model or CSM, as mentioned in the beginning of Chapter 5. An example is the NCAR-GENESIS model (Pollard and Thompson, 1997).

We have emphasized the potential importance of treating the carbon cycle and CO_2 as a free internal variable, in both an EMIC and fuller CSM, throughout this book. In addition, some comment is in order regarding the terrestrial biosphere as a vital fast-response part of such a "supermodel." The importance of the biosphere in long-term climatic evolution was already evident in the discussion of the carbon cycle (Chapter 10). Without going quite to the Gaian limit of treating the whole biosphere as a megaorganism engaged in altering climate for its self-promotion and preservation (e.g., Lovelock, 1979), one can still assert that it is not possible to discuss the evolution of Earth's climate without also considering the ecological evolution of the biosphere. Aside from our discussion of the relevance of this aspect in the carbon cycle and our broad inclusion of the terrestrial biosphere as an essential part of a GCM, we have not really done sufficient justice in this book to this important topic. Complementary references to help fill this void with regard to GCMs are provided in Trenberth (1992) (articles by Dickinson, Aber, and Sellers), and in the excellent collection of papers in Schneider and Boston (1991) with regard to longer term, tectonic-scale global climatic changes. In this last collection an interesting dynamical system study in the context of the Gaian hypothesis is given by Watson and Maddock (1991) [see also Watson and Lovelock (1983)].

In Figs. 5-1 and 5-3 we showed a representation of the fuller CSM system as part of a broad discussion in Chapter 5 outlining the practical difficulties in using such a model as a basis for a theory of paleoclimatic evolution. In essence, it was noted that even if we had such a supermodel at our disposal it would be highly unreasonable to expect that, given an initial condition in the distant past and appropriate boundary conditions and externally imposed forcing, the computer would spew out a solution of the complex nature shown, e.g., in Fig. 1-4 for global ice mass. From basic considerations discussed in Chapter 5, to achieve such a solution the fluxes linking the atmosphere to the ocean, ice sheets, and carbon reservoirs would have to be computed to an unattainable level of accuracy to account for the observed variations. The difficulty is compounded by the degree of instability and nonlinearity implied by a comparison of the

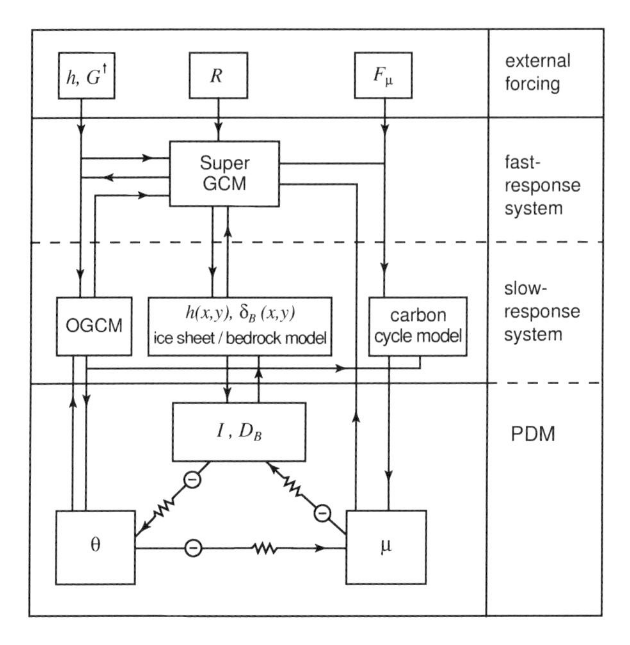

Figure 17-1 Flow diagram pictorializing the connection between a three-dimensional climate system model (CSM) composed of atmospheric/biospheric, ocean, ice-sheet, and carbon cycle components governing the geographic distributions [e.g., $h(x, y)$, $\delta_B(x, y)$, and atmospheric and oceanic state variables] and a low-order, global, paleoclimate dynamics model (PDM) governing global mean values (e.g., I, D_B, θ, μ), all under the influence of external forcing due to solar radiation (R) and tectonic influences (h, $G^{\uparrow}$, F_{μ}).

ice record and the known Milankovitch forcing (Fig. 1-7). Thus, it is strongly suggested that some guidance from observation in the form of a more phenomenological, low-order dynamical component model (i.e., a PDM) will be necessary to complete the theory. Although such a PDM can give only a skeletal representation of the full-three dimensional system, it can provide the global integral constraints by which to tune the CSM governing the temporal evolution of the system, using a minimal number of free parameters. To pictorialize this connection between the CSM and PDM a more complete schematic representation of the full system than was shown in Fig. 5-1 is now shown in Fig. 17-1. In Chapter 15 we considered a low-order dynamical system governing the four global slow-response variables represented in the added lower part of the figure, I, D_B, μ, θ, and S_{φ}. Thus, for example, it is shown in the figure that the evolution of the ice thickness and bedrock depression fields, $h_I(x, y)$ and $\delta(x, y)$, must satisfy the integral constants,

$$I = \iint \rho_i h_I(x, y)\, dx\, dy \tag{17.1}$$

$$D_B = \iint \delta_B(x, y)\, dx\, dy \Big/ \iint dx\, dy \tag{17.2}$$

where the integrals are over the area occupied by all the ice sheets, the extent and height of each of which is to be determined from three-dimensional ice-sheet model coupled to a super-GCM that governs the atmosphere and the oceanic and terrestrial surface state, including the biosphere. With the introduction of such a coupled ice-sheet/GCM model we can proceed with a consideration of the fuller slow-response system $[h_I(x, y, t), \delta_B(x, y, t), \mu(t), \theta(t)]$, where h_I and δ_B are constrained by Eqs. (17.1) and (17.2) to conform with the solution for I, D_B, μ, and θ obtained from the global dynamical model. The execution of the complete scheme shown in this figure is a formidable but necessary undertaking if we are to advance the dynamical theory of climate. At the least, the low-order dynamical system part shown in the lower section of Fig. 17-1 will expose the minimum number of free parameters needed in the more complex three-dimensional model to achieve the observed paleoclimatic evolution.

As indicated in the figure, the free and tectonically forced variations of CO_2, and the variations of θ, would still be governed by the global equations of a form similar to Eqs. (12.5) and (12.6), allowing free internal oscillations arising from a possible instability of the carbon cycle, for example. Thus, whereas in other models (e.g., Marsiat, 1994; DeBlonde and Peltier, 1993) the CO_2 variations have been prescribed from the Vostok ice core observations, CO_2 would be treated as a free, interactive, variable to be deduced along with the ice variations. In accordance with both observations (e.g., Saltzman and Verbitsky, 1994a) and the global theory described in Chapter 15, it would not be possible to exclude this kind of free internal behavior and still be able to account for the record, including a sequence of 100-ky-period oscillations, its onset near 900 ka, and the key bifurcation representing the onset of the ice age near 2.5 My.

On a more fundamental level the proposed scheme constitutes an inversion of the usual approach to the problem, wherein it is often taken as an article of faith that the CSM provides the ultimate physical basis for calculating the long-term, time-dependent evolution of climate as an initial value problem. Here, we have suggested that because of the impracticality (if not impossibility!) of calculating the fluxes of mass and energy involved in the long-term changes to the accuracy required by the geologic evidence (Saltzman, 1984a,b), a more central need is to formulate, phenomenologically, the structure of the slow-response system (i.e., the center manifold) and then use the constraints from such a model to tune the output of a "super-GCM" in order to recover the details in which we are interested. Thus, such a lower order dynamical system model becomes the "control center" of the broader theory (see Fig. 5-4). It is already being recognized that there are many complex natural systems, other than the climate system, for which such a phenomenological low-order dynamical systems component will be appropriate, particularly in the biological sphere.

Conversely, the more detailed full three-dimensional GCM/ice-sheet/carbon cycle/deep ocean components in the upper part of Fig. 17-1 are only a part of the full set of statements constituting a "theory" of paleoclimatic variability (e.g., the ice ages): an additional set of statements is needed to provide the constraints needed to guide the full solution, which are represented by the lower part of Fig. 17-1. In other terms, it would be unreasonable to expect that, without the guidance of the lower order dynamical system, one could obtain a solution agreeing with the observed evolution of all the variables.

17.2 EPILOGUE: THE "ICE AGES" AND "PHYSICS"

The paleoclimatological (e.g., ice-age) problem we have been discussing here is but one example of a class of problems that is as difficult and important as any considered more generally in physics. It is, in short, a problem in ultraslow, complex evolution in which the rates of change, and the fluxes of mass, momentum, and energy that accompany and drive them, are too small to be calculable directly or in some cases even measurable, though we are sure they occurred and are still occurring.

One further example of which we are all aware is the human "aging" process wherein a very slow, unmeasurable process of combined biological, chemical, and physical change occurs on a relatively long time scale (unfortunately, monotonically). From day to day this slow change (perhaps analogous to climatic change) is masked by much higher frequency and amplitude processes (analogous to weather changes). In this case, also, we have very limited prospects of measuring or calculating the statistical-average net fluxes within an organism that must be leading to the inexorable net changes.

Our view is that such a problem in ultraslow, complex, physics is likely to require a "phenomenological" approach of the kind we have illustrated here. The framework of dynamical systems analysis seems especially appropriate, particularly through the construction of low-order models in which the full behavior is projected onto the dynamics of a reduced number of key slow-response, macroscopic or highly aggregated variables (e.g., $I, \mu, \theta, \ldots$). It has been argued that such low-order dynamical systems models are merely "toys" that can only suggest what should be obtainable from a full GCM-type "supermodel" in which an increasing amount of explicit physical detail and resolution is introduced (Lorenz, 1970; Gates, 1981); the tuned parameters of a low-order model are to be determined ultimately from this more explicit model. A typical statement of this view is given by Oerlemans (1982a,b), who states that "few-component systems should be validated against, or better, be derived from more sophisticated models before they are used extensively." From a practical viewpoint, I would suggest the opposite: namely, that the essential slow physics is to be sought in the low-order model, and in the end it is the more explicit supermodel that must be "tuned" to satisfy the constraints imposed by the results of the "best available" low-order phenomenological theory. In this way a supermodel can indeed be achieved that accounts not only for the fast physics and three-dimensional geographic detail but also for the long-term trends that constitute paleoclimatic change.

A question that arises immediately concerns the criterion for the "best available" model. Here I believe we must fall back on the criterion used in all areas of theoretical physics, discussed in Section 5.1; namely, that the best model is the one that requires the lowest number of free (or adjustable) parameters to account for the same set of observations. A wider acceptance of this rule would greatly advance progress toward a theory of paleoclimatic variations, providing the measure by which to judge the relative merit of competing models. Thus, although a fully satisfactory, unique, explanation of paleoclimatic variability (e.g., the ice ages) may never be attainable because of the unmeasurably low rates of the fluxes involved that require the assignment of free parameters, this does not mean that there is no criterion by which one can, as a continuing process, judge which model offers the best explanation up to any time. It is therefore important that a detailed exposure of the free parameters used in any model be a part of any discussion purporting to account for paleoclimatic variability, and by the same token it would be helpful if there were some agreement concerning a target set of time-series observations to be explained at any time (which inevitably will be a "moving target" as measurements are continually improved).

As a corollary to the above, it would also advance the pursuit of such an explanation if it were more widely recognized that whatever theoretical model is proposed, at whatever level of complexity, any such model is no more than a hypothesis to be continually tested, primarily against emerging data and secondarily against other models, by invoking the above criterion (see Section 1.2). In the spirit of these rules of the game we can go forward toward a deeper explanatory understanding of the wonderfully rich and remarkable record of paleoclimatic variability.

Bibliography

Adams, J. M., H. Faure, L. Faure-Denard, J. M. McGlade, and F. I. Woodward (1990). Increases in terrestrial carbon storage from the last glacial maximum to the present. *Nature* **348**, 711–714.

Adhémar, J. F. (1842). "Les Revolutions de la Mer Deluges Periodiques," Paris.

Agassiz, L. (1840). "Etudes Sur les Glaciers." Privately published, Neuchatel.

Algeo, T. J., R. A. Berner, J. B. Maynard, and S. E. Scheckler (1995). Late Devonian oceanic anoxic events and biotic crises: "Rooted" in the evolution of vascular land plants? *GSA Today* **5**, 64–66.

Alley, R. (1992). Flow-law hypothesis for ice sheet modeling. *J. Glaciol.* **38**, 245–256.

Alley, R., and D. MacAyeal (1994). Ice-rafted debris associated with the binge-purge oscillations of the Laurentide Ice Sheet. *Paleoceanography* **9**, 503–511.

Alley, R. B., D. D. Blankenship, S. T. Rooney, and C. R. Bentley (1987a). Till beneath ice stream B, 3. Till deformation: Evidence and implications. *J. Geophys. Res.* **92**, 8921–8929.

Alley, R. B., D. D. Blankenship, S. T. Rooney, and C. R. Bentley (1987b). Till deformation beneath ice stream B, 4. A coupled ice-till flow model. *J. Geophys. Res.* **92**, 8931–8940.

Alvarez, W., L. W. Alvarez, F. Asaro, and H. V. Michel (1984). Impact theory of mass extinctions and the invertebrate fossil record. *Science* **223**, 1135–1141.

Ambio (1997), Vol. **26** (entire volume).

Andersen, B. G., and H. W. Borns Jr. (1994). "The Ice Age World: An Introduction to Quaternary History and Research with Emphasis on North America and Northern Europe during the Last 2.5 Million Years." Scandinavian University Press, Oslo.

Andrews, J. T., and M. A. W. Mahaffey (1976). Growth rate of the Laurentide ice sheet and sea level lowering (with special emphasis on the 115,000 BP sea level low). *Quatern. Res.* **6**, 167–183.

Andrews, J. T., H. Erlenkeuser, K. Tedesco, A. Aksu, and A. J. T. Jull (1994). Late Quaternary (Stage 2 and 3) meltwater and Heinrich events, northwest Labrador Sea. *Quatern. Res.* **41**, 26–34.

Anklin, M., J. Schwander, B. Stauffer, J. Tschumi, and A. Fuchs (1997). CO_2 record between 40 and 8 ky B.P. from the Greenland Ice Core Project ice core. *J. Geophys. Res.* **102**, 26:539–545.

Archer, D., and E. Maier-Reimer (1994). Effect of deep-sea sedimentary calcite preservations on atmospheric CO^2 concentration. *Nature* **367**, 260–263.

Arrhenius, S. (1896). On the influence of carbonic acid in the air upon temperature on the ground. *Philosophical Mag.* **41**, 237–276.

Arthur, M. A., K. R. Hinga, M. E. Q. Pilson, D. Whitaker, and D. Allard (1991). Estimates of pCO_2 for the last 120 My based on the $\delta\,^{13}C$ of marine phytoplanktic organic matter (abstract). *EOS* **17**, 166.

Ashe, S. (1979). A nonlinear model of the time-average axially asymmetric flow induced by topography and diabatic heating. *J. Atmos. Sci.* **36**, 109–126.

Bacastow, R. B., and C. D. Keeling (1981). Atmospheric carbon dioxide concentration and the observed airborne fraction. *In* "Carbon Cycle Modelling, SCOPE16" (B. Bolin, ed.), John Wiley, Chichester.

Bacastow, R., and E. Maier-Reimer (1990). Ocean-circulation model of the carbon cycle. *Climate Dynam.* **4**, 95–125.

Bagnold, R. A. (1954). "The Physics of Blown Sand and Desert Dunes." Methuen, London.

Barnola, J., D. Raynaud, Y. Korotkevitch, and C. Lorius (1987). Vostok ice core provides 160,000-year record of atmospheric CO_2. *Nature* **329**, 408–414.

Barron, E. J. (1987). Eocene equator-to-pole surface ocean temperature: A significant climate problem? *Paleoceanography* **2**, 729–739.

Barron, E., and P. Fawcett (1995). *In* "The Climate of Pangea: A Review of Climate Model Simulations of the Permian," vol. 1, pp. 37–52. Springer-Verlag, Berlin.

Barron, E., and W. Peterson (1989). Model simulation of the cretaceous ocean circulation. *Science* **244**, 684–686.

Beer, J., M. Andree, H. Oeschger, B. Stauffer, R. Balzer, G. Bonani, Ch. Stoller, M. Suter, W. Wolfi, and R. C. Finkel (1985). ^{10}Be variations in polar ice cores. *In* "Greenland Ice Core: Geophysics, Geochemistry, and the Environment" (C. Langway, H. Oeschger, and W. Dansgaard, eds.), pp. 66–70. Geophysical Monograph 33, Am. Geophys. Union, Washington, D.C.

Benzi, R. A., A. Sutera, and A. Vulpiani (1981). The mechanism of stochastic resonance. *J. Phys. A. Math. Gen.* **14**, L453–L457.

Benzi, R., G. Parisi, A. Sutera, and A. Vulpiani (1982). Stochastic resonance in climatic change. *Tellus* **34**, 10–16.

Berger, A. L. (1978a). Long-term variations of caloric insolation resulting from the earth's orbital elements. *Quatern. Res.* **9**, 139–167.

Berger, A. (1978b). A simple algorithm to compute long-term variations of daily or monthly insolation. *Inst. Astron. Geophys. G. Lemaitre*, Contrib. 18.

Berger, A. (1978c). Long term variations of daily insolation and Quaternary climatic changes. *J. Atmos. Sci.* **35**, 2362–2367.

Berger, W. H. (1982). Increase of carbon dioxide in the atmosphere during deglaciation: The coral reef hypothesis. *Naturwissenschaften* **69**, 87–88.

Berger, A., and M. Loutre (1991). Insolation values for climate of the last ten million years. *Quartern. Sci. Rev.* **10**, 297–317.

Berger, A., and M. F. Loutre (1997). Long-term variations in insolation and their effects on climate: The LLN experiments. *Surveys Geophys.* **18**, 147–161.

Berger, A., T. Fichefet, H. Gallée, I. Marsiat, C. Tricot, and J. P. van Ypersele (1990). Physical interactions within a coupled climate model over the last glacial–interglacial cycle. *Trans. Roy. Soc., Edinburgh Earth Sci.* **82**, 357–369.

Berger, W. H., T. Bickert, H. Schmidt, and G. Wefer (1993). Quaternary oxygen isotope record of pelagic foraminifera: Site 806, Ontong Java Plateau. *Proc. Ocean Drilling Program, Scientific Results* **130**, 381–395.

Berger, A., M. F. Loutre, and J. L. Melice (1998a). Instability of the astronomical periods from 1.5 My BP to 0.5 My AP. *Paleoclimates* **24**, 239–280.

Berger, A., M. F. Loutre, and H. Gallée (1998b). Sensitivity of the LLN climate model to the astronomical and CO_2 forcings over the last 200 ky. *Climate Dynam.* **14**, 615–629.

Berggren, W. A., and C. D. Hollister (1974). Paleogeography, paleobiography and the history of circulation in the Atlantic Ocean. *In* "Studies in Paleooceanography" (W. W. Hay, ed.), pp. 126–186. Soc. Econ. Paleontol. Mineral., Spec. Publ., 20.

Berner, R. A. (1987). Models for carbon and sulfur cycles and atmospheric oxygen: Application to Paleozoic geologic history. *Am. J. Sci.* **287**, 177–196.

Berner, R. A. (1989). Biogeochemical cycles of carbon and sulfur and their effect on atmospheric oxygen over Phanerozoic time. *Global Planet. Change* **1**, 97–122.

Berner, R. A. (1990). Atmospheric carbon dioxide levels over Phanerozoic time. *Science* **249**, 1382–1386.

Berner, R. A. (1991). A model for atmospheric CO_2 over Phanerozoic time. *Am. J. Sci.* **291**, 339–379.

Berner, R. A. (1994). GEOCARB II. A revised model of atmospheric CO_2 over Phanerozoic time. *Am. J. Sci.* **294**, 56–91.

Berner, R. A. (1995). A G. Högbom and the development of the concept of the geochemical carbon cycle. *Am. J. Sci.* **295**, 491–495.

Berner, R. A. (1999). A new look at the long-term carbon cycle. *GSA Today* **9** (11), 1–6.

Berner, R. A., and D. E. Canfield (1989). A new model of atmospheric oxygen over Phanerozoic time. *Am. J. Sci.* **289**, 333–361.

Berner, R. A., and D. M. Rye (1992). Calculation of the Phanerozoic strontium isotope record of the oceans from a carbon cycle model. *Am. J. Sci.* **292**, 136–148.

Berner, R. A., and K. A. Maasch (1996). Chemical weathering and controls on atmospheric O_2 and CO_2: Fundamental principles were enunciated by J. J. Ebelmen in 1845. *Geochem. Cosmochim. Acta* **60**, 1633–1637.

Berner, W., H. Oeschger, and B. Stauffer (1980). Information on the CO_2 cycle from ice core studies. *Radiocarbon* **22**, 227–235.

Bigg, G. R., M. R. Wadley, D. P. Stevens, and J. A. Johnson (1998). Simulations of two last glacial maximum ocean states. *Paleoceanography* **13**, 340–351.

Bills, B. (1994). Obliquity–oblateness feedback: Are climatically sensitive calues of obliquity dynamically unstable. *Geophys. Res. Lett.* **21**, 177–180.

Bills, B. G. (1998). An oblique view of climate. *Nature* **396**, 405–406.

Birchfield, G. E. (1977). A study of the stability of a model continental ice sheet subject to periodic variations in heat input. *J. Geophys. Res.* **82**, 4909–4913.

Birchfield, G. E. (1987). Changes in deep-ocean water $\delta^{18}O$ and temperature from the last glacial maximum to the present. *Paleoceanography* **2**, 431–442.

Birchfield, G. E. (1989). A coupled ocean-atmosphere climate model: Temperature versus salinity effects on the thermohaline circulation. *Climate Dynam.* **4**, 57–71.

Birchfield, G. E., and M. Ghil (1993). Climate evolution in the Pliocene and Pleistocene from marine-sediment records and simulations: Internal variability versus orbital forcing. *J. Geophys. Res.* **98**, 10, 385–399.

Birchfield, G., and R. Grumbine (1985). 'Slow' physics of large continental ice sheets and underlying bedrock and its relation to the Pleistocene ice ages. *J. Geophys. Res.* **90**, 11294–11302.

Birchfield, G. E., and J. Weertman (1978). A note on the spectral response of a model continental ice sheet. *J. Geophys. Res.* **83**, 4123–4125.

Birchfield, G., J. Weertman, and A. Lunde (1981). A paleoclimate model of northern hemisphere ice sheets. *Quartern. Res.* **15**, 126–142.

Birchfield, G. E., H. Wang, and M. Wyant (1990). A bimodal climate response controlled by water vapor transport in a coupled ocean-atmosphere box model. *Paleoceanography* **5**, 383–395.

Birchfield, E. G., H. Wang, and J. J. Rich (1994). Century/millenium internal climate oscillations in an ocean–atmosphere–continental ice sheet model. *J. Geophys. Res.* **99**, 12, 459–470.

Bond, G., W. Broecker, S. Johnson, J. McManus, L. Labeyrie, J. Jouzel, and G. Bonani (1993). Correlations between climate records from North Atlantic sediments and Greenland ice. *Nature* **365**, 143–147.

Boulton, G., and K. Dobbie (1993). Consolidation of sediments by glaciers: Relations between sediment geotechnics, soft-bed glacier dynamics and subglacial groundwater flow. *J. Glaciol.* **39**, 26–44.

Boulton, G. S., and R. C. A. Hindmarsh (1987). Sediment deformation beneath glaciers: Rheology and geological consequences. *J. Geophys. Res.* **92**, 9059–9092.

Boyce, W. E., and R. C. DiPrima (1977). "Elementary Differential Equations and Boundary Value Problems." Wiley and Sons, New York.

Boyle, E. A. (1988a). Cadmium: Chemical tracer of deepwater paleoceanography. *Paleoceanography* **3**, 471–489.

Boyle, E. A. (1988b). The role of vertical chemical fractionation in controlling late Quaternary atmospheric carbon dioxide. *J. Geophys. Res.* **93**, 701–715.

Boyle, E. A., and L. D. Keigwin (1982). Deep circulation of the North Atlantic over the last 200,000 years: Geochemical evidence. *Science* **218**, 784–787.

Boyle, E., and A. Weaver (1994). Ocean circulation—Conveying past climates. *Nature* **373**, 41–42.

Bradley, R. S. (1999). "Paleoclimatology: Reconstructing Climates of the Quaternary," 2nd Ed. Academic Press, San Diego.

Brass, G. W., J. R. Southern, and W. H. Peterson (1982). Warm saline bottom water in the ancient ocean. *Nature* **296**, 620–623.

Brickman, D., W. Hyde, and D. G. Wright (1999). Filtering of Milankovitch cycles by the thermohaline circulation. *J. Climate* **12**, 1644–1658.

Broccoli, A., and S. Manabe (1987). The influence of continental ice, atmospheric CO_2 and land albedo on the climate of the last glacial maximum. *Climate Dynam.* **1**, 87–99.

Broccoli, A., and E. Marciniak (1996). Comparing simulated glacial climate and paleodata: A reexamination. *Paleoceanography* **11**, 3–14.

Broecker, W. S. (1982). Glacial to interglacial changes in ocean chemistry. *Prog. Oceanogr.* **11**, 151–197.

Broecker, W. S. (1994). Massive iceberg discharges as triggers for global climatic change. *Nature* **372**, 421–424.

Broecker, W., and G. Denton (1989). The role of ocean–atmosphere reorganisations in glacial cycles. *Geochem. Cosmochim. Acta* **53**, 2465–2501.

Broecker, W.S., and T.-H. Peng (1982). "Tracers in the Sea." Eldigio Press, Lamont-Doherty Geological Observatory, Palisades, New York.

Broecker, W. S., and T.-H. Peng (1989). The cause of the glacial to interglacial atmospheric CO_2 change: A polar alkalinity hypothesis. *Global Biogeochem. Cycles* **3**, 215–239.

Broecker, W. S., and J. van Donk (1970). Insolation changes, ice volumes, and the O^{18} record in deep-sea cores. *Rev. Geophys. Space Phys.* **8**, 169–198.

Broecker, W., G. Bond, M. Klas, E. Clark, and J. McManus (1992). Origin of the North Atlantic's Heinrich events. *Climate Dynam.* **6**, 265–273.

Bryan, F. (1986). High-latitude salinity effects and interhemispheric thermohaline circulations. *Nature* **323**, 301–304.

Budd, W., and D. Jenssen (1989). The dynamics of the Antarctic ice sheet. *Ann. Glaciol.* **12**, 16–22.

Budd, J. F., and B. McInnes (1975). Modelling of periodically surging glaciers. *Science* **186**, 925–927.

Budd, W. F., and P. Rayner (1990). Modelling global ice and climate changes through the ice ages. *Ann. Glaciol.* **14**, 23–27.

Budd, W. F., and I. N. Smith (1981). The growth and retreat of ice sheets in response to orbital radiation changes. "Sea Level, Ice, and Climatic Change," pp. 369–409. IAHS Publ. No. 131.

Budd, W. F., B. Coutts, and R. C. Warner (1998). Modelling the Antarctic and Northern Hemisphere ice-sheet changes with global climate through the glacial cycle. *Ann. Glaciol.* **27**, 153–160.

Budyko, M. I. (1969). The effect of solar radiation variations on the climate of the earth. *Tellus* **21**, 611–619.

Budyko, M. (1974). *In* "Climate and Life," p. 508. Academic Press, New York.

Budyko, M. I. (1977). On present-day climatic changes. *Tellus* **29**, 193–204.

Budyko, M. (1982). *In* "The Earth's Climate: Past and Future," p. 307. Academic Press, New York.

Budyko, M. I., A. B. Ronov, and A. L. Yanshin (1987). "History of the Earth's Atmosphere." Springer, Berlin.

Bush, A., and S. Philander (1997). The late Cretaceous: Simulation with a coupled atmosphere–ocean general circulation model. *Paleoceanography* **12**, 495–516.

Cahalan, R. F., and G. R. North (1979). A stability theorem for energy-balance climate models. *J. Atmos. Sci.* **36**, 1205–1216.

Caldeira, K. (1992). Enhanced Cenozoic chemical weathering and the subduction of pelagic carbonate. *Nature* **357**, 578–581.

Calder, N. (1974). Arithmetic of ice ages. *Nature* **252**, 216–218.

Callendar, G. (1938). The artificial production of carbon dioxide and its influence on temperature. *Q. J. R. Meteorol. Soc.* **64**, 223–237.

Calov, R., and I. Marsiat (1998). Simulations of the Northern Hemisphere through the last glacial–interglacial cycle with a vertically integrated and a three-dimensional thermomechanical ice-sheet model coupled to a climate model. *Ann. Glaciol.* **27**, 169–176.

Cerling, T. E. (1991). Carbon-dioxide in the atmosphere: Evidence from Cenozoic and Mesozoic paleosols. *Am. J. Sci.* **291**, 377–400.

Cerling, T. E., J. M. Harris, B. J. MacFadden, M. G. Leakey, J. Quade, V. Eisenman, and J. R. Ehleringer (1997). Global vegetation change through the Miocene/Pliocene boundary. *Nature* **389**, 153–158.

Cessi, P., and W. R. Young (1992). Multiple equilibria in two-dimensional thermohaline circulation. *J. Fluid Mech.* **241**, 291–309.

Chalikov, D. V., and M. Y. Verbitsky (1984). A new Earth's climate model. *Nature* **308**, 609–612.

Chalikov, D. V., and M. Y. Verbitsky (1990). Modelling the pleistocene ice age. *Adv. Geophys.* **32**, 75–131.

Chamberlin, T. C. (1899). An attempt to frame a working hypothesis of the cause of glacial periods on an atmospheric basis. *J. Geology* **7**, 545–584, 667–685, 751–787.

Chamberlin, T. C. (1906). On a possible reversal of deep-sea circulation and its influence on geologic climates. *J. Geology* **14**, 363–373.

Chapman, M. R., and N. J. Shackleton (1999). Global ice-volume fluctuations, North Atlantic ice-rafting events, and deep-ocean circulation changes between 130 and 70 ka. *Geology* **27**, 795–798.

Chapman, M. R., and N. J. Shackleton (1986). Oxygen isotopes and sea level. *Nature* **324**, 137–140.

Chen, F., and M. Ghil (1995). Interdecadal variability of the thermohaline circulation and high-latitude surface fluxes. *J. Phys. Oceanogr.* **25**, 2547–2568.

Clark, P. (1994). Unstable behavior of the Laurentide Ice Sheet over deforming sediment and its implications for climate change. *Quartern. Res.* **41**, 19–25.

Clark, P. V., and D. Pollard (1998). Origin of the middle Pleistocene transition by ice sheet erosion of regolith. *Paleoceanography* **13**, 1–9.

Clark, P. V., R. B. Alley, and D. Pollard (1999). Northern hemisphere ice-sheet influences on global climate change. *Science* **286**, 1104–1111.

CLIMAP Project Members (1976). The surface of the ice-age earth. *Science* **191**, 1131–1137.

Collins, L. S., A. G. Coates, W. A. Berggren, M.-P. Aubrey, and J. Zhang (1996). The late-Miocene Panama isthmian strait. *Geology* **24**, 687–690.

Crawford, E. (1996). "Arrhenius: From Ionic Theory to the Greenhouse Effect." Watson, Canton, Mass.

Croll, J. (1864). On the physical cause of the change of climate during the geological epochs. *Philosophical Mag.* **28**, 121–137.

Croll, J. (1875). *In* "Climate and Time in Their Geologic Relations. A Theory of Secular Change of the Earth's Climate," p. 577. Stanford, London.

Cronin, T. M. (1999). "Principles of Paleoclimatology." Columbia University Press, New York.

Crowley, T. J. (1991). Modelling Pliocene warmth. *Quatern. Sci. Rev.* **10**, 275–282.

Crowley, T. (1994). *In* "Pangean Climates," pp. 25–39. Geological Society of America.

Crowley, T., and S. Baum (1992). Modeling late paleozoic glaciation. *Geology* **20**, 507–510.

Crowley, T. J., and S. K. Baum (1995). Reconciling Late Ordovician (440 Ma) glaciation with very high (14X) CO_2 levels. *J. Geophys. Res.* **100**, 1093–1101.

Crowley, T. J., and K. C. Burke, eds. (1998). "Tectonic Boundary Conditions for Climate Reconstructions." Oxford University Press, New York.

Crowley, T. J., and G. R. North (1988). Abrupt climate change and extinction events in earth history. *Science* **240**, 996–1002.

Crowley, T., and G. North (1991). *In* "Paleoclimatology," p. 339. Oxford University Press, New York.

Crowley, T. J., D. A. Short, J. G. Mengel, and G. R. North (1986). Role of seasonality in the evolution of climate during the last 100 million years. *Science* **231**, 579–584.

Dansgaard, W., S. J. Johnson, H. B. Clausen, and C. C. Langway, Jr. (1971). Climatic record revealed by the Camp Century ice core. *In* "The Late Cenozoic Glacial Ages" (K. K. Turekian, ed.), pp. 37–56. Yale Univ. Press, New Haven.

DeBlonde, G., and W. Peltier (1991a). A one-dimensional model of continental ice-volume fluctuations throught the pleistocene: Implications for the origin of the mid-pleistocene climate transition. *J. Climate* **4**, 318–344.

DeBlonde, G., and W. Peltier (1991b). Simulations of continental ice sheet growth over the last glacial–interglacial cycle: Experiments with a one level seasonal energy balance model including realistic geography. *J. Geophys. Res.* **96**, 9189–9215.

DeBlonde, G., and W. Peltier (1993). Late pleistocene ice age scenarios based on observational evidence. *J. Climate* **6**, 709–727.

DeBlonde, G., W. Peltier, and W. Hyde (1992). Simulations of continental ice sheet growth over the last glacial–interglacial cycle: Experiments with a one level seasonal energy balance model including seasonal ice albedo feedback. *Palaeogeogr. Palaeoclimatol. Palaeoecol.* **98**, 37–55.

Delmas, R. J., J.-M. Ascencio, and M. Legrand (1980). Polar ice evidence that atmospheric CO_2 20,000 yr BP was 50% of present. *Nature* **284**, 155–157.

Denton, G. H., and T. J. Hughes (1981). "The Last Great Ice Sheets." J. Wiley & Sons, New York.

Denton, G. H., C. J. Heusser, T. V. Lowell, P. I. Moreno, B. G. Andersen, L. E. Heusser, C. Schluchter, and D. R. Marchant (1999). Interhemispheric linkage of paleoclimate during the last glaciation. *Geograf. Ann.* **81A** (2), 107–153.

Dodge, R. E., R. G. Fairbanks, L. K. Benninger, and F. Maurrasse (1983). Pleistocene sea levels from raised coral reefs of Haiti. *Science* **219**, 1423–1425.

Ebelmen, J. J. (1845). Sur les produits de la décomposition des especes minérales de la famille des silicates. *Ann. Mines* **7**, 3–66.
Ebelmen, J. J. (1847). Sur la décomposition des roches. *Ann. Mines* **12**, 627–654.
Ebelmen, J. J. (1855). Recherches sur les altérations des roches stratifées par les agents atmosphériques et les eaux d'infiltration. "Recueil des Travaux Scientifiques de M. Ebelman" (posthumous), 2, pp. 1–79. Mallet-Bachelier.
Egger, J. (1999). Internal fluctuations in an ocean-atmosphere box model with sea-ice. *Climate Dynam.* **15**, 595–604.
Emiliani, C. (1955). Pleistocene temperatures. *J. Geol.* **63**, 538–578.
Engebretson, D. C., K. P. Kelley, H. J. Cashman, and M. A. Richards (1992). 180 million years of subduction. *GSA Today* **2**, 93–95, 100.
Engelhardt, H., and B. Kamb (1997). Basal hydraulic system of a West Antarctic ice stream: Constraints from borehole observations. *J. Glaciol.* **43**, 207–230.
Engelhardt, H., and B. Kamb (1998). Basal sliding of ice stream B, West Antarctica. *J. Glaciol.* **44**, 223–230.
Epstein, S., R. Buchsbaum, H. A. Lowenstam, and H. C. Urey (1953). Revised carbonate-water isotopic temperature scale. *Geol. Soc. Am. Bull.* **64**, 1315–1326.
Eriksson, E. (1963). Possible fluctuations in atmospheric carbon dioxide due to changes in the properties of the sea. *J. Geophys. Res.* **68**, 3871–3876.
Eriksson, E. (1968). Air-ocean-ice cap interactions in relation to climatic fluctuations and glaciation cycles. *Meteorol. Monogr.* **8**, No. 30, 68–92.
Eriksson, E., and P. Welander (1956). On a mathematical model of the carbon cycle in nature. *Tellus* **8**, 155–175.
Fairbanks, R. (1989). A 17,000-year glacio-eustatic sea-level record: Influence of glacial melting rates on younger dryas event and deep-ocean circulation. *Nature* **342**, 637–642.
Farrell, J. W., and W. L. Prell (1991). Pacific $CaCO_3$ preservation and $\delta^{18}O$ since 4 Ma: Paleoceanic and paleoclimatic implications. *Paleoceanography* **6**, 485–498.
Fastook, J. L. (1983). Sea-level control of ice sheet disintegration. *In* "Variations in the Global Water Budget" (A. Street-Perrott *et al.*, eds.), pp. 391–401. D. Reidel, London.
Feely, R. A., R. Wanninkhof, C. Goyet, C. Archer, and T. Takahashi (1997). Variability of CO_2 distributions and sea-air fluxes in the central and eastern equatorial Pacific during the 1991–94 El Nino. *Deep-Sea Res. II* **44**, 1851–1867.
Feely, R. A., R. Wanninkhof, T. Takahashi, and P. Tans (1999). Influence of El Niño on the equatorial Pacific contribution to atmospheric CO_2 accumulation. *Nature* **398**, 597–601.
Felzer, B., R. Oglesby, H. Shao, T. Webb III, D. Hyman, and J. Kutzbach (1995). A systematic study of GCM sensitivity to latitudinal changes in solar radiation. *J. Climate* **8**, 877–887.
Felzer, B., R. Oglesby, T. Webb III, and D. Hyman (1996). Sensitivity of a general circulation model to changes in northern hemisphere ice sheets. *J. Geophys. Res.* **101**, 19077–19092.
Felzer, B., T. Webb III, and R. Oglesby (1998). The impact of ice sheets, CO_2 and orbital isolation on late Quarternary climates: Sensitivity experiments with a general circulation model. *Quartern. Sci. Rev.* **17**, 507–534.
Felzer, B., T. Webb III, and R. J. Oglesby (1999). Climate model sensitivity to changes in boundary conditions during the last glacial maximum. *Paleoclimates* **3**, 257–278.
Fichefet, T., S. Hovine, and J.-C. Duplessy (1994). A model study of the Atlantic thermohaline circulation during the Last Glacial Maximum. *Nature* **372**, 252–255.
Flint, R. F. (1974). Three theories in time. *Quartern. Res.* **4**, 1–8.
Fong, P. (1982). Latent heat of melting and its importance for glaciation cycles. *Climate Change* **4**, 199–206.
Fraedrich, K. (1978). Structural and stochastic analysis of a zero-dimensional climate system. *Q. J. R. Meteorol. Soc.* **104**, 461–474.
Fraedrich, K. (1979). Catastrophes and resilience of a zero-dimensional climate system with ice-albedo and greenhouse feedback. *Q. J. R. Meteorol. Soc.* **105**, 147–167.
Frakes, L. (1979). *In* "Climate throughout Geologic Time," p. 310. Elsevier Publishing Company, New York.
Frakes, L. A., J. E. Francis, and J. I. Syktus (1992). "Climate Modes of the Phanerozoic." Cambridge University Press, New York.

Francois, R., M. A. Altabet, E.-F. Yu, D. M. Sigman, M. P. Bacon, M. Frank, G. Bohrmann, G. Bareille, and L. D. Labeyrie (1997). Contribution of Southern Ocean surface-water stratification to low atmospheric CO_2 concentrations during the last glacial period. *Nature* **389**, 929–935.

Frank, M., B. C. Reynolds, and R. K. O'Nions (1999). Nd and Pb isotopes in Atlantic and Pacific water masses before and after closure of the Panama gateway. *Geology* **27**, 1147–1150.

Franzén, L. G. (1994). Are wetlands the key to the ice-age cycle enigma? *Ambio* **23**, 300–308.

Franzén, L. G., D. Chen, and L. F. Klinger (1996). Principles for a climate regulation mechanism during the late Phanerozoic era, based on carbon fixation in peat-forming wetlands. *Ambio* **25**, 435–442.

Freeman, K. H., and J. M. Hayes (1992). Fractionation of carbon isotopes by phytoplankton and estimates of ancient CO_2 levels. *Global Biogeochem. Cycles* **6**, 185–198.

Friedli, H., H. Loetscher, H. Oeschger, U. Siegenthaler, and B. Stauffer (1986). Ice core record of the $^{13}C/^{12}C$ ratio of atmospheric CO_2 in the past two centuries. *Nature* **324**, 237–238.

Friedlingstein, P., I. Fung, E. Holland, J. John, G. Brasseur, D. Erickson, and D. Schimel (1995). On the contribution of CO_2 fertilization to the missing biospheric sink. *Global Geochem. Cycles* **9**, 541–556.

Fultz, D., R. R. Long, G. V. Owens, W. Bohan, R. Kaylor, and J. Weil (1959). Studies of thermal convection in a rotating cylinder with some implications for large-scale atmospheric motions. *Meteorol. Monogr.* **4**, 1–104.

Gaffin, S. (1987). Ridge volume dependence on seafloor generation rate and inversion using long term sealevel change. *Am. J. Sci.* **287**, 596–611.

Gagan, M. K., L. K. Ayliffe, J. W. Beck, J. E. Cole, E. R. M. Druffel, R. B. Dunbar, and D. P. Schrag (2000). New views of tropical paleoclimates from corals. *Quatern. Sci. Rev.* **19**, 45–64.

Gallée, H., J. P. van Ypersele, T. Fichefet, C. Tricot, and A. Berger (1991). Simulation of the last glacial cycle by a coupled, sectorially averaged climate-ice sheet model 1. The climate model. *J. Geophys. Res.* **96**, 13139–13161.

Gallée, H., J. P. van Ypersele, T. Fichefet, I. Marsiat, C. Tricot, and A. Berger (1992). Simulation of the last glacial cycle by a coupled, sectorially averaged climate-ice sheet model 2. Response to insolation and CO_2 variations. *J. Geophys. Res.* **97**, 15713–15740.

Gallimore, R., and J. Kutzbach (1995). Snow cover and sea-ice sensitivity to generic changes in earth orbital parameters. *J. Geophys. Res.* **100**, 1103–1120.

Gammaitoni, L., P. Hänggi, P. Jung, and F. Marchesoni (1998). Stochastic resonance. *Rev. Mod. Phys.* **70**, 223–287.

Ganopolski, A., S. Rahmstorf, V. Petoukhov, and M. Claussen (1998). Simulation of modern and glacial climates with a coupled global model of intermediate complexity. *Nature* **391**, 351–356.

Gates, W. (1976). Modeling the ice-age climate. *Science* **191**, 1138–1144.

Gates, W. L. (1981). Paleoclimatic modeling—A review of problem and prospects for the pre-Pleistocene. *Report No. 27, Climatic Research Institute, Oregon State Univ.*

Ghil, M. (1976). Climate stability for a Sellers-type model. *J. Atmos. Sci.* **33**, 3–20.

Ghil, M. (1984). Climate sensitivity, energy balance models and oscillatory climate models. *J. Geophys. Res.* **89**, 1280–1284.

Ghil, M., and S. Childress (1987). "Topics in Geophysical Fluid Dynamics: Atmospheric Dynamics, Dynamo Theory, and Climate Dynamics." Springer-Verlag, New York.

Ghil, M., and H. LeTreut (1981). A climate model with cryodynamics and geodynamics. *J. Geophys. Res.* **86**, 5262–5270.

Ghil, M., A. Mulhaupt, and P. Pestiaux (1987). Deep water formation and Quarternary glaciations. *Climate Dynam.* **2**, 1–10.

Gill, A. E. (1982). "Atmosphere-Ocean Dynamics." Academic Press, Orlando.

Goosse, H., J. M. Campin, T. Fichefet, and E. Deleersnijder (1997). Sensitivity of a global ice–ocean model to the Bering Strait throughflow. *Climate Dynam.* **13**, 349–358.

Goulden, M. L., S. C. Wofsy, J. W. Harden, S. E. Trumbore, P. M. Crill, S. T. Gower, T. Fries, B. C. Daube, S.-M. Fan, D. J. Sutton, A. Bazzaz, and J. W. Munger (1998). Sensitivity of boreal forest carbon balance to soil thaw. *Science* **279**, 214–217.

Grassberger, P., and I. Procaccia (1983). Measuring the strangeness of strange attractors. *Physica* **9D**, 189–208.

Greenland Summit Ice Cores (1998). Reprinted from *J. Geophys. Res.*, pp. 26315–26886. American Geophysical Union, Washington D. C.

Greve, R., and K. Hutter (1995). Polythermal three-dimensional modelling of the greenland ice sheet with varied geothermal heat flux. *Ann. Glaciol.* **21**, 8–12.

Greve, R., and D. MacAyeal (1996). Dynamic/thermodynamic simulations of laurentide ice sheet instability. *Ann. Glaciol.* **23**, 328–335.

Griffies, S. M., and E. Tziperman (1995). A linear thermohaline oscillator driven by stochastic atmospheric forcing. *J. Climate* **8**, 2440–2453.

Grigoryan, S., M. Krass, and P. Shumskiy (1976). Mathematical model of a three-dimensional non-isothermal glacier. *J. Glaciol.* **17**, 401–417.

Grimm, E. C., G. L. Jacabson, W. A. Watts, B. C. S. Hansen, and K. A. Maasch (1993). A 50,000-year record of climate oscillations from Florida and its temporal correlation with Heinrich events. *Science* **261**, 198–200.

Grousset, F., L. Labeyrie, A. Sinko, M. Cremer, G. Bond, J. Duprat, E. Cortijo, and S. Huon (1993). Patterns of ice-rafted detritus in the glacial North Atlantic (40–55°N). *Paleoceanography* **8**, 175–192.

Haidvogel, D. B., and F. O. Bryan (1992). Ocean general circulation modeling. *In* "Climate System Modeling" (K. E. Trenberth, ed.), pp. 371–412. Cambridge Univ. Press, New York.

Haken, H. (1983). *In* "Synergetics: An Introduction," p. 508. Springer Verlag, Heidelberg.

Hall, N., and P. Valdes (1997). A GCM simulation of the climate 6000 years ago. *J. Climate* **10**, 3–17.

Hallam, A. (1984). Pre-Quaternary sea level changes. *Ann. Rev. Earth Planet. Sci.* **12**, 205–243.

Haney, R. L. (1971). Surface thermal boundary condition for ocean circulation models. *J. Phys. Oceanograp.* **1**, 241–248.

Hargraves, R. B. (1976). Precambrian geologic history. *Science* **193**, 363–371.

Hartmann, D. L. (1994). "Global Physical Climatology." Academic Press, San Diego.

Hasselmann, K. (1976). Stochastic climate models, Part 1. Theory. *Tellus* **28**, 473–485.

Haug, G., and R. Tiedemann (1998). Effect of the formation of the Isthmus of Panama on Atlantic Ocean thermohaline circulation. *Nature* **393**, 673–676.

Hays, J. D., J. Imbrie, and N. J. Shackleton (1976). Variations in the earth's orbit: Pacemaker of the ice ages. *Science* **194**, 1121–1132.

Hecht, A. (1985). *In* "Paleoclimate analysis and modeling," p. 445. J. Wiley and Sons, New York.

Heinrich, H. (1988). Origin and consequences of cyclic ice rafting in the Northeast Atlantic Ocean during the past 130,000 years. *Quartern. Res.* **29**, 143–152.

Held, I. (1983). *In* "Stationary and Quasi-Stationary Eddies in the Extratropical Troposphere: Theory," pp. 127–168. Academic Press, New York.

Herterich, K. (1988). A three-dimensional model of the Antarctic ice sheet. *Ann. Glaciol.* **11**, 32–35.

Hewitt, C. D., and J. F. B. Mitchell (1997). Radiative forcing and response of a GCM to ice age boundary conditions: Cloud feedback and climate sensitivity. *Climate Dynam.* **13**, 821–834.

Hindmarsh, R. (1993). Modelling the dynamics of ice sheets. *Prog. Phys. Geogr.* **17**, 391–412.

Hodell, D. A. (1993). Late Pleistocene paleoceanography of the South Atlantic sector of the southern ocean: Ocean drilling program hole 704A. *Paleoceanography* **8**, 47–67.

Högbom, A. G. (1894). On the probability of secular variations of atmospheric carbon dioxide. *Svensk Kimisk Tidskrift* **6**, 169–176 (in Swedish).

Hollander, D. J., and J. A. McKenzie (1991). CO_2 control on carbon-isotope fractionation during aqueous photosynthesis: A paleo-pCO_2 barometer. *Geology* **19**, 929–932.

Hooghiemstra, H., and T. van der Hammen (1998). Neogene and quaternary development of the neotropical rain forest: The forest refugia hypothesis, and a literature overview. *Earth Sci. Rev.* **44**, 147–183.

Hooke, R. L., B. Hanson, N. R. Iverson, P. Jansson, and U. H. Fischer (1997). Rheology of till beneath Storglaciaren, Sweden. *J. Glaciol.* **43**, 172–180.

Hostetler, S. W., P. U. Clark, P. J. Bartlein, A. C. Mix, and N. J. Pisias (1999). Atmospheric transmission of North Atlantic Heinrich events. *J. Geophys. Res.* **104**, 3947–3952.

Houghton, J., G. Jenkins, and J. Ephraums (eds.) (1990). *In* "Climate Change: The IPCC Scientific Assessment," p. 365. Cambridge University Press, New York.

Hughes, T. (1996). The structure of a pleistocene "glaciation cycle. *In*" The Minerals, Metals and Materials Society: The Johannes Westerman Symposium (R. Arsenalt, ed.), pp. 375–399.

Hughes, T. (1998). *In* "Ice Sheets," p. 399. Oxford University Press, New York.

Hunt, B. (1979). Effects of past variations of the earths rotation rate on climate. *Nature* **281**, 188–191.

Hutter, T. (1983). *In* "Theoretical Glaciology: Material Science of Ice and the Mechanisms of Glaciers and Ice Sheets," p. 399. D. Reidel Publishing Company, Dordrecht.

Huybrechts, P. (1990). A 3-D model of the antarctic ice sheet: A sensitivity study on the glacial–interglacial contrast. *Climate Dynam.* **5**, 79–92.

Huybrechts, P. (1992). The antarctic ice sheet and environmental change: A three dimensional modelling study. *Ber. Polarforch*, p. 99.

Huybrechts, P. (1994). Formation and disintegration of the antarctic ice sheet. *Ann. Glaciol.* **20**, 336–340.

Huybrechts, P., and J. Oerlemans (1988). Evolution of the East Antarctic ice sheet: A numerical study of thermo-mechanical response patterns with changing climate. *Ann. Glaciol.* **11**, 52–59.

Hyde, W. T., and W. R. Peltier (1985). Sensitivity experiments with a model of the ice age cycle: The response to harmonic forcing. *J. Atmos. Sci.* **42**, 2170–2188.

Hyde, W. T., T. J. Crowley, L. Tarasov, and W. R. Peltier (1999). The Pangean ice age: Studies with a coupled climate-ice sheet model. *Climate Dynam.* **15**, 619–629.

Imbrie, J. (1982). Astronomical theory of the Pleistocene ice ages: A brief historical review. *Icarus* **50**, 408–422.

Imbrie, J., and J. Z. Imbrie (1980). Modeling the climatic response to orbital variations. *Science* **207**, 943–953.

Imbrie, J., and N. G. Kipp (1971). A new micropaleontological method for quantitative paleoclimatology: Application to a late Pleistocene Caribbean core. *In* "The Late Cenozoic Glacial Ages" (K. K. Turekian, ed.), pp. 71–179. Yale Univ. Press, New Haven.

Imbrie, J., J. D. Hays, D. G. Martinson, A. McIntyre, A. C. Mix, J. J. Morley, N. G. Pisias, W. L. Prell, and N. J. Shackleton (1984). *In* "The Orbital Theory of Pleistocene Climate: Support from a Revised Chronology of the Marine $\delta^{18}O$ Record. Milankovitch and Climate, Part 1 (A. Berger, J. Imbrie, J. Hays, G. Kukla, and B. Saltzman, eds.), pp. 269–305. Reidel, Dordrecht.

Imbrie, J., A. Boyle, S. Clemens, A. Duffy, W. Howard, G. Kukla, J. Kutzbach, D. Martinson, A. McIntyre, A. Mix, B. Molfino, J. Morley, L. Peterson, N. Pisias, W. Prell, M. Raymo, N. Shackleton, and J. Toggweiler (1992). On the structure and origin of major glaciation cycles. 1. Linear response to Milankovitch forcing. *Paleocenography* **7**, 701–738.

Imbrie, J., A. Berger, E. A. Boyle, S. C. Clemens, A. Duffy, W. R. Howard, G. Kukla, J. Kutzbach, D. G. Martinson, A. McIntyre, A. C. Mix, B. Molfino, J. J. Morley, L. C. Peterson, N. G. Pisias, W. L. Prell, M. E. Raymo, N. J. Shackleton, and J. R. Toggweiler (1993). On the structure and origin of major glaciation cycles 2. The 100,000-year cycle. *Paleoceanography* **8**, 699–735.

IPCC (Intergovernmental Panel of Climate Change) (1996). "Climate Change 1995: The Science of Climate Change" (J. T. Houghton, F. G. Meira Filho, B. A. Callander, N. Harris, A. Kattenberg, and K. Maskell, eds.), Cambridge Univ. Press, Cambridge.

Iverson, N. R., P. Jansson, and R. L. Hooke (1994). *In situ* measurement of the strength of deforming subglacial sediments. *J. Glaciol.* **40**, 497–503.

Iverson, N. R., B. Hanson, R. L. Hooke, and P. Jansson (1995). Flow mechanisms of glaciers on soft beds. *Science* **267**, 559–562.

Jacobson, M. C., R. J. Charlson, H. Rodhe, and G. H. Orians, eds. (2000). "Earth System Science." Academic Press, San Diego.

Jenkins, G. (1993). A general circulation model study of the effects of the effects of faster rotation rate, enhanced CO_2 concentrations and reduced solar forcing — Implications for the faint young sun paradox. *J. Geophys. Res.* **98**, 20803–20811.

Jenkins, G. (1996). A sensitivity study of changes in earth's rotation rate with an atmospheric general circulation model. *Global Planet. Change* **11**, 141–154.

Jenkins, G. S., H. G. Marshall, and W. R. Kuhn (1993). Precambrian climate: The effects of land area and earth's rotation rate. *J. Geophys. Res.* **98**, 8785–8791.

Jenssen, D. (1977). A three-dimensional polar ice-sheet model. *J. Glaciol.* **18**, 373–390.

Jiang, X., and W. R. Peltier (1996). Ten million year histories of obliquity and precession: The influence of the ice-age cycle. *Earth Planet. Sci. Lett.* **139**, 17–32.

Johannessen, O. M., E. V. Shalina, and M. W. Miles (1999). Satellite evidence for an Arctic sea ice cover in transformation. *Science* **286**, 1937–1939.

Jones, P. D., T. M. L. Wigley, and P. B. Wright (1986). Global temperature variations between 1861 and 1984. *Nature* **322**, 430–434.

Jouzel, J., C. Lorius, J. Petit, C. Genthon, N. Barkov, V. Kotlyakov, and V. Petrov (1987). Vostok ice core — A continuous isotope temperature record over the last climatic cycle (160,000 years). *Nature* **329**, 403–408.

Jouzel, J., N. Barkov, J. Barnola, M. Bender, J. Chapellaz, C. Genthon, V. Kotlayakov, V. Lipenlov, C. Lorius, J. Petit, D. Raymond, G. Raisbeck, C. Ritz, T. Sowers, M. Steivenard, F. Fiou, and P. Yiou (1993). Extending the Vostok ice-core record of paleoclimate to the penultimate glacial period. *Nature* **364**, 407–412.

Källén, E., C. Crafoord, and M. Ghil (1979). Free oscillations in a climate model with ice-sheet dynamics. *J. Atmos. Sci.* **36**, 2292–2303.

Karol, I., and E. Rozanov (1982). Radiative convective models of climate. *Izvestia Atm. Oceanic Phys.* **18**, 910–922.

Kasting, J. (1989). Long-term stability of the earths climate. *Global Planet. Change* **75**, 83–95.

Keeling, C. D., R. B. Bacastow, A. F. Carter, S. C. Piper, T. P. Whorf, M. Heinmann, W. G. Mook, and H. Roeloffzen (1989). A three dimensional model of atmospheric CO_2 transport based on observed winds: Analysis of observational data. *In* "Aspects of Climate Variability in the Pacific and the Western Americas" (D. H. Peterson, ed.). *Geogr. Monogr.* **55**, 305–363. AGU, Washington.

Keffer, T., D. Martinson, and B. Corliss (1988). The position of the Gulfstream during Quarternary glaciation. *Science* **241**, 440–442.

Keir, R. S., and W. H. Berger (1983). Atmospheric CO_2 content in the last 120,000 years: The phosphate extraction model. *J. Geophys. Res.* **88**, 6027–6038.

Kellogg, W. W. (1978). Global influences of mankind on the climate. *In* "Climate Change" (J. Gribbin, ed.), pp. 205–227. Cambridge University Press, Cambridge.

Kellogg, T. (1987). Glacial–interglacial changes in global deepwater circulation. *Paleocenography* **2**, 259–271.

Kennett, J. P. (1981). "Marine Geology." Prentice-Hall, Englewood Cliffs, N.J.

Kerrick, D. M., and K. Caldeira (1993). Paleoatmospheric consequences of CO_2 released during early Cenozoic regional metamorphism in the Tethyan orogen. *Chem. Geology* **108**, 201–230.

Khodakov, V. (1965). O zavisimosti summarnoy ablyatsii poverkhnosti lednikov ot temperatury vozdukha [relationship between the sum of ablation of the glacier surface and air temperature]. *Meteorol. Ghidrol.* **7**, 48–50.

Klinger, L. F. (1991). Peatland formation and ice ages: A possible Gaian mechanism related to community succession. *In* "Scientists on Gaia" (S. H. Schneider and P. J. Boston, eds.), pp. 247–255. M.I.T. Press, Cambridge, Mass.

Klinger, L. F., J. A. Taylor, and L. G. Franzén (1996). The potential role of peatland dynamics in ice-age initiation. *Quatern. Res.* **45**, 89–92.

Knox, F., and M. McElroy (1984). Changes in atmospheric CO_2: Influence of marine biota at high latitudes. *J. Geophys. Res.* **89**, 4629–4637.

Komhyr, W. D., R. H. Gammon, T. B. Harris, L. S. Waterman, T. J. Conway, W. R. Taylor, and K. W. Thoning (1985). Global atmospheric CO_2 distribution and variations from 1968–1982 NOAA/GMCC CO_2 flask sampling data. *J. Geophys. Res.* **90**, 5567–5596.

Kortenkamp, S. J., and S. F. Dermott (1998). A 100,000-year periodicity in the accretion rate of interplanetary dust. *Science* **280**, 874–876.

Kothavala, Z., R. Oglesby, and B. Saltzman (1999). Sensitivity of equilibrium surface temperature of CCM3 to systematic changes in atmospheric CO_2. *Geophys. Res. Lett.* **26**, 209–212.

Kothavala, Z., R. Oglesby, and B. Saltzman (2000). Evaluating the climatic response to changes in CO_2 and solar luminosity. *In* "11th Symposium on Global Change Studies," Long Beach, California, pp. 348–351. American Meteorological Society.

Krinsley, D., and W. Wellendorf (1980). Wind velocities determined from the surface textures of sand grains. *Nature* **283**, 372–373.

Kuhn, W. R., J. C. G. Walker, and H. G. Marshall (1989). The effect on earth's surface temperature from variations in rotation rate, continent formation, solar luminosity, and carbon dioxide. *J. Geophys. Res.* **94**, 11,129–11,136.

Kukla, G. (1977). Pleistocene land-sea correlations I: Europe. *Earth Sci. Rev.* **13**, 307–374.

Kump, L. R., S. L. Brantley, and M. A. Arthur (2000). Chemical weathering, atmospheric CO_2, and climate. *Annu. Rev. Earth Planet. Sci.* **28**, 611–667.

Kutzbach, J. (1976). The nature of climate and climatic variations. *Quartern. Res.* **6**, 471–480.

Kutzbach, J. (1980). Estimates of past climate at paleolake Chad, North-Africa, based on a hydrological and energy-balance model. *Quartern. Res.* **14**, 210–223.

Kutzbach, J. (1981). The nature of climate and climatic variations. *Science* **214**, 55–61.

Kutzbach, J. E. (1994). Idealized Pangean climates: Sensitivity to orbital change. *In* "Pangea: Paleoclimate, Tectonics, and Sedimentation during Accretion, Zenith and Breakup of a Supercontinent" (G. D. Klein, ed.), pp. 41–56. Geological Society of America Special Paper 288, Boulder, CO.

Kutzbach, J., and P. Guetter (1986). The influence of changing orbital parameters and surface boundary conditions on climate simulations for the past 18,000 years. *J. Geophys. Res.* **33**, 1726–1759.

Ladurie, E. L. R. (1971). "Times of Feast, Times of Famine: A History of Climate Since the Year 1000." Doubleday and Co.

Lamb, H. H. (1977). "Climate Present, Past and Future," Vol. 2. Methuen & Co., London.

Landau, L. D. (1944). "On the Nature of Turbulence" [reprinted in Collected Papers, 1965 (D. ter Haar, ed.)]. Pergamon Press, Oxford.

Laskar, J. (1988). Secular evolution of the solar system over 10 million years. *Astron. Astrophys.* **198**, 341–362.

Laskar, J. (1999). The limits of earth orbital calculations for geological time-scale use. *Phil. Trans. R. Soc. Lond. A* **357**, 1735–1759.

Lemke, P. (1977). Stochastic climate models, part 3. Application to zonally averaged energy models. *Tellus* **29**, 385–392.

Le Treut, H., and M. Ghil (1983). Orbital forcing, climatic interactions, and glaciation cycles. *J. Geophys. Res.* **88**, 5167–5190.

Liao, X., A. Street-Perrott, and J. F. B. Mitchell (1994). GCM experiments with different cloud parameterization: Comparisons with palaeoclimatic reconstructions for 6000 years B.P. *Palaeoclimates* **1**, 99–123.

Lindstrom, D. R. (1990). The Eurasian ice sheet formation and collapse resulting from natural atmospheric CO_2 concentration variations. *Paleoceanography* **5**, 207–227.

Lindzen, R. S. (1986). A simple model for 100 K-year oscillations in glaciation. *J. Atmos. Sci.* **43**, 986–996.

Liss, P., and L. Merlivat (1986). Air-sea exchange rates: Introduction and synthesis. *In* "The Role of Air-Sea Exchange in Geochemical Cycling" (Baut-Meanrd, ed.), pp. 113–127. Reidel, Dordrecht.

Loehle, C. (1993). Geologic methane as a source for post-glacial CO_2 increases: The hydrocarbon pump hypothesis. *Geophys. Res. Lett.* **20**, 1415–1418.

Lohmann, G., R. Gerdes, and D. Chen (1996). Stability of the thermohaline circulation in a simple coupled model. *Tellus* **48A**, 465–476.

Lorenz, E. N. (1963). Deterministic non-periodic flow. *J. Atmos. Sci.* **20**, 130–141.

Lorenz, E. (1968). Climatic determinism. *Meteorol. Mag.* **5**, 1–3.

Lorenz, E. (1970). Climatic change as a mathematical problem. *J. Appl. Meteorol.* **9**, 325–329.

Lorenz, E. (1975). Climate predictability. *In* "The Physical Basis of Climate and Climate Modelling," pp. 132–136. Geneva, Switzerland.

Lorenz, E. (1993). "The Essence of Chaos." Univ. of Washington Press, Seattle.

Lorius, C., J. Jouzel, and D. Raynaud (1993). Glacials–interglacials in Vostok: Climate and greenhouse gases. *Global Planet. Change* **7**, 131–143.

Lovelock, J. E. (1979). "Gaia, a New Look at Life on Earth." Oxford University Press, Oxford.

Lowe, J., and M. Walker (1997). *In* "Reconstructing Quarternary Environments," 2nd Ed., p. 446. Longman, New York.

Maasch, K. A. (1989). Calculating climate attractor dimension from $\delta^{18}O$ records by the Grassberger–Procaccia algorithm. *Climate Dynam.* **4**, 45–55.

Maasch, K. (1992). Ice age dynamics. *Encycl. Earth Syst. Sci.* **2**, 559–569.

Maasch, K. A., and B. Saltzman (1990). A low-order dynamical model of global climatic variability over the full Pleistocene. *J. Geophys. Res.* **95**, 1955–1963.

MacAyeal, D. (1993). A low-order model for the heinrich-event cycle. *Paleoceanography* **8**, 767–773.

MacAyeal, D., and D. R. Lindstrom (1990). Effects of glaciation on methane-hydrate stability. *Ann. Glaciol.* **14**, 183–185.

MacCracken, M., and S. Ghan (1988). Design and use of zonally-averaged climate models. *In* "Physically-based Modelling and Simulation of Climate and Climatic Change," Part II (M. Schlesinger, ed.), pp. 755–809. Kluwer Academic, Dordrecht.

MacDonald, G. J. F. (1964). Tidal friction. *Rev. Geophys.* **2**, 467–541.

Mahaffy, M. (1976). A three-dimensional numerical model of ice-sheets: Tests on the barnes ice-cap, northwest territories. *J. Geophys. Res.* **81**, 1059–1066.

Maier-Reimer, E., U. Mikolajewicz, K. Hasselmann (1993). Mean circulation of the Hamburg LSG OGCM and its sensitivity to the thermohaline surface forcing. *J. Phys. Oceanogr.* **23**, 731–757.

Manabe, S., and A. Broccoli (1985). The influence of continental ice-sheets on the climate of an ice age. *J. Geophys. Res.* **90**, 2167–2190.

Manabe, S., and K. Bryan, Jr. (1985). CO_2-induced change in a coupled ocean-atmosphere model and its paleo-climatic implications. *J. Geophys. Res.* **90**, 11689–11707.

Manabe, S., and D. Hahn (1977). Simulation of tropical climate of an ice age. *J. Geophys. Res.* **82**, 3889–3911.

Manabe, S., and R. J. Stouffer (1988). Two stable equilibria of a coupled ocean-atmosphere model. *J. Climate* **1**, 841–866.

Marotzke, J., P. Welander, and J. Willebrand (1988). Instability and multiple steady states in a meridional-plane model of the thermohaline circulation. *Tellus* **40A**, 162–172.

Marshall, S., R. Oglesby, J. Larson, and B. Saltzman (1994). A comparison of GCM sensitivity to changes in CO_2 and solar luminosity. *Geophys. Res. Lett.* **21**, 2487–2490.

Marsiat, I. (1994). Simulation of the northern hemisphere continental ice-sheets over the last glacial–interglacial cycle: Experiments with a latitude-longitude vertically integrated ice-sheet model coupled to a zonally-averaged climate model. *Paleoclimates* **1**, 59–98.

Martin, J. H. (1990). Glacial–interglacial CO_2 change: The iron hypothesis. *Paleoceanography* **5**, 1–13.

Maslin, M. A., N. J. Shackleton, and U. Pflaumann (1995). Surface water temperature, salinity, and density changes in the northeast Atlantic during the last 45,000 years: Heinrich events, deep water formation, and climatic rebounds. *Paleoceanography* **10**, 527–544.

Matteucci, G. (1989). Orbital forcing in a stochastic resonance model of the Late-Pleistocene climatic variations. *Climate Dynam.* **3**, 179–190.

Matteucci, G. (1991). A study of the climatic regimes of the Pleistocene using a stochastic resonance model. *Climate Dynam.* **6**, 67–81.

McGuffie, K., and A. Henderson-Sellers (1997). *In* "A Climate Modelling Primer," p. 217. J. Wiley and Sons, New York.

McManus, J. F., D. W. Oppo, and J. L. Cullen (1999). A 0.5-million-year record of millenial-scale climate variability in the North Atlantic. *Science* **283**, 971–975.

Mikolajewicz, U., E. Maier-Reimer, T. J. Crowley, and K.-Y. Kim (1993). Effect of Drake and Panamanian gateways on the circulation of an ocean model. *Paleoceanography* **8**, 409–426.

Milankovitch, M. (1930). Mathematische Klimalehre. *In* "Koppen-Geiger," p. 176. Gebruder Barutrager, Berlin.

Milankovitch, M. (1941). "Kanon der Erdbestrahlung und seine Anvendung auf des Eiszeitproblem." Royal Servian Academy, Belgrade. [English transl., 1969, U. S. Dept. Commerce, Clearinghouse for Fed. Scientific and Technical Inf., Springfield, Va.]

Mitchell, J. M. (1976). An overview of climatic variability and its causal mechanisms. *Quatern. Res.* **6**, 481–493.

Monin, A., and A. Yaglom (1971). *In* "Statistical Fluid Mechanics," p. 769. MIT Press, Cambridge.

Moon, F. C. (1992). "Chaotic and Fractal Dynamics." J. Wiley and Sons, New York.

Mora, C. I., S. G. Driese, and P. G. Seager (1991). Carbon dioxide in the Paleozoic atmosphere: Evidence from carbon-isotope compositions of pedogenic carbonate. *Geology* **19**, 1017–1020.

Moritz, R. E. (1979). Nonlinear analysis of a simple sea-ice ocean temperature oscillator model. *J. Geophys. Res.* **84**, 4916–4920.
Moritz, R. E., and A. Sutera (1981). The predictability problem: Effects of stochastic perturbations in multiequilibrium systems. *Adv. Geophys.* **23**, 345–383.
Moss, F. (1991). Stochastic resonance. *Ber. Bunsenges. Phys. Chem.* **95**, 303–310.
Muller, R., and G. MacDonald (1995). Glacial cycles and orbital inclinations. *Nature* **377**, 107–108.
Munhoven, G. (1997). Modelling glacial–interglacial atmospheric CO_2 variations: The role of continental weathering. Doctoral Dissertation, Univ. of Liege, Belgium.
Munhoven, G., and L. M. Francois (1996). Glacial–interglacial variability of atmospheric CO_2 due to changing continental silicate rock weathering: A model study. *J. Geophys. Res.* **101**, 21423–21437.
Munk, W. H. (1950). On the wind-driven ocean circulation. *J. Meteorol.* **7**, 79–93.
Muszynski, I., and G. E. Birchfield (1987). A coupled marine ice-stream-ice-shelf model. *J. Glaciol.* **33**, 3–15.
Najjar, R. G. (1992). Marine biogeochemistry. *In* "Climate System Modeling" (K. E. Trenberth, ed.). Cambridge Univ. Press, Cambridge.
Nakamura, M., P. H. Stone, and J. Marotzke (1994). Destabilization of the thermohaline circulation by atmospheric eddy transports. *J. Climate* **7**, 1870–1882.
Neeman, B. U., G. Ohring, and J. H. Joseph (1988). The Milankovitch theory and climate sensitivity, 1. Equilibrium climate model solutions for the present surface conditions. *J. Geophys. Res.* **93**, 11153–11174.
Neftel, A., H. Oeschger, J. Schwander, B. Stauffer, and R. Zumbrum (1982). Ice core sample measurements give atmospheric CO_2 content during the past 40,000 yr. *Nature* **295**, 220–223.
Neftel, A., E. Moor, H. Oeschger, and B. Stauffer (1985). Evidence from polar ice cores for the increase in atmospheric CO_2 in the past two centuries. *Nature* **315**, 45–47.
Newell, R. E. (1974). Changes in the poleward energy flux by the atmosphere and ocean as a possible cause for ice ages. *Quatern. Res.* **4**, 117–127.
Newman, M., and R. Rood (1977). Implications of solar evolution for the earth's early atmosphere. *Science* **198**, 1035–1037.
Nicolis, C. (1982). Stochastic aspects of climatic transitions-response to a periodic forcing. *Tellus* **34**, 1–9.
Nicolis, C. (1984). Self-oscillations and predictability in climate dynamics-periodic forcing and phase locking. *Tellus* **36A**, 217–227.
Nicolis, C. (1987). Long-term climatic variability and chaotic dynamics. *Tellus* **39A**, 1–9.
Nicolis, C. (1993). Long-term climatic transitions and stochastic resonance. *J. Statistic. Phys.* **70**, 3–14.
Nicolis, C., and G. Nicolis (1984). Is there a climatic attractor? *Nature* **311**, 529–532.
Nicolis, C., and G. Nicolis (1986). Reconstruction of the dynamics of the climatic system from time-series data. *Proc. Natl. Acad. Sci. U.S.A.* **83**, 536–540.
Nicolis, C., and G. Nicolis (1995). From short-scale atmospheric variability to global climate dynamics: Toward a systematic theory of averaging. *J. Atmos. Sci.* **52**, 1903–1913.
Niiler, P. P. (1992). The ocean circulation. "Climate System Modeling" (K. E. Trenberth, ed.), pp. 117–148. Cambridge Univ. Press, Cambridge.
North, G. R. (1975a). Analytical solution to a simple climate model with diffusive heat transport. *J. Atmos. Sci.* **32**, 1301–1307.
North, G. R. (1975b). Theory of energy-balance climate models. *J. Atmos. Sci.* **32**, 2033–2043.
North, G., and J. A. Coakley (1979). Differences between seasonal and mean annual energy balance model calculations of climate and climate sensitivity. *J. Atmos. Sci.* **36**, 1189–1204.
North, G. R., L. Howard, D. Pollard, and B. Wielicki (1979). Variational formulation of Budyko–Sellers climate models. *J. Atmos. Sci.* **36**, 255–259.
North, G. R., R. F. Cahalan, and J. A. Coakley (1981). Energy balance climate models. *Rev. Geophys. Space Phys.* **19**, 91–121.
North, G., J. Mengel, and D. Short (1983). Simple energy balance model resolving the seasons and the continents: Application to the astronomical theory of ice ages. *J. Geophys. Res.* **88**, 6576–6586.
Oerlemans, J. (1979). A model of a stochastically driven ice sheet with planetary wave feedback. *Tellus* **31**, 469–477.
Oerlemans, J. (1980a). Continental ice sheets and the planetary radiation budget. *Quatern. Res.* **14**, 349–359.

Oerlemans, J. (1980b). Model experiments on the 100,000-yr glacial cycle. *Nature* **287**, 430–432.
Oerlemans, J. (1981a). Modelling of Pleistocene European ice sheets: Some experiments with simple mass balance parameterization. *Quartern. Res.* **15**, 77–85.
Oerlemans, J. (1981b). Some basic experiments with a vertically-integrated ice sheet model. *Tellus* **33**, 1–11.
Oerlemans, J. (1982a). A model of the Antarctic ice sheet. *Nature* **297**, 550–553.
Oerlemans, J. (1982b). Glacial cycles and ice-sheet modelling. *Climatic Change* **4**, 353–374.
Oerlemans, J., and C. Van der Veen (1984). *In* "Ice-Sheets and Climate," p. 217. D. Reidel, Dordrecht.
Oglesby, R. (1989). A GCM study of Antarctic glaciation. *Climate Dynam.* **3**, 135–156.
Oglesby, R., and B. Saltzman (1990a). Sensitivity of the equilibrium surface temperature of a GCM to systematic changes in atmospheric carbon dioxide. *Geophys. Res. Lett.* **17**, 1089–1092.
Oglesby, R. J., and B. Saltzman (1990b). Extending the EBM: The effect of deep ocean temperature on climate with applications to the Cretaceous. *Paleogeogr. Paleoclimatol. Paleoecol.* (*Global Planet. Change Sect.*) **82**, 237–259.
Oglesby, R., and B. Saltzman (1992). Equilibrium climate statistics of a general circulation model as a function of atmospheric carbon dioxide. Part I: Geographic distributions of primary variables. *J. Climate* **5**, 66–92.
Oglesby, R., B. Saltzman, and H. Hu (1997). Sensitivity of GCM simulations of paleoclimate to the initial state. *Paleoclimates* **2**, 33–45.
Ohmura, A., M. Wild, and L. Bengtsson (1996). Present and future mass balance of the ice-sheets simulated with a GCM. *Ann. Glaciol.* **23**, 187–193.
Opdyke, B. N., and J. C. G. Walker (1992). Return of the coral reef hypothesis: Basin to shelf partioning of $CaCO_3$ and its effect on atmospheric CO_2. *Geology* **20**, 733–736.
Otto-Bliesner, B. (1996). Initiation of a continental ice sheet in a global climate model (GENESIS). *J. Geophys. Res.* **101**, 16909–16920.
Otto-Bliesner, B., G. Branstator, and D. Houghton (1982). A global low-order spectral general-circulation model 1. Formulation and seasonal climatology. *J. Atmos. Sci.* **39**, 929–948.
Oxburgh, R. (1998). Variations in the osmium isotope composition of sea water over the past 200,000 years. *Earth Planet. Sci. Lett.* **159**, 183–191.
Pagani, M., M. A. Arthur, and K. H. Freeman (1999). Miocene evolution of atmospheric carbon dioxide. *Paleoceanography* **14**, 273–292.
Paillard, D. (1995). The hierarchical structure of glacial climatic oscillations: Interactions between ice-sheet dynamics and climate. *Climate Dynam.* **11**, 162–177.
Paillard, D. (1998). The timing of Pleistocene glaciations from a simple multiple-state climate model. *Nature* **391**, 378–381.
Paillard, D., and L. Labeyrie (1994). Role of the thermohaline circulation in the abrupt warming after Heinrich events. *Nature* **372**, 162–164.
Parrish, J. T. (1998). "Interpreting Pre-Quaternary Climate from the Geologic Record." Columbia Univ. Press, New York.
Paterson, W. (1994). *In* "The Physics of Glaciers," 3rd Ed., p. 480. Pergamon Press, Oxford.
Pearman, G. I. (1988). Greenhouse gases: Evidence for atmospheric changes and anthropogenic causes. *In* "Greenhouse, Planning for Climate Change" (G. I. Pearman and E. J. Brill, eds.). Leiden, The Netherlands.
Pearman, G. I., D. M. Atheridge, F. de Silva, and P. J. Fraser (1986). Evidence of changing concentrations of atmospheric CO_2, N_2O, and CH_4 from air bubbles in Antarctic ice. *Nature* **320**, 248–250.
Pedlosky, J. (1979). "Geophysical Fluid Dynamics." Springer-Verlag, New York.
Pedlosky, J. (1987). "Geophysical Fluid Dynamics." Springer-Verlag, New York.
Peixoto, J., and A. Oort (1992). *In* "Physics of Climate," p. 520. American Institute of Physics, New York.
Peltier, W. (1982). Dynamics of the ice age earth. *Adv. Geophys.* **24**, 1–146.
Peltier, W. (1994). Ice age paleotopography. *Science* **265**, 195–201.
Peltier, W. R., and W. Hyde (1984). A model of the ice age cycle. *In* "Milankovitch and Climate," Part 2 (A. Berger, J. Imbrie, J. Hays, G. Kukla, and B. Saltzman, eds.), pp. 565–580. Reidel, Amsterdam.
Peltier, W., and X. Jiang (1996). Mantle viscosity from the simulations inversion of multiple data sets pertaining to post glacial rebound. *Geophys. Res. Lett.* **23**, 503–506.

Peltier, W., and S. Marshall (1995). Coupled energy-balance/ice-sheet simulations of the glacial cycle: A possible connection between terminations and terrigeneous dust. *J. Geophys. Res.* **100**, 14269–14289.

Petit, J. R., J. Jouzel, D. Raynaud, N. I. Barkov, J.-M. Barnola, I. Basile, M. Bender, J. Chappellaz, M. Davis, G. Delaygue, M. Delmotte, V. M. Kotlyakov, M. Legrand, V. Y. Lipenkov, C. Lorius, L. Pépin, C. Ritz, E. Saltzman, and M. Stievenard (1999). Climate and atmospheric history of the past 420,000 years from the Vostok ice core, Antarctica. *Nature* **399**, 429–436.

Phillips, P., and I. Held (1994). The response to orbital perturbations in an atmospheric model coupled to a slab ocean. *J. Climate* **7**, 767–782.

Plass, G. N. (1956). The carbon dioxide theory of climate change. *Tellus* **8**, 140–154.

Plass, G. N. (1961). The influence of infrared absorptive molecules on the climate. *Ann. N.Y. Acad. Sci.* **95**, 61–71.

Pollard, D. (1978). An investigation of the astronomical theory of the ice ages using a simple climate-ice sheet model. *Nature* **272**, 233–234.

Pollard, D. (1980). A simple parameterization of ice sheet ablation rate. *Tellus* **32**, 384–388.

Pollard, D. (1982). A simple ice sheet model yields realistic 100 ky glacial cycles. *Nature* **296**, 334–338.

Pollard, D. (1983a). Ice-age simulations with a calving ice-sheet model. *Quartern. Res.* **20**, 30–48.

Pollard, D. (1983b). A coupled climate-ice sheet model applied to the Quaternary ice ages. *J. Geophys. Res.* **88**, 7705–7718.

Pollard, D., and S. L. Thompson (1997). Driving a high resolution dynamic ice sheet model with GCM climate: Ice sheet initiation at 116,000 BP. *Ann. Glaciol.* **25**, 296–304.

Pollard, D., A. P. Ingersoll, and J. G. Lockwood (1980). Response of a zonal climate-ice sheet model to the orbital perturbations during the Quaternary ice ages. *Tellus* **32**, 301–319.

Pond, S., and G. L. Pickard (1983). "Introductory Dynamical Oceanography," 2nd Ed. Pergamon Press, Oxford.

Prahl, F. G., and S. G. Wakeham (1987). Calibration of unsaturation patterns in long-chain ketone compositions for palaeotemperature assessment. *Nature* **330**, 367–369.

Raich, J. W., and C. S. Potter (1995). Global patterns of carbon dioxide emissions from soils. *Global Biogeochem. Cycles* **9**, 23–36.

Ramanathan, V., and J. Coakley (1978). Climate modeling through radiative-convective models. *Rev. Geophys. Space Phys.* **16**, 465–489.

Ramsay, W. (1925). The probable solution of the climate problem in geology. *Smithsonian Rep.* **1924**, Washington D.C., pp. 237–248.

Raymo, M. E., and G. Rau (1994). On glaciations and their causes (abstract). *EOS Trans. Am. Geophys. Union* **75**, 52.

Raymo, M. E., and W. F. Ruddiman (1992). Tectonic forcing of the late Cenozoic. *Nature* **359**, 117–122.

Raymo, M. E., W. F. Ruddiman, and P. N. Froelich (1988). Influence of late Cenozoic mountain building on ocean geochemical cycles. *Geology* **16**, 649–653.

Reusch, D. N., and K. A. Maasch (1998). The transition from arc volcanism to exhumation, weathering of young Ca, Mg, Sr silicates, and CO_2 drawdown. *In* "Tectonic Boundary Conditions for Climate Reconstructions" (T. J. Crowley and K. C. Burke, eds.), pp. 261–276. Oxford University Press, New York.

Reusch, D. R., J. D. Wright, K. A. Maasch, and G. Ravizza (1996). Miocene seawater $^{187}Os/^{186}Os$ ratios inferred from metalliferous carbonates. *EOS Trans. (AGU)* **79F**, 325.

Robock, A. (1978). Internally and externally caused climate change. *J. Atmos. Sci.* **35**, 1111–1122.

Rooth, C. G. H., C. Emiliani, and H. W. Poor (1978). Climate response to astronomical forcing. *Earth Planet. Sci. Lett.* **41**, 387–394.

Rothrock, D. A., Y. Yu, and G. A. Maykut (1999). Thinning of the Arctic sea-ice cover. *Geophys. Res. Lett.* **26**, 3469–3472.

Royer, D. L. (2001). Stomatal density and stomatal index as indicators of paleoatmospheric CO_2 concentration. *Rev. Paleobotany Palynology* **114**, 1–28.

Royer, D. L., R. A. Berner, and D. J. Beerling (2001). Phanerozoic atmospheric CO_2 change: Evaluating geochemical and paleobiological approaches. *Earth Sci. Rev.*, in press.

Ruddick, B., and L. Zhang (1996). Qualitative behavior and nonoscilation of Stommel's thermohaline box model. *J. Climate* **9**, 2768–2777.

Ruddiman, W. F., and A. McIntyre (1977). Late Quaternary surface ocean kinematics and climate change in the high latitude North Atlantic. *J. Geophys. Res.* **82**, 3877–3887.

Ruddiman, W. F., and A. McIntyre (1981). The mode and mechanism of the last deglaciation: Oceanic evidence. *Quatern. Res.* **16**, 125–134.

Ruddiman, W. F., and A. McIntyre (1984). Ice-age thermal response and climatic role of the surface Atlantic Ocean, 40°N to 63°N. *Geol. Soc. Am. Bull.* **95**, 381–396.

Ruddiman, W., A. McIntyre, V. Nieblerhunt, and J. Durazzi (1980). Oceanic evidence for the mechanism of rapid northern hemisphere glaciation. *Quartern. Res.* **13**, 33–64.

Sakai, K., and W. R. Peltier (1995). A simple model of the Atlantic thermohaline circulation: Internal and forced variability with paleoclimatological implications. *J. Geophys. Res.* **100**, 13455–13479.

Sakai, K., and W. R. Peltier (1996). A multibasin reduced model of the global thermohaline circulation: Paleoceanographic analyses of the origins of ice-age climate variability. *J. Geophys. Res.* **101**, 22535–22562.

Sakai, K., and W. R. Peltier (1997). Dansgaard–Oeschger oscillations in a coupled atmosphere–ocean climate model. *J. Climate* **10**, 949–970.

Salmon, R. (1998). "Lectures on Geophysical Fluid Dynamics." Oxford Univ. Press, New York.

Saltzman, B. (1962). Finite amplitude free convection as an initial value problem—I. *J. Atmos. Sci.* **19**, 329–341.

Saltzman, B. (1968). Surface boundary effects on the general circulation and macroclimate: A review of the theory of the quasi-stationary perturbations in the atmosphere. *Meteorol. Magazine* **5**, 4–19.

Saltzman, B. (1978). A survey of statistical-dynamical models of the terrestrial climate. *Adv. Geophys.* **20**, 183–304.

Saltzman, B. (1979). A simple sea-ice/ocean-temperature oscillator model. Proceedings of the JOC Study Conference on Climate Models: Performance, Intercomparison, and Sensitivity Studies. GARP Publ. Ser. No. 22, pp. 917–933. World Meteorological Organization, Geneva.

Saltzman, B. (1982). Stochastically-driven climatic fluctuations in the sea-ice, ocean temperature, CO_2 feedback system. *Tellus* **34**, 97–112.

Saltzman, B. (1983). Climatic system analysis. *Adv. Geophys.* **25**, 173–233.

Saltzman, B. (1984a). On the role of equilibrium atmospheric climate models in the theory of long period glacial variations. *J. Atmos. Sci.* **41**, 2263–2266.

Saltzman, B. (1984b). Modeling the late-quaternary glacial variations with multi-component climatic systems. *Ann. Glaciol.* **5**, 225–227.

Saltzman, B. (1985). Paleoclimatic modeling. *In* "Paleoclimate Analysis and Modeling" (A. D. Hecht, ed.), Chapter 8, pp. 341–396. J. Wiley and Sons, New York.

Saltzman, B. (1986). Climatic "equilibrium" for the Quaternary. *J. Atmos. Sci.* **43**, 109–110.

Saltzman, B. (1987a). Carbon dioxide and the $\delta^{18}O$ record of late-Quaternary climatic change: A global model. *Climate Dynam.* **1**, 77–85.

Saltzman, B. (1987b). *In* "Modeling the $\delta^{18}O$-Derived Record of Quarternary Climatic Change with Low Order Dynamical Systems," pp. 355–380. D. Reidel, Dordrecht.

Saltzman, B. (1988). Modelling the slow climatic attractor. *In* "Physically-Based Modelling and Simulation of Climate and Climatic Change," Part II (M. E. Schlesinger, ed.), pp. 737–754. Reidel, Dordrecht.

Saltzman, B. (1990). Three basic problems of paleoclimatic modeling: A personal perspective and review. *Climate Dynam.* **5**, 67–78.

Saltzman, B., and S. Ashe (1976). Variance of surface temperature due to diurnal and cyclone-scale forcing. *Tellus* **28**, 308–322.

Saltzman, B., and K. Maasch (1988). Carbon cycle instability as a cause of the late Pleistocene ice age oscillations: Modeling the asymmetric response. *Global Biogeochem. Cycles* **2**, 177–185.

Saltzman, B., and K. Maasch (1990). A first-order global model of late Cenozoic climate change. *Trans. R. Soc. Edinb.* **81**, 315–325.

Saltzman, B., and K. A. Maasch (1991). A first-order global model of late Cenozoic climatic change. *Climate Dynam.* **5**, 201–210.

Saltzman, B., and R. Moritz (1980). A time-dependent climatic feedback system involving sea-ice extent, ocean temperature, and CO_2. *Tellus* **32**, 93–118.

Saltzman, B., and J. A. Pollack (1977). Sensitivity of the diurnal temperature range to changes in physical parameters. *J. Appl. Meteorol.* **16**, 614–619.
Saltzman, B., and A. Sutera (1984). A model of the internal feedback system involved in later Quarternary climatic variations. *J. Atmos. Sci.* **41**, 736–745.
Saltzman, B., and A. Sutera (1987). The mid-Quaternary climatic transition as the free response of a three-variable dynamical model. *J. Atmos. Sci.* **44**, 236–241.
Saltzman, B., and C.-M. Tang (1975). Formation of meanders, fronts, and cut-off thermal pools in a baroclinic ocean current. *J. Phys. Oceanogr.* **5**, 86–92.
Saltzman, B., and M. Verbitsky (1992). Asthenospheric ice load effects in a global dynamical system model of the pleistocene climate. *Climate Dynam.* **8**, 1–11.
Saltzman, B., and M. Verbitsky (1993). Multiple instabilities and modes of glacial rythmicity in the Plio-Pleistocene: A general theory of late Cenozoic climatic change. *Climate Dynam.* **9**, 1–15.
Saltzman, B., and M. Verbitsky (1994a). CO_2 and the glacial cycles. *Nature* **367**, 418.
Saltzman, B., and M. Verbitsky (1994b). Late Pleistocene climatic trajectory in the phase space of global ice, ocean state, and CO_2: Observations and theory. *Paleoceanography* **9**, 767–779.
Saltzman, B., and M. Verbitsky (1995a). Heinrich-scale surge oscillations as an internal property of ice sheets. *Ann. Glaciol.* **23**, 348–351.
Saltzman, B., and M. Verbitsky (1995b). Predicting the Vostok CO_2 curve. *Nature* **377**, 690.
Saltzman, B., and M. Y. Verbitsky (1996). Heinrich-scale surge oscillations as an internal property of ice sheets. *Ann. Glaciol.* **23**, 348–351.
Saltzman, B., and A. Verneker (1971). An equilibrium solution for the axially symmetric component of the earth's macroclimate. *J. Geophys. Res.* **76**, 1498–1524.
Saltzman, B., and A. Verneker (1975). A solution for the northern hemisphere climatic zonation during a glacial maximum. *Quartern. Res.* **5**, 307–320.
Saltzman, B., A. Sutera, and A. Evenson (1981). Structural stochastic stability of a simple autooscillatory climate feedback system. *J. Atmos. Sci.* **38**, 494–503.
Saltzman, B., A. Sutera, and A. R. Hansen (1982). A possible marine mechanism for internally generated long-period climate cycles. *J. Atmos. Sci.* **39**, 2634–2637.
Saltzman, B., A Sutera, and A. R. Hansen (1984). Long-period free oscillations in a three-component climate model. *In* "New Perspectives in Climate Modelling" (A. Berger and C. Nicolis, eds.), pp. 289–298. Elsevier, Amsterdam.
Saltzman, B., K. A. Maasch, and M. Y. Verbitsky (1993). Possible effects of anthropogenically-increased CO_2 on the dynamics of climate: Implications for ice age cycles. *Geophys. Res. Lett.* **20**, 1051–1054.
Saltzman, B., H. Hu, and R. Oglesby (1997). Transitivity properties of surface temperature and ice cover in the CCM1. *Dynam. Atmos. Oceans* **27**, 619–629.
Sarmiento, J. L., and M. Bender (1994). Carbon biogeochemistry and climate change. *Photosynth. Res.* **39**, 209–234.
Sarmiento, J., and J. Toggweiler (1984). A new model for the role of the oceans in determining atmospheric CO_2. *Nature* **308**, 621–624.
Savin, S. M. (1977). The history of the earth's surface temperature during the past 100 million years. *Annu. Rev. Earth Planet. Sci.* **5**, 319–355.
Schlesinger, M. E. (ed.) (1988). "Physically-Based Modelling and Simulation of Climate and Climatic Change," Parts 1 and 2. Kluwer Acad. Publ., Dordrecht.
Schlesinger, M., and J. Mitchell (1987). Climate model calculations of the equilibrium climatic response to increased carbon dioxide. *Rev. Geophys.* **25**, 760–798.
Schneider, S. H., and P. J. Boston (eds.) (1991). "Scientists on Gaia." MIT Press, Cambridge, Mass.
Schneider, S., and R. Dickinson (1974). Climate modeling. *Rev. Geophys. Space Phys.* **12**, 447–493.
Schneider, S. H., and S. L. Thompson (1979). Ice ages and orbital variations: Some simple theory and modeling. *Quatern. Res.* **12**, 188–203.
Schuss, Z. (1980). Singular perturbation methods in stochastic differential equations of mathematical physics. *SIAM Rev.* **22**, 119–155.
Scotese, C. R. (1997). "Paleogeographic Atlas." PALEOMAP Program Report 90-0497, Department of Geology, University of Texas at Arlington, Arlington, Texas.

Scotese, C. R., and J. Golonka (1992). "Paleogeographic Atlas." Paleomap Project, Univ. of Texas at Arlington, Arlington, Texas.

Seidov, D., and M. Maslin (1999). North Atlantic deep water circulation collapse during Heinrich events. *Geology* **27**, 23–26.

Sellers, W. D. (1969). A global climatic model based on the energy balance of the earth–atmosphere system. *J. Appl. Meteorol.* **8**, 392–400.

Semtner, A., and R. Chervin (1988). A simulation of global ocean circulation with resolved eddies. *J. Geophys. Res.* **93**, 15502–15522.

Semtner, A. J., Jr., and R. M. Chervin (1992). Ocean general circulation from a global eddy-resolving simulation. *J. Geophys. Res.* **97**, 5493–5550.

Sergin, V. Y. (1979). Numerical modeling of the glaciers-ocean-atmosphere global system. *J. Geophys. Res.* **84**, 3191–3204.

Sergin, V. Y., and S. Y. Sergin (1976). *In* "The Simulation of the 'Glaciers-Ocean-Atmosphere' Planetary System" (S. Y. Sergin, ed.), pp. 5–51. Far East Science Center, U.S.S.R. Academy of Science, Vladivostok (in Russian).

Shackleton, N. J., and J. Imbrie (1990). The δ^{18}O spectrum of oceanic deep water over a five-decade band. *Climate Change* **16**, 217–230.

Shackleton, J., and J. Kennett (1975). "Initial Reports of the Deep Sea Drilling Project XXIX," pp. 743–755. U. S. Govt. Printing Office, Washington, D.C.

Shackleton, N. J., and N. D. Opdyke (1973). Oxygen isotope and paleomagnetic stratigraphy of equatorial Pacific core V28-238: Oxygen isotope temperature and ice volumes on a 10^5 and 10^8 year scale. *Quatern. Res.* **3**, 39–55.

Shackleton, N. J., and N. G. Pisias (1985). Atmospheric carbon dioxide, orbital forcing, and climate. Geophysical Monograph. 32, pp. 303–317. AGU, Washington, D.C.

Shackleton, N. J., J. Backman, H. Zimmerman, D. V. Kent, M. A., Hall, D. G. Roberts, D. Schnitker, J. G. Baldauf, A. Desprairies, R. Homrighousen, P. Huddlestun, J. B. Keene, A. J. Kaltenback, K. A. O. Krumsiek, A. C. Morton, J. W. Murray, and J. Westberg-Smith (1984). Oxygen isotope calibration of the onset of ice-rafting and history of glaciation in the North Atlantic region. *Nature* **307**, 620–623.

Shackleton, N. J., J. Imbrie, and N. G. Pisias (1988). The evolution of oceanic oxygen-isotope variability in the North Atlantic over the past three million years. *Phil. Trans. R. Soc. Lond.* **3**, 318, 679–688.

Shackleton, N. J., J. Le, A. Mix, and M. A. Hall (1992). Carbon isotope records from Pacific surface waters and atmospheric carbon dioxide. *Quatern. Sci. Rev.* **11**, 387–400.

Shaffer, G. (1989). A model of biogeochemical cycling of phosphorus, nitrogen, oxygen and sulphur in the ocean: One step toward a global climate model. *J. Geophys. Res.* **94**, 1979–2004.

Shaffer, G. (1990). A non-linear climate oscillator controlled by biogeochemical cycling in the ocean: An alternative model of Quaternary ice age cycles. *Climate Dynam.* **4**, 127–143.

Shumskiy, P. (1955). "Principles of Structural Glaciology." USSR Academy of Sciences, Moscow [English translation by D. Kraus and Dover Publications, 1964].

Shumskiy, P. (1975). O zakone techeniya pollikristallicheskogo l'da [on the flow law of polycrystalline ice]. *Trudy Inst. Mekh.* **42**, 54–68.

Siegenthaler, V. (1990). El Nino and atmospheric CO_2. *Nature* **345**, 295–296.

Sloan, L., and D. Rea (1995). Atmospheric carbon dioxide and early Eocene climate: A general circulation modeling sensitivity study. *Palaeogeogr. Palaeoclimatol. Palaeoecol.* **119**, 275–292.

Sloan, L., J. Walker, and T. Moore (1995). Possible role of oceanic heat-transport in early eocene climate. *Palaeogeogr. Palaeoclimatol. Palaeoecol.* **10**, 347–356.

Smith, T. M., and H. H. Shugart (1993). The transient response of terrestrial carbon storage to a perturbed climate. *Nature* **361**, 523–526.

Snieder, R. K. (1985). The origin of the 100,000-year cycle in a simple ice age model. *J. Geophys. Res.* **90**, 5661–5664.

Sowers, T., M. Bender, D. Raynaud, Y. S. Korotkevich, and J. Orchardo (1991). The δ^{18}O of the atmospheric O_2 from an inclusion in the Vostok ice core: Timing of CO_2 and ice volume changes during the penultimate deglaciation. *Paleoceanography* **6**, 679–696.

Sowers, T., M. Bender, L. Labeyrie, D. Martinson, J. Jouzel, D. Raymond, J. Pinchon, and Y. Korotkivich (1993). A 135,000-year vostok-specmap common temporal framework. *Paleoceanography* **8**, 737–766.

Stern, M. E. (1975). "Ocean Circulation Physics." Academic Press, New York.
Stockton, C. W., W. R. Boggess, and D. M. Meko (1985). Climate and tree rings. *In* "Paleoclimate Analysis and Modeling" (A. D. Hecht, ed.), pp. 71–161. Wiley and Sons, New York.
Stockwell, J. N. (1875). Memoir on the secular variations of the elements of the orbits of the eight principle planets. *Smithsonian Contrib. Knowledge* **232**.
Stokes, W. L. (1955). Another look at the ice age. *Science* **122**, 815–821.
Stommel, H. (1948). The westward intensification of wind-driven ocean currents. *Trans. Am. Geophys. Union* **99**, 202–206.
Stommel, H. (1961). Thermohaline convection with two stable regimes of flow. *Tellus* **13**, 224–228.
Stone, P., and J. Risbey (1990). On the limitations of General Circulation Models. *Geophys. Res. Lett.* **17**, 2173–2176.
Stott, L. D. (1992). Higher temperatures and lower oceanic pCO_2: A climate enigma at the end of the Paleocene epoch. *Paleoceanography* **7**, 395–404.
Suarez, M., and I. Held (1976). Modelling climatic response to orbital parameter variations. *Nature* **263**, 46–47.
Sutera, A. (1980). Stochastic perturbation of a pure convective motion. *J. Atmos. Sci.* **37**, 245–249.
Sutera, A. (1981). On stochastic perturbation and long-term climate behaviour. *Q. J. Roy. Meteorol. Soc.* **107**, 137–153.
Sverdrup, H. V. (1947). Wind-driven currents in a baroclinic ocean, with application to the equatorial currents of the eastern Pacific. *Proc. Natl. Acad. Sci. U.S.A.* **33**, 318–326.
Syktus, J., H. Gordon, and J. Chappell (1994). Sensitivity of a coupled atmosphere-dynamic upper ocean GCM to variations of CO_2, solar constant, and orbital forcing. *Geophys. Res. Lett.* **21**, 1599–1602.
Syktus, J., J. Chappell, R. Oglesby, J. Larson, S. Marshall, and B. Saltzman (1997). Latitudinal dependence of signal-to-noise patterns from two general circulation models with CO_2 forcing. *Climate Dynam.* **13**, 293–302.
Takahashi, T. (1989). The carbon dioxide puzzle. *Oceanus* **32**, 22–29.
Takahashi, T., J. Olafsson, J. Goddard, D. Chipman, and S. Sutherland (1993). Seasonal variation of CO_2 and nutrients in the high-latitude surface oceans: A comparative study. *Global Biogeochem. Cycles* **7**, 843–878.
Tanner, W. F. (1965). Cause and development of an ice age. *J. Geol.* **73**, 413–430.
Tarasov, L., and W. R. Peltier (1997a). A high-resolution model of the 100 ka ice-age cycle. *Ann. Glaciol.* **25**, 58–65.
Tarasov, L., and W. R. Peltier (1997b). Terminating the 100 ky ice age cycle. *J. Geophys. Res.* **102**, 21665–21693.
Tarasov, L., and W. R. Peltier (1999). Impact of thermomechanical ice sheet coupling on a model of the 100 ky ice age cycle. *J. Geophys. Res.* **104**, 9517–9545.
Taylor, N. K. (1992). The role of the ocean in the global carbon cycle. *Weather* **47**, 146–241; 237–241.
Taylor, K. (1994). *In* "Climate Models for the Study of Paleoclimates," pp. 21–41. Springer-Verlag, Berlin.
TEMPO (1996). Potential role of vegetation feedback in the climate sensitivity of high-latitude regions: A case study at 6000 years B. P. *Global Biogeochem. Cycles* **10**, 727–736.
Thomas, R. (1973a). The creep of ice shelves: Interpretation of observed behaviour. *J. Glaciol.* **12**, 55–70.
Thomas, R. (1973b). The creep of ice shelves: Theory. *J. Glaciol.* **12**, 45–53.
Thomas, R. (1979). Dynamics of marine ice sheets. *J. Glaciol.* **24**, 167–177.
Thompson, J. M. T., and H. B. Stewart (1986). Nonlinear dynamics and chaos. J. Wiley and Sons, New York.
Toggweiler, J., and B. Samuels (1995). Effect of drake passage on the global thermohaline circulation. *Deep Sea Res.* **42**, 477–500.
Toggweiler, J. R., and J. L. Sarmiento (1985). Glacial to interglacial changes in atmospheric carbon dioxide: The critical role of ocean surface water in high latitudes. Geophysical Monograph, 32, pp. 163–184. AGU, Washington, D.C.
Torbett, M. (1989). Solar system and galactic influences on the stability of the earth. *Global Planet. Change* **75**, 1–33.
Trenberth, K. E. (1992). "Climate System Modeling." Cambridge University Press, Cambridge.
Turekian, K. K. (1976). "Oceans." Prentice-Hall, Englewood Cliffs, N.J.

Turekian, K. K. (1996). "Global Environmental Change." Prentice-Hall, Englewood Cliffs, N.J.

Tyndall, J. (1861). On the absorption and radiation of heat by gases and vapours, and on the physical connection of radiation, absorption, and conduction. *Philosoph. Mag.* **22**, 273–285.

Urey, H. C. (1947). The thermodynamic properties of isotopic substances. *J. Chem. Soc.* **1974**, 562–581.

Vail, P. R., R. M. Mitchum, and S. Thompson (1977). Seismic stratigraphy and global changes in sea level, 4. Global cycles of relative changes of sea level. *Am. Assoc. Petrol. Geol. Mem.* **26**, 83–97.

van der Veen, C. (1985). Response of marine ice sheet to changes at the grounding line. *Quartern. Res.* **24**, 257–267.

van der Veen, C., and J. Oerlemans (1984). Global themodynamics of a polar ice sheet. *Tellus* **36**, 228–235.

van Kreveld, S. A., M. Knappertsbusch, J. Ottens, G. M. Ganssen, and J. E. van Hinte (1996). Biogenic carbonate and ice-rafted debris (Heinrich layer) accumulation in deep sea sediments from a Northeast Atlantic piston core. *Marine Geol.* **131**, 21–46.

Verbitsky, M. (1992). Equilibrium ice-sheet scaling in climate modeling. *Climate Dynam.* **7**, 105–110.

Verbitsky, M., and R. Oglesby (1995). The CO_2-induced thickening/thinning of the greenland and antarctic ice sheets as simulated by a GCM and an ice-sheet model. *Climate Dynam.* **11**, 247–253.

Verbitsky, M. Y., and B. Saltzman (1993). Heinrich-type events in a low-order dynamical model of global climatic change. EOS 74 (Fall Meeting Suppl.), p. 357. AGU, Washington, D.C.

Verbitsky, M., and B. Saltzman (1994). Heinrich-type glacial surges in a low-order dynamical climate model. *Climate Dynam.* **10**, 39–47.

Verbitsky, M., and B. Saltzman (1995). A diagnostic analysis of Heinrich glacial surge events. *Paleoceanography* **10**, 59–65.

Verbitsky, M., and B. Saltzman (1995b). Behavior of the East Antarctic ice sheet as deduced from a coupled GCM/ice sheet model. *Geophys. Res. Lett.* **22**, 2913–2916.

Verbitsky, M., and B. Saltzman (1997). Modeling the antarctic ice sheet. *Ann. Glaciol.* **25**, 259–268.

Vernekar, A. D. (1972). Long-period global variations of incoming solar radiation. *Meteorol. Monogr.* **12** (No. 34).

Verneker, A., and H. Chang (1978). A statistical-dynamical model for stationary perturbations in the atmosphere. *J. Atmos. Sci.* **35**, 433–444.

Vinnikov, K. Y., A. Robock, R. J. Stouffer, J. E. Walsh, C. L. Parkinson, D. J. Cavalieri, J. F. B. Mitchell, D. Garrett, and V. F. Zakharov (1999). Global warming and northern hemisphere sea ice extent. *Science* **286**, 1934–1937.

Volk, T. (1989a). Rise of angiosperms as a factor in long-term climatic cooling. *Geology* **17**, 107–110.

Volk, T. (1989b). Sensitivity of climate and atmospheric CO_2 to deep-ocean and shallow-ocean carbonate burial. *Nature* **337**, 637–640.

Volk, T., and M. I. Hoffert (1985). Ocean carbon pumps: Analysis of relative strengths and efficiencies in ocean-driven atmospheric CO_2 changes. *In* "The Carbon Cycle and Atmospheric CO_2: Natural Variations Archean to Present" (E. T. Sundquist and W. S. Broeker, eds.), Geophysical Monograph 32, pp. 99–110. Am. Geophys. Union, Washington D.C.

Waelbroeck, C., P. Monfray, W. C. Oechel, S. Hastings, and G. Vourlitis (1997). The impact of permafrost thawing on the carbon dynamics of tundra. *Geophys. Res. Lett.* **24**, 229–232.

Walder, J., and A. Fowler (1994). Channelized subglacial drainage over a defromable bed. *J. Glaciol.* **40**, 3–15.

Walker, J. C. G. (1978). "Evolution of the Atmosphere." Macmillan Publ. Co., New York.

Walker, J., P. Hays, and J. Kasting (1981). A negative feedback mechanism for the long-term stabilisation of earth's surface temperature. *J. Geophys. Res.* **86**, 9776–9782.

Washburn, A. L. (1973). "Periglacial Processes and Environments." Edward Arnold, London.

Washington, W., and C. Parkinson (1986). *In* "An Introduction to Three Dimensional Climate Modelling," p. 422. University Science, Mill Valley, CA.

Watson, A. J., and J. E. Lovelock (1983). Biological homeostasis of the global environment: The parable of daisy world. *Tellus* **35B**, 284–289.

Watson, A. J., and L. Maddock (1991). A geophysiological model for glacial–interglacial oscillations in the carbon and phosphorus cycles. *In* "Scientists on Gaia" (S. H. Schneider and P. J. Boston, eds.), pp. 240–246. The M.I.T. Press, Cambridge, Mass.

Watson, R. T., H. Rodhe, H. Oeschger, and U. Siegenthaler (1990). Greenhouse gases and aerosols. *In* "Climate Change, The IPCC Scientific Assessment" (J. T. Houghton, G. J. Jenkins, and J. J. Ephraums, eds.), p. 8. Cambridge University Press, Cambridge.

Watts, R. G. (1981). Comparing zero- and one-dimensional climate models. *J. Atmos. Sci.* **38**, 2333–2336.

Watts, R. G., and M. E. Hayder (1983). The origin of the 100-kiloyear ice sheet cycle in the Pleistocene. *J. Geophys. Res.* **88**, 5163–5166.

Watts, R. G., and E. Hayder (1984). A two-dimensional, seasonal, energy balance climate model with continents and ice sheets: Testing the Milankovitch theory. *Tellus* **36A**, 120–131.

Weart, S. R. (1997). The discovery of the risk of global warming. *Phys. Today* **January**, 34–40.

Weertman, J. (1957). On the sliding of glaciers. *J. Glaciol.* **3**, 33–38.

Weertman, J. (1964). Rate of growth or shrinkage of nonequilibrium ice sheets. *J. Glaciol.* **6**, 145–158.

Weertman, J. (1966). Effect of a basal water layer on the dimensions of ice sheets. *J. Glaciol.* **6**, 191–207.

Weertman, J. (1969). Water lubrication mechanism of glacier surges. *Can. J. Earth Sci.* **6**, 929–942.

Weertman, J. (1976). Milankovitch solar radiation variations and ice age ice sheet sizes. *Nature* **261**, 17–20.

Weiss, R. F. (1974). Carbon dioxide in water and seawater: The solubility of a non-ideal gas. *Marine Chem.* **2**, 203–215.

Wells, J. (1963). Coral growth and geochronometry. *Nature* **197**, 948–950.

Wenk, T., and U. Siegenthaler (1985). The high latitude ocean as a control of atmospheric CO_2. Geophysical Monographs 32, pp. 185–194. AGU, Washington, D.C.

Whitehead, J. A. (1998). Topographic control of oceanic flows in deep passages and straits. *Rev. Geophys.* **36**, 423–440.

Williams, J., R. Barry, and W. Washington (1974). Simulation of the atmospheric circulation using the NCAR global circulation model with ice-age boundary conditions. *J. Appl. Meteorol.* **13**, 305–317.

Williams, M., D. Dunkerley, P. de Decker, P. Kershaw, and J. Chappell (1998). "Quaternary Environments." Arnold, London.

Wilson, A. T. (1964). Origin of ice ages: An ice shelf theory for Pleistocene glaciation. *Nature* **201**, 147–148.

Wolfe, J. A. (1978). A paleobotanical interpretation of tertiary climates in the Northern Hemisphere. *Am. Sci.* **66**, 694–703.

Wright, D. G., and T. F. Stocker (1991). A zonally averaged ocean model for the thermohaline circulation. Part I: Model development and flow dynamics. *J. Phys. Oceanogr.* **21**, 1713–1724.

Yang, J., and J. D. Neelin (1993). Sea-ice interaction with the thermohaline circulation. *Geophys. Res. Lett.* **20**, 217–220.

Yang, J., and J. D. Neelin (1997). Sea-ice interaction and the stability of the thermohaline circulation. *Atmos. Ocean* **35**, 433–469.

Zhang, S., C. A. Lin, and R. J. Greatbatch (1995). A decadal oscillation due to the coupling between an ocean circulation model and a thermodynamic sea-ice model. *J. Marine Res.* **53**, 79–106.

Index

AGCMs, *see* Atmospheric general circulation models
Atmosphere
carbon doxide, *see* Carbon dioxide
composition assumptions in climate modeling, 13
ideal gas equation of state, 49
variables affecting surface temperature, 45
Atmospheric general circulation models (AGCMs), time averaging, 55–56
Attractor set, dynamical systems modeling, 96, 108–109

Bedrock depression
depression-calving hypothesis of sea level change, 173–176, 178–179, 271
ice inertia modeling
bedrock-calving effects, 271
direct bedrock effects, 270–271
ice sheet modeling, 148, 172–173
Bifurcation theory, dynamical systems modeling
bifurcation diagrams, 87–90
energy balance model, 126–127
Hopf bifurcation, 91, 100
pitchfork bifurcation, 87–89
saddle node bifurcation, 90
structural stability, 87
transcritical bifurcation, 90
Binge–purge model, ice surge behavior, 307
Bolide, forced evolution of tectonic-mean climatic state, 253

Carbon-13
GEOCARB model formulation, 204–205
ratiometric methods, 23, 25
water content analysis, 25
carbonate content analysis, 25
atmospheric carbon dioxide change estimation, 25–26
Carbon-14, half-life and geochronometry, 27, 29
Carbon dioxide
air–sea flux
bulk gas-exchange coefficient, 184
carbonate alkalinity, 185–186
carbon balance
deep ocean, 191–192
upper layer, 190–191
dissolved inorganic carbon, 184
factors affecting mass flux
circulation of ocean, 187, 189
dust-borne depositions, 190
overview, 185–186
sea level change, 189–190
temperature of ocean, 186–187
feedback-loop representation in unified dynamical theory, 240
influx equations, 191
mass flux equations, 183–184
mixed layer concentration, 184
oxygen levels, 191–192
parameterization for flux, 192–193
surface water layer concentration, 184
atmospheric mixing assumptions, 149
change estimation with isotopes, 25–26, 150
continuity equation for atmospheric carbon, 181
determinants of atmospheric content, 181–182
equivalent carbon dioxide, 149
glacial effects, 151
global carbon cycle, 206
global dynamical equation, 201
greenhouse effect and global temperature response, 44–45
historical perspective of climate studies, 275–276
multimillion-year evolution
first-order response of global ice mass and deep ocean temperature, 259–260
GEOCARB solution, 255–259
overview, 255
outgassing processes, 197–198
paleoclimate dynamics model
equations, 236, 244
structural stability as function of level, 290–291

Carbon dioxide (*continued*)
partial sensitivity of surface temperature to carbon dioxide, 140–141
phase-space trajectory with temperature and carbon dioxide, 154, 156–157
Pleistocene ice age hypothesis, 276–277, 300–302
positive feedback in carbon cycle, 150–151
rock weathering downdraw, 198–200
seasonal changes, 182
slow response variables in structured dynamical approach, 77
small changes and model accuracy, 70–71
sources, 151
tectonically forced variations, long-term process modeling
GEOCARB model, 202–206
oceanic carbon balance, 202
overview, 201–202
terrestrial organic carbon exchange
components, 193–194
global mass balance, 197
ice cover effects, 195–196
long-term organic burial, 196–197
sea level change effects, 195
seasonal variation, 194–195
temperature effects, 195
thermal response function, 137–139
volume mixing ratio, 181
Cenozoic ice-age departures, *see* Ice-age problem
Center manifold, elimination of fast-response variables, 241–242
Chaos, dynamical systems modeling, 108–109
Climate, definition, 3
Climate system, definition, 3–4
Climate system model (CSM)
paleoclimate dynamics model role, 80–82
prediction limitations
heterogeneity and nonlinearity, 69
instability, 69
unknown long-term forcing, 69
prospects for more complete theory, 317–320
schematic representations, 68, 78–79
small changes and model accuracy, 69–71
Cloud
rate of condensation of drops, 51
rate of melting, 51
Conservation equations
energy, 49
mass, 49
momentum, 49
symbols, 50
Coral
layer application in climate change studies, 19
nutrient-shelf hypothesis, 190
sea level change effects on reef formation and carbon dioxide levels, 189–190
Cosmic bombardment, climate effects, 12
Cosmic dust, forced evolution of tectonic-mean climatic state, 253–254
CSM, *see* Climate system model

Dansgaard–Oeschger oscillations
dynamical equations and specialization, 315–316
overview of features, 303
variable set, 314–315
Data sources, climate changes
direct measurement, 17–18
qualitative accounts, 18
Day, length changes over time, 12
Deacon cell, formation, 213–214
Depression-calving hypothesis
ice inertia modeling, 271
sea level change, 173–176, 178–179
Dynamical approach, paleoclimate modeling
adiabatic elimination, 78
best available model criteria, 321–322
Dansgaard–Oeschger oscillations, 315–316
diagnostic versus prognostic variables, 78
dynamical systems theory
deterministic chaos, 108–109
generic cubic nonlinearity, 86–87
local stability, 84–86
multivariable systems, two-variable phase plane, 92, 94
prototype two-variable model
attractor sets, 96
equilibria sensitivity to changes in parameters, 97–99
net downward flux of energy at ocean surface, 96
ocean temperature/sea ice-extent model overview, 95–96
random forcing effects, 103–108
sample deterministic solution near stable equilibrium, 96
structural stability, 99–103
structural stability and bifurcation theory, 87–92
external forcing function
astronomical/cosmic forcing, 82
tectonic forcing, 83
ground rules, 71–72
Heinrich oscillations, 311–314
minimization of free parameters, 79
notational simplifications for resolving total climate variability, 74–76

paleoclimate dynamics model role in climate system model, 80–82
problems in long-term climate change theory, 73–74
prospects for more complete theory, 317–320
response time subsets, 76–78
small changes and model accuracy, 69–71
strategy, 72–74
time averaging, 72–73
transivity properties of atmospheric and surface climatic state, 132–134
two-box thermohaline circulation model analysis, 223–226
unified model, *see* Paleoclimate dynamics model

Earth orbit, *see* Orbit, Earth
Earth rotation, *see* Rotation, Earth
EBM, *see* Energy balance model
Eccentricity, Earth orbit changes over time, 9–11
Energy, conservation equations, 49
Energy balance
flux vectors, 65
sensible heat per unit horizontal area of atmosphere, 65
surface heat balance condition, 65
water balance integration, 66–67
Energy balance model (EBM)
applications, 123
assumptions and critical approximations, 121–122
formal feedback analysis of fast-response equilibrium state, 142–143
one-dimensional model
ice-baroclinicity feedback, 131
ice line latitude, 129–130
overview, 123
solar constant requirements for ice line maintenance, 130
temperature variation as function of latitude, 131
stochastic resonance, 127–129
temperature change calculation, 122–123
transivity properties of atmospheric and surface climatic state, 132–134
zero dimensional, global average model
bifurcation theory, 126–127
effective planetary radiative equilibrium temperature, 124
equilibrium dependence on solar constant, 126–127
global mean thermal state, 123, 125–126
time constant, 124–125
transient departures of temperature, 124

Equilibrium portraits, dynamical systems modeling, 98
External forcing
astronomical forcing
cosmic bolide bombardments, 12
Earth orbital changes, 9–12
external forcing function in paleoclimate dynamics model, 82
miscellaneous factors, 12
rotation of Earth, 12
solar luminosity changes, 9
combination of factors in climate change, 45–46
forced evolution of tectonic-mean climatic state
bolide, 253
carbon dioxide multimillion-year evolution
first-order response of global ice mass and deep ocean temperature, 259–260
GEOCARB solution, 255–259
overview, 255
cosmic dust, 253–254
general circulation model simulations, 261
global property summary for tectonic states, 252–253
land–ocean distribution and topography, 249–253
rotation of Earth, 249
salinity-driven instability of tectonic mean state, 260–261
set of variables, 247
solar luminosity, 248
volcanic activity, 253–254
paleoclimate dynamics model effects, 245–246
steady forcing, 9
tectonic forcing
external forcing function in paleoclimate dynamics model, 83
geothermal heat flux, 12
miscellaneous factors, 13
plate tectonic crustal movements, 12
volcanic activity, 12–13

Feedback
carbon cycle positive feedback, 150–151
energy balance model
formal feedback analysis of fast-response equilibrium state, 142–143
ice-baroclinicity feedback, 131
general circulation model, formal feedback analysis of fast-response equilibrium state
feedback factors, 141
partial sensitivity of surface temperature to carbon dioxide, 140–141
perturbation parameters, 139–140
Heinrich oscillation feedback loop, 312

Feedback (*continued*)
  paleoclimate dynamics model feedback loop representation, 238–240
  sea level dynamic feedback on ice sheets, 174–175
Fossil, climate change evidence, 22–23

GCM, *see* General circulation model
General circulation model (GCM)
  closure relationships based on sensitivity experiments
    carbon dioxide thermal response function, 137–139
    external forcing, 136
    fast-response variables, 134
    sensitivity functions, 135, 138
    slow-response variables, 135, 139
    surface temperature sensitivity, 134–139
  complete time-averaged state, 119
  equilibrium modeling, 114–115
  external forcing simulations, 261
  fast-response weather variables, 77–78
  formal feedback analysis of fast-response equilibrium state
    feedback factors, 141
    partial sensitivity of surface temperature to carbon dioxide, 140–141
    perturbation parameters, 139–140
  foundations, 51
  paleoclimate dynamics model component, 237, 245
  paleoclimatic simulations, 143–145
  parameterizations, 114–115
  prognostic equations, 114
  prospects for more complete theory, 318–320
  response time of variables, 113–114
  small changes and model accuracy, 69–71
  super model, 59, 68, 78
  transivity properties of atmospheric and surface climatic state, 132–134
Generic cubic nonlinearity, dynamical systems theory, 86–87
GEOCARB model
  applications, 206
  balance equations, 203–204
  carbon-13 formulation, 204–205
  multimillion-year carbon dioxide evolution solution, 255–259
  outgassing, 202–204
  parameters, 204
  silicate weathering, 203
Geophysical fluid dynamics (GFD), foundations, 51
Geothermal heat flux, climate effects, 12
GFD, *see* Geophysical fluid dynamics
Glen's rheological law, 159
Global circulation model, *see* General circulation model
Greenhouse gases, *see also* Carbon dioxide
  equivalent carbon dioxide, 149
  types, 149

Heinrich oscillations
  basal temperature
    melting point relationships, 306–307, 311–312
    scale analysis of influencing factors, 308–309
  binge–purge model, 307
  carbon dioxide fluctuations, 304
  diagnostic analysis, 309–311
  dynamical analysis and model, 311–314
  ice-rafted debris events, last ice age, 37–38
  overview of features, 303–304
  sea surface temperature variation, 303–304
Human activity, climate effects, 13
Hydrostatic balance equation, 49

Ice-age problem
  carbon dioxide hypothesis, 276–277, 300–302
  complexity of system, 15–16
  coupled ice-sheet/atmospheric climate model enhancement, 272–273
  forced ice-line models
    classification
      stable equilibria models, 268
      unstable equilibria and stochastic resonance models, 268
    coupling with slow-response variables
      deep ocean temperature, 274–275
      regolith mass, 273
      salinity gradient, 275
    inertia models
      basal meltwater and sliding, 271
      bedrock-calving effects, 271
      continental ice-sheet movement, 271–272
      direct bedrock effects, 270–271
      historical perspective, 269–270
      ice stream and shelf effects, 271
      simple forms, 268–269
      three-dimensional models, 272
    overview, 266
  forced versus free models
    Earth-orbital forcing in models, 263–265
    hierarchical classification in terms of increasing physical complexity, 266
    historical perspective, 262–263
    instability-driven models, 265–266
  overview, 14–15, 262–263

paleoclimate dynamics model
carbon dioxide hypothesis, 300–302
complete solution, 292–297
Milankovitch forcing of free oscillation, 288–290
100-ky oscillation as free response
adjustable parameter determination, 284–288
internal stability analysis, 286–288
nondimensionalization of system, 285–286
physically constrained parameters, 284–285
parameter overview, 280–281, 284
predictions and verification, 297–298
robustness, 299
sensitivity analysis, 299–300
sequence of processes implied by model, 300–302
specialization of model, 281–284
structural stability as function of tectonic carbon dioxide level, 290–291
time-dependent solution, 293–295
Ice core
beryllium content studies, 25
cadmium analysis, 26–27
deuterium content studies, 24
dust layer analysis, 27
isotopic analysis, 23–26
overview of studies, 20–21
oxygen-18 analysis, 24
trapped gas analysis, 27, 36
Ice mass
basal sliding velocity, 162
basal temperature
overview, 148
melting temperature relationship, 169–171, 306–307, 311–312
scale analysis of influencing factors, 308–309
basic equations, 158–160
bedrock depression, 148, 172–173
binge–purge model, 307
boundary conditions, 160–163
deformable basal regolith, 171–172
depression-calving hypothesis of sea level change, 173–176, 178–179, 271
dissipative rate constants and sources of instability, 242–244
equation of state, 49
evidence of retreat from terrestial signs, 20
flotation condition, 163
frictional heating per unit mass, 158, 160
general circulation models, paleoclimatic simulations, 143–145
global dynamical equations, 177–180
global variability estimates using oxygen-18, 5–7, 14, 148–149
grounding line junction of sheet and shelf, 162–163
mass equations
global mass, 147
sheet, 147
mean thickness of sheet, 147
net melting rate, 179
paleoclimate dynamics model equations, 236, 244
phase-space trajectory with temperature and carbon dioxide, 154, 156–157
rate of melting equation, 51
response time, 148
rheological coefficient, 160
scale analysis
areas and volumes of ice sheets at present and last glacial maximum, 165–166
notation, 164
physical parameters of ice sheets, 165
regions of ice sheet, 164
thickness equations, 163–164
vertical and horizontal temperature difference, 165
shelves of ice, 172
slow response variables in structured dynamical approach, 77
streams of ice, 172
stress relation to shear, 159–160
surface mass balance, 168–169
surge flux, 148, 178–179
terrestrial organic carbon exchange effects, 195–196
variation over Phanerozoic Eon
Cenozoic Era, 34–35
eras, periods, and epochs, 31–33
ice-rafted debris events during last ice age, 37–38
last glacial maximum, 38, 40
Late Precambrian, 33
Miocene Epoch, 34
Ordovician Period, 33
past 100 years, 43–44
Permo–Carboniferous Period, 34
Plio–Pleistocene Epoch, 35–36
postglacial changes, 40–42
vertically integrated ice-sheet model
overview, 166–168
paleoclimatic applications, 176–177
viscosity equation, 160
Inclination, Earth orbit changes over time, 11
Iron, ocean deposition and carbon dioxide draw-down, 190

Lake level, change application in climate change studies, 19
Land mass
  changes and climate effects, 12, 249–250
  Phanerozoic Eon changes in continent–ocean distribution, 31–33
Local stability, dynamical systems theory, 84–86
Lyapunov exponent, definition, 109

Magnetic field, reversal in Earth history, 13
Mass
  conservation equations, 49
  specialization of mass continuity equation, 51
Methane, equivalent carbon dioxide, 149
Millenial-scale variations, *see* Dansgaard–Oeschger oscillations; Heinrich oscillations
Mollusks, application in climate change studies, 19
Momentum, conservation equations, 49

Navier–Stokes equations, 48

Obliquity, Earth axis changes over time, 9–10
Ocean
  baroclinic eddy circulations, 216
  box-thermodynamic models, overview, 207–208
  carbon dioxide content, *see* Carbon dioxide
  deep ocean
    equations
      salinity, 229
      temperature, 227–228, 231–232
    temperature and ice sheet modeling coupling, 274–275
  Ekman layer balance, 213–214
  equation of state, 49, 207
  flow components
    baroclinically unstable waves, 208–209
    buoyancy-driven convective circulation cells, 208
    thermohaline circulation, 208
    wind-driven circulation, 208, 212–216
  goal of modeling, 207
  level, *see* Sea level
  local convective overturnings, 216
  paleoclimate dynamics model state equations, 237
  slow response variables in structured dynamical approach, 77
  surface layer wind-driven gyre circulation, 215–216
  Sverdrup balance, 214–215
  temperature, *see* Sea surface temperature; Temperature
  thermohaline state, *see* Thermohaline ocean state
Orbit, Earth
  eccentricity changes over time, 9–11
  forcing in models, 263–265
  inclination changes over time, 11
  precessional index changes over time, 9–11
Osmium, isotopic evaluation of sedimentary samples, 26, 205
Oxygen-18
  fossil distribution, 23–24
  fractionation in nature, 23–24
  global ice mass variability estimates, 5–7, 14
  ratiometric methods, 23

Paleoclimate dynamics model (PDM)
  approach, 235–236
  carbon dioxide equations, 236
  center manifold for elimination of fast-response variables, 241–242
  damping rate constants, 242
  dissipative rate constants and sources of instability, 242–244
  external forcing effects, 245–246
  feedback-loop representation, 238–240
  formal separation into tectonic equilibrium and departure equations
    carbon dioxide, 244
    ice mass, 244
    thermohaline ocean state, 244–245
  general circulation model component, 237, 245
  goals, 235
  ice sheet equations, 236
  late Cenozoic ice age modeling
    carbon dioxide hypothesis, 300–302
    complete solution, 292–297
    Milankovitch forcing of free oscillation, 288–290
    100-ky oscillation as free response
      adjustable parameter determination, 284–288
      internal stability analysis, 286–288
      nondimensionalization of system, 285–286
      physically constrained parameters, 284–285
    parameter overview, 280–281, 284
    predictions and verification, 297–298
    robustness, 299
    sensitivity analysis, 299–300
    sequence of processes implied by model, 300–302
    specialization of model, 281–284
    structural stability as function of tectonic carbon dioxide level, 290–291
    time-dependent solution, 293–295
  ocean state equations, 237

prospects for more complete theory, 317–320
unification concepts, 278–279
PDM, *see* Paleoclimate dynamics model
Plate tectonics
land mass changes and climate effects, 12, 249–250
Phanerozoic Eon changes in continent–ocean distribution, 31–33
Precessional index, Earth orbit changes over time, 9–11

Radiative–convective models (RCMs), overview, 119–120
RCMs, *see* Radiative–convective models
Regolith mass
deformable basal regolith, 171–172
ice sheet modeling coupling, 273
Residence density portrait, dynamical systems modeling, 101
Response time
center manifold for elimination of fast-response variables, 241–242
climatic-mean state of atmosphere, 59
definition, 56–57
departure from equilibrium, 57–58
dissipative term in equations, 57
formal feedback analysis of fast-response equilibrium state, 139–143
slow-response control variables
carbon dioxide, 149–151
ice sheets, 147–149
phase-space trajectory, 154, 156–157
prescribed variables, 146
prognostic systems, 146
schematic of main components, 146–147
thermohaline ocean state, 151–154
thermal response time, 58
Rock weathering
carbon dioxide downdraw, 198–200, 203
weatherability, 203
Rotation, Earth
external forcing, 12
forced evolution of tectonic-mean climatic state, 249

Salinity gradient, *see* Thermohaline ocean state
Sand formations, application in climate change studies, 19–20
SDMs, *see* Statistical-dynamical models
Sea level
carbon dioxide effects of change
air–sea flux, 189–190
terrestrial organic carbon exchange, 195
causes of change, 174
climate change studies, 19, 176
depression-calving hypothesis of sea level change, 173–176, 178–179, 271
dynamic feedback on ice sheets, 174–175
eustatic change equation, 174
Phanerozoic Eon changes, 250–251
Sea surface temperature (SST)
core K708-1 studies, 153–154
Heinrich oscillations, 303–304
Sedimentary rock
fossil evidence, 22–23
ice-rafated debris in marine cores, 21, 36
isotopic evaluation, 23–26
overview of studies, 20–21
physical indicators of climate variation, 21–22
Separatrices, dynamical systems modeling, 94
Slow climatic attractor, 156
Soil, application in climate change studies, 19–20
Solar constant
equilibrium dependence, 126–127
requirements for ice line maintenance, 130
Solar luminosity
changes over time, 9, 14
forced evolution of tectonic-mean climatic state, 248
Spatial averaging
East–West variations, 60–61
global mass-weighted average, 62–63
mass per unit area, 61
North–South variations, 60–61
vertical dimensions, 61–62
SST, *see* Sea surface temperature
Statistical-dynamical models (SDMs), *see also* Energy balance model
axially asymmetric models, 117–119
friction function, 116
heating function, 116
thermodynamic models
radiative–convective models, 119–120
vertically averaged models, 120
time averaging, 55–56, 116
trace constituent source function, 115
zonal-averaged model, 116–117
Stochastic resonance, energy balance model, 127–129
Stochastic-dynamical system
flowchart of climatic variability model, 103–104
Ito form, 104
limitations, 103
stochastic amplitude, 104
structural stochastic stability, 104–108
Strontium, isotopic evaluation of sedimentary samples, 26, 200, 205

Structural stability
bifurcation theory, 87–92
function of tectonic carbon dioxide level, 290–291
prototype two-variable model, 99–103
structural stochastic stability, 104–108

Temperature, *see also* Thermohaline ocean state
atmospheric thermal variance at midlatitude, 4–5, 44
combination of factors in climate change, 45–46
mean global temperature over age of Earth, 5
phase-space trajectory with ice and carbon dioxide, 154, 156–157
small changes and model accuracy, 70
surface temperature variation over Phanerozoic Eon
Cenozoic Era, 34–35
eras, periods, and epochs, 31–33
ice-rafted debris events during last ice age, 37–38
last glacial maximum, 38, 40
Late Precambrian, 33
Miocene Epoch, 34
Ordovician Period, 33
past 100 years, 43–44
Permo–Carboniferous Period, 34
Plio–Pleistocene Epoch, 35–36
postglacial changes, 40–42
thermal response time, 58
variance spectra, 44
Theory, connecting external forcing and initial conditions with paleoclimatic variability observations, 7–8
Thermohaline ocean state
boundary conditions for flux, 210
bulk ocean temperature global dynamical equations, 230–232
cold glacial model, 153
deep ocean
circulation, 152–153
equations
salinity, 229
temperature, 151, 227–228, 231–232
dissipative rate constants and sources of instability, 243–244
four-box model, 211–212
general equations for modeling, 209–210
gradients, 152
paleoclimate dynamics model, 244–245
phase-space trajectory with ice, carbon dioxide, and temperature, 154, 156–157
salinity
driven instability of tectonic mean state, 260–261
global dynamical equations, 230–232
gradient, ice sheet modeling coupling, 275
sea surface temperature from core K708-1, 153–154
two-box thermohaline circulation model
cross-section of circulation, 217
dynamical analysis, 223–226
freshwater flux, 220
global mean salinity, 219
gyre/eddy volume exchange, 223
heat fluxes, 219
meridional fluxes, 222–223
net thermal and salinity fluxes, 219
overview, 212, 217
simple model of thermohaline circulation, 220–221
three-box model reduction, 218
Time averaging
complete time-averaged state, 119
consequences, 53–54
dynamical approach to paleoclimate modeling, 72–73
periods, 52, 56
physical phenomena over periods, 52–53
random forcing, 54–55
Reynolds conditions, 52–53
time-dependent ordinary differential equations, 55–56
Tree ring, application in climate change studies, 18–19

Unified paleoclimate dynamics model, *see* Paleoclimate dynamics model
Urey reaction, reverse reaction and carbon balance, 197

Volcanic activity
climate effects, 12–13
forced evolution of tectonic-mean climatic state, 253–254

Water mass balance
energy balance integration, 66–67
evaporation rate, 64
flux vectors, 63–64
rate of change of mass
column mass, 63
surface mass, 64

International Geophysics Series

EDITED BY

RENATA DMOWSKA
Division of Applied Science
Harvard University
Cambridge, Massachusetts

JAMES R. HOLTON
Department of Atmospheric Sciences
University of Washington
Seattle, Washington

H. THOMAS ROSSBY
Graduate School of Oceanography
University of Rhode Island
Narragansett, Rhode Island

Volume 1 BENO GUTENBERG. Physics of the Earth's Interior. 1959*

Volume 2 JOSEPH W. CHAMBERLAIN. Physics of the Aurora and Airglow. 1961*

Volume 3 S. K. RUNCORN (ed.). Continental Drift. 1962*

Volume 4 C. E. JUNGE. Air Chemistry and Radioactivity. 1963*

Volume 5 ROBERT G. FLEAGLE AND JOOST A. BUSINGER. An Introduction to Atmospheric Physics. 1963*

Volume 6 L. DEFOUR AND R. DEFAY. Thermodynamics of Clouds. 1963*

Volume 7 H. U. ROLL. Physics of the Marine Atmosphere. 1965*

Volume 8 RICHARD A. CRAIG. The Upper Atmosphere: Meteorology and Physics. 1965*

Volume 9 WILIS L. WEBB. Structure of the Stratosphere and Mesosphere. 1966*

Volume 10 MICHELE CAPUTO. The Gravity Field of the Earth from Classical and Modern Methods. 1967*

Volume 11 S. MATSUSHITA AND WALLACE H. CAMPBELL (eds.). Physics of Geomagnetic Phenomena (In two volumes). 1967*

Volume 12 K. YA. KONDRATYEV. Radiation in the Atmosphere. 1969*

Volume 13 E. PAL MEN AND C. W. NEWTON. Atmosphere Circulation Systems: Their Structure and Physical Interpretation. 1969*

Volume 14 HENRY RISHBETH AND OWEN K. GARRIOTT. Introduction to Ionospheric Physics. 1969*

Volume 15 C. S. RAMAGE. Monsoon Meteorology. 1971*

Volume 16 JAMES R. HOLTON. An Introduction to Dynamic Meteorology. 1972*

Volume 17 K. C. YEH AND C. H. LIU. Theory of Ionospheric Waves. 1972*

Volume 18 M. I. BUDYKO. Climate and Life. 1974*

Volume 19 MELVIN E. STERN. Ocean Circulation Physics. 1975*

Volume 20 J. A. JACOBS. The Earth's Core. 1975*

Volume 21 DAVID H. MILLER. Water at the Surface of the Earth: An Introduction to Ecosystem Hydrodynamics. 1977*

Volume 22 JOSEPH W. CHAMBERLAIN. Theory of Planetary Atmospheres: An Introduction to Their Physics and Chemistry. 1978*

Volume 23 JAMES R. HOLTON. An Introduction to Dynamic Meteorology, Second Edition. 1979*

Volume 24 ARNETT S. DENNIS. Weather Modification by Cloud Seeding. 1980*

Volume 25 ROBERT G. FLEAGLE AND JOOST A. BUSINGER. An Introduction to Atmospheric Physics, Second Edition. 1980

Volume 26 KUG-NAN LIOU. An Introduction to Atmospheric Radiation. 1980*

Volume 27 DAVID H. MILLER. Energy at the Surface of the Earth: An Introduction to the Energetics of Ecosystems. 1981*

Volume 28 HELMUT G. LANDSBERG. The Urban Climate. 1991

Volume 29 M. I. BUDKYO. The Earth's Climate: Past and Future. 1982*

Volume 30 ADRIAN E. GILL. Atmosphere-Ocean Dynamics. 1982

Volume 31 PAOLO LANZANO. Deformations of an Elastic Earth. 1982*

Volume 32 RONALD T. MERRILL AND MICHAEL W. MCELHINNY. The Earth's Magnetic Field. Its History, Origin, and Planetary Perspective. 1983*

Volume 33 JOHN S. LEWIS AND RONALD G. PRINN. Planets and Their Atmospheres: Origin and Evolution. 1983

Volume 34 ROLF MEISSNER. The Continental Crust: A Geophysical Approach. 1986

Volume 35 M. U. SAGITOV, B. BODKI, V. S. NAZARENKO, AND K. G. TADZHIDINOV. Lunar Gravimetry. 1986*

Volume 36 JOSEPH W. CHAMBERLAIN AND DONALD M. HUNTEN. Theory of Planetary Atmospheres, 2nd Edition. 1987

Volume 37 J. A. JACOBS. The Earth's Core, 2nd Edition. 1987*

Volume 38 JOHN R. APEL. Principles of Ocean Physics. 1987

Volume 39 MARTIN A. UMAN. The Lightning Discharge. 1987*

Volume 40 DAVID G. ANDREWS, JAMES R. HOLTON, AND CONWAY B. LEOVY. Middle Atmosphere Dynamics. 1987

Volume 41 PETER WARNECK. Chemistry of the Natural Atmosphere. 1988

Volume 42 S. PAL ARYA. Introduction to Micrometeorology. 1988

Volume 43 MICHAEL C. KELLEY. The Earth's Ionosphere. 1989*

Volume 44 WILLIAM R. COTTON AND RICHARD A. ANTHES. Storm and Cloud Dynamics. 1989

Volume 45 WILLIAM MENKE. Geophysical Data Analysis: Discrete Inverse Theory, Revised Edition. 1989

Volume 46 S. GEORGE PHILANDER. El Niño, La Niña and the Southern Oscillation. 1990

Volume 47 ROBERT A. BROWN. Fluid Mechanics of the Atmosphere. 1991

Volume 48 JAMES R. HOLTON. An Introduction to Dynamic Meteorology, Third Edition. 1992

Volume 49 ALEXANDER A. KAUFMAN. Geophysical Field Theory and Method. 1992
Part A: Gravitational, Electric, and Magnetic Fields. 1992
Part B: Electromagnetic Fields I. 1994
Part C: Electromagnetic Fields II. 1994

Volume 50 SAMUEL S. BUTCHER, GORDON H. ORIANS, ROBERT J. CHARLSON, AND GORDON V. WOLFE. Global Biogeochemical Cycles. 1992

Volume 51 BRIAN EVANS AND TENG-FONG WONG. Fault Mechanics and Transport Properties of Rocks. 1992

Volume 52 ROBERT E. HUFFMAN. Atmospheric Ultraviolet Remote Sensing. 1992

Volume 53 ROBERT A. HOUZE, JR. Cloud Dynamics. 1993

Volume 54 PETER V. HOBBS. Aerosol-Cloud-Climate Interactions. 1993

Volume 55 S. J. GIBOWICZ AND A. KIJKO. An Introduction to Mining Seismology. 1993

Volume 56 DENNIS L. HARTMANN. Global Physical Climatology. 1994

Volume 57 MICHAEL P. RYAN. Magmatic Systems. 1994

Volume 58 THORNE LAY AND TERRY C. WALLACE. Modern Global Seismology. 1995

Volume 59 DANIEL S. WILKS. Statistical Methods in the Atmospheric Sciences. 1995

Volume 60 FREDERIK NEBEKER. Calculating the Weather. 1995

Volume 61 MURRY L. SALBY. Fundamentals of Atmospheric Physics. 1996

Volume 62 JAMES P. MCCALPIN. Paleoseismology. 1996

Volume 63 RONALD T. MERRILL, MICHAEL W. MCELHINNY, AND PHILIP L. MCFADDEN. The Magnetic Field of the Earth: Paleomagnetism, the Core, and the Deep Mantle. 1996

Volume 64 NEIL D. OPDYKE AND JAMES CHANNELL. Magnetic Stratigraphy. 1996

Volume 65 JUDITH A. CURRY AND PETER J. WEBSTER. Thermodynamics of Atmospheres and Oceans. 1998

Volume 66 LAKSHMI H. KANTHA AND CAROL ANNE CLAYSON. Numerical Models of Oceans and Oceanic Processes. 1999

Volume 67 LAKSHMI H. KANTHA AND CAROL ANNE CLAYSON. Small Scale Processes in Geophysical Fluid Flows. 1999

Volume 68 RAYMOND S. BRADLEY. Paleoclimatology, Second Edition. 1999

Volume 69 LEE-LUENG FU AND ANNY CAZANAVE. Satellite Altimetry. 1999

Volume 70 DAVID A. RANDALL. General Circulation Model Development. 1999

Volume 71 PETER WARNECK. Chemistry of the Natural Atmosphere, Second Edition. 1999

Volume 72 M. C. JACOBSON, R. J. CHARLESON, H. RODHE, AND G. H. ORIANS. Earth System Science: From Biogeochemical Cycles to Global Change. 2000

Volume 73 MICHAEL W. MCELHINNY AND PHILLIP L. MCFADDEN. Paleomagnetism: Continents and Oceans. 2000

Volume 74 ANDREW E. DESSLER. The Chemistry and Physics of Stratospheric Ozone. 2000

Volume 75 BRUCE DOUGLAS, MICHAEL KEARNEY, AND STEPHEN LEATHERMAN. Sea Level Rise: History and Consequences. 2000

Volume 76 ROMAN TEISSEYRE AND EUGENIUSZ MAJEWSKI. Earthquake Thermodynamics and Phase Transformations in the Interior. 2001

Volume 77 GEROLD SIEDLER, JOHN CHURCH, AND JOHN GOULD. Ocean Circulation and Climate: Observing and Modelling The Global Ocean. 2001

Volume 78 ROGER PIELKE. Mesoscale Meteorological Modeling, 2nd Edition. 2001

Volume 79 S. PAL ARYA. Introduction to Micrometeorology. 2001

Volume 80 BARRY SALTZMAN. Dynamical Paleoclimatology: Generalized Theory of Global Climate Change. 2002

ISBN 0-12-617331-1